Nanoscience and Soil-Water Interaction in Agroecosystem

Editors

Kamel A. Abd-Elsalam

Plant Pathology Research Institute
Agricultural Research Center
Giza, Egypt

Muhammad Zahid

Department of Chemistry
University of Agriculture
Faisalabad, Pakistan

CRC Press
Taylor & Francis Group
Boca Raton London New York

CRC Press is an imprint of the
Taylor & Francis Group, an **informa** business

A SCIENCE PUBLISHERS BOOK

First edition published 2025
by CRC Press
2385 NW Executive Center Drive, Suite 320, Boca Raton FL 33431

and by CRC Press
4 Park Square, Milton Park, Abingdon, Oxon, OX14 4RN

CRC Press is an imprint of Taylor & Francis Group, LLC

Library of Congress Cataloging-in-Publication Data (applied for)

ISBN: 978-1-032-55156-2 (hbk)
ISBN: 978-1-032-55157-9 (pbk)
ISBN: 978-1-003-42928-9 (ebk)

DOI: 10.1201/9781003429289

Typeset in Times New Roman
by Prime Publishing Services

Preface

Nanotechnology integration into agricultural practices demonstrates a transformative era to achieve sustainable environmental protection, food production, and efficient resource management. This book, "Nanoscience and Soil-Water Interaction in Agroecosystem," represents a comprehensive collection of cutting-edge research, techniques and innovations that explore the possible ways in which nanotechnology can improve and revolutionize modern agriculture practices. This book contains 13 chapters written by prominent authors from Bangladesh, Czech Republic, Egypt, India, Malaysia, Oman, Pakistan, Slovakia, and the United States of America.

Significant challenges are being faced by Agriculture such as pest infestations, nutrient deficiencies, microbial infections, and worsening the quality of soil and water resources. Conventional methods, and the use of chemical fertilizers and pesticides, have led to soil degradation, biodiversity loss and ecological imbalances. To deal with these challenges, the scientific community has turned its attention to the promising field of nanotechnology, which offers sustainable solutions.

This volume brings together contributions from established researchers and experts across the globe, who are involved in different applications of nanotechnology in agriculture. Each chapter is precisely designed to present both the theoretical background and practical implications of nanotechnology to offer a complete understanding of this field.

The book begins with the potential impacts of nanoparticles in agriculture and highlights their roles such as nano-fertilizers, nano-pesticides, and nano-sensors. This chapter explains how nanoparticles can improve plant growth and protection in comparison with traditional agricultural practices. The chapter offers insights into specific applications and their impacts by summarizing recent research in this domain.

The use of nanoparticles in enhancing soil moisture retention is discussed in subsequent chapters. These sections give insights into the intricate dynamics between nanoparticles and soil ecosystems, emphasizing the importance of soil moisture retention and carbon sequestration. The discussion extends to how nanoparticles deal with pest control, nutrient cycling, and soil microbial activity, reporting a comprehensive overview of nanoparticles in sustainable agriculture. The high precision and sensitivity of nanosensors with detailed investigations explain how nanosensors revolutionize soil-water system monitoring and management. These devices can detect a wide range of parameters and provide real-time data that enhances resource management and decision-making in agriculture. The nanosensor's role in measuring water quality, soil health, and nutrient dynamics is explored to highlight their potential to improve agricultural efficiency and sustainability.

The desalination of water using nanomaterials is analyzed in another chapter, addressing both their economic feasibility and advantages. It highlights the practical challenges in desalination and their solutions, which a critical issue in many parts of the world realizing water scarcity. Environmental consequences of various nanoparticles are critically assessed to emphasize sustainable practices to mitigate their toxic effects. This chapter specifies a detailed overview of various metal oxide nanoparticles regarding their toxicity in aquatic and soil ecosystems. Therefore, the need for careful use and management of these toxic materials is recommended.

Further chapters explore the use of hydrogels in sandy soils to improve crop production by seizing the constraints of sandy soils. The conversion of nano-agrochemicals in soil-plant systems is also examined, discussing the complex interactions and possible impacts of these materials in the environment. The impacts of nano-agrochemicals on soil microbiomes are considered, highlighting both the possible risks and benefits. This chapter highlights the importance of managing and understanding the effects of nanotechnology on microbial communities in soil, which are essential for soil fertility and health.

Promising applications of hybrid magnetic-nanostructures and nanofibers for wastewater treatment along with the related issues are considered. These discussions underline the need for thorough research to manage the environmental effects of nanotechnology while connecting its benefits to pollution management. The progress and application of nano-biofertilizers for agriculture are underlined, highlighting their potential to increase crop productivity and nutrient management. This chapter explains the ways of combining nanotechnology with biofertilizers to enhance plant growth, thereby contributing to more viable agricultural practices.

The final chapter presents the use of biomass-derived carbon quantum dots to emphasize the possible solutions for wastewater treatment, thus addressing water pollution problems. By exploring the fabrication, characterization and application of quantum dots, the book concludes with a forward-looking perspective of nanotechnology to solve some of the most challenging environmental issues.

As editors, we are deeply grateful to the authors who have contributed their insights and invaluable research, making this book a rich resource for students, researchers, scientists, environmentalists, agronomists, and policymakers. We hope that this collection will stimulate further development and exploration of nanotechnology in agriculture. This will certainly pave the way for future sustainable practices to ensure environmental health and food security.

We would like to thank all contributing authors for their excellent work and unwavering support in creating these thought-provoking and scholarly chapters. We also want to express our gratitude to each reviewer for giving up their valuable time to complete the process. Lastly, we would like to thank our friends, family, and colleagues for their continuous support during this effort. We dedicate this book to the relentless pursuit of knowledge and innovation and to the countless individuals working tirelessly to create a more sustainable and prosperous world.

Kamel A. Abd-Elsalam
(Agricultural Research Center, Egypt)
Muhammad Zahid
(University of Agriculture Faisalabad, Pakistan)

Contents

Preface   iii

1. **Effect of Nanoparticles on Agriculture Production**   1
   Rahat Javaid and *Umair Yaqub Qazi*

2. **Use of Nanoparticles in Moisture Retention and Soil Health Management**   21
   J. C. Tarafdar and *Indira Rathore*

3. **Nanosensor in Soil-Water Agroecosystem**   35
   Ghulam Mustafa, Muafia Akbar, Nauman Sadiq, Bushra Iqbal and *Muhammad Zahid*

4. **Nanomaterials in Water Desalination**   62
   Rahat Javaid and *Umair Yaqub Qazi*

5. **Nanotoxicity Effects of Metal Oxide Nanoparticles on Aquatic and Soil Ecosystems**   79
   Josef Jampílek and *Katarína Kráľová*

6. **Hydrogel Materials in Sandy Soil**   130
   Mujeebat Bashiru and *Noureen Siraj*

7. **Nano-agrochemicals' Transformation in the Soil-plant System**   141
   Iqra Naseer, Sumera Javad, Ajit Singh, Nimrah Azam, Khajista Jabeen and *Mohammad Faizan*

8. **Effect of Nano-based Agrochemicals on Soil Microbiome**   163
   Tayyaba Samreen, Muhammad Ahmad, Sehar Rasool, Muhammad Zulqernain Nazir, Samia Arshad, Faisal Nadeem and *Sehrish Kanwal*

9. **Nanomaterial Pollution in Agricultural Soils**   178
   Farah Noshin Chowdhury and *Md. Mostafizur Rahman*

10. **Hybrid Magnetic Nanostructures for Wastewater Treatment**   208
    Nimra Nadeem, Muhammad Zahid, Usman Zubair, Palwasha Tehseen and *Zulfiqar Ahmad Rehan*

11. **Nutrient Management through Nano-biofertilizers for Smart Agriculture**   228
    Tayyaba Samreen, Sehar Rasool, Aimen Tahir, Umair Riaz, Muhammad Zulqernain Nazir, Sehrish Kanwal and *Sidra-Tul-Muntaha*

12. Nanofibers with Functional Properties for Water Purification   244
Kamel A. Abd-Elsalam, Toka E. Abdelkhalek, Rawan K. Hassan and *Chandra S. Seth*

13. Utilizing Biomass-Sourced Carbon Quantum Dots: An Eco-Friendly Strategy for Wastewater Treatment   267
Muhammad Usama Ghafoor, Kashif Ali, Muhammad Zahid, Saima Noreen, Ghulam Mustafa and *Zulfiqar Ahmad Rehan*

Index   291

1

Effect of Nanoparticles on Agriculture Production

Rahat Javaid[1,*] and *Umair Yaqub Qazi*[2,*]

Introduction

Agriculture is a significant building block defining any country's economy, especially for developing countries. Increasing food production is important in enhancing a country's Gross Domestic Production (GDP). World's population is estimated to be ten billion by 2050, increasing food requirements by 50% (Farooq et al. 2022). Available conventional agricultural technologies are incapable of fulfilling food requirements. The food production rate greatly depends on agricultural technologies and agrochemicals such as fertilizers and pesticides. As an estimation, pesticide, and fertilizer usage globally crossed 4.1 and 125 million tons, respectively (Liu et al. 2021, Ajmal et al. 2022). Applying conventional chemical-based fertilizers, pesticides, and other products harms soil fertility, biodiversity, and ecosystems. Therefore, it is necessary to regulate and modify conventional agricultural technologies to increase food production, which is required to meet the needs of the increasing world's population in a sustainable manner. Moreover, conventional agricultural production is inefficient. For example, conventional pesticides are generally synthetic organic compounds with highly hydrophobic and less stable characteristics, which cause the loss of a significant portion of the applied dosage through volatilization, degradation, and photolysis. According to an estimation, less than 1% of the applied dosage serves the purpose (Huang et al. 2018, Camara et al. 2019). Similarly, conventional fertilizers used for providing essential plant growth nutrients such as potassium (K), nitrogen (N), magnesium (Mg), sulfur (S), phosphorus (P), calcium (Ca), and silicon (Si), known as macronutrients, and boron (B), molybdenum (Mo), iron (Fe), manganese (Mn), nickel (Ni), zinc (Zn), copper (Cu), sodium (Na), and chlorine (Cl), known as micronutrients, are inefficient (Rai et al. 2015). Volatilization and leaching cause an estimated 50% loss of N in conventional fertilizers, which leads to increased production costs and drastically affects the ecosystem and human health (Kalia and Sharma 2020, Mejias et al. 2021). Therefore, it is necessary to regulate and modify conventional agricultural technologies.

[1] Department of Chemical Engineering, University of South Carolina, Columbia, SC, USA.

[2] Department of Chemistry, Benedict College, Columbia, SC, USA.

* Corresponding authors: RQAZI@mailbox.sc.edu; umair.qazi@benedict.edu

Nanotechnology has revolutionarily improved human life by promising breakthroughs in various fields such as electronics, optics, pharmaceuticals, energy, etc. (Javaid and Nanba 2020, 2023, Fatima et al. 2021, Qazi et al. 2022, Qazi and Javaid 2023). The application of nanoparticles in agriculture has been trending for many years to achieve a smart and precise approach to efficient agrotechnology by controlling and regulating release behavior, long-term chemical stability, increased solubility, and target capability (Duhan et al. 2017, Kalia and Sharma 2020). Nanoparticles' structural characteristics and composition can be monitored to attain desired functional efficiency, stability, precision characteristics, and sustainability (Duhan et al. 2017, Kalia and Sharma 2020). Application of various metals, metal oxides, carbon nanotubes, and polymer-based nanoparticles as nano-pesticides and nano-fertilizers showed their potential for sustainably efficient agriculture production (Chhipa 2019, Cruz-Luna et al. 2021, Mahaletchumi 2021, Beig et al. 2022). In addition, applying nanoparticles as nano-sensors has helped in developing smart agriculture production through soil condition detection (Chhipa 2019). Along with various positive impacts of nanoparticles, negative influences such as toxicity to plants and ecosystems have also been reported (Rizwan et al. 2017). Fabrication of nanoparticles through environmentally benign methods greatly enhanced the potential applications of metal and metal oxides-based nanoparticles in agriculture by reducing toxicological impact and enhancing stability (Chhipa 2019). Though nanoparticles can be fabricated to attain desired results and productivity, applying nanoparticles to agriculture is still a relatively new field requiring comprehensive exploration. This chapter will focus on recent trends in using nanoparticles for agricultural production.

Nanoparticles in agriculture

Nano-fertilizers

One of the most significant disadvantages of conventional fertilizers is that only a small concentration, lesser than the minimum requirement, approaches the targeted site due to various factors, including photolytic and microbial degradation, leaching/removal of fertilizers, drift, hydrolysis, and evaporation. According to an estimation, up to 70% N, 90 % P, and 90% P in the applied dose of the fertilizers are lost, causing sustainable and economical losses (Duhan et al. 2017, Bijay-Singh and Craswell 2021, Tara Meghana et al. 2021). Due to the loss of this large concentration, fertilizers are used repeatedly in more significant amounts, which drastically damages the inherent nutrient composition of the soil and causes environmental pollution by affecting flora and fauna. It has been described in the literature that excessive application of conventional fertilizers and pesticides causes a reduced microflora, increased bioaccumulation of pesticides, lack of nitrogen fixation, and affects the birds' habitat (Rajak et al. 2023). Therefore, replacing conventional chemical fertilization with smart, sustainable, and eco-friendly fertilizers is imperative to efficiently fulfill the required nutrient demands and to reduce the negative impacts on the ecosystem (Shaji et al. 2020, Harman et al. 2021).

Fertilizers based on nanoparticles (particles in size of a few nanometers) are called nano-fertilizers. Nano-fertilizers enable site-targeted efficient delivery, increased nutrient uptake, and controlled release of agriculturally required chemicals, minimizing environmental contamination and enhancing cost-effectiveness (Babu et al. 2022, Nongbet et al. 2022). These characteristics of nano-fertilizers are because of their higher surface area to volume ratios, controlled targeting due to nanometer size, high solubility, increased mobility, and relatively low toxic effect of nanoparticles (Babu et al. 2022, Nongbet et al. 2022). For example, the deposition of nanoparticles on the surface of fertilizer particles causes a firm hold of the material because of higher surface tension resulting in controlled release (Duhan et al. 2017). Figure 1 shows the novel characteristics of nano-fertilizers. The application of nanoparticles influences the elongation of roots and shoots, biomass, amount of chlorophyll, and germination of seeds. It has been suggested that nano-fertilizers could move faster inside the plant as compared to conventional fertilizers. This phenomenon was associated with

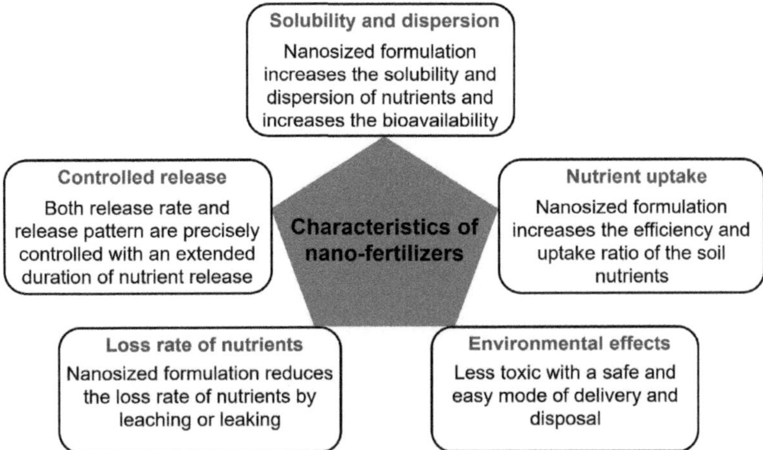

Figure 1. Characteristics of nano-fertilizers.

thermodynamics and higher entropy of nano-fertilizers due to colloidal suspension. The correlation between Gibbs energy and entropy enables quick movement and simple penetration through the cell membrane (Abdel-hakim et al. 2023).

Nano-fertilizers should be fabricated in such a way as to attain all required characteristics, including high solubility, durability, controlled nutrient release, enhanced targeted approach with higher concentration, and less toxicity (Kumar et al. 2023). Various natural and synthetic materials were used to create multiple formats of nano-fabricated fertilizers. Nano-fertilizers can be categorized into three types: (1) fertilizers in which nanoparticles are used as supports/carriers to control the fertilizer or nutrient release are called nano-supported fertilizers, (2) those in which the fertilizers themselves are made in nanoscale are nano-sized fertilizers, and (3) fertilizers in which nanoparticles are used for coating of the regular sized fertilizers (Shao et al. 2022).

Nano-supported fertilizers constitute the extensively studied nano-fertilizers that can be prepared by various methods. There are two basic concepts of fabrication: (1) entrapment, in which nutrients are entrapped or encapsulated in nano-carriers hindering direct exposure; (2) adsorption method, in which nutrients are deposited on nano-carriers by physical adsorption (Shao et al. 2022). The entrapment method is a very common technique to fabricate nano-fertilizers. Entrapped nanoparticles show a slow and controlled release rate. A slower release rate of nutrients promotes efficient usage of nutrients with reduced wastage and accumulation in the soil. Kottegoda et al. (Kottegoda et al. 2017) fabricated urea-deposited hydroxyapatite (HA) nanoparticles using a urea/HA ratio of 6/1 by adding H_3PO_4 to a suspension of $Ca(OH)_2$ and urea. Due to the large surface area of HA, a large concentration of urea could be deposited on the surface. The formation of strong N–C–N bond between the carbonyl groups of HA and the amine groups of urea contributed to the slower and controlled release of N for one week in the soil, increasing the plant growth rate. Moreover, HA is a good source of P and is bio-compatible; therefore, it can be a potential source as the smart carrier of nutrients. Likewise, Tarafder et al. (Tarafder et al. 2020) fabricated nano-fertilizer by incorporating nanoparticles of HA with urea, $Fe(OH)_2$, $Zn(OH)_2$, and $Cu(OH)_2$, that slowly and continuously released the nutrients for more than two weeks. This research claimed 1% reduced usage of nano-fertilizer as compared to conventional fertilizer. Some researchers applied nanoparticles of amorphous calcium phosphate (ACP) to control the release rate of urea (Ramírez-Rodríguez et al. 2020, Carmona et al. 2021, 2022) by depositing urea onto ACP. Due to the relatively larger surface area of ACP, simultaneous release of P and Ca was possible at monitored release rates. Chitosan is a promising natural nano additive. As chitosan gains positive charges in an acidic medium due to the amino groups, it can be nano-sized by the ionic gelation. Enzymatic hydrolysis and mass transfer resistance properties of chitosan help in the controlled release of nutrients

(Minh et al. 2019, Dhlamini et al. 2020). Liposome nanoparticles have also been employed as fertilizer supports (Karny et al. 2018). Nutrients deposited in liposome could be prepared by lipid-film hydration/extrusion or solvent injection techniques. Integrity disruption by osmotic pressure causes an intracellular release of nutrients. Researchers have also used nanofibers and polymers to fabricate nano-fertilizers by entrapment method. For example, Nooeaid et al. (Nooeaid et al. 2021) deposited N, P, and K onto the nanofibers through electrospinning, where PVA was applied in the core and hydrophobic PLA was used in the shell. According to the studies, the stimulated release could monetize the release of nutrients more than the passive release (Li et al. 2019). Considering the fluctuation in the requirement of Fe by crops with a change in temperature conditions, Chi et al. (Chi et al. 2018) used copolymer (ethylene oxide/propylene oxide) to capture Fe. Due to the temperature-sensitive characteristics of ethylene oxide/propylene oxide copolymer, the Fe release rate was monitored according to the crop requirement. Another study reported the fabrication of pH-sensitive nanoparticles by depositing Fe on carboxyl cellulose though chelation. These nanoparticles could be broken down in acidic conditions (Wang et al. 2016). The adsorption process is preferred for synthesis of nano-fertilizers from porous or carbonaceous feedstocks. Synthesis of biochar as nano-support from low-value biomass or agriculture wastes has gained researchers' attention because of its relatively lower cost and higher physical adsorption abilities (Lateef et al. 2019). Moreover, biochars are excellent soil conditioner as they have exceptional swelling capacity which can retain water benefiting the soil (Lateef et al. 2019, Khan et al. 2021). Researchers observed slow and controlled release of various nutrients such as K, P, N, Ca, Na, Mg, and Zn using nano-fertilizers prepared from biochar-based adsorption method. Zn-based mesoporous materials like Zn aluminosilicate (Naseem et al. 2020), Zn-layered hydroxide-nitrate and Zn-layered hydroxide phosphate (Khadiran et al. 2021) were used as supports for the release of nutrient such as P, Zn and N. Zeolites have high porosity and have been used for the fabrication of nanocomposites for Fe release (Saleh and Webster 2020).

Nano-sized fertilizers are synthesized to enhance the adsorption of insoluble nutrients to the plants. HA nanoparticles are known for higher P uptake efficiency (Xiong and Wang 2018), which was further modified with organic acids (Jeon and Adamiano 2020) or urea and starch (Ribeiro 2015) to increase its potential as a nano-fertilizer. Ceischi et al. (Cieschi et al. 2019) fabricated an iron-humic nano-fertilizer showing increased Fe uptake by soybean plants in iron-deficient soil. Nano-sized fertilizers generally have smaller sizes than the leaf stomata enabling them efficient and favorable as foliar fertilizers. Direct uptake of nano-fertilizers by plants through leaves helps in avoiding the shortcomings of soil applications (Kaboli et al. 2021). Shebl et al. (Shebl et al. 2020) synthesized foliar nano-fertilizer from manganese zinc ferrite nanoparticles through the microwave-assisted hydrothermal process, which increased the yield of the squash plant (*Cucurbita pepo* L). Many research studies used metal-organic frameworks (MOFs) to synthesize nano-sized fertilizers (Abdelhameed et al. 2019, Wu et al. 2019).

The process of wrapping the fertilizers is used to avoid water dissolution, reducing the loss or wastage of nutrients. Conventional petroleum-based wrapping materials are hydrophobic, whereas biodegradable materials are mostly hydrophilic. Therefore, it isn't easy to control the release of nutrients by using conventional wrapping materials (Liu et al. 2018). The use of nanomaterials as wrapping agents helps in controlling the characteristics and release rate of the nutrients. Yang et al. used nano-silica (SiO_2) along with methylene diphenyl diisocyanate (MDI) and bio-polyol to increase the stability and reduce the release rate of the urea from the wrapped tablet (Liu et al. 2018). They further refined the wrapping by using nano-SiO_2 (Liu et al. 2018) and nano lauric acid copper (Zhang et al. 2019). The nanoparticles blocked the holes, avoiding the dissolution of urea, thus resulting in sustained and controlled release.

Nano-pesticides

In agriculture, crop pests are a significant concern because they cause diseases that harm crop development and production and lower farmers' yields. According to an estimation, crop diseases cost $220 billion in lost crops each year globally (Chhipa 2019). Different pathogens, including fungi, bacteria, insects, viruses, nematodes, parasites, and protozoa, are responsible for plant diseases. Different conventional chemical and biological pesticides have been used to control them, but their inappropriate use damaged the ecological balance and promoted pests' resistance. Therefore, developing novel approaches to control environmental imbalance is highly required to lessen the severe effects on agriculture caused by resistance of crop pathogens. The application of nanoparticles in crop protection products has improved crop productivity (Chhipa 2019). For this purpose, many types of metal nanoparticles have been reported as pesticides. Nanoparticles of silver (Ag), copper (Cu), SiO_2, and nano-formulations of chemical pesticides, including hexaconazole and sulfur (S), showed effective pesticide activity against a variety of insects and fungi. Nano-formulation results in enhanced solubility of active chemicals, shelf life, and nano-carrier efficiency via controlled release (Chhipa 2019). Currently, nano-emulsion is a growing research area in the field of nano-pesticides. Due to nano-size and higher absorption, nano-emulsion improved the ability of active ingredients to be retained by plants. For example, the nano-emulsion of Permethrin with neem oil demonstrated greater efficacy for larvicidal application than the active component (Anjali et al. 2010, 2012). Additionally, the effective targeted distribution offered by nano-emulsion reduced the effects on unintended species (Chhipa 2019). Figure 2 shows nano-pesticides' impact on bioavailability and efficiency (Huang et al. 2018).

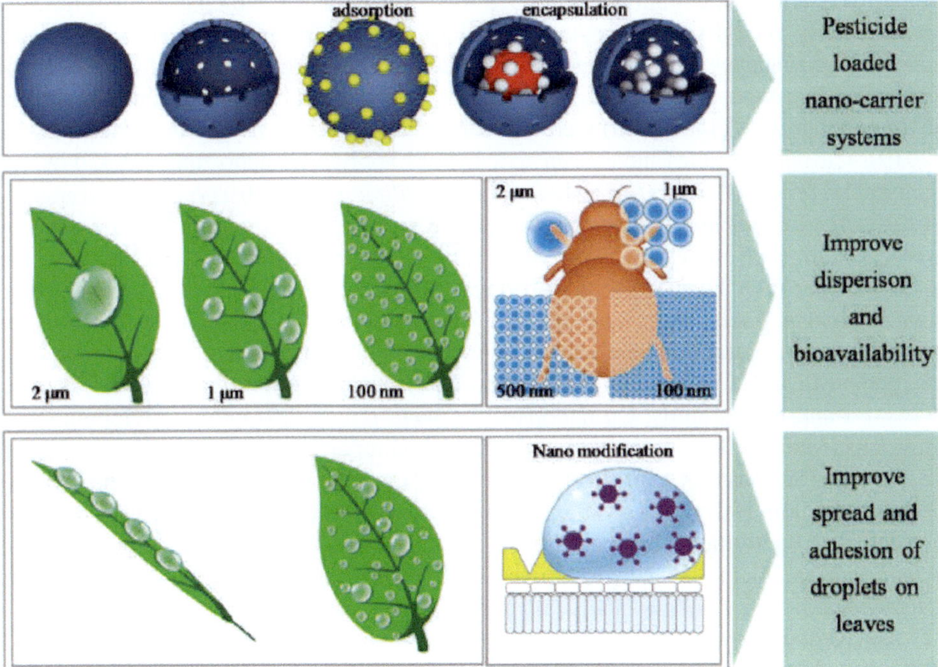

Figure 2. Nano-pesticides enhance bioavailability and efficiency. Reprinted from (Huang et al. 2018) (open access publication).

Nano-fungicide

Fungal infections account for severe problems in crop growth, responsible for more than 70% of agricultural diseases reducing crop yield drastically (Baker et al. 2017). Many metal nanoparticles and nano-formulations have been proposed to combat fungal infections as nano-fungicides. Due to the advantageous characteristics of Ag, such as antimicrobial, antifungal, and antioxidant efficiency, easy production, non-toxicity, and biocompatibility, these nanoparticles have been extensively studied and applied in agrochemical products, especially to control phytopathogenic fungi (Hazarika et al. 2022a). Ag nanoparticles' size, shape, and structure greatly influence the antifungal activity. Mostly Ag nanoparticles applied as antifungal agents were spherical, possibly because spherical-shaped nanoparticles are easy to synthesize. In general, Ag nanoparticles prepared by different methods are polydisperse; therefore, the effect of size cannot be studied in detail (Hazarika et al. 2022a). Singh et al. reported Ag nanoparticle-based fungicides as more efficient than conventional chemical fungicides (Singh et al. 2021). Several studies have reported the efficacy of Ag nanoparticles against various phytopathogenic fungi (Jebril et al. 2020, Nguyen et al. 2020, Sahayaraj et al. 2020, Al-Otibi et al. 2021). Elamawi and El-Shafey (Elamawi et al. 2013) reported that Ag nanoparticles decreased the growth of a fungal pathogen called *Magnaporthe grisea*, causing rice blast disease.

Cu nanoparticles demonstrated high antifungal efficacy against many phytopathogenic fungi, affecting them intracellularly and extracellularly. As a result, Cu nanoparticles could be an ideal choice for controlling and managing various agronomical diseases. Smaller nanoparticles of Cu (10–70 nm) penetrate the cell membrane more readily, resulting in a rupture and loss of cell contents (Malandrakis et al. 2020). At the same time, larger Cu nanoparticles (80 to > 100 nm) are reported to exhibit antifungal activity by limiting mycelium and spore formation (Pariona et al. 2019). Cu nanoparticles are most commonly produced in spherical shape, which can easily penetrate the membrane gaining access to the enzymes that activate cellular inhibition (Bramhanwade et al. 2016). Among other shapes, faceted nanoparticles (200–500 nm) demonstrated high efficacy against *F. oxysporum*, *F. solani*, and *Neofusicoccum* sp. (Pariona et al. 2019). Truncated octahedron-shaped Cu nanoparticles (14–37 nm) were also effective against *F. oxysporum* (Hermida-Montero et al. 2019). Cu nanoparticles have been extensively applied as nano-fungicides. For example, Cu nanoparticles produced from *Citrus medica Linn. (Idilimbu)* were observed for antifungal activity against *Fusarium oxysporum*, *Fusarium culmorum*, *Fusarium graminearum*, *Fusarium solani*, and *Neofusicoccum* sp. (Pariona et al. 2019). According to Bramhanwade et al. (Bramhanwade et al. 2016), Cu nanoparticles have a strong antifungal effect on the most common pathogenic fungi *Fusarium culmorum*, *Fusarium oxysporum*, and *Fusarium equiseti*. Malandrakis et al. compared the influence of dose concentration of Cu nanoparticles by utilizing different concentrations ranging from 50–1000 ppm against *A. alternata*, *B. cinerea*, *F. solani*, *C. gloeosporioides*, *V. dahlia*, *M. fructicola*, and *F. oxysporum*. In this study, Cu nanoparticles were reported to inhibit phytopathogens at all concentrations (Malandrakis et al. 2019). Other metal nanoparticles investigated as nano-fungicide include selenium (Se) (Nandini et al. 2017), nickel (Ni) (Jagana et al. 2017), magnesium (Mg) (Jagana et al. 2017), palladium (Pd) (Osonga et al. 2020), and iron (Fe) (Asghar et al. 2018), demonstrating efficient antifungal activities. Nanoparticles of Se were biologically synthesized using six strains of *Trichoderma* sp. including *T. harzianum*, *T. atroviride*, *T. asperellum*, *T. longibrachiatum*, *T. brevicompactum*, and *T. virens*, and evaluated against *S. graminicola*. Nano-fungicide produced from *T. asperellum* showed the highest efficiency as antifungal agent (Nandini et al. 2017). Chitosan nanoparticles were reported efficient to reduce blast infections in rice (Manikandan and Sathiyabama 2016). Oh et al. reported efficient treatment of phytopathogenic fungi including *Phytophthoracapsici*, *Colletotrichum gelosporidies*, *Fusarium oxysporum*, *Gibberella fujikuori*, and *Sclerotinia sclerotiorum* in tomato plants using chitosan nanoparticles (Oh et al. 2019).

Nano-bactericide

Nanoparticles have also been employed for synthesis of nano-bactericides. *Ralstonia solanacearum* (Rsol), *Clavibactermichiganensis* (Cm), *Erwinia amylovora* (Eam), *Xanthomonas campestris pv.*, *Dickeyasolani* (Dsol), and *Campestris* (Xcc) are a few examples of pathogenic bacteria. According to Mansfield et al. (Mansfield et al. 2012) and Kamber et al. (Kamber et al. 2012), Eam is a Gram (–) bacterial phytopathogen affecting the *Pomoideae* and *Maloideae* subfamilies of Rosaceae plants. On the other hand, Cm is a Gram (+) bacterial pathogen causing bacterial canker on tomatoes (Hausbeck et al. 2000). Contrarily, Rsol is a Gram (–) bacterial pathogen that impacts many plant species from over 44 families (Hazarika et al. 2022b). The rod-shaped Gram (–) bacterium Xcc frequently inhabits the leaves of kale, cauliflower, cabbage, radish, broccoli, and Brussels sprouts (Vicente and Holub 2013). Another Gram (–) rod-shaped bacteria is Dsol. According to Toth et al. (Toth et al. 2011), dsol causes blackleg (softening and blackening of the stem base) and soft rot (collapse of the internal tuber tissue).

Nanoparticles have also been applied as nano-bactericide or antibacterial agents against plants' bacterial pathogens. Ag nanoparticles are referred to as commonly used antibacterial agents. Dzimitrowicz et al. (Dzimitrowicz et al. 2018) prepared silver nanoparticles, effective against bacterial phytopathogens such as *Xanthomonas campestris pv. Campestris, Clavibactermichiganensis, Ralstonia Erwinia amylovora, solanacearum,* and *Dickeyasolani.* Another study reported the fabrication of Ag nanoparticles from *Piper nigrum* extract and their application as antibacterial agents against *Erwinia cacticida* and *Citrobacter freundii* (Paulkumar et al. 2014). Aravinthan et al. (Aravinthan et al. 2015) found that Ag nanoparticles derived from the extract of sun root tuber showed antibacterial efficacy against *Ralstonia solanacearum* and *Xanthomonas axonopodis.* Cu is another metal reported for its efficiency and potential as nano-bactericides. According to reports, Cu nanoparticles can effectively inhibit the *Xanthomonas oryzae* and *Xanthomonas campestris* pathogens causing leaf spot and rice blast disease. In addition, Mondal and Mani (Mondal and Mani 2012) observed that Cu nanoparticles-based nano-bactericide was effective against the bacterial blight disease in pomegranates at a concentration of 0.2 ppm, which was 60,000 times lower than the concentration required for the conventional bactericide, i.e., copper oxychloride. A reduction in the required amount of the pesticide increases the cost-effectiveness with greater environmental balance preservation. Kalaba et al. (Kalaba et al. 2021) fabricated zinc oxide (ZnO)-based nanoparticles effective against *Erwinia amylovora* bacteria. Cai et al. (Cai et al. 2018) found magnesium oxide (MgO)-based nanoparticles efficient against agricultural antibacterial agent *Ralstonia solanacearum.* There is very limited research on the use of nanoparticles as antimicrobial agents. There is a great scope in the development and analysis of various nanoparticles as nano-bactericides.

Nano-insecticide

Pests pose an increasing threat to agriculture by lowering crop production and, consequently, crop quality. Conventional pesticides applied to crops or soil harm the ecosystem and unintended species (Jafir et al. 2021). In contrast, nano-pesticides have less environmental impact and cause fewer losses than conventional pesticides. They thereby lower expenses, toxicity to humans, phytotoxicity, and harm to both the plant and unintended species (Hosamani et al. 2019) with efficient delivery systems (Hazarika et al. 2022a).

Nanoparticles based on Ag have been extensively studied and reported as efficient nano-insecticide. Ag nanoparticles prepared from the leaf extract of *Borago officinalis* demonstrated efficient larvicidal activity against cotton leafworm *Spodoptera littoralis* (Hazaa et al. 2021). Likewise, Ag nanoparticles prepared from the leaves of *E. prostrata* were reported to be effective in controlling *Sitophilusoryzae.* According to a study by Jafir et al. (Jafir et al. 2021), Ag nanoparticles fabricated from the leaf extract of *Ocimum basilicum* were more efficient against *S. litura* than conventional insecticides. Hosamani et al. (Hosamani et al. 2019) investigated the effectiveness of Ag nanoparticles as an insecticide prepared from the seeds of soya beans. Kasmara et al. (Kasmara et al. 2018) fabricated Ag nanoparticles from the leaf extract of *Lantana camara* and applied them as an efficient nano-insecticide against *Spodoptera Litura*. Ag nanoparticles prepared from the pomegranate peel extract demonstrated higher efficiency as an insecticide against *S. litura* (Bharani et al. 2017). Other than Ag nanoparticles, titanium oxide (TiO_2)-based nanoparticles have also been reported efficient as insecticides against specific pests (Singh et al. 2021). Werdin González et al. (Werdin González et al. 2015) synthesized polyethylene glycol (PEG)-based nanoparticles with essential oils and utilized them against *Blatella germanica*. The effectiveness of impregnated Cu as a nano-insecticide was reported to increase with the production of chitosan-based nano-gels (Brunel et al. 2013). El Bendary and El Helaly (El-Bendary and El-Helaly 2013) found SiO_2-based nanoparticles efficient against *Spodoptera littoralis*, and broken beetle *C. chinensis*. In this study, the researchers observed that nanoparticles of SiO_2 decreased the fertility of *Spodoptera littoralis*, reducing insect population and lessening crop loss (El-Bendary and El-Helaly 2013). Nanoparticles of SiO_2 have been reported as effective nano-insecticides against various insect pests, including the coconut mite, white fly, mustard weevil, and rice weevil (Chhipa 2019). According to research by Barik, Kamaraju, and Gowswami (Barik et al. 2012), the insecticidal effect of SiO_2 nanoparticles is because of physio sorption in cuticle lipids of the insects, promoting insect killing. According to Stadler et al. (Stadler et al. 2010), *Rhyzoperthadominica (F.)*, and *Sitophilus oryzae* L., two prominent insects found in preserved food commodities, were successfully controlled by alumina nanoparticle-based pesticides.

Nano-sensors

To reduce environmental damage, care must be taken when using fertilizers, pesticides, and heavy metals in agricultural techniques. Advanced technologies should be established to regularly check the condition of the soil and any potential disease outbreaks. For the real-time monitoring of vast field regions, accurate sensing technologies with convenience and portable applications must be developed. The use of nanoparticles in biosensors improved their efficiencies for agricultural applications. Extensive research and remarkable advancements in nanotechnology and analytic techniques with enhanced sensitivity enabled the development of responsive sensors. Nano-sensors have shown their potential in wide applications, including monitoring soil parameters such as nutrients and pH, crop cultivation and harvesting, and detecting pathogens. The potential of nanoparticles to develop advanced sensors is because of their distinct surface characteristics, thermal and electrical properties, higher sensitivity, and detection limits (Idris et al. 2023). In agriculture, nanoparticles-based sensors are used for multiple applications, such as detecting residual pesticides, pathogens, soil characteristics, and plant sensors monitoring plant health, stress, etc.

Nano-biosensors for the detection of residual pesticides

Monitoring the amounts of used pesticides to attain a sustainable and smart agriculture approach is vital. Gas chromatograph/Mass spectrometer (GC/MS) and liquid chromatography are two conventional methods for the detection of pesticides. These techniques result in high reproducibility, have low detection limits, and require intensive efforts (collection and extraction of samples and then analysis) and highly equipped labs. Various nanoparticles, including metal-based nanoparticles, carbon nanotubes (CNTs), graphene, and quantum dots (QDs), have been applied for the detection of heavy metals, pesticides, and other pollutants in water and soil (Niu et al. 2023). Adding nanoparticles with biosensors provides the benefits of fast detection, low sample requirement, and enhanced sensitivity. The array-based method is commonly used for nano-biosensors in which biomolecules are attached to a substrate enabling the simultaneous detection of numerous analytes

(Xu et al. 2023). Different measurement methods used to analyze the binding of the target molecule with biosensors include observing changes in color, fluorescence, magnetic sensing, electrical potential, etc. Cadmium telluride (CdTe) nanoparticles, also known as QDs, were used to detect 2,4-dichlorophenoxyactic acid (Vinayaka et al. 2009). Due to their fluorescent characteristics, QDs can be used to analyze other dangerous pesticides, including atrazine (Cummins et al. 2006, Vinayaka and Thakur 2010), methiocarb (Hua et al. 2010), and pyrethroid (Kranthi et al. 2009). Several kinds of nanoparticles, specifically QDs, and gold (Au) were applied in optically based immunosensors to improve signal amplification. To identify the pesticide chlorpyrifos in drinking water, Chen et al. used QDs conjugated with streptavidin (Chen et al. 2010). According to Lisha and Pradeep (Lisha et al. 2009), Au nanoparticles can be used to identify onsite water pesticides from agricultural lands and other resources. Au nanoparticles were used in immunochromatographic test strips to detect various pesticides. These Au nanoparticles-based immunoassay strips can simultaneously detect carbofuran and triazophos. A critical enzyme for human health called acetylcholinesterase (AChE) is inhibited by most pesticides. This enzymatic condition can be a powerful tool in making biosensors. For example, this enzymatic reaction was utilized for localized surface plasmon resonance fiberoptic biosensor that tracks AChE inhibition to quantify the concentration of a pesticide (Guo et al. 2009). The enzyme, mounted on the surface of Au nanoparticles, modifies the light when inhibited by a pesticide (such as paraoxon), making it possible to quantify the enzyme's activity.

Nano-sensors for the detection of pathogens

Nanoparticles can be used as a quick diagnostic biomarker for direct or indirect detection of different plant pathogens or particular diseases. Singh et al. (Singh et al. 2010) reported the application of surface plasmon resonance with Au nanoparticle-based immunosensors to identify the Karnal bunt (Tilletia indica) disease in wheat. Fluorescence Si nanoparticles were used to detect *Xanthomonas axonopodis pv. vesicatoria*, causing bacterial spot disease in Solanaceae plants (Yao et al. 2009). Although this field is new and still developing, it has potential for practical field applications. Another crucial tactic is utilizing nano-chips, a type of microarray that uses fluorescent oligo probes to detect hybridization (López et al. 2007). *Phytoplasma aurantifolia*, which causes witches broom disease of lime, was detected using QD fluorescence resonance energy transfer-based immunosensors demonstrating great sensitivity of 100% (Rad et al. 2012). Plants develop distinctive volatile signatures whenever they become infected with any pathogen and identifying these signature molecules helps confirm the existence of pathogens. TiO_2 and SnO_2 nanoparticles were used in an electrochemical sensing device to identify p-ethylguaiacol, which was emitted by plants after they became infected with the fungus *Phytophthora cactorum* (Fang et al. 2014). Both TiO_2 and SnO_2 nanoparticles demonstrated high sensitivity.

Nano-sensors for the plant sensing system

Plant sensing system efficiently manages expensive agrochemicals such as nutrients, pesticides, and fertilizer and reduces plant stress caused by the changing climate. For boosting total crop output, it is necessary to investigate and apply novel and efficient plant sensing techniques, which can enhance the tolerance of the plant against stress and the effective use of nutrients and water. Sensors based on nanoparticles can interact via optical signals to monitor the real-time plant health status. These sensors are utilized to monitor crop growth, efficient usage of limited resources, and precise detection of stress (Giraldo et al. 2019). The conventional methods to analyze plant health and the possibility of any disease involve the application of Infrared and Raman spectroscopic techniques, which are expensive and give results with low signal-to-noise ratios (Wilson et al. 2023). Recent advancements based on smart plant sensing systems enhanced the signal-to-noise ratio and can be directly accessed with smartphones and imaging cameras (Wolfert et al. 2017, Padilla et al. 2018). A smart sensing system based on nanoparticles can monitor plant resource shortages or stressors in real time using key signaling molecules like NO, Ca, sucrose, glucose, and ROS, as well as plant hormones like jasmonic acid, ABA, methyl salicylate, and ethylene (Giraldo et al. 2014, 2015). Because of special characteristics such as thermal and electrical properties, surface characteristics, increased detection limits, and sensitivities, nanoparticles have potential applicability to develop efficient sensing systems (Yao et al. 2014, Hong et al. 2015). Application of nano-sensors in plant sensing systems provides fluorescence in transparent-background windows with lesser photobleaching of tissue allowing the measurements of analytes with high resolution at the molecular level within milliseconds (Guo et al. 2014, Hong et al. 2015). QDs-based nano-sensors are fluorescent and used within visible and near-infrared regions to assess plant health (Li et al. 2018). Molecules with short lives (NO, ROS, Ca, nitroaromatics, glucose) at specific sites of the leaf could be monitored using single-walled carbon nanotubes (SWCNTs)-based nano-sensors with high resolution (Giraldo et al. 2014, 2015, Wong et al. 2017). A changed or different amount of analyte gives rapid changes in the spatiotemporal pattern with time and locations (Kruss et al. 2014). Kinetic simulations for diffusion and stochastic help to manage complex data under various stress conditions and resource deficiencies. Lee et al. (Lee et al. 2014) reported the application of SWCNT and graphite-based electrodes on the leaf's surface to monitor gaseous components present in traces (less than 5 ppm). Esser et al. (Esser et al. 2012) reported nano-sensors based on Cu complexes and SWCNT to detect ethylene coming from plant fruits. Nano-sensors based on CNT are available in markets and are used for the measurement of ethylene, which is a hormone of plants and an essential parameter for fruit ripping. Moreover, the application of SWCNTs combined with bombolitin peptide has been reported to detect the presence of picric acid, which acts as explosives for plants (Wong et al. 2017). Likewise, Ca sensors based on green fluorescent protein serve as plant stress detectors. Nano-sensors enable real-time monitoring through meteorological stations, hyperspectral cameras, or smartphones to enhance high yielding plant traits and remove stress. To protect the plant from salinity and heat, Graham et al., and Wu et al. (Graham et al. 2015, Wu et al. 2017, 2018) reported the delivery of CeO_2 and ZnO nanoparticles by scavenging reactive oxygen. Likewise, the delivery of CuO and nanocrystals was reported to protect plants from fungal disease (*Citrullus lanatus*) and cold damage (Alhamid et al. 2018, Borgatta et al. 2018), respectively. Though nano-sensors provide ease and efficiency for monitoring plant health, applying them to the actual agriculture field conditions and improving their communication with electronic devices is still required. If the field tests are successful, the nano-sensor can offer real-time monitoring for fertilizers, nutrients, water, and pesticides according to the plant requirements.

Uptake of nanoparticles by plants

To achieve desired results and benefits from nanoparticles, it is essential to understand the mechanism of their uptake by plants and the parameters, which can affect the fate of nanoparticles after their application to plants. Nanoparticles enter the plant through the root, shoot, leaf tissues, wounds, and root junctions (Jia-Yi et al. 2022). Nanoparticles can travel through the plant's system by phloem loading, bulk flow, or diffusion. Figure 3 shows the transportation of nanoparticles in the plant cell. Designing efficient nanoparticles for agricultural usage requires a thorough understanding of nanoparticles' absorption and transport behavior in plants. Likewise, knowledge of the plant uptake and bioaccumulation of nanoparticles can serve for biological safety and provide guidelines for the safe application of nanoparticles.

Uptake by plant leaves (Foliar uptake)

The uptake of nanoparticles could be done by plant leaves, known as foliar uptake. Nanoparticles are often sprayed onto the leaf surface during agricultural treatments, where they settle and get absorbed by plants via the stomata or cuticle. The cuticles on the leaf epidermis are mostly covered with wax, cutin, and pectin, which prevents water loss from the leaves during plant growth (Wang et al. 2023). It is a critical natural barrier to stop nanoparticles from entering plant leaves (Yang

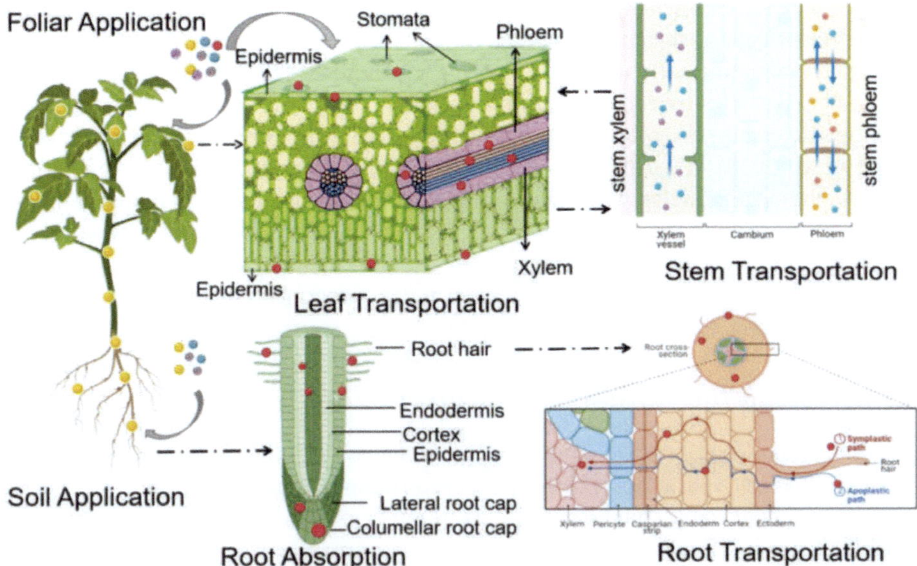

Figure 3. Mechanism of nanoparticles' uptake by the plant. Reprinted from (Wang et al. 2023) (open access publication).

et al. 2015, Pérez-de-Luque 2017). On the surface of stratum corneum, there are two types of channels: one is lipophilic, and the other is hydrophilic (Avellan et al. 2019). Both hydrophilic and lipophilic channels have a diameter range of 0.6–4.8 nm (Eichert et al. 2008). The hydrophilic channels transport nanoparticles smaller than 4.8 nm through diffusion (Banerjee et al. 2019), whereas leaves can take up nanoparticles with lipophilic characteristics via diffusion through the lipophilic channels on the surface of the cuticle (Bussières 2014). Applying confocal fluorescence microscopy, Hu et al. demonstrated that smaller carbon dots (2 nm) infiltrated cotton leaves from cuticles (Hu et al. 2020). However, plants' ability to absorb nanoparticles through the epidermis is constrained by the smaller diameter of the channels in the cuticles. The researchers suggested uptake by the stomatal channel as another way for nanoparticle uptake by the leaf. In plants, the stomata on the surface of the leaves regulate the exchange of gases and water and range in size from 10 to 100 μm. The precise size exclusion limit (SEL) of the stomata is still unknown because of the stomata's distinctive geometric design and physiological function. After entering mesophyll cells through the cuticle and stomata, depending on the particle size and surface charge, nanoparticles can be transported in plants by the extracellular or plastid route which includes the cell wall, channels within the cell walls, lamella, and xylem (Chen et al. 2010, Miralles et al. 2012). In the protoplast pathway, the transportation occurs through intercellular channels called plasmodesmata, having a diameter of 2–20 nm (Cornu et al. 2020). Nanoparticles that cross the plasmodesmata enter the cytoplasm and go to the Casparian strip and plant endothelium layer (Miralles et al. 2012) (Figure 3). The studies show that nanoparticles smaller than 50 nm are transported via the plastid route, whereas the apoplast route transports nanoparticles within a range of 50–200 nm (Raliya et al. 2018). Nanoparticle transportation from the plant leaf to the root occurs through the vascular system–phloem (Bussières 2014).

Uptake by plant roots

The uptake on the surface of roots takes place by absorption. As the root hair has a negative charge due to the discharge of mucus or organic acids, the positively charged nanoparticles easily accumulate on the roots (Zhao et al. 2012, Lv et al. 2019) (Figure 3). Plant root epidermis functions like that of the plant leaf surface. The presence of semipermeable epidermal cells with small pores blocks the absorption of large nanoparticles (Pérez-de-Luque 2017). Nanoparticles can enter the central column or xylem of the root when the exodermis is absent (Péret et al. 2009). After getting access to plant tissue, plant cells absorb nanoparticles through different pathways including the ion route, endocytosis, and interacting cell membrane proteins (Lv et al. 2019). Some researchers reported the absorption of nanoparticles by roots, which then entered cells by hydrophilic route. However, the entry of nanoparticles into cells through this channel is relatively constrained because of the small pore size (Li et al. 2020). Absorption of nanoparticles by endocytosis is another important way. In this process, the nanoparticles with sizes less than 1 μm enter the cell through the invagination of the plasma membrane; therefore, this process has no selectivity for particle size (Ma and Yan 2018). After crossing the cell wall, the symplast or exoplast pathway allows nanoparticles to reach the endothelium layer from the epidermis. These particles subsequently go through xylem vessels to the plant's above-ground portion (Tripathi et al. 2017).

Factors affecting the uptake of nanoparticles

Various factors, including the size of nano particles and charge on the surface, duration and amount of exposure, and plant type affect nanoparticle absorption and transportation. The influence of nanoparticle size is considered extensively in plant uptake. The optimum size of nanoparticles for efficient uptake depends on the specific site of the plant, such as leaves, roots, etc., the composition of nanoparticles, and the type of plant. According to the literature, nanoparticles based on metal having less than 50 nm size may be able to infiltrate plant leaves through the stomatal channel. The ability of leaves to absorb nanoparticles reduced as particle size increased (Avellan et al. 2019). Likewise, it has been found that plant roots may absorb nanoparticles with less than 10 nm diameter. For example, Au nanoparticles with an average diameter of 3.5 nm were reported to be absorbed by the roots of Viciafaba L. (Hu et al. 2022), and nanoparticles of CeO_2 with an average of 8 nm diameter (Zhao et al. 2012) were absorbed by the roots of maize. Another study on wheat showed the absorption of TiO_2-based nanoparticles having a particle size range of 36–140 nm through wheat roots. The absorption rate decreased with an increase in the nanoparticle size, and TiO_2 nanoparticles larger than 140 nm could not be absorbed (Larue et al. 2012). It is commonly accepted that metal-based nanoparticles larger than 100 nm diameter are difficult for plants to absorb through the roots (Banerjee et al. 2019). On the other hand, it is intriguing to learn that nanoparticles with particle sizes greater than 100 nm made of Si or natural polymers could be absorbed by plant roots. For instance, the Arabidopsis roots absorbed Si nanoparticles with a 200 nm diameter (Slomberg and Schoenfisch 2012). By application of transmission electron microscopy and confocal microscopy, it was demonstrated that sugarcane roots could absorb zein nanoparticles (135 nm) (Prasad et al. 2018).

In addition to size, nanoparticles' penetration into plant mesophyll tissue is also influenced by their shape and charge. Different structures of nanoparticles attribute to different contact angles and surface areas affecting the absorption properties (Sun et al. 2021, Zhu et al. 2021). When comparing gold nanoparticles with different shapes, Zhang et al. discovered that nanoparticles with rod-like shapes were more readily absorbed by Arabidopsis leaves (Zhang et al. 2022). Plant leaves can absorb nanoparticles that are both positively and negatively charged. Sun et al. observed the absorption of both positively and negatively charged graphene quantum dots (GQDs) on maize leaves via stomata (Sun et al. 2022). Similarly, Zhu et al. (Zhu et al. 2021) confirmed by confocal microscopy that both positively charged F-P-ZnO nanoparticles and negatively charged F-N-ZnO nanoparticles gathered at the stomata of wheat leaves. Additionally, positively charged nanoparticles were more easily absorbed than negatively charged ones. This observation was supported by the presence of electrostatic attraction between the negatively charged plant cell walls and positively charged nanoparticles. Compared to plant leaves, the phenomenon of nanoparticle absorption is slightly different in plant roots (Spielman-Sun et al. 2017). The plant roots usually have negative charge on cell walls. Due to the presence of electrostatic affinity between nanoparticles with positive charge and the cell wall with negative charge, nanoparticles tend to accumulate on the surface of the roots but are unable to penetrate the root tissues (Bosker et al. 2019).

Conclusion

Nanoparticles have demonstrated substantial applications in agrochemical products such as nano-pesticides, nano-sensors, and nano-fertilizers. Due to controllable nano-sized release and site-specific delivery, nanoparticles can potentially contribute to the development of effective smart agriculture systems. Moreover, due to the extensive research in synthesis techniques of nanoparticles, their characteristics such as shape, size, and composition and their efficiencies can be controlled and modified according to the mode and site of application. Nanoparticles have also demonstrated potential applications as nano-sensors in detecting residual pesticides or pathogens and ensuring consumer food safety. Nanoparticle encapsulation methods have been proved very efficient in alleviating nutrient wastage. Although the application of nanoparticles in agriculture is increasing, it is still highly required to discover the threshold amounts for each crop system to avoid toxicity. Despite the enormous potential applications of nanoparticles in agricultural production, their possible unforeseeable risks should also be considered. For the efficient and cost-affect application of nanoparticles, it is highly required to realize the fate of nanoparticles in the environment and the mechanism of uptake by plants. Nanoparticles should be designed in such a way that they can anchor to the plants' roots, to the soil, or the organic substances in the soil. For this purpose, it is important to realize the involved mechanisms between the structures of the targets and the nanoscale delivery system. This advancement will reduce the usage amounts of various agrochemical products and waste production, resulting in sustainable agriculture.

References

Abdel-hakim, S.G., Shehata, A.S.A., Moghannem, S.A., Qadri, M., El-ghany, M.F.A., Abdeldaym, E.A. et al. 2023. Nanoparticulate fertilizers increase nutrient absorption efficiency and agro-physiological properties of lettuce plant. Agronomy, 13: 691.

Abdelhameed, R.M., Abdelhameed, R.E., and Kamel, H.A. 2019. Iron-based metal-organic-frameworks as fertilizers for hydroponically grown *Phaseolus vulgaris*. Materials Letters, 237: 72–79.

Ajmal, S., Bui, H.T.D., Bui, V.Q., Yang, T., Shao, X., Kumar, A. et al. 2022. Accelerating water reduction towards hydrogen generation via cluster size adjustment in Ru-incorporated carbon nitride. Chemical Engineering Journal, 429: 132282.

Al-Otibi, F., Perveen, K., Al-Saif, N.A., Alharbi, R.I., Bokhari, N.A., Albasher, G. et al. 2021. Biosynthesis of silver nanoparticles using *Malva parviflora* and their antifungal activity. Saudi Journal of Biological Sciences, 28(4): 2229–2235.

Alhamid, J.O., Mo, C., Zhang, X., Wang, P., Whiting, M.D., and Zhang, Q. 2018. Cellulose nanocrystals reduce cold damage to reproductive buds in fruit crops. Biosystems Engineering, 172: 124–133.

Anjali, C.H., Khan, S.S., Margulis-Goshen, K., Magdassi, S., Mukherjee, A., and Chandrasekaran, N. 2010. Formulation of water-dispersible nanopermethrin for larvicidal applications. Ecotoxicology and Environmental Safety, 73: 1932–1936.

Anjali, C.H., Sharma, Y., Mukharjee, A., and Chandrasekaran, N. 2012. Neem oil (*Azadirachta indica*) nanoemulsion—a potent larvicidal agent against Culex quinquefasciatus. Pest Management Science, 68: 158–163.

Aravinthan, A., Govarthanan, M., Selvam, K., Praburaman, L., Selvankumar, T., Balamurugan, R. et al. 2015. Sunroot mediated synthesis and characterization of silver nanoparticles and evaluation of its antibacterial and rat splenocyte cytotoxic effects. International Journal of Nanomedicine, 10: 1977–1983.

Arvind Bharani, R., and Karthick Raja Namasivayam, S. 2017. Biogenic silver nanoparticles mediated stress on developmental period and gut physiology of major lepidopteran pest *Spodoptera litura* (Fab.) (Lepidoptera: Noctuidae) - An eco-friendly approach of insect pest control. Journal of Environmental Chemical Engineering, 5(1): 453–467.

Asghar, M.A., Zahir, E., Shahid, S.M., Khan, M.N., Asghar, M.A., Iqbal, J. et al. 2018. Iron, copper and silver nanoparticles: Green synthesis using green and black tea leaves extracts and evaluation of antibacterial, antifungal and aflatoxin B1 adsorption activity. Lwt, 90 (December 2017): 98–107.

Avellan, A., Yun, J., Zhang, Y., Spielman-Sun, E., Unrine, J.M., Thieme, J. et al. 2019. Nanoparticle size and coating chemistry control foliar uptake pathways, translocation, and Leaf-to-Rhizosphere transport in wheat. ACS Nano, 13(5): 5291–5305.

Babu, S., Singh, R., Yadav, D., Singh, S., Raj, R., Avasthe, R. et al. 2022. Chemosphere Nanofertilizers for agricultural and environmental sustainability. Chemosphere, 292 (December 2021): 133451.

Baker, S., Volova, T., Prudnikova, S.V., Satish, S., and Prasad, M.N.N. 2017. Nanoagroparticles emerging trends and future prospect in modern agriculture system. Environmental Toxicology and Pharmacology, 53: 10–17.

Banerjee, K., Pramanik, P., Maity, A., Joshi, D.C., Wani, S.H., and Krishnan, P. 2019. Methods of using nanomaterials to plant systems and their delivery to plants (Mode of Entry, Uptake, Translocation, Accumulation, Biotransformation and Barriers). Advances in Phytonanotechnology: From Synthesis to Application.

Barik, T.K., Kamaraju, R., and Gowswami, A. 2012. Silica nanoparticle: A potential new insecticide for mosquito vector control. Parasitology Research, 111(3): 1075–1083.

Beig, B., Niazi, M.B.K., Sher, F., Jahan, Z., Malik, U.S., Khan, M.D. et al. 2022. Nanotechnology-based controlled release of sustainable fertilizers. A review. Environmental Chemistry Letters, 20(4): 2709–2726.

Bijay-Singh, and Craswell, E. 2021. Fertilizers and nitrate pollution of surface and ground water: an increasingly pervasive global problem. SN Applied Sciences, 3(4): 1–24.

Borgatta, J., Ma, C., Hudson-smith, N., Elmer, W., David, C., Pe, P. et al. 2018. Copper based nanomaterials suppress root fungal disease in watermelon (*Citrullus lanatus*): role of particle morphology, composition and dissolution behavior. ACS Sustainable Chemistry & Engineering, 6: 14847–14856.

Bosker, T., Bouwman, L.J., Brun, N.R., Behrens, P., and Vijver, M.G. 2019. Microplastics accumulate on pores in seed capsule and delay germination and root growth of the terrestrial vascular plant *Lepidium sativum*. Chemosphere, 226: 774–781.

Bramhanwade, K., Shende, S., Bonde, S., Gade, A., and Rai, M. 2016. Fungicidal activity of Cu nanoparticles against Fusarium causing crop diseases. Environmental Chemistry Letters, 14(2): 229–235.

Brunel, F., El Gueddari, N.E., and Moerschbacher, B.M. 2013. Complexation of copper(II) with chitosan nanogels: Toward control of microbial growth. Carbohydrate Polymers, 92(2): 1348–1356.

Bussières, P. 2014. Estimating the number and size of phloem sieve plate pores using longitudinal views and geometric reconstruction. Scientific Reports, 4: 1–11.

Cai, L., Chen, J., Liu, Z., Wang, H., Yang, H., and Ding, W. 2018. Magnesium oxide nanoparticles: Effective agricultural antibacterial agent against *Ralstonia solanacearum*. Frontiers in Microbiology, 9(APR): 1–19.

Camara, M.C., Campos, E.V.R., Monteiro, R.A., Do Espirito Santo Pereira, A., De Freitas Proença, P.L., and Fraceto, L.F. 2019. Development of stimuli-responsive nano-based pesticides: Emerging opportunities for agriculture. Journal of Nanobiotechnology, 17(1): 1–19.

Carmona, F.J., Dal Sasso, G., Ramírez-Rodríguez, G.B., Pii, Y., Delgado-López, J.M., Guagliardi, A. et al. 2021. Urea-functionalized amorphous calcium phosphate nanofertilizers: optimizing the synthetic strategy towards environmental sustainability and manufacturing costs. Scientific Reports, 11(1): 1–14.

Carmona, F.J., Guagliardi, A., and Masciocchi, N. 2022. Nanosized calcium phosphates as novel macronutrient nano-fertilizers. Nanomaterials, 12(15).

Chen, R., Ratnikova, T.A., Stone, M.B., Lin, S., Lard, M., Huang, G. et al. 2010. Differential uptake of carbon nanoparticles by plant and mammalian cells. Small, 6(5): 612–617.

Chen, Y., Ren, H. ling, Liu, N., Sai, N., Lu, X., Liu, Z. et al. 2010. A fluoroimmunoassay based on quantum dot—streptavidin conjugate for the detection of chlorpyrifos. Journal of Agricultural and Food Chemistry, 58: 8895–8903.

Chhipa, H. 2019. Applications of nanotechnology in agriculture. Methods in Microbiology, 46: 115–142.

Chi, Y., Zhang, G., Xiang, Y., Cai, D., and Wu, Z. 2018. Fabrication of reusable temperature-controlled-released fertilizer using a palygorskite-based magnetic nanocomposite. Applied Clay Science, 161 (April): 194–202.

Cieschi, M.T., Polyakov, A.Y., Lebedev, V.A., Volkov, D.S., Pankratov, D.A., Veligzhanin, A.A. et al. 2019. Eco-friendly iron-humic nanofertilizers synthesis for the prevention of iron chlorosis in soybean (*Glycine max*) grown in calcareous soil. Frontiers in Plant Science, 10: 1–17.

Cornu, J.Y., Bussière, S., Coriou, C., Robert, T., Maucourt, M., Deborde, C. et al. 2020. Changes in plant growth, Cd partitioning and xylem sap composition in two sunflower cultivars exposed to low Cd concentrations in hydroponics. Ecotoxicology and Environmental Safety, 205 (July).

Cruz-Luna, A.R., Cruz-Martínez, H., Vásquez-López, A., and Medina, D.I. 2021. Metal nanoparticles as novel antifungal agents for sustainable agriculture: Current advances and future directions. Journal of Fungi, 7(12).

Cummins, C.M., Koivunen, M.E., Stephanian, A., Gee, S.J., Hammock, B.D., and Kennedy, I.M. 2006. Application of europium (III) chelate-dyed nanoparticle labels in a competitive atrazine fluoroimmunoassay on an ITO waveguide. Biosensors and Bioelectronics, 21: 1077–1085.

Dhlamini, B., Paumo, H.K., and Katata-seru, L. 2020. Sulphate-supplemented NPK nanofertilizer and its effect on maize growth Sulphate-supplemented NPK nanofertilizer and its effect on maize growth. Material Research Express, 7: 095011.

Duhan, J.S., Kumar, R., Kumar, N., Kaur, P., Nehra, K., and Duhan, S. 2017. Nanotechnology: The new perspective in precision agriculture. Biotechnology Reports, 15 (March): 11–23.

Dzimitrowicz, A., Motyka-Pomagruk, A., Cyganowski, P., Babinska, W., Terefinko, D., Jamroz, P. et al. 2018. Antibacterial activity of Fructose-stabilized silver nanoparticles produced by direct current atmospheric pressure glow discharge towards quarantine pests. Nanomaterials, 8(10).

Eichert, T., Kurtz, A., Steiner, U., and Goldbach, H.E. 2008. Size exclusion limits and lateral heterogeneity of the stomatal foliar uptake pathway for aqueous solutes and water-suspended nanoparticles. Physiologia Plantarum, 134(1): 151–160.

El-Bendary, H.M., and El-Helaly, A.A. 2013. First record nanotechnology in agricultural: Silica nano-particles a potential new insecticide for pest control. Applied Science Reports, 4(3): 241–246.

Elamawi, R.M.A., and El-Shafeyl, R.A. 2013. Inhibition effects of silver nanoparticles against rice blast disease caused by *Magnaporthe Grisea*. Egyptian Journal of Agricultural Research, 91(4): 1271–1283.

Esser, B., Schnorr, J.M., and Swager, T.M. 2012. Selective detection of ethylene gas using carbon nanotube-based devices: utility in determination of fruit ripeness. Angewandte Chemie - International Edition, 51: 5752–5756.

Fang, Y., Umasankar, Y., and Ramasamy, R.P. 2014. Electrochemical detection of p-ethylguaiacol, a fungi infected fruit volatile using metal oxide nanoparticles. Analyst, 139: 3804–3810.

Farooq, M.A., Hannan, F., Islam, F., Ayyaz, A., Zhang, N., Chen, W. et al. 2022. The potential of nanomaterials for sustainable modern agriculture: present findings and future perspectives. Environmental Science: Nano, 9(6): 1926–1951.

Fatima, N., Qazi, U.Y., Mansha, A., Bhatti, I.A., Javaid, R., Abbas, Q. et al. 2021. Recent developments for antimicrobial applications of graphene-based polymeric composites: A review. Journal of Industrial and Engineering Chemistry, 100: 40–58.

Giraldo, J.P., Landry, M.P., Faltermeier, S.M., Mcnicholas, T.P., Iverson, N.M., Boghossian, A.A. et al. 2014. Plant nanobionics approach to augment photosynthesis and biochemical sensing. Nature Materials, 13: 400–408.

Giraldo, J.P., Landry, M.P., Kwak, S., Jain, R.M., Wong, M.H., Iverson, N.M. et al. 2015. A ratiometric sensor using single chirality near- infrared fluorescent carbon nanotubes : application to *in vivo* monitoring. Small, 11: 3973–3984.

Giraldo, J.P., Wu, H., Newkirk, G.M., and Kruss, S. 2019. Nanobiotechnology approaches for engineering smart plant sensors. Nature Nanotechnology, 14: 541–553.

Graham, J.H., Johnson, E.G., Myers, M.E., Science, W., and Departments, P.P. 2015. Potential of nano-formulated zinc oxide for control of citrus canker on grapefruit trees. Plant Disease, 100(12): 2442–2447.

Guo, Y., Liu, S., Gui, W., and Zhu, G. 2009. Gold immunochromatographic assay for simultaneous detection of carbofuran and triazophos in water samples. Analytical Biochemistry, 389(1): 32–39.

Guo, Z., Park, S., Yoon, J., and Shin, I. 2014. Recent progress in the development of near-infrared fluorescent probes for bioimaging applications. Chemical Society Reviews, 43(1): 1–488.

Harman, G., Khadka, R., Doni, F., and Uphoff, N. 2021. Benefits to plant health and productivity from enhancing plant microbial symbionts. Frontiers in Plant Science, 11(April).

Hausbeck, M.K., Bell, J., Medina-Mora, C., Podolsky, R., and Fulbright, D.W. 2000. Effect of bactericides on population sizes and spread of *Clavibacter michiganensis* subsp. michiganensis on Tomatoes in the greenhouse and on disease development and crop yield in the field. Phytopathology, 90(1): 38–44.

Hazaa, M., Alm-Eldin, M., Ibrahim, A.-E., Elbarky, N., Salama, M., Sayed, R. et al. 2021. Biosynthesis of Silver Nanoparticles using *Borago officinslis* leaf extract, characterization and larvicidal activity against cotton leaf worm, *Spodoptera littoralis* (Bosid). International Journal of Tropical Insect Science, 41: 145–156.

Hazarika, A., Yadav, M., Yadav, D.K., and Yadav, H.S. 2022a. An overview of the role of nanoparticles in sustainable agriculture. Biocatalysis and Agricultural Biotechnology, 43(April): 102399.

Hazarika, A., Yadav, M., Yadav, D.K., and Yadav, H.S. 2022b. An overview of the role of nanoparticles in sustainable agriculture. Biocatalysis and Agricultural Biotechnology, 43(April): 102399.

Hermida-Montero, L.A., Pariona, N., Mtz-Enriquez, A.I., Carrión, G., Paraguay-Delgado, F., and Rosas-Saito, G. 2019. Aqueous-phase synthesis of nanoparticles of copper/copper oxides and their antifungal effect against *Fusarium oxysporum*. Journal of Hazardous Materials, 380(December 2018): 120850.

Hong, G., Diao, S., Antaris, A.L., and Dai, H. 2015. Carbon nanomaterials for biological imaging and nanomedicinal therapy. Chemical Reviews, 115: 10816–10906.

Hosamani, B.G., Patil, R.R., Benagi, V.I., Chandrashekhar, S.S., and Nandihali, B.S. 2019. Synthesis of green silver nanoparticles from soybean seed and its bioefficacy on *Spodoptera litura* (F.). International Journal of Current Microbiology and Applied Sciences, 8(09): 610–618

Hu, D., Bai, Y., Li, L., Ai, M., Jin, J., Song, K. et al. 2022. Phytotoxicity assessment study of gold nanocluster on broad bean (*Vicia Faba* L.) seedling. Environmental Pollutants and Bioavailability, 34(1): 284–296.

Hu, P., An, J., Faulkner, M.M., Wu, H., Li, Z., Tian, X. et al. 2020. Nanoparticle charge and size control foliar delivery efficiency to plant cells and organelles. ACS Nano, 14(7): 7970–7986.

Hua, X., Qian, G., Yang, J., Hu, B., Fan, J., Qin, N. et al. 2010. Biosensors and bioelectronics development of an immunochromatographic assay for the rapid detection of chlorpyrifos-methyl in water samples. Biosensors and Bioelectronics, 26(1): 189–194.

Huang, B., Chen, F., Shen, Y., Qian, K., Wang, Y., Sun, C. et al. 2018. Advances in targeted pesticides with environmentally responsive controlled release by nanotechnology. Nanomaterials, 8(2): 102.

Idris, A.O., Akanji, S.P., Orimolade, B.O., Omobola, F., Olorundare, G., Azizi, S. et al. 2023. Using nanomaterials as excellent immobilisation layer for biosensor design. Biosensors, 13: 192.

Jafir, M., Ahmad, J.N., Arif, M.J., Ali, S., and Ahmad, S.J.N. 2021. Characterization of *Ocimum basilicum* synthesized silver nanoparticles and its relative toxicity to some insecticides against tobacco cutworm, *Spodoptera litura* Feb. (Lepidoptera; Noctuidae). Ecotoxicology and Environmental Safety, 218: 112278.

Jagana, D., R. Hegde, Y., and Lella, R. 2017. Green nanoparticles - a novel approach for the management of Banana Anthracnose Caused by *Colletotrichum musae*. International Journal of Current Microbiology and Applied Sciences, 6(10): 1749–1756.

Javaid, R. and Nanba, T. 2020. mgfe2o4-supported ru catalyst for ammonia synthesis: promotive effect of chlorine. ChemistrySelect, 5(14): 4312–4315.

Javaid, R., and Nanba, T. 2023. Ru/CeO2/MgO catalysts for enhanced ammonia synthesis efficiency. Topics in Catalysis, (0123456789).

Jebril, S., Khanfir Ben Jenana, R., and Dridi, C. 2020. Green synthesis of silver nanoparticles using *Melia azedarach* leaf extract and their antifungal activities: *In vitro* and *in vivo*. Materials Chemistry and Physics, 248 (February).

Jeon, J., and Adamiano, A. 2020. Synergistic Release of Crop Nutrients and Stimulants from Hydroxyapatite Nanoparticles Functionalized with Humic Substances: Toward a Multifunctional Nanofertilizer.

Jia-Yi, Y., Meng-Qiang, S., Zhi-Liang, C., Yu-Tang, X., Hang, W., Jian-Qiang, Z. et al. 2022. Effect of foliage applied chitosan-based silicon nanoparticles on arsenic uptake and translocation in rice (*Oryza sativa* L.). Journal of Hazardous Materials, 433 (January): 128781.

Kaboli, H., Azizi, M., Teymouri, M., Reza, A., and Reza, M. 2021. Scientia Horticulturae synthesis and characterization of nanoliposome containing Fe 2 + element : A superior nano-fertilizer for ferrous iron delivery to sweet basil. Scientia Horticulturae, 283 (February): 110110.

Kalaba, M.H., Moghannem, S.A., El-Hawary, A.S., Radwan, A.A., Sharaf, M.H., and Shaban, A.S. 2021. Green synthesized zno nanoparticles mediated by streptomyces plicatus: Characterizations, antimicrobial and nematicidal activities and cytogenetic effects. Plants, 10(9).

Kalia, A., and Sharma, S.P. 2020. Multifunctional Hybrid Nanomaterials for Sustainable Agri-food and Ecosystems. 1st ed. Elsevier.

Kamber, T., Smits, T.H.M., Rezzonico, F., and Duffy, B. 2012. Genomics and current genetic understanding of Erwinia amylovora and the fire blight antagonist *Pantoea vagans*. Trees - Structure and Function, 26(1): 227–238.

Karny, A., Zinger, A., Kajal, A., Shainsky-roitman, J., and Schroeder, A. 2018. Therapeutic nanoparticles penetrate leaves and deliver nutrients to agricultural crops. Scientific Reports, 8: 7589.

Kasmara, H., Melanie, Nurfajri, D.A., Hermawan, W., and Panatarani, C. 2018. The toxicity evaluation of prepared Lantana camara Nano Extract Against *Spodoptera litura* (Lepidoptera: Noctuidae). In: AIP Conference Proceeding.

Khadiran, N.F., Zobir, M., Rozita, H., Tumirah, A., and Zulkarnain, K. 2021. Preparation and properties of zinc layered hydroxide with nitrate and phosphate as the counter anion , a novel control release fertilizer formulation. Journal of Porous Materials, 28(6): 1797–1811.

Khan, H.A., Naqvi, S.R., Mehran, M.T., Khoja, A.H., Khan Niazi, M.B., Juchelková, D. et al. 2021. A performance evaluation study of nano-biochar as a potential slow-release nano-fertilizer from wheat straw residue for sustainable agriculture. Chemosphere, 285 (July).

Kottegoda, N., Sandaruwan, C., Priyadarshana, G., Siriwardhana, A., Rathnayake, U.A., Berugoda Arachchige, D.M. et al. 2017. Urea-Hydroxyapatite Nanohybrids for Slow Release of Nitrogen. ACS Nano, 11(2): 1214–1221.

Kranthi, K.R., Davis, M., Mayee, C.D., Russell, D.A., Shukla, R.M., Satija, U. et al. 2009. Development of a colloidal-gold based lateral-flow immunoassay kit for 'quality-control' assessment of pyrethroid and endosulfan formulations in a novel single strip format. Crop Protection, 28: 428–434.

Kruss, S., Landry, M.P., Ende, E. Vander, Lima, B.M.A., Reuel, N.F., Zhang, J. et al. 2014. Neurotransmitter detection using corona phase molecular recognition on fluorescent single-walled carbon nanotube sensors. Journal of the American Chemical Society, 136: 713–724.

Kumar, N., Ram, S., Venkatesh, K., and Tripathi, S.C. 2023. Soil & tillage research global trends in use of nano-fertilizers for crop production : Advantages and constraints – A review. Soil & Tillage Research, 228 (December 2022): 105645.

Larue, C., Veronesi, G., Flank, A.-M., Surble, S., Herlin-Boime, N., and Carriere, M. 2012. Comparative uptake and impact of TiO$_2$ nanoparticles in wheat and rapeseed. Journal of Toxicology and Environmental Health, Part A, 75: 722–734.

Lateef, A., Nazir, R., Jamil, N., Alam, S., Shah, R., Khan, M.N. et al. 2019. Synthesis and characterization of environmental friendly corncob biochar based nano-composite—A potential slow release nano-fertilizer for sustainable agriculture. Environmental Nanotechnology, Monitoring and Management, 11 (January): 100212.

Lee, K., Park, J., Lee, M., Kim, J., Hyun, B.G., Kang, D.J. et al. 2014. *In-situ* synthesis of carbon nanotube—graphite electronic devices and their integrations onto surfaces of live plants and insects. Nano Letters, 14: 2647–2654.

Li, J., Wu, H., Santana, I., Fahlgren, M., and Giraldo, J.P. 2018. Standoff optical glucose sensing in photosynthetic organisms by a quantum dot fluorescent probe. Applied Materials and Interfaces, 10: 28279–28289.

Li, L., Luo, Y., Li, R., Zhou, Q., Peijnenburg, W.J.G.M., Yin, N. et al. 2020. Effective uptake of submicrometre plastics by crop plants via a crack-entry mode. Nature Sustainability, 3(11): 929–937.

Li, T., Lü, S., Yan, J., Bai, X., Gao, C., and Liu, M. 2019. An environment-friendly fertilizer prepared by layer-by-layer self-assembly for pH-responsive nutrient release. ACS Applied Materials and Interfaces, 11(11): 10941–10950.

Lisha, K.P., Paradeep, A., and Pradeep, T. 2009. Enhanced visual detection of pesticides using gold nanoparticles. Journal of Environmental Science and Health, Part B, 44(7): 697–705.

Liu, B., Fan, Y., Li, H., Zhao, W., Luo, S., Wang, H. et al. 2021. Control the entire journey of pesticide application on superhydrophobic plant surface by dynamic covalent trimeric surfactant coacervation. Advanced Functional Materials, 31(5).

Liu, L., Shen, T., Yang, Y., Gao, B., Li, Y.C., Xie, J. etal. 2018. Bio-based large tablet controlled-release urea: synthesis, characterization, and controlled-released mechanisms. Journal of Agricultural and Food Chemistry, 66(43): 11265–11272.

López, M.M., Llop, P., and Olmos, A. 2007. Are Molecular Tools Solving the Challenges Posed by Detection of Plant Pathogenic Bacteria and Viruses ? 13–46.

Lv, J., Christie, P., and Zhang, S. 2019. Uptake, translocation, and transformation of metal-based nanoparticles in plants: recent advances and methodological challenges. Environmental Science: Nano, 6(1): 41–59.

Ma, X. and Yan, J. 2018. Plant uptake and accumulation of engineered metallic nanoparticles from lab to field conditions. Current Opinion in Environmental Science and Health, 6: 16–20.

Mahaletchumi, S. 2021. Review on the Use of nanotechnology in fertilizers. Journal of Research Technology and Engineering, 2(1): 60–72.

Malandrakis, A.A., Kavroulakis, N., and Chrysikopoulos, C.V. 2019. Use of copper, silver and zinc nanoparticles against foliar and soil-borne plant pathogens. Science of the Total Environment, 670: 292–299.

Malandrakis, A.A., Kavroulakis, N., and Chrysikopoulos, C.V. 2020. Synergy between Cu-NPs and fungicides against *Botrytis cinerea*. Science of the Total Environment, 703: 135557.

Manikandan, A., and Sathiyabama, M. 2016. Preparation of Chitosan nanoparticles and its effect on detached rice leaves infected with *Pyricularia grisea*. International Journal of Biological Macromolecules, 84: 58–61.

Mansfield, J., Genin, S., Magori, S., Citovsky, V., Sriariyanum, M., Ronald, P. et al. 2012. Top 10 plant pathogenic bacteria in molecular plant pathology. Molecular Plant Pathology, 13(6): 614–629.

Mejias, J.H., Salazar, F., Pérez Amaro, L., Hube, S., Rodriguez, M., and Alfaro, M. 2021. Nanofertilizers: A cutting-edge approach to increase nitrogen use efficiency in grasslands. Frontiers in Environmental Science, 9 (March): 1–8.

Minh, N., Ha, C., Huyen, T., San, N., and Wang, L. 2019. Preparation of NPK nanofertilizer based on chitosan nanoparticles and its effect on biophysical characteristics and growth of coffee in green house. Research on Chemical Intermediates, 45(1): 51–63.

Miralles, P., Church, T.L., and Harris, A.T. 2012. Toxicity, uptake, and translocation of engineered nanomaterials in vascular plants. Environmental Science and Technology, 46(17): 9224–9239.

Mondal, K.K., and Mani, C. 2012. Investigation of the antibacterial properties of nanocopper against *Xanthomonas axonopodis* pv. *punicae*, the incitant of pomegranate bacterial blight. Annals of Microbiology, 62(2): 889–893.

Nandini, B., Hariprasad, P., Prakash, H.S., Shetty, H.S., and Geetha, N. 2017. Trichogenic-selenium nanoparticles enhance disease suppressive ability of Trichoderma against downy mildew disease caused by *Sclerospora graminicola* in pearl millet. Scientific Reports, 7(1): 1–11.

Naseem, F., Zhi, Y., Farrukh, M.A., Hussain, F., and Yin, Z. 2020. Mesoporous ZnAl2Si10O24 nanofertilizers enable high yield of *Oryza sativa* L. Scientific Reports, 10(1): 1–11.

Nguyen, D.H., Vo, T.N.N., Nguyen, N.T., Ching, Y.C., and Thi, T.T.H. 2020. Comparison of biogenic silver nanoparticles formed by *Momordica charantia* and *Psidium guajava* leaf extract and antifungal evaluation. PLoS ONE, 15 (9 September): 1–13.

Niu, C., Yao, Z., and Jiang, S. 2023. Science of the Total Environment Synthesis and application of quantum dots in detection of environmental contaminants in food : A comprehensive review. Science of the Total Environment, 882 (April): 163565.

Nongbet, A., Mishra, A.K., Mohanta, Y.K., Mahanta, S., Ray, M.K., Khan, M. et al. 2022. Nanofertilizers: a smart and sustainable attribute to modern agriculture. Plants, 11: 2587.

Nooeaid, P., Chuysinuan, P., Pitakdantham, W., and Aryuwananon, D. 2021. Eco-friendly polyvinyl alcohol/polylactic acid core/shell structured fibers as controlled - release fertilizers for sustainable agriculture. Journal of Polymers and the Environment, 29(2): 552–564.

Oh, J.W., Chun, S.C., and Chandrasekaran, M. 2019. Preparation and *in vitro* characterization of chitosan nanoparticles and their broad-spectrum antifungal action compared to antibacterial activities against phytopathogens of tomato. Agronomy, 9(1): 21.

Osonga, F.J., Kalra, S., Miller, R.M., Isika, D., and Sadik, O.A. 2020. Synthesis, characterization and antifungal activities of eco-friendly palladium nanoparticles. RSC Advances, 10(10): 5894–5904.

Padilla, F.M., Gallardo, M., Pena-Fleitas, M.T., Souza, R. de, and Thompson, R.B. 2018. Proximal optical sensors for nitrogen management of vegetable crops : a review. Sensors, 18: 2083.

Pariona, N., Mtz-Enriquez, A.I., Sánchez-Rangel, D., Carrión, G., Paraguay-Delgado, F., and Rosas-Saito, G. 2019. Green-synthesized copper nanoparticles as a potential antifungal against plant pathogens. RSC Advances, 9(33): 18835–18843.

Paulkumar, K., Gnanajobitha, G., Vanaja, M., Rajeshkumar, S., Malarkodi, C., Pandian, K. et al. 2014. *Piper nigrum* leaf and stem assisted green synthesis of silver nanoparticles and evaluation of its antibacterial activity against agricultural plant pathogens. The Scientific World Journal, 2014.

Péret, B., De Rybel, B., Casimiro, I., Benková, E., Swarup, R., Laplaze, L. et al. 2009. Arabidopsis lateral root development: an emerging story. Trends in Plant Science, 14(7): 399–408.

Pérez-de-Luque, A. 2017. Interaction of nanomaterials with plants: What do we need for real applications in agriculture? Frontiers in Environmental Science, 5 (APR): 1–7.

Prasad, A., Astete, C.E., Bodoki, A.E., Windham, M., Bodoki, E., and Sabliov, C.M. 2018. Zein nanoparticles uptake and translocation in hydroponically grown sugar cane plants. Journal of Agricultural and Food Chemistry, 66(26): 6544–6551.

Qazi, U.Y., and Javaid, R. 2023. Graphene utilization for efficient energy storage and potential applications: Challenges and Future Implementations. Energies, 16: 2927.

Qazi, U.Y., Javaid, R., Ikhlaq, A., Khoja, A.H., and Saleem, F. 2022. A comprehensive review on zeolite chemistry for catalytic conversion of biomass/waste into green fuels. Molecules, 27(23).

Rad, F., Mohsenifar, A., Tabatabaei, M., Safarnejad, M.R., Shahryari, F., Safarpour, H. et al. 2012. Detection of *Candidatus phytoplasma* aurantifolia with a quantum dots fret-based biosensor. Journal of Plant Pathology, 94: 525–534.

Rai, M., Ribeiro, C., Mattoso, L., and Duran, N. 2015. Nanotechnologies in food and agriculture. Nanotechnologies in Food and Agriculture, (May), 1–347.

Rajak, P., Roy, S., Ganguly, A., Mandi, M., Dutta, A., Das, K. et al. 2023. Agricultural pesticides – friends or foes to biosphere? Journal of Hazardous Materials Advances, 10 (February): 100264.

Raliya, R., Saharan, V., Dimkpa, C., and Biswas, P. 2018. Nanofertilizer for precision and sustainable agriculture: current state and future perspectives. Journal of Agricultural and Food Chemistry, 66(26): 6487–6503.

Ramírez-Rodríguez, G.B., Dal Sasso, G., Carmona, F.J., Miguel-Rojas, C., Pérez-De-Luque, A., Masciocchi, N. et al. 2020. Engineering biomimetic calcium phosphate nanoparticles: a green synthesis of slow-release Multinutrient (NPK) Nanofertilizers. ACS Applied Bio Materials, 3(3): 1344–1353.

Ribeiro, C. 2015. RSC Advances nanoparticles incorporated into biodegradable, 104179–104186.

Rizwan, M., Ali, S., Qayyum, M.F., Ok, Y.S., Adrees, M., Ibrahim, M. et al. 2017. Effect of metal and metal oxide nanoparticles on growth and physiology of globally important food crops: A critical review. Journal of Hazardous Materials, 322: 2–16.

Sahayaraj, K., Balasubramanyam, G., and Chavali, M. 2020. Green synthesis of silver nanoparticles using dry leaf aqueous extract of *Pongamia glabra* Vent (Fab.), Characterization and phytofungicidal activity. Environmental Nanotechnology, Monitoring and Management, 14 (July): 100349.

Saleh, B., and Webster, T.J. 2020. Green Synthesis of Zeolite/Fe_2O_3 Nanocomposites : Toxicity & Cell Proliferation Assays and Application as a Smart Iron Nanofertilizer, 1005–1020.

Shaji, H., Chandan, V., and Mathew, L. 2020. Organic fertilizers as a route to controlled release of nutrients. pp. 231–245. *In*: Lewu, F.B., Volova, T., Thomas, S., and Rakhimol, K.R. (eds.). Conrolled Release Fertilizers for Sustainable Agriculture. Elsevier.

Shao, C., Zhao, H., and Wang, P. 2022. Recent development in functional nanomaterials for sustainable and smart agricultural chemical technologies. Nano Convergence, 9(1).

Shebl, A., Hassan, A.A., Salama, D.M., El-aziz, M.E.A., and Abd, M.S.A. 2020. Heliyon Template-free microwave-assisted hydrothermal synthesis of manganese zinc ferrite as a nanofertilizer for squash plant (*Cucurbita pepo* L). Hcliyon, 6(3): e03596.

Singh, R.P., Handa, R., and Manchanda, G. 2021. Nanoparticles in sustainable agriculture: An emerging opportunity. Journal of Controlled Release, 329 (July 2020): 1234–1248.

Singh, S., Singh, M., Varun, V., and Kumar, A. 2010. An attempt to develop surface plasmon resonance based immunosensor for Karnal bunt (*Tilletia indica*) diagnosis based on the experience of nano-gold based lateral fl ow immuno-dipstick test. Thin Solid Films, 519(3): 1156–1159.

Slomberg, D.L., and Schoenfisch, M.H. 2012. Silica nanoparticle phytotoxicity to *Arabidopsis thaliana*. Environmental Science and Technology, 46(18): 10247–10254.

Spielman-Sun, E., Lombi, E., Donner, E., Howard, D., Unrine, J.M., and Lowry, G.V. 2017. Impact of surface charge on cerium oxide nanoparticle uptake and translocation by wheat (*Triticum aestivum*). Environmental Science and Technology, 51(13): 7361–7368.

Stadler, T., Buteler, M., and Weaver, D.K. 2010. Novel use of nanostructured alumina as an insecticide. Pest Management Science, 66(6): 577–579.

Sun, H., Lei, C., Xu, J., and Li, R. 2021. Foliar uptake and leaf-to-root translocation of nanoplastics with different coating charge in maize plants. Journal of Hazardous Materials, 416 (February): 125854.

Sun, H., Wang, M., Wang, J., and Wang, W. 2022. Surface charge affects foliar uptake, transport and physiological effects of functionalized graphene quantum dots in plants. Science of the Total Environment, 812: 151506.

Tara Meghana, K., Wahiduzzaman, M., and Vamsi, G. 2021. Nanofertilizers in agriculture. Acta Scientific Agriculture, 5(3): 35–46.

Tarafder, C., Daizy, M., Alam, M.M., Ali, M.R., Islam, M.J., Islam, R. et al. 2020. Formulation of a hybrid nanofertilizer for slow and sustainable release of micronutrients. ACS Omega, 5(37): 23960–23966.

Toth, I.K., van der Wolf, J.M., Saddler, G., Lojkowska, E., Hélias, V., Pirhonen, M. et al. 2011. Dickeya species: An emerging problem for potato production in Europe. Plant Pathology, 60(3): 385–399.

Tripathi, D.K., Shweta, Singh, S., Singh, S., Pandey, R., Singh, V.P., Sharma, N.C. et al. 2017. An overview on manufactured nanoparticles in plants: Uptake, translocation, accumulation and phytotoxicity. Plant Physiology and Biochemistry, 110: 2–12.

Vicente, J.G., and Holub, E.B. 2013. *Xanthomonas campestris* pv. Campestris (cause of black rot of crucifers) in the genomic era is still a worldwide threat to brassica crops. Molecular Plant Pathology, 14(1): 2–18.

Vinayaka, A.C., Basheer, S., and Thakur, M.S. 2009. Biosensors and bioelectronics bioconjugation of CdTe quantum dot for the detection of 2 , 4-dichlorophenoxyacetic acid by competitive fluoroimmunoassay based biosensor. Biosensors and Bioelectronics, 24: 1615–1620.

Vinayaka, A.C., and Thakur, M.S. 2010. Focus on quantum dots as potential fluorescent probes for monitoring food toxicants and foodborne pathogens. Analytical and Bioanalytical Chemistry, 397: 1445–1455.

Wang, M., Zhang, G., Zhou, L., Wang, D., Zhong, N., Cai, D. et al. 2016. Fabrication of pH-controlled-release ferrous foliar fertilizer with high adhesion capacity based on nanobiomaterial. ACS Sustainable Chemistry and Engineering, 4(12): 6800–6808.

Wang, X., Xie, H., Wang, P., and Yin, H. 2023. Nanoparticles in Plants: Uptake, Transport and Physiological Activity in Leaf and Root, 1–21.

Werdin González, J.O., Stefanazzi, N., Murray, A.P., Ferrero, A.A., and Fernández Band, B. 2015. Novel nanoinsecticides based on essential oils to control the German cockroach. Journal of Pest Science, 88(2): 393–404.

Wilson, R.H., Nadeau, K.P., Jaworski, F.B., Tromberg, B.J., and Durkin, A.J. 2023. Review of short-wave infrared spectroscopy and imaging methods for biological tissue characterization. Journal of Biomedical Optics, 20(3): 030901.

Wolfert, S., Ge, L., Verdouw, C., and Bogaardt, M. 2017. Big data in smart farming – a review. Agricultural Systems, 153: 69–80.

Wong, M.H., Giraldo, J.P., Kwak, S., Koman, V.B., Sinclair, R., Thomas, T. et al. 2017. Nitroaromatic detection and infrared communication from wild-type plants using plant nanobionics. Nature Materials, 16 (February): 264–274.

Wu, H., Shabala, L., Shabala, S., and Giraldo, J.P. 2018. Hydroxyl radical scavenging by cerium oxide nanoparticles improves Arabidopsis salinity tolerance by enhancing leaf mesophyll potassium retention. Environmental Science Nano, 5: 1567–1583.

Wu, H., Tito, N., and Giraldo, J.P. 2017. Anionic cerium oxide nanoparticles protect plant photosynthesis from abiotic stress by scavenging reactive oxygen species. ACS Nano, 11: 11283–11297.

Wu, K., Du, C., Ma, F., Shen, Y., and Zhou, J. 2019. Optimization of metal-organic (citric acid) frameworks for controlled release of nutrients. RSC Advances, (55): 32270–32277.

Xiong, L., and Wang, P. 2018. Environmental Science Nano, 2888–2898.

Xu, R., Ouyang, L., Chen, H., and Zhang, G. 2023. Recent Advances in Biomolecular Detection Based on Aptamers and Nanoparticles, 1–31.

Yang, C., Powell, C.A., Duan, Y., Shatters, R., and Zhang, M. 2015. Antimicrobial nanoemulsion formulation with improved penetration of foliar spray through citrus leaf cuticles to control citrus huanglongbing. PLoS ONE, 10(7): 1–14.

Yao, J., Yang, M., and Duan, Y. 2014. Chemistry, biology, and medicine of fluorescent nanomaterials and related systems : new insights into biosensing, bioimaging, genomics, diagnostics, and therapy. Chemical Reviews, 114: 6130–6178.

Yao, K.S., Li, S.J., Tzeng, K.C., Cheng, T.-C., Chang, C.-Y., Chiu, C.Y. et al. 2009. Fluorescence silica nanoprobe as a biomarker for rapid detection of plant pathogens. Advanced Materials Research, 79–82: 513–516.

Zhang, H., Goh, N.S., Wang, J.W., Pinals, R.L., González-Grandío, E., Demirer, G.S. et al. 2022. Nanoparticle cellular internalization is not required for RNA delivery to mature plant leaves. Nature Nanotechnology, 17(2): 197–205.

Zhang, S., Gao, N., Shen, T., Yang, Y., Gao, B., Li, Y.C. et al. 2019. One-step synthesis of superhydrophobic and multifunctional nano copper-modified bio-polyurethane for controlled-release fertilizers with "multilayer air shields": new insight of improvement mechanism. Journal of Materials Chemistry A, 7(16): 9503.

Zhao, L., Peralta-Videa, J.R., Ren, M., Varela-Ramirez, A., Li, C., Hernandez-Viezcas, J.A. et al. 2012. Transport of Zn in a sandy loam soil treated with ZnO NPs and uptake by corn plants: Electron microprobe and confocal microscopy studies. Chemical Engineering Journal, 184: 1–8.

Zhao, L., Peralta-Videa, J.R., Varela-Ramirez, A., Castillo-Michel, H., Li, C., Zhang, J. et al. 2012. Effect of surface coating and organic matter on the uptake of CeO_2 NPs by corn plants grown in soil: Insight into the uptake mechanism. Journal of Hazardous Materials, 225–226: 131–138.

Zhu, J., Wang, J., Zhan, X., Li, A., White, J.C., Gardea-Torresdey, J.L. et al. 2021. Role of charge and size in the translocation and distribution of zinc oxide particles in wheat cells. ACS Sustainable Chemistry and Engineering, 9(34): 11556–11564.

2

Use of Nanoparticles in Moisture Retention and Soil Health Management

J. C. Tarafdar[1],* and *Indira Rathore*[2]

Introduction

Nanosize particles have the potential to play an important role in agriculture, including soil moisture retention and soil health management (Tarafdar 2021). There is a lot of research going on globally to find out the effect of nano-based materials on soil moisture holding and maintaining soil health. The use of nanoclays and hydrogels has been found to be very effective in improving moisture retention capacity in arid agroecosystems. Nanoparticles were also found to be very effective materials for water treatment due to their larger surface area, lower volume, and ability to make chemical and biological reactions easier. In general, four important factors that are responsible for maintaining soil health are carbon transformations, nutrient cycling, soil structure maintenance, and the control of pests and diseases. Nanoparticles have the potential to control all factors and play a significant role in soil health management. It can also play an important role in water purification techniques (Kundru et al. 2017). Innovative solutions for water treatment may be contributed through nanotechnology due to their high aspect ratio, reactivity, and tunable pore volume, electrostatic, hydrophilic, and hydrophobic interactions that can be important in adsorption, catalysis, sensoring, and optoelectronics. Frequently, nanomembranes are used for softening the water and removing contaminants. Due to their large surface area and intense sorption properties, they can minimize the chance of losses due to runoff and decrease release kinetics. Peteu et al. (2010) found that soil application of micronutrients trapped in nanomaterials helps slow the release of nutrients and improve plant growth and soil health. It has also been noticed that nanomaterials have the potential to rapidly equilibrate between the aqueous and solid phases. Their aggregation rates in soil are positively correlated with ionic strength, zeta potential, and soil pH. Nanoadsorbents are

[1] ICAR-Central Arid Zone Research Institute, Jodhpur-342008 Rajasthan, India.
[2] Nano Research Lab, IFFCO, Aonla unit, Bareilly 243403 Utter Pradesh, India.
 Email: rathoreindusamrpan@gmail.com
* Corresponding author: jctarafdar@ yahoo.in

now widely used for the removal of organics, heavy metals, and bacteria from the soil (Gehrke et al. 2015). The industry now uses nano absorbents for bioprocessing and environmental remediation due to their excellent capacity of biocompatibility, high reactivity, more surface to volume ratio and metal binding. Nanomembranes are globally used as a wastewater treatment process. Nanometals and metal oxides are widely used to remove heavy metals in media filters and slurry reactors. Nanomaterials like nanoclay composite have been found to be very effective in water retention capacity and nontoxic (Jatav and De 2013), which may be used in agricultural and horticultural applications to increase input use efficiency. The present chapter suggests how nanoparticles or nanomaterials can help in soil moisture retention and soil health management.

Microbial polysaccharides

They are mainly biopolymers, which are easily soluble in water. They have massive commercial value as binders, emulsifiers, gelling agents, coagulants, stabilizers, lubricants, and suspension agents due to their rheological characteristics. The important microbial polysaccharides are curdlan, pollulan, xanthan, pectin, cellulose, chitosan, dextran, gellan, cellulose, sodium alginate, and Arabic gum. Microorganisms generally release polysaccharides extracellularly in the presence of surplus carbon sources. Microbial extracellular exopolysaccharides offer significant benefits to soil properties and crop productivity. It has characterized by unique properties, including an increase in soil water-holding capacity, improvement in soil aggregate formation due to its binding nature, and regulation of nutrient and water flows to plant roots through biofilm formation (Şengör 2019). Additionally, microbial polysaccharides have been shown to enhance plant nutrient uptake in numerous plant species, leading to better growth and development (Rolli et al. 2015). Microbial polysaccharides may be neutral or acidic. Acidic polysaccharides (e.g., xanthan, gellan) are more important as they have ionized groups such as carboxyl that may function as polyelectrolytes. Microbial polysaccharide has the potential to alleviate abiotic and biotic stresses on plants, such as drought, salinity, and heavy metal toxicity, by protecting plants through osmoprotection and sequestering toxic metals (Bramhachari et al. 2018). It may also act as a biofertilizer, promoting plant growth and yield by increasing the availability of nutrients to plants. Polysaccharides are of interest in agriculture due to their environmental compatibility. Unlike chemically synthesized polymers, microbial extracellular polysaccharides are fully biodegradable and can be fully degraded and mineralized to CO_2 and water under the influence of soil microbial activity. Furthermore, microbial polysaccharides containing sugar moieties as structural components can serve as an excellent carbon source for soil microbes, promoting their growth and activity (Han et al. 2015).

Nutrient and oxygen supply may affect the polysaccharide release. They are large molecular weight carbohydrates that can be produced by fungi, bacteria, algae and yeast. They can be released and gathered inside the cells such as glycogen where they function as energy and carbon reserves. The cell wall of polysaccharides is mainly chitin,which is mainly responsible in stabilization of the unity of the cell. They may form capsular polysaccharides (associated with the surface of the cell) or exopolysaccharides, which are loosely attached to the cell surface (Freitus et al. 2021). Polysaccharides in capsular form are mainly associated with pathogenicity; on the other hand, exopolysaccharides are usually responsible for overcoming the environmental stress and generally responsible for water and carbon storage. Exopolysaccharides may be homopolysaccharides or heteropolysaccharides. Homopolysaccharides are made of only one type of monosaccharide unit such as cellulose, dextran or pollulan whereas heteropolysaccharides are composed of two or more different polysaccharide units such as xanthan or hyaluronic acid. Generally, microorganisms produce extra polysaccharide to protect against environmental stresses. Polysaccharide production by the microorganisms can be triggered by nanosized particles of Zn, Fe, and Mg. It has been noticed that more than 100 enzymatic reactions (directly or indirectly) may be involved in polysaccharide synthesis. Their structural diversity depends on the number of different sugars present in them. The utility of microbial polysaccharides depends on their rheological features and wide range

Table 1. Commercial application of some important microbial polysaccharides.

Name of microbial polysaccharide	Commercial importance
Xanthan	Soil aggregation, soil moisture retention, soil carbon build up, food additive and stabilization, ice cream, cheese, oil recovery, tooth paste, water based paints
Dextran	Soil moisture retention, soil carbon build up, used as adsorbent, food additive, blood plasma expander
Pollulan	Soil aggregation, soil carbon build up, food coating and packaging
Curdlan	Soil moisture retention, soil carbon build up, gelling agent
Gellan	Used in food industry as a thickener, binder, stabilizer in food application
Alginate	Used in food industry as thickener, emulsifier, stabilizer and in medical industry as pharmaceutical additives

of adaptation to temperature and pH. Due to this, xanthan gum is the most popular microbial polysaccharide in the industry. Normally, microbial polysaccharides are secreted from the cell and form a layer over the surface of the organism. They are generally designated as exopolysaccharides because of their position. Microbial polysaccharides may be precipitated after using salts, organic solvents, and acids and can be recovered with appropriate techniques. They may be linked either with the cell surface or ejected into the growth medium. It has enormous commercial importance in different industries. Besides agriculture, they may be used in the food, pharmaceutical, and oil industries as thickening and gelling agents. Most important polysaccharide-producing organisms are fungi and bacteria. A list of commercially important microbial polysaccharides and their uses is presented in Table 1.

Nano-induced production of microbial polysaccharides

Nanoparticles like Zn, Fe, and Mg may be able to enhance polysaccharide production by different polysaccharide-producing fungi and bacteria. The nano-induced extracellular release of polysaccharides can be powdered (Tarafdar et al. 2018a) and used for soil aggregation, moisture retention, and soil carbon buildup in harsh arid environments. In general, there was 33–81% improvement in soil aggregation, 10–14% improvement in moisture retention, and a 3-5% improvement in soil carbon buildup with the application of nano-induced polysaccharide powder originating from different fungi and bacterial sources being observed (Tarafdar 2022). The higher production of microbial polysaccharides was noticed when the C:N ratio was maintained at 10:1, the temperature was between 30 and 35°C, the moisture condition was between 50 and 75%, and the pH was between 6 and 7.5. Nanosized Zn, Fe, and Mg are found to be the most effective nanoparticles to trigger polysaccharide secretion from microbial sources, which may be due to their structural presence in polysaccharide molecules. They are nothing but complex carbohydrates consisting of a greater number of monosaccharides joined together by glycosidic linkages. The important polysaccharide-producing organisms that can be triggered by different nanoparticles and their type of polysaccharide production are listed in Table 2.

The major fungal polysaccharides identified from nano-induced microbial gum were xanthan, curdlan, and pollulan. It has been found that these microbial polysaccharides are very useful in minimizing soil erosion and soil health management in arid environments, as they help with more than 80% of soil aggregation, up to 14% more soil moisture retention, and soil carbon buildup, which rises to almost 5%. They are mainly hydrocolloids and, therefore, disperse easily in water. They may very well be utilized as stabilizers, emulsifiers, thickeners, and gelling agents (Koocheki et al. 2009). It can also be used in the pharmaceutical industry for capsule formation. It has been observed that many organisms have a great ability to extracellularly secrete polysaccharides, and some may produce more than 40 g per liter under stress conditions (Lin et al. 1984, Ravella et al. 2010). This production can be induced up to 16 times with the application of different nanoparticles

Table 2. List of some nano-induced polysaccharide producing organisms and quantum of increase in their production.

Name of the organism	Nanoparticles responded in polysaccharide release	Quantum of nanoparticle effect
Agrobacterium tumefaciens	Fe and Zn	8–11 folds
Aspergillus flavus	Zn	7–12 folds
Aspergillus fumigates	Fe and Zn	8–14 folds
Aspergillus terreus	Mg and Zn	8–11 folds
Bacillus subtilis	Zn	6–9 folds
Caulobacter vibrioides	Mg and Zn	5–8 folds
Chaetomium globossum	Zn	9–16 folds
Carvularia lunata	Fe and Zn	5–12 folds
Paenibacillus illinoisensis	Fe and Zn	7–8 folds
Rhizobium sp.	Fe, Mg, Zn	4–12 folds
Sinorhizobium sp.	Zn	4–8 folds
Sporosarcina ginsengissoli	Zn	6–10 folds
Stenotrophomonas maltophilia	Fe and Zn	8–12 folds

(Tarafdar 2021). That way, the main problem of production cost can be minimized in several ways for industrial, agricultural, and pharmaceutical uses. Nano-induced polysaccharides may contain only one sugar (homo) or two or more sugar (hetero) components. It has linear molecules and may have different lengths of side chains in their structure. They are more likely to form helical strands (xanthan, gellan) or triple strands (curdlan) (Sutherland 1994). The monosaccharides present under nano-induced polysaccharides are mainly D-glucose, D-galactose, and D-mannose. Figure 1 shows nanosized Zn particle-induced microbial polysaccharide powder (both from fungi and bacteria) that can be used in agriculture for soil aggregation, moisture retention, and soil carbon buildup.

Nano-induced polysaccharides may be conjugated, cross-linked, or functionally modified and then used as nanocarrier materials. It can also be used in tissue engineering, wound healing and delivery of drugs (Gong et al. 2022) and many other fields (Peng et al. 2023). A large amount of nano-induced extracellular polysaccharides can be made through endophytic bacteria, microalgae, cyanobacteria, extremophiles as well as from mixed microbial communities. Nano particle induced

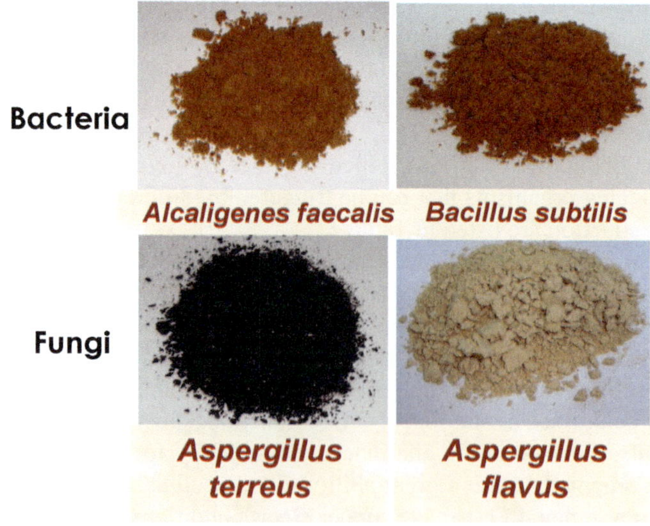

Figure 1. Nano-induced microbial polysaccharide powder (modified after Tarafdar 2021).

polysaccharides may be the future in coming days as they have many desired properties for industrial use. They can also meet the present demand of large scale use in polysaccharides under various industrial sectors. It may also overcome challenges in production that the polysaccharide industry is now facing, along with help in purification and improvement of product quality, which could assist in the path to their commercialization. Nano-induced polysaccharides may overcome the problem of microbial contamination, batch-to-batch variations, reduced viscocity during storage and thickening of the material. It may overcome all the problems now being suffered by the polysaccharide industries and may help to reduce the overall production cost.

Factors affecting nano-induced polysaccharide production

Important factors that are affecting the nano-induced polysaccharide secretion are temperature, pH, incubation time, and sucrose concentration. Temperature plays an important role, as more polysaccharide secretion was observed between 30 and 35°C. It has been noticed that nano-induced polysaccharide release is adversely affected below 25°C and above 40°C (Tarafdar 2021). Increase in temperature from 40°C increases the degradation of polysaccharide to oligosaccharides and monosaccharides. The increased temperature also affected the intermolecular associations of polysaccharides as well as glycosidic branching. In general, nanoparticle-induced bacterial polysaccharide secretion was greater between pH 6.0 and 7.5, while more polysaccharide secretion from fungi was observed between pH 5.0 and 6.0. A negligible increase in microbial polysaccharide release due to the application of nanoparticles was perceived when pH was below 4.0 and above 10.0. Generally, the variation in pH may affect the molecular weight as well as production of exopolysaccharide by the microorganisms. The best incubation time period to release maximum nano-induced polysaccharides from different organisms was observed between 72 and 96 hours. Sucrose plays a major role in releasing nano-induced polysaccharides, as a large amount of polysaccharide production was visualized between 20 and 25% sucrose concentration. Bacteria release more polysaccharide in presence of sucrose than fungi. The presence of oxygen as well as carbon sources, which play a substrate, also affects the release of nano-induced polysaccharides. Excess of carbon presence always plays a positive role in polysaccharide secretion.

Nanoparticles for water management

Nanoparticles are now very popular for water management due to their large surface area and reactivity. Nanofiltration is now a very popular and eco-friendly technology for water treatment. It has also been observed that many recent problems in maintaining water quality can easily be solved after using nanocatalysts, nanoabsorbents, magnetic nanoparticles, bioactive nanoparticles, nanostructured catalytic membranes, and nanotubes (Mamadou and Savage 2005). Moreover, nanoparticles may be transported productively by the groundwater flow and may remain in suspension for *in situ* water treatment (Zhang 2003). Nanoparticles such as iron oxide (Nowack and Buchelli 2007), TiO_2, and cerium oxide (Peng et al. 2005) play a vital role as efficient sorbents to remove organic and inorganic pollutants from contaminated water. TiO_2 nanoparticles are also able to eradicate arsenic under acidic conditions (Peng et al. 2005). Nanofiltration is now very popular for drinking water and waste water treatment. They are very effective at removing turbidity, inorganic ions like Na and Ca, as well as microorganisms. Nanofiltration can very well remove cations, natural organic matter, organic pollutants, and biological contaminants both from groundwater and surface water (Bruggen and Vandercasteele 2003). It can also remove organic pollutants as well as nitrates and arsenic. It can also be used to desalinate water (Mohsen et al. 2003) and make it potable. Nanostructured mixed oxides are being applied effectively for ground water treatment. Carbon nanotube filters may be effective for water management (Srivastava et al. 2004) and have been found to be very effective in removing bacteria from contaminated water. It has been noticed that chemically modified nanomaterials like cerium oxide supported on carbon nanotubes

(CeO$_2$-CNTs) can be used as novel sorbent (Peng et al. 2005). The tubes can also very well wash through ultrasonication and autoclaving. Nanoceramic filters can very effectively remove viruses and bacteria and can chemisorb dissolved heavy metals (Shah and Ahmed 2011). Researchers have demonstrated that mixed oxides nano-agglomerates like iron–cerium, iron–manganese, iron–zirconium, iron–titanium, iron–chromium, cerium-manganese, etc., can be successfully employed for pollutant removal (i.e., arsenic, fluoride, etc.) from aqueous solutions. For nano-remediation, iron nanoparticles are very popular because of their large surface area and large number of reactive sites. It is also noted that nanoscale metallic iron is more potent to overcome contaminants like polychlorinated hydrocarbons, pesticides, and dyes (Zhang 2003). Nanoparticles, due to their greater potential, are an ideal candidate to solve the challenges of various water treatments and show a lot of promise for future water management. It can be used as photocatalyst to trap the energy from light such as UV radiation which may help to degrade wide variety of organic materials like organic acids, estrogens, pesticides, dyes, crude oil, microbes (including viruses and chlorine resistant organisms), and inorganic molecules such as nitrous oxides (NOx). They actually oxidize these organic materials into readily biodegradable compounds and convert them to carbon dioxide and water (Kuwahara et al. 2010). It has massive scope and enormous opportunities for the removal of heavy metals, organic pollutants and microorganisms from the waste water due to their versatile biological and physiochemical properties.

Nanoparticles for soil health management

Soil health is the capacity of the soil to promote growth and productivity by sustaining and intensifying water and solute transport, nutrient cycling, physical and chemical stability, and the capability to reclaim toxic materials (Lehman et al. 2015, Stevens 2018). But the health of the global soil is degrading day by day because of intensive cropping, pollution, and environmental stresses like salinity, acidity, and heavy metal contamination (Liu et al. 2017). Nanoparticles have been found to be effective in more catalytic degradation, have higher absorptive capacity with higher carbon sequestration, help minimize soil-borne greenhouse gas emissions, and increase both soil water and nutrient retention (Raliya et al. 2014, Tan et al. 2016). Silver nanoparticles are found to be very effective in improving soil physico-chemical properties (Das et al. 2018). The best soil pH range for plant growth is between 5 and 7. Nanoparticles like Cu, Ag, Zn, Au, etc., may regulate the soil pH. It was observed that the pH of the soil also guided the toxicity of the nanoparticles on the soil microorganisms and nematodes. Scientists have noted more toxicity of Zn nanoparticles than dissolved Zn in bacterial community (Read et al. 2016). It was found that CuO nanoparticles help to increase the pH of paddy soils, affecting on soil properties. Silver nanoparticles' accumulation by insects is also reported to influence the soil pH (Pappa et al. 2018). Figure 2 represents the important parameters that can be controlled through nanoparticle introduction for soil health management. They include soil physical properties like aggregate stability, water holding capacity, soil compaction and penetration resistance. The chemical properties include nutrient status in the soil, electrical conductivity, pH and salinity. The biological properties include soil organic matter, carbon content, enzyme activity, microbial load and diversity, and flora and faunal population. The important soil storage that affects soil health are water, nutrient and microbial biomass. Nanosize particles of Fe and Zn help in more soil aggregate stability. Ag nanoparticles were found to be effective to retain water content in the soil. Nano Cu and nano gypsum are very effective for soil compaction. Lignin nanoparticles may assist in soil penetration resistance after decreasing or increasing swelling, shrinkage, and the elasticity index. Se and cerium oxide nanoparticles are found to be effective against soil salinity stress. Nanobiochar along with copperoxide nanoparticles are reported to improve soil nutrient availability as well as crop production (Rashid et al. 2023). Nanotechnology is reported to be very effective to show the pathways to enhance biological nitrogen fixation successfully (Li et al. 2023). It has been pointed out that positive effects of nanoparticles was observed on soil microbial diversity, nutrient cycling, and overall soil health and crop yields (Rafiq et al. 2017). Soil microbial

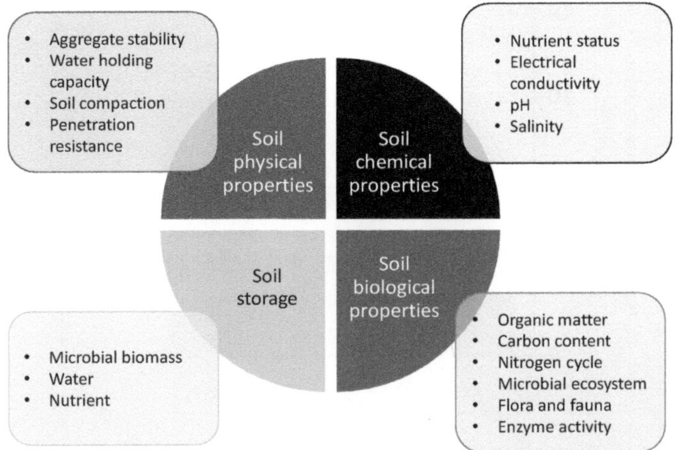

Figure 2. Conceptual framework on parameters responsible to maintain good soil health.

communities can also be altered by Ag and Cu-based nanoparticles (McKee and Filser 2016). Nano-clay, nano-chitosan, and nano-zeolite were reported to be very effective for soil microbial growth and health improvement (Khati et al. 2017). It has been found that carbon nanomaterials such as grapheme, grapheme oxide, and carbon nanotubes increase the population and diversity of microarthropods, resulting in improved soil health (Bai et al. 2017). Nanonutrients like N, P, K, Ca, Mg, S, B, Cu, Fe, Mn, Mo, and Zn can influence soil enzymatic activities (Tarafdar 2021), which play a proactive role in sustaining mineralization and the transformation of nutrients that improve soil health. The trigger of the microbial population by the nanonutrients can secrete more extracellular enzymes in the soil environment that help to generate more humic acid from plant decay, which ultimately boosts the weathering potential of the soil (Shen 2017). Therefore, to maintain soil health for sustainable agricultural production, nanonutrients may help as agricultural amendments. They may help to increase important enzymes like phosphatases, urease, esterase, dehydrogenase, amylase, cellulase,chitinase, lipase, glucosidases, sulfatases, and proteases (Tarafdar et al. 2018b). The application of nanoparticles as soil amendment resulted in positive stimulatory effects on crops due to increased enzymatic activity and management of the soil nutrient pool (Das et al. 2016). Nanocompounds such as nanoclay, nanochitosan, and nanozeolite have been found to help increase soil N, P, K, and organic carbon status (Khati et al. 2017). Nanocompounds are also found to be good chelaters of many macro and micronutrients, with slower release and improvements in nutrient use efficiency. Soil pore water also affected the earthworm population in presence of Ag nano particles (Schlich et al. 2013). A similar situation was noticed at higher concentration of Zn nanoparticles. The overall results pointed out the poisonous quality of most of the metallic nanoparticles at higher concentration, which also rests on soil pH and that may create a danger under acidic soil condition. Nanoparticles may help to reduce salt concentration in soil solution after enhancing Na^+ removal and working as polymeric carriers through clay binding process. It can also improve soil drainage and change carbonate chemistry after forming nanoorganic carbonates. Nano-Si is reported to be a strong candidate to mitigate salt stresses due to their ability to counteract the negative effects of salts on plant growth. It also has positive effect on xylem humidity, and water translocation, which help to increase plant water use efficiency. Authors found that carbon sequestration potential in soil (Figure 3) has been tremendously improved with the continuous application of nanofertilizers as compared to chemical, organic fertilizers or natural farming. Almost 73% more carbon sequestration was observed in soil with the continuous application of nanofertilizer (for five years) to crop production as compared to the no input soil (natural farming) or chemical fertilizer-applied soil. The results clearly indicated the role of nanofertilizers or nanonutrients to reduce environmental carbon dioxide concentration and help to store more carbon in the native soil for better plant production.

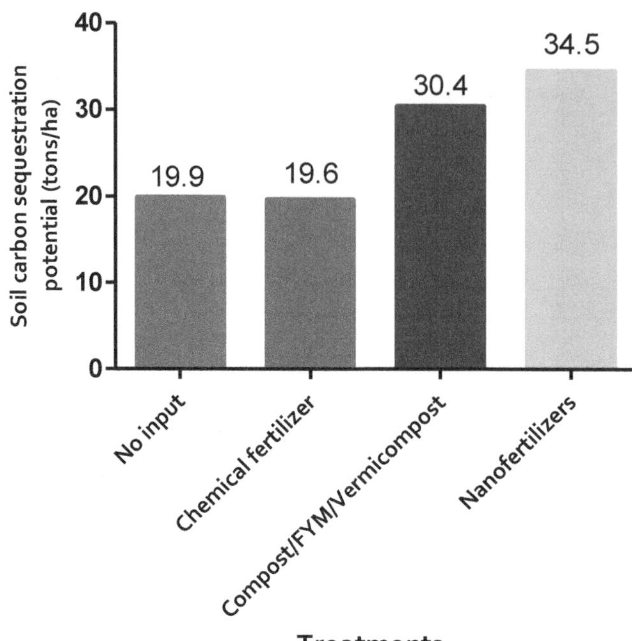

Figure 3. Role of nanofertilizers on carbon sequestration potential under different input application.

Moreover, the application of nanoparticles helps in increasing the population of polysaccharide releasing organisms by 10–12% that may also help in more polysaccharide release to restore more carbon and soil aggregates, besides helping in moisture retention in the soil.

Nanoparticles may be used as agricultural amendments for the restoration of soil acidity, increasing soil biodiversity, and increasing productivity (Liu et al. 2017). Salt stress in soil can be minimized with the application of nanoparticles (Khan et al. 2017). Zn nanoparticles' application showed that they could overcome salt stress. Moreover, nanosized Si, ZnO, Fe_2O_3, and gold nanoparticles have been reported to mitigate the adverse effects of salinity. It has also been reported that carbon-based nanotubes could effectively minimize the salt stresses on plants (Pandey et al. 2018). They have been found to remove toxic sodium from the soil solution, thereby helping in sustainable soil health management. Due to their unusual properties and renewable nature, nanocellulose-based nanocomposite hydrogels have proven to be an effective material for soil health management (Nascimento et al. 2018). They have also been found to have more biodegradability, water absorptive capacity, and nutrient retention potential. It has also been noted that methyl cellulose nanocomposite hydrogel has 2000 times more water retention capacity. In general, nanocomposite hydrogel may be used as a water, moisture, and nutrient reservoir in soil, leading to an improvement in soil granular structure, which ultimately helps in the cultivation capacity of soil. Tarafdar (2021) reported the soil health index (Figure 4) with the continuous application of different generations of fertilizer for five consecutive years. The first generation is considered as chemical fertilizer, the second generation is organic fertilizer, the third generation is growth stimulator and the nanofertilizer has been considered as fourth generation of fertilizers. The average soil health results from the five different crops growing under three different kinds of soil, which clearly indicated the ill effect of chemical fertilizer on soil health whereas biofertilizer, growth stimulator and nanofertilizers can help to maintain good to excellent soil health. Nanofertilizer may be considered as good as biofertilizer to maintain sustainable soil health condition. Nanobioformulations may help further to improve soil health. This may be due to the increase in diversity with the application of biofertilizer and nanofertilizer that can break disease cycles, revive plant growth and give shelter to the habitat responsible for pollination in the soil. It may affect the transportability of soil pollutants as it has

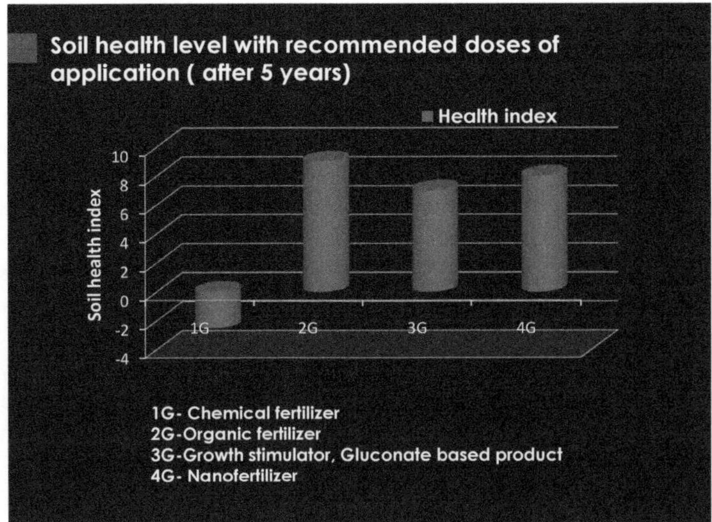

Figure 4. Soil health index with continuous application of different generations of fertilizers (Source: Tarafdar 2021).

been reported that particle size distribution as well as the composition of organic matter change microbial population in the contaminated soils (Calvarro et al. 2014).

Engineered nanomaterial has the deciding factors to maintain soil quality and health (Dotaniya et al. 2019). It can help to maintain soil security and is capable of managing the soil adverse conditions. It has come out as a promising material for sustaining agricultural production under increasing constraints (Bandala and Berli 2018). They are responsible to mitigate soil erosion, compaction, acidification, salinization, loss of organic matter, and heavy metal contamination, by enhancing mineralization, nutrient cycling, and soil microbial diversity which could potentially strengthen soil health and fertility toward sustainable agriculture. Researchers showed that it helps to improve soil bulk density, structural stability, water-holding capacity, CEC, nutrient cycling, and soil acidity,which have influential effects on the sustainability of agricultural development. Nanoparticles may lead many agricultural controlling processes and help in enhancement of food quality and safety besides reduction of agricultural inputs.

Nanoparticles for maintaining soil conditions and plant growth

Nanoparticles have unique properties in physical strength, electrical conductance, magnetism, and optical effects that result in better soil conditions than bulk materials. They may exist in different shapes and can be placed under different groups, such as metal, metal oxide, carbon-based, and based on polymer compounds. The function of nanoparticles depends on their properties, concentration, and method of application. Different techniques, such as seed treatment, foliar application, soil application, drip, sprinkler, hydroponic, aeroponic, and aquaponic, can be used for nanoparticle application (Tarafdar 2021). The accumulation, translocation, and uptake of nanoparticles mainly depend on their size, type, and stability, plant species, and their interaction with the roots, soil, and microbes (Ali et al. 2021). It has been observed that microbes have a role in the mobility of nanoparticles when applied through foliar spray, soil, or root application. Absorption of nanoparticles through the leaf surface depends on the morphology of the leaves and the presence of trichomes, waxes, and exudates on the leaf surface (Larue et al. 2014). Moreover, the size of the cell wall pores plays a significant role in allowing the nanoparticles to enter the plant cells. After entering the plant body, nanoparticles induce morphological, molecular, physiological, and biochemical changes in the crops (Khan et al. 2019). These changes appreciably improve the plant's growth and development. Nanoparticle application (for example, ZnO) significantly decreases MDA and H_2O_2 accumulation, which results in membrane stability and reduces the loss of crucial osmolytes (Sun et al. 2020). Nanoparticle application may also alter the root morphology with the increase in the lateral roots and root biomass that results in better water status under drought conditions. SiO_2 nanoparticles were found to be very effective in reducing the negative impact of drought after increasing photosynthesis, transpiration, water uptake, and relative water content (Sutuliene et al. 2021). Nanoparticles also help to increase root hydraulic conductivity, gene expression, and hormonal signaling, resulting in better plant water status in adverse environments. Metal and non-metal oxides nanoparticles like Au, Ag, ZnO, Se, TiO_2, Fe_2O_3, Al_2O_3, carbon nanotubes and quantum dots may play an important role in plant stress management and growth and development (Aqueel et al. 2022). The overall effect of nanoparticles on plant growth and management is represented as Figure 5. They may enter through foliage (cuticle, stomata, hydothades, stigma), can be fed through stems and may be through roots (root tips, rhizodermis, lateral root junction and any wounding). After entrance, they may translocate through the cell sap and trigger the different plant co-enzyme system for more release of beneficial enzymes to mobilize plant nutrients from native soil for better uptake and nutrition.

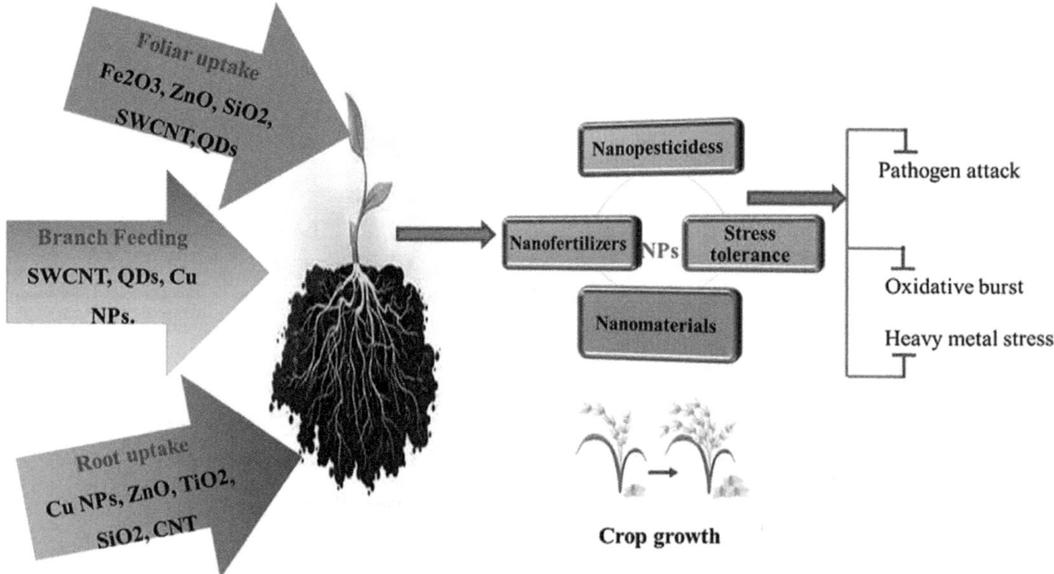

Figure 5. Overall effect of nanoparticles on plant growth and development (Sources: Aqueel et al. 2022).

Nanoparticles have found the capability to change molecular mechanisms after altering gene expression. Nanoparticles play a crucial role in improving nutrient uptake, their translocation and allocation to different plant parts, and their use efficiency (Tarafdar 2021). Foliar or soil application of nanoparticles significantly improves the uptake of N, P, K, and Zn and can also overcome the adverse effects of drought. Nanoparticles were also found to be very effective in overcoming both abiotic (heat, cold, drought, salinity, heavy metals, flooding, UV rays) and biotic (pathogenic, insect, herbivore) stresses plants are facing during their growth period. The most effective nanoparticles to overcome these stresses are Ag, TiO_2, SiO_2, Zn, chitosan, Fe, Se, CeO_2, Mn, Al_2O_3, MgO, and CuO. Nanoparticles may activate both enzymatic and non-enzymatic antioxidant defense mechanisms such as APOX, CAT, SOD, GPOX, GR, etc., (enzymatic) and glutathione, ascorbate, thiols, phenolics, etc., (non-enzymatic), resulting in better plant defense. Multiwalled carbon nanotubes have been found to be very effective in the seed germination of many crop species. Nanomaterials have proven to penetrate the seed coat, increase the absorption and utilization of water, act as a tonic for the enzymatic system, and improve seed germination and seedling growth. Aside from this, a large number of nanomaterials are reported to increase crop growth, quality enhancement, and yields (Tarafdar 2021). Nanoparticles like Ag, ZnO, $CuSO_4$, Fe, and TiO_2 help in the enhancement of secondary metabolite production in plants after generating reactive oxygen species that may help in flowering and fruit setting. Nanoparticles can help reduce heavy metal toxicity, improve plant growth, chlorophyll, and leaf relative water content, cause cooling of leaves after inducing stomatal opening, and upgrade the expression of stress-related genes. It can help overcome seed dormancy and enhance the expression of Rubisco and chlorophyll-binding protein genes. It also helps in increasing the proline content in leaves, reducing lipid peroxidation in roots, and increasing the net CO_2 assimilation rate. Many nanoparticles are found to be very effective as pesticidal agents; they can also be used as nanocarriers of different pesticides. Nanoherbicides are an eco-friendly alternative for weed control, as no toxic residues in the soil were observed (Pérez-de-Luque 2017). Nanocapsules are very popular carriers of different pesticides and herbicides. Nanoparticles were found to have the potential for gradual and controlled release of nutrients to the targeted sites and can be used as a potential material for precision farming and nutrient use efficiency. Application of nanosized nutrients can reduce nutrient loss, minimize the risk of environmental degradation, and improve crop quality and yield. Tarafdar et al. (2012) demonstrated that the foliar application of these nanonutrients as fertilizer can diminish plant stresses. Nanosized particles like calcium peroxide are reported to be very effective for degrading organic compounds. Similarly, nanozerovalent iron is useful for destroying halogenated organic compounds. Generally, nanoscale metal oxide can be very effectively used for metal adsorption. However, systematic evaluation of the potential adverse effect of nanoparticles is needed under long term basis for critical assessment towards adverse effects.

Soil restoration

Nanomaterials and nanonutrients can be used to restore the degraded soils and may prevent the losses due to soil biodiversity, productivity and fertility. The main causes of soil degradation are mainly due to industrial and agricultural pollution, commercial activities, overgrazing and urban development, climate change and inappropriate agricultural practices. Nanomaterials or nanonutrients like CuO, Se, SiO_2, ZnO, FeO, nano-biochar, etc., are found to be very effective in restoring degraded soils like sandy soils, waterlogged soils, polluted soils, mined soils as well as salt effected soils. Plants may take up nanoparticles through foliage or soil and use them in the process of remediation of pollutants presentboth in soil and water (Vélez et al. 2021). Nanoparticles may prove to be a good material for soil restoration which may alleviate the plant stresses (Gelaw and Sanan-Mishra 2022). Halophytic nanoparticles may be used for remediation of saline soils (Munir et al. 2021). But one should use the optimum prescribed doses of nanomaterials; otherwise, overdoses of nanoparticles during remediation process may help in nano-toxicity to the cultivated crops and may advance through the food chain resulting in negative health impact to humans as well as environment. Therefore, further studies are needed on nano-restoration to sustain fertility and productivity of soils. More research also should be aimed to the supportive role of plant, nanoparticles and soil microbes in the integrated way. The general mechanism of nano-toxicity may be due to the oxidative stress as well as destruction in biochemical and morpho-physiological changes in edible plants that adversely affect human health as well as soil microbial biota (Júnior et al. 2022), although the adverse effect depends on the type of nanomaterials, the applied doses and plant species as well as soil type. Moreover, size, surface and mineralogy of the particles also matter. Flow rate and flow conditions of nanoparticles and also way of delivery of nano nutrients may have an impact on nano-toxicity. It has also been noticed that some of the native microorganisms are hyper accumulator of nanoparticles that can be used as nano phytoremediation or microbial mediated nano remediation. The nanobioremediation is accountable for adsorption, redox reaction as well as mobility, aggregation and stability of nanoparticles. Nanomaterials like carbon nanotubes, metal oxide nanoparticles, titanium oxides as well as various nanocomposites may be used to immobilize soil contaminants to restore soil degradation. They have the capacity to immobilize heavy metals like cadmium and arsenic. TiO_2 nanoparticles can be used as photocatalyst with a UV light source to enhance degradation of organic materials as well as pesticides. Both carbon and cellulose nanomaterials as well as metal and metal oxide nanoparticles can found to deliver well of specific nutrients to the soil from seeds, roots and leaves, resulting in minimizing the occurrence of unwanted soil contaminants. Actually soil needs enough fertility and diversified properties to keep them healthy and produce healthy food to support human and environmental health. Soil restoration can reverse the degradation process that may be achieved through nanoparticles with recommended doses of application. Nano-restoration can very well be possible through the sustainable production of nanoparticles using plant and microbes. The plant, microbe and nanoparticles consortia may be an important issue for soil fertility and maintenance of soil health. It depends on the bioavailability of nutrients and control by large number of factors that are responsible for soil fertility and plant health conditions.

Conclusion

The present chapter brings out the important role nanosized particles are playing in soil moisture retention and their health management. The different problems posed by modern agriculture may be solved with the introduction of nanoparticles or nanomaterials due to their effectiveness and efficiency as compared to traditional sources. Nanoinduced polysaccharide powder provides better moisture retention, soil aggregation, and soil carbon buildup. More soil carbon sequestration was observed with the application of nanosized nutrient particles in agricultural fields. Nanoparticles also help plants to overcome abiotic and biotic stresses. It helps to boosts the stability of membranes as well as nutrient and water uptake. Moreover, may assist more photosynthesis under drought situations. The use of nanoparticles also activates the different enzymes involved in mobilization of native nutrients and induce the production of plant hormones needed to overcome adverse conditions. Nanoparticles have a great impact on water treatment and purification systems, but more studies are needed on the environmental fate and toxicity of the nanomaterials. However, nanoparticles showed an extremely high potential and are suitable for use in soil, water and health management, including groundwater.

References

Ali, S., Mehmood, A., and Khan, N. 2021. Uptake, translocation, and consequences of nanomaterials on plant growth and stress adaptation. J. Nanomater., 2021: 6677616.

Aqeel, U., Aftab, T., Khan, M.H.A., Naeem, M., and Khan, M.N. 2022. A comprehensive review of impacts of diverse nanoparticles on growth, development and physiological adjustments in plants under changing environment. Chemosphere, 291: 132672.

Bai, X., Zhao, S.L., and Duo, L. 2017. Impacts of carbon nanomaterials on the diversity of microarthropods in turfgrass soil. Sci. Rep., 7.

Bandala, E.R., and Berli, M. 2018. Nanomaterials: new agrotechnology tools to improve soil quality? pp. 127140. *In*: López-Valdez, F., and Fernañdez-Luqueñõ, F. (eds.). Agricultural Nanobiotechnology: Modern Agriculture for a Sustainable Future. Springer International Publishing, Cham, Switzerland.

Bruggen, V.D.B., and Vandercasteele, C. 2003. Removal of pollutants from surface water and groundwater by nanofiltration: Overview of the possible applications in the drinking water industry. Environ. Pollut., 22: 435–445.

Calvarro, L.M., de Santiago-Martín, A., Gómez, J.Q., González-Huecas, C., Quintana, J.R. et al. 2014. Biological activity in metal-contaminated calcareous agricultural soils: the role of the organic matter composition and the particle size distribution. Environ. Sci. Pollut. Res., 21: 6176–6187.

Das, P., Sarmah, K., Hussain, N., Pratihar, S., Das, S., Bhattacharyya, P. et al. 2016. Novel synthesis of an iron oxalate capped iron oxide nanomaterial: a unique soil conditioner and slow release eco-friendly source of iron sustenance in plants. RSC Adv., 6: 103012–103025.

Das, P., Barua, S., Sarkar, S., Karak, N., Bhattacharyya, P., Raza, N. et al. 2018. Plant extract mediated green silver nanoparticles: efficacy as soil conditioner and plant growth promoter. J. Hazard. Mater., 346: 62–72.

Dotaniya, M.L., Aparna, K., Dotaniya, C.K., Singh, M., and Regar, K.L. 2019. Role of soil enzymes in sustainable crop production. Enzymes in Food Biotechnology. Academic Press, pp. 569589.

Freitas, F., Torres, C.A.V., Araújo, D., Farinha, I., Pereira, J.R., Concórdio-Reis, P. et al. 2021. Advanced microbial polysaccharides. Biopolym. Biomed. Biotechnol. Appl. Chapter 2: 19–62.

Gehrke, I., Geiser, A., and Somborn-Schulz, A. 2015. Innovations in nanotechnology for water treatment. Nanotech. Sci. Appl., 8: 1–17.

Gelaw, T.A., and Sanan-Mishra, N. 2022. Nanomaterials coupled with microRNAs for alleviating plant stress: A new opening towards sustainable agriculture. Physiol. Mol. Biol. Plants., 28: 791–818.

Gong, H.X., Li, W.A., Sun, J.L., Jia, L., Guan, Q.X., Guo, Y.Y. et al. 2022. A review on plant polysaccharide based on drug delivery system for construction and application, with emphasis on traditional Chinese medicine polysaccharide. Int. J. Biol. Macromol., 211: 711–728.

Han, Y., Liu, E., Liu, L., Zhang, B., Wang, Y., Gui, M. et al. 2015. Rheological, emulsifying and thermostability properties of two exopolysaccharides produced by *Bacillus amyloliquefaciens* LPL061. Carbohydr. Polym., 115: 230–237.

Jatav, G.K., and De, N. 2013. Application of nanotechnology in soil plant system. Asian J. Soil Sci., 8(1): 176–184.

Júnior, A.H.D.S., Mulinari, J., de Oliveira, P.V., de Oliveira, C.R.S., and Júnior, F.W.R. 2022. Impacts of metallic nanoparticles application on the agricultural soils microbiota. J. Hazard. Mater. Adv., 7: 100103.

Khan, I., Saeed, K., and Khan, I. 2019. Nanoparticles: Properties, applications and toxicities. Arab. J. Chem., 12: 908–931.

Khan, M.N., Mobin, M., Abbas, Z.K., AlMutairi, K.A., and Siddiqui, Z.H. 2017. Role of nanomaterials in plants under challenging environments. Plant Physiol. Biochem., 110: 194–209.

Khati, P., Sharma, A., Gangola, S., Kumar, R., Bhatt, P., and Kumar, G. 2017. Impact of agriusable nano compounds on soil microbial activity: an indicator of soil health. Clean Soil Air Water, 45: 1600458.

Koocheki, A., Mortazavi, S.A., Shahidi, F., Razavi, S.M.A., and Taherian, A.R. 2009. Rheological properties of mucilage extracted from alyssum homolocarpum seed as a new source of thickening agent. J. Food Eng., 91: 490–496.

Kunduru, K.R., Nazarkovsicy, M., Farah, S., Pawar, R.S., Basu, A., and Domb, A.J. 2017. Nanotechnology for water purification : Application of nanotechnology method in waste water treatment. pp. 33–73. *In*: Nanotechnology for water purification, Chapter 2, Elsevier Inc.

Kuwahara, Y., Kamegawa, T., Mori, K., and Yamashita, H. 2010. Design of new functional Titanium oxide-based photocatalysts for degradation of organics diluted in water and air. Curr. Org. Chem., 14(7): 616–629.

Larue, C., Castillo-Michel, H., Sobanska, S., Cécillon, L., Bureau, S., Barthès, V. et al. 2014. Foliar exposure of the crop *Lactuca sativa* to silver nanoparticles: evidence for internalization and changes in Ag speciation. J. Hazard. Mater., 264: 98–106.

Lehman, R.M., Cambardella, C.A., Stott, D.E., Acosta-Martinez, V., Manter, D.K., Buyer, J.S. et al. 2015. Understanding and enhancing soil biological health: the solution for reversing soil degradation. Sustainability, 7: 988–1027.

Li, M., Gao, L., White, C.J., Haynes, C.L., O'Keefe, T.L., Zhiling, G. et al. 2023. Nano enabled strategies to enhance biological nitrogen fixation. Nature Nanotech. 18: 688–691.

Lin, C.C., Cassida, L.E., Jr., Lin, C.C., and Cassida, Jr.,., L.E. 1984. Gelrite as a gelling agent for the growth of thermophilic microorganisms. Appl. Environ. Microbiol., 47: 427–429.

Liu, S.K., Qi, X., Han, C., Liu, J.M., Sheng, X.B., Li, H. et al. 2017. Novel nano submicronmineral-based soil conditioner for sustainable agricultural development. J. Clean. Prod., 149: 896–903.

Mamadou, S.D., and Savage, N. 2005. Nanoparticles and water quality. J. Nano Res., 7: 325–330.

McKee, M.S., and Filser, J. 2016. Impacts of metal-based engineered nanomaterials on soil communities. Environ. Sci. Nano, 3: 506–533.

Mohsen, M.S., Jaber, J.O., and Afonso, M.D. 2003. Desalination of brakish water by nanofiltration and reverse osmosis. Desalination, 157(1): 167–167.

Munir, N., Hanif, M., Dias, D.A., and Abideen Z. 2021. The role of halophytic nanoparticles towards the remediation of degraded and saline agricultural lands. Environ. Sci. Pollut. Res., 28:60383–60405.

Nascimento, D.M., Nunes, Y.L., Figueiredo, M.C.B., de Azeredo, H.M.C., Aouada, F.A., Feitosa, J.P.A. et al. 2018. Nanocellulose nanocomposite hydrogels: technological and environmental issues. Green Chem., 20: 2428–2448.

Nowack, B., and Bucheli, T.D. 2007. Occurrence, and effects of nanoparticles in the environment. Environ. Pollut. 150: 5–22.

Pandey, K., Lahiani, M.H., Hicks, V.K., Hudson, M.K., Green, M.J., and Khodakovskaya, M. 2018. Effects of carbon-based nanomaterials on seed germination, biomass accumulation and salt stress response of bioenergy crops. Plos One, 13: e0202274.

Pappa, A.M., Parlak, O., Scheiblin, G., Mailley, P., Salleo, A., Owens, R.M. 2018. Organic electronics for point-of-care metabolite monitoring. Trends Biotechnol. 36: 45–59.

Peng, X., Luan, Z., Ding, J., Di, Z., Li, Y., and Tian, B. 2005. Ceria nanoparticles supported nanotubes for the removal of arsenate from water. Mater. Lett., 59: 399–403.

Peng, W., Guo, X., Xu, X., Zou, D., Zou, H., and Young, X. 2023. Advances in polysaccharide production based on co-culture of microbes. Polymers, 15(13): 2847.

Pérez-de-Luque, A. 2017. Interaction of nanomaterials with plants: what do we need for real applications in agriculture. Front. Environ. Sci., 5: 12.

Peteu, S.F., Oancea, F., and Sicuia, O.A. 2010. Constantinescu F, Dinu S. Responsive polymers for crop protection. Polymers, 2: 229–251.

Rafiq, M.K., Joseph, S.D., Li, F., Bai, Y.F., Shang, Z.H., Rawal, A. et al. 2017. Pyrolysis of attapulgite clay blended with yak dung enhances pasture growth and soil health: characterization and initial field trials. Sci. Total Environ., 607: 184–194.

Raliya, R., Tarafdar, J.C., Mahawar, H., Kumar, R., Gupta, P., Mathur, T. et al. 2014. ZnO nanoparticles induced exopolysaccharide production by *B. subtilis* strain JCT 1 for arid soil applications. Int. J. Biol. Macromol., 65: 362–368.

Rashid, M.I., Shah, A.G., Sadiq, M., Amin, N.U., Ali, A.M., Ondrasek, G. et al. 2023. Nanobiochar and copper oxide nanoparticles mixture synergistically increases soil nutrient availability and improves wheat production. Plants, 12(6): 1312.

Ravella, S.R., Quinones, T.S.R., Retter, A., Heiermann, M., Amon, T., and Hobbs, P.J. 2010. Extracellular polysaccharide (EPS) production by a novel strain of yeast-like fungus aureobasidium pullulans. Carbohydr. Polm., 82: 728–732.

Read, D.S., Matzke, M., Gweon, H.S., Newbold, L.K., Heggelund, L., Ortiz, M.D. et al. 2016. Soil pH effects on the interactions between dissolved zinc, non-nano-and nano-ZnO with soil bacterial communities. Environ. Sci. Pollut. Res., 23: 4120–4128.

Rolli, E., Marasco, R., Vigani, G., Ettoumi, B., Mapelli, F., Deangelis, M.L. et al. 2015. Improved plant resistance to drought is promoted by the root-associated microbiome as a water stress-dependent trait. Environ. Microbiol., 17: 316–331.

Schlich, K., Klawonn, T., Terytze, K., and Hund-Rinke, K. 2013. Effects of silver nanoparticles and silver nitrate in the earthworm reproduction test. Environ. Toxicol. Chem., 32: 181–188.

Şengör, S.S. 2019. Review of current applications of microbial biopolymers in soil and future 493 perspectives. pp. 275–299. *In*: Introduction to Biofilm Engineering, ACS Publications.

Shah, M.A., and Ahmed, T. 2011. Principles of nanoscience and nanotechnology. Narosa Publishing House, New Delhi, India, pp. 34–47.

Shen, Y.F. 2017. Rice husk silica derived nanomaterials for sustainable applications. Renew. Sust. Energ. Rev., 80: 453–466.

Srivastava, A., Srivastava, O.N., Talapatra, S., Vajtai, R., and Ajayan, P.M. 2004. Carbon nanotube filters. Nat. Mater., 3: 610–614.

Sun, L., Song, F., Guo, J., Zhu, X., Liu, S., Liu, F. et al. 2020. Nano-ZnO induced drought tolerance is associated with melaton in synthesis and metabolism in maize. Int. J. Mol. Sci., 21: 782.

Sutuliene, R., Rageliene, L., Samuoliene, G., Brazaityte, A., Urbutis, M., and Miliauskiene, J. 2021. The response of antioxidant system of drought-stressed green pea (*Pisumsativum* l.) affected by watering and foliar spray with silica nanoparticles. Hortic., 8: 35.

Stevens, A.W. 2018. Review: the economics of soil health. Food Policy, 80: 1–9.

Sutherland, I.W. 1994. Structure-function relationships in microbial exopolysaccharides. Biotech. Adv., 12: 393–448.

Tan, X.F., Liu, Y.G., Gu, Y.L., Xu, Y., Zeng, G.M., Hu, X.J. et al. 2016 . Biochar-based nano-composites for the decontamination of wastewater: a review. Bioresour. Tech., 212: 318–333.

Tarafdar, J.C., Xiang, Y., Wang, W.N., Dong, Q., and Biswas, P. 2012. Standardization of size, shape and concentration of nanoparticle for plant application. Appl. Biol. Res., 14: 138–144.

Tarafdar, J.C., Raliya, R., and Praveen-Kumar. 2018a. Nanoinduced bacterial polysaccharide production. Indian Patent No. 304253.

Tarafdar, J.C., Rathore, I., Kaur, R., and Jain, A. 2018b. Biosynthesis of nanonutrients for agricultural applications. pp. 15–30. *In*: Singh, B. (ed.). Nano Agroceuticals and NanoPhytoChemicals. CRC Press: Boca Raton, FL, USA.

Tarafdar, J.C. 2021. Nanofertilizers: Challenges and Prospects. Scientific publisher, India, pp. 363.

Tarafdar, J.C. 2022. Nanotechnology application in agriculture: current developments of nanofertilizers. pp. 19–29. *In*: Singh, P., Anam, Srivastava, T.K., and Verma, R.R. (eds.). Nanoparticles Application in Agriculture, Chapter 2, Scientific Publishers.

Vélez, Y.S.P., Carrillo-González, R., and González-Chávez, M.D.C.A. 2021. Interaction of metal nanoparticles–plants–microorganisms in agriculture and soil remediation. J. Nanopart. Res. 23: 206.

Zhang, W.X. 2003. Nano-scale iron particles for environmental remediation: an overview. J. Nanopart. Res., 5: 323–332.

3

Nanosensor in Soil-Water Agroecosystem

Ghulam Mustafa,[1,*] *Muafia Akbar,*[1] *Nauman Sadiq,*[1,2]
Bushra Iqbal[1] and *Muhammad Zahid*[3]

Introduction

Natural ecosystems are self-sustaining, complex adaptive systems (CASs), having a flow of matter, free energy and information (Levin 1998). On the other hand, agroecosystems, often known as agricultural ecosystems, are not CAS. Self-sustainability cannot be achieved with artificially fewer system components. Therefore, the system's ability to produce food, fiber, and biomass as well as maintain its structural integrity over time depends on ongoing human contributions of free energy, matter, and information in the equipment, infrastructure, irrigation water use, human or robotic workforce, weed and management of soil, fertilizer, and pesticide use, among other things (Cochran et al. 2016). Although almost all agronomic practices have an impact on the environment, fertilizers and pesticides have received a lot of interest and concern because of their negative effects. This is probably the case because these two elements have a direct impact on food safety in addition to modifying the soil and water chemistry and biodiversity. The utilization of nanoparticles (NPs) of metals and semimetals as well as organic (like chitosan) and other materials like nanoclay is currently presenting a potentially significant opportunity in the commercial niche of pesticides and fertilizers, which includes a wide range of products like insecticides, acaricides, fungicides, nematicides, and so on. These NPs and nanomaterials (NMs), which include nanosensors, nanopesticides, nanocarriers of pesticides, nanofertilizers, soil supplements, and others, provide a possible replacement to reduce the utilization of fertilizers and pesticides (Feregrino-Perez et al. 2018). To meet societal expectations and a growing population, agriculture must today focus on maximizing output (better yields and quality) while minimizing input consumption (water, fertilizer, pesticides, and energy) (Godfray et al. 2010). Nanotechnology has a huge potential to change agriculture's production and related industries, and it may provide fresh ideas for more effective

[1] Department of Chemistry, University of Okara, Okara 56300, Pakistan.
[2] Department of Chemistry, Quaid-i-Azam University, Islamabad, 45320, Pakistan.
[3] Department of Chemistry, University of Agriculture Faisalabad, Pakistan.
* Corresponding author: ghulam.mustafa@uo.edu.pk

and environmentally friendly farming. Nanopesticides, nanofertilizers, and nanosensors have been developed for testing soil and water to improve plant growth, seed germination and quality. In addition, their use may enhance nutraceutical properties, require fewer inputs, and increase crop value. Additionally, it is anticipated that agricultural nanotechnology will produce more effective fertilizer materials, reducing the quantity of the various elements used while ensuring increased production and fewer losses from leaching, volatilization, and immobilization by precipitation and soil fixation. Research, development, and application of nanoproducts in agriculture have all extended postharvest life (Feregrino-Perez et al. 2018, Mesci-Haftaci et al. 2014). By using materials that are more reactive and efficient due to a high surface-to-volume ratio, such as NMs with biostimulant potential, which boost the crop plants' natural defenses, it is also possible to minimize the number of pesticides employed. The aforementioned is also made possible by the increased potency of pesticides and the regulators connected to nanocarriers that enable their action on particular targets (Juárez-Maldonado et al. 2019). The application of nanomaterials for human benefit is known as "nanotechnology". Nanosensors are used to monitor size and special qualities at the nanoscale and for innovative sensing. Nanosensors' primary function is to track any alterations in chemical, mechanical, or physical properties that are connected to an interest indication. Nanosensors are tiny electronic components designed to recognize specified chemical, biological, or environmental components. These sensors are incredibly precise, portable, economical, and determined at a lower level than their macroscale counterparts.

Components of nanosensors for monitoring agroecosystem

The three fundamental parts of a typical nanosensor device are as follows:

- **Sample preparation:** It could consist of a homogenous or intricate mixture of solid, liquid, or gas phases. Agroecosystem sample preparation is highly difficult because of contaminants and interferents. The sensor can target specific molecules in the sample, functional groups of chemicals, or organisms.

  The analyte is a term used for the targeted molecules or organisms, it can comprise any of the following: molecules such as colors/dyes, hormones, pesticides, toxicants, antibodies and vitamins; biomolecules like allergens, DNA/RNA and enzymes and gases/vapors of CO_2, O_2, water vapors, and organisms.

- **Recognition:** Specific molecules or substances can identify the analyte present in the sample. These recognition molecules, which include antibodies, aptamers, polymers, and others, must exhibit high specificity, sensitivity, and selectivity toward their analytes to quantify them to an acceptable level.

- **Signal transduction:** Using different transduction techniques, these simple devices have been divided into many categories including optical, electrochemical, pyroelectric, piezoelectric, electronic and gravimetric biosensors. They transform events of recognition into signals that may be analyzed further to provide data (Figure 1).

Nanosensors can be categorized as optical, electrical, or mechanical based on the type of signal they produce (Nguyen et al. 2014). The two main kinds of nanosensors are those that operate at the nanoscale and those that are used to measure properties at the nanoscale. The first kind of nanosensor requires cheap materials, little weight, and little power. These designed NMs might have harmful effects on the second class of nanosensors (Simon et al. 2016). Due to their selectivity, speed, and sensitivity as compared to conventional methods, nanosensors are frequently employed in the agriculture, food, and medical industries. Monitoring of microorganisms, toxins, and pollutants may be done with nanosensors (Omanović-Mikličanina et al. 2016). Aptasensors are biosensors made of nanomaterials (the signal transducers) and aptamers (the target recognition component). Single-stranded peptides or nucleic acids with a size of less than 25 kDa are known as aptamers.

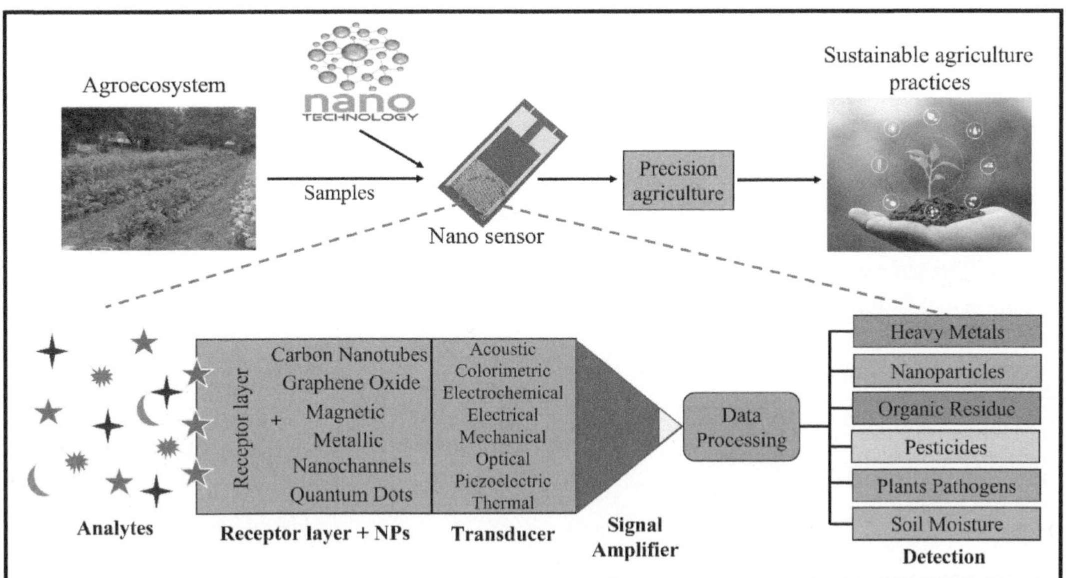

Figure 1. A streamlined illustration of the components of nanosensors for monitoring agroecosystem.

Due to their 3D structures, they are highly specific toward their target compounds, such as proteins, toxins, microbes, viruses, etc. For the detection of diverse analytes, a broad range of NMs, such as metal nanoparticles and nanoclusters, carbon nanoparticles, magnetic nanoparticles, etc., can be utilized in aptasensors (Lewinski et al. 2008).

Nanosensors are proving to be an effective instrument for use in agricultural and food production. Compared to conventional chemical and biological approaches, they offer considerable advantages in selectivity, speed, and sensitivity. The assessment of microorganisms, toxins, pollutants, and food freshness may be done with nanosensors. These nanosensors have the power to completely change how we perceive and take care of soil-water agroecosystems. These sensors may be used to keep an eye on several variables, including toxins or pollutants as well as soil moisture, nutrients, pH, and salt. Nanosensors can assist farmers and researchers in making knowledgeable decisions about irrigation, fertilization, and pest control, among other things, by providing real-time and precise data (Omanović-Mikličanina et al. 2016). There is an urgent need to ensure global food security as a result of several issues from previous decades. Agriculture ecosystems have faced many obstacles as a result of rising food production, including the persistence of lingering pesticide particles, the accumulation of heavy metals, and contamination with dangerous components, all of which have impacted the agricultural environment. Numerous health effects, including disorders of the nervous system and metabolism, infertility, disturbance of cellular biological processes and bone marrow, and respiratory and immune disorders, result from the ingestion of such toxic elements through agricultural products. When considering the approximately 220,000 annual deaths brought on by the hazardous effects of residual pesticides, it is clear that there is a need for monitoring agroecosystems (Akoto et al. 2015). Currently, techniques like high-performance liquid chromatography (HPLC), gas chromatography (GC), mass spectroscopy, etc., are utilized to monitor agroecosystems, which have several drawbacks, including being expensive, time consuming, and requiring complicated equipment as well as skilled personnel (Dhouib et al. 2016). Over the past few decades, the science of nanotechnology has made major advancements, and these advances have greatly aided in the development of low-cost, quick, and useful bio and nanosensors for identifying various contaminants in natural agroecosystems, which have the advantage of being safe for human health (Sahani et al. 2021). The rapid advancement of nanotechnology has enabled the development of nanosensors and biosensors for the detection of various composites, ranging from the detection of

Figure 2. Schematic diagram illustrating the variation between traditional and advanced monitoring techniques.

proteins, numerous metal ions, and pesticides to the detection of whole microbes (Akoto et al. 2015). Figure 2 shows the schematic diagram illustrating the variation between traditional and advanced monitoring techniques.

The current chapter focuses on numerous biosensors and nanosensors used to monitor ecosystems in agriculture in order to emphasize the aspects impacting their application from the proof of concept to commercialization stages.

Background of polluted soils and wastewater

Despite the fact that water is a renewable resource, many parts of the world may experience a shortage over the next 20 years, especially in the case of portable water, which is rapidly running out and affects 1.2 billion people annually (Kim et al. 2010). One of the main problems the world is now experiencing is the conservation of water for future generations. The development of a society with an unfavorable environmental imbalance brought on by high production of unwanted pollutants, such as wastewater effluents from industries, heavy metals, nonbiodegradable materials, hazardous materials, organic materials, fertilizers, and pesticides is significantly impacted by industrial and agricultural revaluations. By mingling with them, these materials eventually contaminate soil and water sources. These contaminants affect the vast population residing close to the contaminated locations, as well as the strength of the soil and the quality of the water. If the side effects of the process were not addressed, even cleanup therapy would have an impact on the population and workforce. In addition to plant nutrients (N, P, and K), wastewater contains various components such as pathogenic organisms (viruses, protozoa, bacteria, and helminths), heavy metals (Cd, Cr, Cu, Hg, Ni, Pb, and Zn), organic pollutants (polychlorinated biphenyls, pesticides, biodegradable organics, and poly-aromatics), as well as micro-plants such as medication (Raja et al. 2020). Due to its vast range and intricate method of management and governance, soil contamination was formerly not seen as essential as air and water pollution. However, in the last several decades, the scope of soil pollution treatment has significantly increased (Hu et al. 2020). In many civil engineering and agricultural practices, the quality of the soil is a crucial factor. The main sources of soil contaminants, including heavy metals, organic material, and nonbiodegradable chemicals, are manufacturing processes, construction methods, agriculture, fertilizer industries, paper industries, and the food sector. The fertilizer companies produce natural radionuclides and heavy metals such as Hg, Cd, Cu, Pb, Ni, that pollute soil (Sonmez et al. 2007). In the case of heavy metal-contaminated soil, self-purification techniques are either exceedingly slow or unsuccessful. Certain heavy metal-contaminated soils may require 100 or 200 years to self-remediate (Umabharathi et al. 2023). The soil microorganisms and microfauna that encourage soil toxicity are negatively impacted by the continuous use of fertilizers (Prashar et al. 2016). Analyzing soil microbiology is an expensive and time-consuming operation; as an alternative, sensing devices are used to show the status of a site, monitor soil attributes and pH levels, observe how pesticides and fertilizers affect crops, and determine the soil's toxicity (Tothill 2001).

Role of nanomaterials in managing water and soil quality

Due to deforestation, industrialization, wildfires, the agricultural revolution, and inappropriate solid waste treatment, the entire globe is currently experiencing an environmental catastrophe. Pollutants that pollute air, water, and soil include excessive use of pesticides, herbicides, and fertilizers, as well as significant emissions of hazardous particulate matter, greenhouse gases, heavy metals, industrial effluents, and small particles. The environmental balance and ecology are being slowly destroyed by these pollutants, which is one of the largest problems we are now facing and has to be addressed with careful planning and laws. The environmental issue has become worse over the past ten years due to the various government entities' patchy or insufficient remedial activity. Since remediation is not an emergency reaction to a catastrophe, it usually does not require quick action. Because of this, it promotes a thorough understanding of contaminated areas, danger to the populace from the use of water and land, risk to workers engaged in remediation work, production of waste, and net benefits of remediation, in addition to the formulation of target-specific goals, sound planning, moral consideration, and ongoing research towards new technologies (Vaseashta et al. 2007). Finding new materials to use in environmental rehabilitation prompts the examination of a wide variety of novel ways in comparison to traditional remediation techniques. Nanotechnology is one of these techniques that provides an array of chances for the degradation of pollutants, which is a challenging task given the complexity of the compound mixture, low reactivity, and high volatility (Tratnyek et al. 2006). The unusual form, high surface area, and porosity of NMs allow for effective environmental cleanup through surface activities with contaminants. In terms of environmental remediation, microbes, chemicals produced by microbes, and synthetic NMs mediated via plant extracts all play a vital role. The use of these biological agents in the creation of NMs is thought to be more economical and ecologically benign than using traditional chemical processes (Husen et al. 2014). In this type of remediation, harmful metals and organic chemicals are detected and removed from the various environmental components using NMs and NPs, nanopowders, and nanomembranes. To create a green and safe atmosphere and, eventually, encourage people to adopt a certain technology, the material utilized shouldn't be a pollutant, and there shouldn't be any trash left over after the corrective activity. The most crucial requirement is that the NMs utilized to degrade contaminants be made of biodegradable materials. Furthermore, due to their great efficiency, novel technologies that depend on the task-oriented capture of harmful pollutants stand out (Guerra et al. 2018).

As a result, some researchers have been inspired to incorporate surface chemical and physical modification with nanotechnology to create unique materials that can overcome a variety of difficult obstacles throughout the restorative process (Kamat et al. 2003). Different NMs have so far been successful in removing harmful metals and organic chemicals through a variety of processes, including adsorption, transformation, photocatalysis, and catalytic reduction (Li et al. 2014).

Development of nanosensors in agroecosystem

One of the main things nanotech devices do is make it easier to use self-distributed nanosensors in crop fields, along with a global positioning system for accurate time and a thorough survey of crop progress. This gives high-quality and important data, like the best time to plant and harvest crops, which leads to better agronomic practices (Initiative 2012). The Intel Corporation has created processors with nanoscale features and deployed large wireless sensor nodes (referred to as "motes") throughout an Oregon vineyard. The sensors are the first steps towards the vineyard's complete automation and record temperature once per minute. Motes created by Crossbow Technologies may be used on agricultural fields to calculate water quality, diagnose frost damage, notify users of the need to use pesticides, estimate irrigation needs, and estimate harvest times. Agricultural manufacturing might be monitored using nanoprocessing and nanobarcodes. Crop diseases were efficiently analyzed and identified using the notion of grocery barcoding. Nanobarcodes were developed that could identify different infections in a field of crops that could be observed by using fluorescence-based monitoring equipment (Li et al. 2005).

Nanosensors for detection of conditions and plant growth hormone

The advancement of nanotechnology enables real-time monitoring of crop development and farm conditions, including soil fertility, moisture content, temperature, crop nutrition capacity, pathogens, plant diseases, etc. Farmers may use inputs more effectively by screening for infections, estimating nitrogen absorption, and properly measuring soil criteria including pH, minerals, residual pesticides, and soil moisture with nanosensors. This practice promotes sustainable agriculture (Parisi et al. 2015). To fulfill the goal of precision agriculture, these nanosensors may be utilized to monitor the period, volume, and delivery of pesticide treatments. Modern research is being done to develop accurate irrigation water distribution technologies in the future. Water storage, the capacity to store water in place, the supply of water closest to plant roots, plant ability to consume water, and contained water released under crop needs are all vital components for this advancement. Interactions between roots and microorganisms in the rhizosphere of wheat or other crops play a significant role in chemical signaling networks. As a result, metabolites produced by microorganisms or plants are released into the soil. According to some theories, nanofertilizer incorporates nanobiosensors suspended in a biopolymer that coats fertilizer particles. The primary step in the process of nutrient release is the identification and joining of a particular plant SOS (salt-overly sensitive) by a nanobiosensor suspended in a biopolymer film that covers Zn-fertilizer NPs (Yadav et al. 2023).

To improve micronutrient usage efficiency and decrease fertilizer loss to the environment, such nanofertilizer delivery methods must be developed first and then evaluated in diverse crops, climates, and farming environments. The ability of nanotechnocrates to study the regulation of phytohormones in plants that are responsible for root growth and seedling organization will help researchers better understand the mechanism by which plant roots adapt to their surroundings. Nanotechnology-based microelectromechanical system sensors use electronic circuits to recognize and react to environmental changes. Researchers have successfully created wireless nanosensors composed of micromachined MEMS cantilever beams wrapped in moisture-optimized water-sensitive nanopolymers (Prasad et al. 2017). Nanosensors are produced at the nanoscale yet have the same arrangement as regular sensors. As a result, a nanosensor is a very small device that can attach to anything that is being detected and give back a signal. These small sensors can detect and react to physical, chemical, and biological signals, and they can also detect and respond to biological signals, converting their reaction into a signal or output that people can utilize. Nanosensors provide many advantages over conventional sensors, including high sensitivity and selectivity, near real-time recognition, inexpensive, and mobility, as well as other important characteristics that are enhanced by incorporating nanomaterials in their fabrication. There are several methods for the construction of nanosensors, including bottom-up assembly methods, molecular self-assembly, and top-down lithography. Currently, available nanosensors may be categorized into materials with nanostructures, such as porous silicon, nanoparticle-based sensors, nanoprobes, nanosystems: cantilevers, nanowire nanosensors, and nanoelectromechanical systems (NEMS). Nanosensors may be categorized depending on their uses in food analysis, such as nanoparticle-based nanosensors and electrochemical nanosensors and nanooptical sensors (Omanović-Miklićanina et al. 2016).

Nanotechnology delivery systems for plant hormones and nutrients

It is simple to employ agricultural natural resources like water, fertilizers and chemicals as nanosensors in farming. It utilizes nanomaterials, global positioning systems, and satellite imaging of fields to identify crop pests and other facts of stress like drought. Plant viruses and the amount of soil nutrients may be detected using field-distributed nanosensors (Karn et al. 2009). They reduce pollution to the environment and fertilizer use. Fertilizers with the nano-encapsulated slow release are often utilized (DeRosa et al. 2010). Nanoprocessing and nanobarcoding might be used to assess the manufacturing quality of agricultural products utilizing the concept of supermarket barcodes for efficient, effective, quick and easy decoding and illness detection. They developed nanobarcodes that may easily be recognized by any fluorescent-based instruments and may flag many diseases on a farm. Nanotechnocrates can examine how plants control hormones like auxin, which are responsible for development of root and seedling organization. Auxin-reactive nanosensors have been created which is a development in auxin research since it enlightens researchers on how plant roots adapt to their surroundings, particularly to marginal soils (Prasad et al. 2017).

Nanosensors for detection of soil moisture

The soil moisture nanosensors also known as nanobiosensors combine knowledge from biology, chemistry, and nanotechnology. The existence of gaseous smells, chemical pollutants, infections, and even changes in the environment may all be detected using nanosensors. Numerous uses for soil moisture sensors exist, mostly in agriculture. A lot of water and human resources waste may be avoided by monitoring soil moisture levels and managing irrigation systems. These sensor types make farming automation simpler. Additionally, they are utilized in controlled environments. Engineering, geological, agronomic, ecological, bioorganic, and hydrological properties of the soil mass are strongly influenced by the soil's moisture content. Soil water acts as an intermediary in the absorption of minerals by roots. Soil and water are the fundamental requirements for plant existence and development. Numerous laboratory research is now being conducted on the significance of soil moisture and its measures. The findings of this study will be put to use in the real world. By combining nanosensors with a basic soil moisture detector (SMD), new and creative methods for accurate component detection are made possible. To meet the various needs for soil inspection, numerous kinds of nanosensors have been created (Chajanovsky et al. 2021, Paul 2007).

Recent developments in soil moisture analysis using nanosensors

Many of us take extended trips away from home, leaving the plants unattended while we are gone. Plants can sometimes perish due to lack of water. We have thus included instructions for a straightforward soil moisture detection circuit that will look after your plants by watering them when necessary, allowing you to rest at ease when you step outside knowing that they are secure. If the amount of moisture in the soil is insufficient, this circuit will automatically turn on the pump that is connected to the water supply through the relay (Peng and Loew 2017).

This straightforward soil moisture detecting circuit is built using a relay, an IC 7404, and a few other parts that are readily accessible as shown in Figure 3. The CD7404 is an inverter IC that has six separate NOT Gates on one IC. When the input is high, the output decreases. Its output turns high when the input is low, which is how it earns the name "inverter IC", In the soil, embedded probes measure the soil's moisture content. Be cautious when inserting the probes; they should be isolated

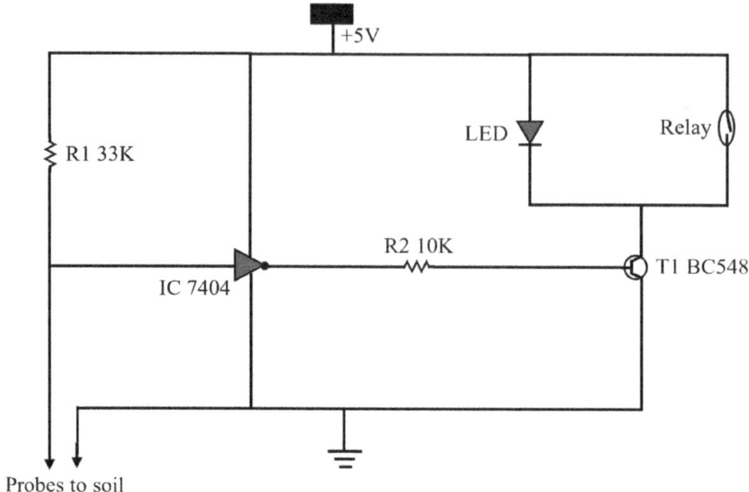

Figure 3. Circuit for Soil moisture detection (Umasankareswari et al. 2022).

from one another and placed at a sufficient depth. Pin 1 of IC1 drops low in the absence of moisture in the soil, which causes output pin 2 to go high.

Components of project

- IC 74 L S0$_4$
- LED
- Relay switch
- Resistor

Because the motor attached to the relay turns on and the plant receives water, the transistor connected to R1 conducts current because the motor connected with the relay starts and water is provided to the plant. The motor attached to the relay stops operating when the plant receives enough water, causing pin 1 of IC1 to go high and pin 2 of IC1 to go low. This is because the transistor is now in the cut-off state and does not conduct. Whenever you leave your residence, you can quickly construct this straightforward circuit and place it in your yard (Umasankareswari et al. 2022).

With the help of modern technology, nanotechnology is a groundbreaking revelation that has the potential to reform rural communities. By using engineered nanoparticles, nanotechnology is employed in agriculture to increase crop output and meet rising food demand. To develop nutrition, moisture, and physiological determination of plant life, nanosensors may be used. This aids in the adoption of suitable and timely remedial actions. Precision agriculture may need the use of intelligent nanosensors to administer fertilizers while monitoring the demands of the crops.

As a result, nanosensors that may be used in agricultural areas have the potential to track soil nutrients and moisture. Nanosensors can also be used in agriculture to understand interactions between soil and root organisms, nutrient processes, and crop stature maintenance. We must make an effort to increase the nanosensors' speed, precision, and range. The use of nanotechnology in consumer items has sparked several ethical and sociological issues, including worries about health and the environment as well as the rights to intellectual property. Consequently, regardless of the level of skill in nanogeneration in farming, it may be necessary to verify its relevance to both humans and the environment (Peng et al. 2017).

Micro electromechanical system (MEMS) for detecting soil moisture

MEMS that are based on nanotechnology are made up of a variety of microsensors, nanosensors, and actuators that employ microcircuit control to perceive their surroundings. The micro-cantilever beam and microsensor chip, which combines a customized nano-polymer detection element with Wheatstone's bridge piezoresistor circuit, make up the MEMS utilized for measuring soil moisture. The change on the cantilever surface may be investigated using the nano-resistor's shear stress and stress sensitivity. On its upper surface, the MEMS cantilever has to expand, and nanomoisture vapor polymer films installed. When the thin film is in contact with soil moisture molecules, the cantilever beam bends downward and expands until the stress in the cantilever beam equals the stress in the thin film brought on by the moisture molecules. This results in shear constraint, which forces the cantilever beam to deflect. The shear stress and moisture molecule concentration are directly proportional to the deflection, which is detected as a change in resistance in the embedded strain gauges and converted into a proportionate differential voltage change. An on-chip temperature sensor is used to measure the temperature of the moisture vapor to monitor temperature. A theoretical and practical investigation on the viability of employing low-cost MEMS-based nanotechnology-based devices for the field detection of soil moisture was carried out by Jackson (Jackson et al. 2008). The cantilever beam's thickness, elasticity modulus, and moisture content are what the researchers identified to be the key determinants of how the resistance of the sensor changes as a result of moisture. The scientists have assumed that only the thickness and stiffness of the cantilever beam have an impact on the MEMS sensitivity which is unaffected by the length of the cantilever and the shear stress at the polymer/cantilever contact. The actual use of this device and the impact of various soil elements on MEMS responsiveness, however, need further research (Molden 2013).

Nanosensors for detection of pesticides

To increase agroecosystem productivity, pesticides are frequently employed in agricultural systems to prevent, control, or remove weeds, fungi, and other pests. The usage of pesticides is continuously rising, and they may account for about one-third of all agricultural goods produced worldwide (Rhouati et al. 2018, Sharma et al. 2020). Additionally, the widespread use of pesticides in the field has harmed non-target species including people and animals by accumulation in food and ground water.

Pesticide exposure can harm people's health in many ways, including via mutagenicity, carcinogenicity, neurotoxicity, and genotoxicity (Nsibande et al. 2016). When used in low quantities, certain pesticides, such as organophosphates, build up in the bodies of animals, and exposure to larger doses inhibits enzymes like acetylcholinesterase, posing major health risks to humans as shown in Figure 4 (Liang et al. 2013). Therefore, it is essential to create improved methods of detecting pesticide residues to ensure the safety of food.

High performance liquid chromatography, as well as colorimetric assays, enzyme-linked immune sorbent assays, liquid/gas chromatography mass spectrometry electrophoresis, and flourimetric methods for testing, have all been used for detecting pesticide residues for a very long time (Biparva et al. 2020, Gascon et al. 1997, Guan et al. 2021, López et al. 2014, Madianos et al. 2018, Yetim et al. 2020). Although the majority of these procedures are single-single assays that demand costly equipment, expert technicians and thorough sample preparation, some of them are particularly sensitive to environmental changes (Christopher et al. 2020, Verdian 2018). Additionally, on-site pesticide residue detection is not appropriate for using such detection methods. Additionally, they are believed to be inappropriate for real-time detection, which limits their application in emergencies (Lei et al. 2018). As a result, the convenience and reliability of the analysis are improved by detection methods that use multiple signals. For instance, techniques that combine

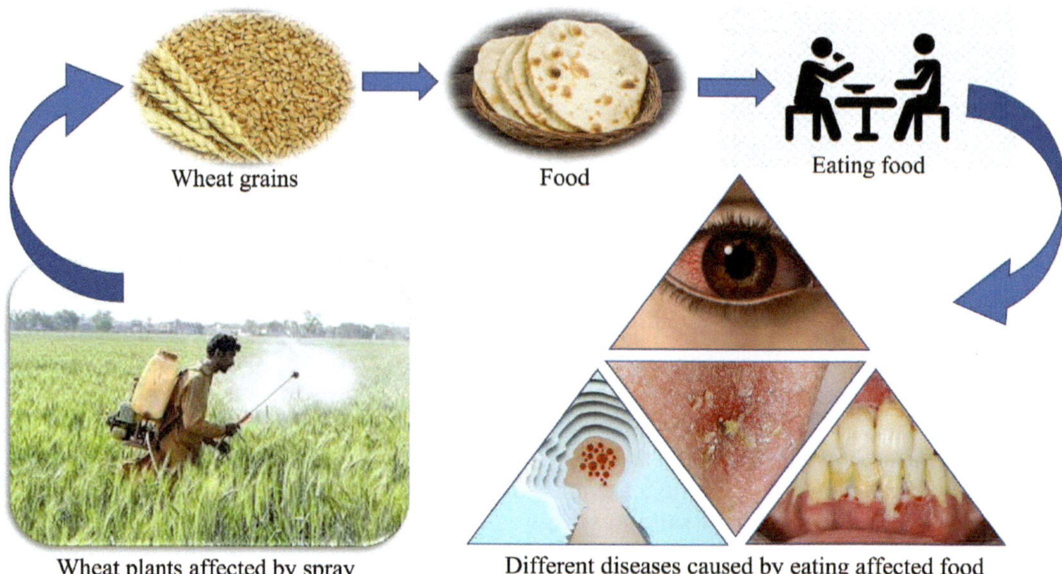

Figure 4. Harmful impacts of pesticides on human health

a multi-signal fluorescent approach with colorimetric analysis can supplement naked-eye sensing in a variety of real-world situations while avoiding the effects of background in multidimensional structures (H. Xie et al. 2018). The development of optical sensors to search for pesticide traces came about as a result of more effort being put into analyzing various approaches for the quick, easy, delicate, selective, exact, and understandable detection of pesticides (Verdian 2018). The sensing of pesticides has been made possible by many optical approaches that made use of recognition factors such as enzymes, antibodies, molecularly imprinted polymers, aptamers, and host-guest recognizers. Such methods may unwaveringly identify and find the specific pesticide particle (Li et al. 2018, Qu et al. 2009, Yan et al. 2012). Furthermore, the combination of nanomaterials and identification elements gives higher levels of sensitivity and remarkable precision for fast implementation, which is a vital necessity for quick and efficient pesticide detection (López et al. 2014). Therefore, the introduction of nanosensors as a leading alternative to traditional methods for detecting pesticide residue has resulted from the need for a rapid, sensitive, specific, and user-friendly method. This is because it is inexpensive, compact, very sensitive, and quickly detectable (Majdinasab et al. 2017).

An optical sensor generally consists of a transducer used to transmit the signal necessary for a particular pesticide residue to attach to the sensor and a recognition element, which is unique to that pesticide particle. To enhance the diagnostic performance of any sensor, the scientific community is paying special attention to the recognition components. Based on the forms of output signals, the four main types of optical probes could be identified. These optical sensors use surface plasmon resonance (SPR), surface enhanced Raman scattering (SERS), colorimetric (CL), and fluorescence (FL) (Yan et al. 2018). Another well-known type of nanosensor is the immunochromatographic strip (ICTS) nanosensor, which is utilized in point-of-care analytical equipment (Huang et al. 2016). Due to their behavior as point-of-care tests, immunochromatographic assessments have also been reported for their involvement in controlling agroecosystems. For example, the reported immunochromatographic technique for the detection of genetically modified crops had a visual colorimetric readout technique, which usually had insufficient sensitivity and merely produced a yes/no answer (Emslie et al. 2007, Rowland et al. 1961, Zhou et al. 2015). Similar to this, it has been noticed that gold nanoparticle based ICTS sensors are limited in their usage due to their poor detection sensitivity and comparatively lesser color density (Wang et al. 2017). However, some amplification strategies have been suggested to increase their sensitivity, including boosting the reagent's potency, enhancing the detection signal's intensity, enhancing labeling methods, and changing strip device shapes (Zhou et al. 2019). As a result, the enhanced ICTS nanosensors may potentially prove to be a practical and affordable instrument for detecting pesticide residue in agroecosystems. The combination of several electrochemical methods and nanotechnology gives the sensor a larger operational surface area and provides a reasonable level of control over the electrode microenvironment. Nanoparticles have the ability to perform a wide range of activities in electrochemical sensing structures, including accelerating the transport of electrons, tagging, and serving as a reactant in electrochemical reactions (Jeevanandam et al. 2020). Therefore, it seems that electrochemical nanosensors could be a beneficial tool for pesticide detection. Electrochemical biosensors based primarily on the enzyme cholinesterase have recently become effective tools for detecting residual pesticide fragments, especially those from the class of carbamates and organophosphates, due to their excellent perception, caution, and straightforward method of production (Cesarino et al. 2012, Zhou et al. 2017). Enzyme-based biosensors, however, are subject to some limitations, including high cost, decreased enzyme activity, and limited repeatability (Ahmad et al. 2017). Enzymes also frequently exhibit intrinsic instability and are susceptible to denaturation in severe environments, which limits the effectiveness and robustness of biosensors (Zhou et al. 2014). Additionally, the sensitivity and selectivity of the enzyme during detection might be affected by the presence of various contaminants, such as the presence of specific heavy metals in biological samples, which increases the likelihood of false-positive results (Khairy et al. 2018). As a result, non-enzymatic electrochemical biosensors become necessary. To create non-enzymatic electrochemical sensors, nanomaterials seem to be a potential contender (Paul 2007). It has been discovered that many different types of nanomaterials, including nanoparticles (such as CuO, CuO-TiO$_2$, and NiO), nanocomposites (for example, nanocomposite of molybdenum) and nanotubes (such as carbon nanotubes and peptides) are used to electrochemically determine the remaining pesticidal particles (Du et al. 2008, Qu et al. 2008, Wang et al. 2014).

Such NMs' explicit and thorough examination of leftover pesticide particles is made possible by their very tiny size, increased surface area, and unique electrical and chemical characteristics. Numerous researches have been reported regarding the sensitivity and specificity of different nanosensors for detecting pesticides. For example, it was discovered that metribuzin and dimethoate could be detected at a level of 6.8 × 10^{-8} M and 0.002 ppm, respectively, by two distinct optical sensors based on upconverting nanoparticles and silver nanodendrites (Pham et al. 2020, Wang et al. 2014). The electrochemical aptasensors made of chitosan-iron oxide nanocomposite detected malathion at the level of 0.01 nM, but malathion was detected at a concentration of 0.02 nM by the electrochemical nanosensor fabricated with Cuo NPs covered with a three-dimensional graphene nanocomposite (Xie et al. 2018).

Based on reduced graphene quantum dots (rGQDs) and multi-walled carbon nanotubes (MWCNTs), an optical apta-nanosensor was created and developed. It was used to detect the organophosphorus pesticide diazinon, one of the most often used in the world. Due to the strong fluorescence emission and optical stability of GQDs and rGQDs, effective optical transducers could be created, and based on these transducers, accurate detection techniques could be created. A straightforward, cost-effective fluorescence procedure has been developed to determine diazinon with a detection limit of 0.4 nM (0.1 g/L) in the range of 4–31 nM, meeting the requirements and standards set by the European Union and the World Health Organization. This method uses rGQDs, a diazinon-specific aptamer, and MWCNTs. The practical application of the unique apta-nanosensor was subsequently verified by determining diazinon in real samples, such as tap water, river water, and agricultural runoff water, to evaluate apta-nanosensor performance. The obtained results showed that the emission recoveries were maintained in the existence of other pesticides due to the diazinon-specific aptamer's lack of binding affinity. As a result, the designed apta-nanosensor was able to rapidly detect diazinon pesticides with increased accuracy and selectivity in real samples. As a result, the precise apta-nanosensor formed in this work offers a sensitive, quick, affordable, and portable instrument for the selective determination of diazinon over other pesticides in polluted areas. The newly developed technology offered a useful on-site detection strategy for actual samples of several pesticides, and it could also be used for additional pesticides (Talari et al. 2021).

Nanosensors for detection of heavy metals

Pb^{2+}, Hg^{2+}, Ag^+, Cd^{2+}, and Cu^{2+} are just a few of the heavy metal ions that may be found in our environment and have dangerous effects on both people and their environments. Heavy metal ion accumulation in various ecosystems is facilitated by the constant development of agricultural and industrial successes, insufficient removal of heavy metals from wastewater, and residential emissions (Li et al. 2013, Sadiq et al. 2023). To maintain environmental security and health analysis, it is therefore highly desired to remove trace heavy metal ions employing effective practices. Investigating various analytical systems will help one better comprehend heavy metals (O'Connell et al. 2001), examples include inductively coupled plasma mass spectrometry (ICP-MS), atomic absorption spectrometry (AAS), X-ray fluorescence spectroscopy (XRF), and atomic emission spectrometry (AES), but their application is severely constrained by expensive equipment, labor-intensive processes, and other issues. A variety of electrochemical, optical, and colorimetric stratagems have been thoroughly researched in order to get over these limitations and offer low-cost, efficient approaches to understand sensitive, rapid and discerning analysis of heavy metal ions (Aragay et al. 2011). A specific kind of chemical sensor that frequently generates a diagnostic signal in the transduction element using electromagnetic radiation are grouped together with optical sensor, which is frequently used to detect heavy metals. A certain optical factor that might be connected to the concentration of an analyte is altered by the sample's interaction with the radiation (Ullah et al. 2018). For example, nanohybrid CdSe quantum dots were used to build cadmium detection optical nanosensors that recovered their green photoluminescence (Wang et al. 2016). The working principle of optical chemical sensors is based on apparent changes in optical properties (emission, absorption, transmission, lifespan, etc.) that appear from the delayed indicator's binding to the analyte. The strategy of alluring graphene-based nanotechnology begins as a creditable tool that resolves these issues and improves the performance of the sensing platform. Because of the prospective advantages of its reverent construction and conscious appreciation of some various metal ions, graphene-based optical approaches have recently been proposed as one of the motivating strategies for detecting heavy metal ions (Zhang et al. 2018). When coupled with graphene, noble NPs like Ag, Au, and Pd have a special capacity to mimic peroxidase activity, increasing their durability and catalytic efficiency. Numerous sensors are used for the detection of several heavy metal ions as a result of this feature. In order to make nanohybrids that imitated the function of the peroxidase enzyme and could discriminate between double- and single-stranded DNA molecules, silver nanoparticles and graphene oxide were mixed. By changing its form to either a quadruplex configuration or a hairpin-like structure when they are present, Pb^{2+} and Hg^{2+} can be detected using colorimeter (Xia et al. 2019). Such colorimetric techniques also have the benefit of easy-to-use applications, transportable equipment, inexpensive viability, and simple operation. The removal of the targeted species is discovered to be hampered by chemosensors for detecting heavy metals since they would cause secondary contamination. As a result, combining fluorescent and magnetic capabilities into a single nanocomposite particle appears to be an effective replacement (Mossoba et al. 2010). However, the presence of magnetic nanoparticles strongly quenches the fluorescent moiety's photoluminescence, making the synthesis of these types of nanocomposites more difficult. To address this issue, nanocomposite synthesis frequently focuses on a variety of molecular interactions, including electrostatic and hydrophobic interactions, hydrogen bonds, and covalent bonds. For instance, the Fe_2O_3 globules with shallow polymer layers were coated with quantum dots using thiol chemical techniques. Gold nanoparticles that have been electrostatically connected to silica microspheres and other materials, such as Fe_2O_3 nanoparticles, have also been prepared (Neuberger et al. 2005). The approach that uses nanochemistry to synthesize multimodal nanosensors not only effectively identifies but also extracts heavy metal ions from the water is more attractive. Satapathi et al. fabricated multimodal nanosensor included a fluorescent quantum dot for simultaneous determination and removal of the spotted mercury ion, a fixed spacer arm, and a thin silica shell housing magnetic (Fe_2O_3) nanoparticle. With a limit of detection as small as 1 nm, this nanosensor's extraordinary sensitivity may be observed in its capacity to detect Hg^{2+} at the nanomolar level (Satapathi et al. 2018). It is possible to highlight the eco-friendliness of nanosensors by demonstrating their singular capacity to remove the identified analyte by an external bar magnet and leave no residual contaminants. In order to improve the qualities of nanosensors, a variety of chemicals, such as polysaccharides citrates, some polymers, and proteins are utilized to stabilize them (Yong et al. 2009). Even in the existence of other monitoring metal ions, the silver NPs stabilized with epicatechin can be employed for precise Pb^{2+} detection. Because of its low detection limit, simplicity in synthesis, superior discernment, and affordable production, epicatechin-coated silver nanoparticles (ECAgNPs) are a useful sensor created for recurrent monitoring of Pb^{2+} concentrations in ecological models (Ikram et al. 2019). Fluorescent quantum dots are an effective instrument for sensing a variety of metal ions due to their favorable photophysical and chemical characteristics. The main drawback of using quantum dots, however, is how difficult and time-consuming it is to separate and recover them for use in practical applications. Magnetic nanoparticles (Fe_2O_3) are included into quantum dot-based fluorescence sensors to alleviate the issue and offer a number of additional advantages because of their large particular surface area, distinctive magnetic characteristics, magnetic operability, and other advantages. Yang et al. developed multifunctional magnetic-fluorescent nanoparticles based on fluorescent quantum dots, carboxymethyl chitosan, and magnetic nanomaterials and demonstrated simultaneous detection and separation of Hg^{2+} at a detection limit of 9.1×10^{-8} mol/L. However, the straightforward and novel method of nanotechnology offers a perspective on the development of heavy metal sensors in the field in the future, which at the moment seem to be a difficult task with a number of restrictions (Yang et al. 2018).

Nanosensors for detection of plant pathogens

Pathogen detection and evaluation are crucial for scientific clarification, ecological surveillance, and the management of food safety. For successful investigative outcomes, the intricate component of biological source, which is a component of biological provenance, must interact with the chemical species in the test. Some comprehensive, reliable, and quick recognition elements such as aptamers, lectin, phage, antibodies, and bacterial imprints of cell receptors have been discovered for exposing bacteria (Chen et al. 2017). Bacterial receptors, antibodies, and lectins are the most often employed biosensing components for examining infections. Due to their versatility in combining with other components to form biosensors, these substances are widely used as biosensing components to examine pathogens (Liu et al. 2014). Aptamers, which are single-stranded nucleic acids, are more practical economically and chemically than antibody-based recognition elements for classifying bacteria (Kim et al. 2014). However, they do have a number of limitations, including batch-to-batch variability, resilience in complex materials, and relative preparation complexity. The method referred to as the "chemical nose" is a recently developed tool for the detection of infection. To enable their rating, it allocates many selective receptors, each of which generates a unique response configuration for each objective. It operates similarly to the way our minds work when we smell something (Verma et al. 2016). This method involves establishing a reference database by training sensors with competent bacterial samples. By comparing them to the reference catalog, bacterial pathogens are identified. The surface of the nanoparticle is often modified with a variety of ligands, each of which is responsible for a distinct communication with the target in "chemical nose" biosensors (Sun et al. 2018). The size variation and external makeup of the nanoparticles are chosen in such a manner that each set of particles can react to various classes of bacteria uniquely, adding more characteristics to the absorption spectra. The incorporation of nanoparticles to the bacterium results in the agglomeration encasing the bacteria because of electrostatic interactions between the anionic parts of the cell walls of bacteria and the cationic cetyltrimethylammonium bromide (CTAB). This agglomeration mechanism encourages a shift in localized surface plasmon resonance-induced coloration. By obtaining an absorption spectrum when numerous bacteria are present, the color change is further indicated (Verma et al. 2014). The components of the bacterial cell wall that causes this form of aggregation are teichoic acids in gram-positive bacteria and lipopolysaccharides and phospholipids in gram-negative bacteria (Sun et al. 2012). These peculiar aggregation patterns are caused by extracellular polymeric molecules that are present on the bacterial surface. It is because of these various aggregation patterns that distinguishable colorimetric responses are provided. Therefore, it would be possible to recognize mixtures of different bacterial species using the "chemical nose" established by nanoparticles. The "chemical nose" is powerful enough to distinguish between polymicrobial and monomicrobial instances during infections, enabling greater effectiveness and triggering antimicrobial medication without the need for laborious and drawn-out sample examination. Species and strains of bacteria found in biofilms can be readily identified by the very sensitive multichannel nanosensors. In order to locate and recognize biofilms based on their physiochemical features, Li et al. developed a multichannel sensor composed of gold NPs (AuNPs). The nanosensor's capacity to distinguish between six different biofilms is a strong argument for its sensitivity (Li et al. 2014). Another sensor developed by Phillips et al. based on hydrophobically used gold nanoparticles recognized three different *E. coli* strains very quickly (Phillips et al. 2008). In the sensor devices, pathogenic cells finally replaced the conjugated polymers with negative charges and differentially restored the polymer fluorescence. Nanotechnology provides us with new opportunities for altering the constraints of human perception. Humans have an extraordinary capacity to detect volatile organic chemicals that exist in a wide range of environments at very low levels through the development of their olfactory system (Sela et al. 2010). Olfaction is a promising platform for a variety of biotechnological applications due to human beings' high sensitivity and flexibility in differentiating among more than a trillion olfactory stimuli (Bushdid et al. 2014, Hellwig and Henle 2014). For revealing microorganisms, several effective sensors that mostly rely on olfaction have been proposed. Three distinct components make up the system of these nanosensors: pro-smell fragments, surface-functionalized nanoparticles, and enzymes that cut pro-fragrances to produce the olfactory output. These three components have been modified to develop a sensitive sensory system that allows for the fast identification of bacteria at concentrations as low as 102 CFU/mL. By adding magnetic nanoparticles, it is also feasible to separate, clean, and detect diseases in difficult circumstances. The "enzyme nose" nanosensor, which is based on nanomaterials, is also a practical tool for doing research that is designed to find targets in natural samples that are toxicologically significant. A novel enzyme nanosensor for pathogen detection built on non-covalent centers was created by Sun (Sun et al. 2018). With an accuracy of 90.7% and a speed of 30 minutes, the use of magnetic nanoparticle-urease sensors allowed for the thorough identification of bacteria at a concentration of 102 CFU/LL. Similar to this, several other unique forms of electrochemical, optical, and immunosensors have been created for identifying a variety of plant pathogenic bacteria. In contrast, the Fe_2O_4/SiO_2-based immunosensor revealed the existence of bean pod mottle virus, tomato ringspot virus, and Arabis mosaic virus at concentrations of 10^{-4} mg/mL (Zhang et al. 2013). For example, the optic particle plasmon resonance immunosensor synthesized using gold nanorods successfully determined Cymbidium mosaic virus (CymMV) or Odontoglossum ringspot virus at concentrations of 48 and 42 pg/mL. Therefore, it is possible to modify human annotations about their environments in a way that would otherwise seem unmanageable by influencing the molecular-level performance of accessible nanomaterials (Welbaum et al. 2004).

Nanosensors for detection of nanoparticles (NPs)

Natural NMs include things like humic acids and clay minerals, etc. As a result of significant human activity, many nanomaterials may be accidentally synthesized in the environment, such as through the emission of welding fumes or diesel oil, or they may be purposefully made to exhibit unmatched chemical, electrical, optical, or physical capabilities (Picó et al. 2014). These properties are used in a wide range of consumable products, including medicines, meals, sunscreen, cosmetics, paints, electronics, and companies who directly release NPs into the environment, such as cleaning up polluted areas (O'Brien et al. 2008). Additionally, metal NPs may pose environmental risks because activities used to produce, transport, use, and dispose of them ultimately release them into the environment. The widespread use of NPs in many systems has given rise to several worries. Since carbon-based NMs are very resistant to deterioration, they accumulate in the environment (Farré et al. 2009). Due to their larger surface area, it is easier for NPs to adhere to the surface of cells. They have a number of effects on the cell, such as obstructing the pathway for proteins to cross the membrane, decreasing the cell membrane's permeability, or further hindering the operation of the vital organelles (Kumar et al. 2011).

Several synthetic NPs currently in use for various industrial and ecological goals or produced as a result of various human activities are eventually discharged into soil systems. The most common types of NPs used are metal-engineered NPs (elementary Ag, Fe, Au, etc.), metal oxides (SiO_2, FeO_2, TiO_2, CuO, Al_2O_3, ZnO, etc.), composite compounds (Co-Zn-Fe oxide), fullerenes (including Buckminster fullerenes, nanocones, CNTs, etc.), and quantum dots frequently covered with organic polymers. Different plant growth-promoting rhizobacteria (PGPR) (including *Pseudomonas aeruginosa*, *Bacillus subtilis*, *P. putida*, and *P. fluorescens*, as well as various microorganisms) involved in the transformation of soil nitrogen (Mishra et al. 2009). Free radicals, which are recognized for their extreme toxicity to microorganisms, are said to be the cause, and are produced when the copper and iron-based NPs interact with the local peroxides (Saliba et al. 2006). Therefore, there is an urgent need to keep an eye on the various NPs that end up in the soil, particularly in agroecosystems. One of the methods that might be investigated for sensing NPs is the application of microcavity sensors, which have gotten a lot of attention in the form of whispering gallery resonators. The optical characteristics are disrupted in this case by the particle binding on the outside of the microactivity, leading to a resonant wavelength swing whose size is determined by the particle's polarizability. Both the particle size and the binding activities may be monitored in real-time using the measure of the change. For applications outside of homeland security, such as environmental monitoring, there is projected to be a considerable demand for optical sensing that is equipped with the high sensitivity of individual nanoscale species. Raman lasers, which are also extremely sensitive optical sensors, can detect a single nanoparticle in split-mode microactivity. According to reports, NPs can be detected down to a radius of 29 nm, and by observing the particular variations in the Raman lasers' beat frequency, it is possible to confirm their presence (Li et al. 2014).

Nanosensors for detection of organic residues and pH

Globally, people utilize water for drinking, irrigation, wastewater treatment, and other purposes. The quality and management of water used for agriculture are of highest importance due to the possibility of heavy metals, organic residues, and other contaminants building up in crops, which may subsequently affect their growth. For many years, rivers and subterranean sources provided the majority of the world's water for irrigation of agricultural land (Molden 2013). However, excessive exploitation has resulted in severe shortages, forcing policymakers to reevaluate strategies for managing water supply that are more environmentally friendly and sustainable. There has been a lot of effort done recently to fill reservoirs with rainwater, especially in dry and semi-arid areas, due to the great variation in rainfall (Chajanovsky et al. 2021). This method of water management has a lot of potential because it is inexpensive, relatively easy to use, and improves agricultural output (Ogilvie et al. 2019). However, numerous studies have found high concentrations of contaminants in both precipitation and runoffs that collect in reservoirs, calling for regular water quality monitoring before any use is made of it. For instance, it was found that the rains from Polish reservoirs contained various contaminants with lower load limits, such as heavy metals, as well as a high amount of salt, unsuitable quantities of phosphorus, and chloride ions (Yosef et al. 2015). Even trace levels of heavy metals are known to be critically harmful and can linger in the environment, which can seriously harm both the ecosystem and human health, despite these lower percentages (Milik et al. 2018). Even a tiny reservoir needs to be managed, according to Lee et al., to lessen chemical and microbiological pollution. For instance, flush filters may be installed to address the development of various pollutants that have an influence on the water quality when precipitation stagnates (Lee et al. 2010). Schets et al.'s research demonstrated that the proliferation of incredibly dangerous bacteria in reservoirs was caused by the fecal and coliform contamination of captured rainwater (Schets et al. 2010). In general, pollutants such as heavy metals, bacteria and organic debris are indications of water quality and possible health hazards. Spectrometry analysis, which can also involve fluorescence, is the most popular technique for determining the presence of contaminants and their concentration (Philp et al. 2003, Tokalioglu et al. 2000). These technologies cannot be used in real-time since they are expensive and unable to react rapidly in the field (Soylemez et al. 2018). Sensors are another frequent technology, and nanosensors in particular are becoming more and more popular since they could tackle the previously mentioned problems. They also offer a number of benefits, including speed, real-time analysis, mobility, high sensitivity, miniaturization, and ease of preparation (Ullah et al. 2018). Nanosensors are being used increasingly frequently to identify and analyze water pollutants as a consequence of these factors (Ballen et al. 2021). Shtenberg et al. designed a horseradish peroxidase inhibition-based optical biosensor mounted onto nanostructured porous silicon to identify and measure a range of metallic ions in polluted water (Patel et al. 2011). Additionally, nanosensors, particularly those based on carbon nanotubes (CNTs), have improved in sensitivity and selectivity, making them a unique technique for the determination of analytes (Vashist et al. 2011). Wang et al. presented a glucose biosensor that relies on multi-wall carbon nanotubes (MWCNTs), polypyrrole, and glucose oxidase. They also noted the extremely high sensitivity of 2.33 nA/mM and excellent linearity (Wang and Musameh 2005). The simultaneous determination of three various types of organic materials was accomplished by Savk et al. by utilizing a ZnNi bimetallic nanoalloy-based MWCNT-based sensor with excellent electrochemical properties. Since, they can monitor the pH of water in real-time to find whether a water sample is polluted or not, optical pH-based sensors are an option (Savk et al. 2019). Dye indicators can be employed to monitor the color changes brought on by pH variations. When utilizing dyes, the sensor design must take into account additional chemical or physical immobilizing procedures (Wencel et al. 2014). Recently created pH-sensitive conductive polymers allow for the avoidance of dyes. Today's most effective polymers are thought to be intrinsically conducting polymers (CPs), as these materials' exceptional electrical and optical characteristics make them ideal for a variety of real-world uses, including sensing (Gupta et al. 2006, Soylemez et al. 2018). The most common choice among these CPs is polyaniline (PANI), which is easy to synthesize, has high environmental stability, and can be modified to undergo a protonation/deprotonation process in order to change its electric characteristics from an insulator to a metallic conductor. When PANI is protonated in an acidic liquid, it conducts, but when it is deprotonated in a basic medium, it becomes an insulator (Kuswandi et al. 2012, Liu et al. 2017). Inverted emulsion polymerization is one method of making PANI (Soares et al. 2006), which has the benefit of resolving potential viscosity and heat problems while keeping PANI's strong chemical characteristics (Rao et al. 2002). Free-radical oil-in-water emulsion polymerization technique requires an initiator, monomers, a dispersion medium, and a surfactant. While the monomers are present in the organic phase, the surfactant is present in a continuous aqueous phase. The emulsion is formed when the surfactant aggregates to form spherical micelles, which are subsequently polymerized to create a stable colloidal dispersion. The monomer is retained in the continuous phase during the water-in-oil polymerization process, also referred to as inverse emulsion polymerization (Li et al. 2012).

PANI, on the other hand, has a low processing ability, which has a negative impact on its organic solvent solubility. Charge-transfer doping, which integrates dopant molecules like dodecylbenzene sulfonic acid (DBSA) into the PANI chain, is utilized to address this problem (Han et al. 2005). This increases the solubility of the PANI chain, a novel family of materials recently described that includes both organic polymeric molecules and inorganic NPs. The usage of CNTs and graphene as nanomaterials for sensors is quite common (Rowley-Neale et al. 2018), and their combination with PANI gives nanocomposite synergetic capabilities (Rowley-Neale et al. 2018). The exceptional structure and characteristics of CNTs (high stiffness, tensile strength, amazing flexibility, high thermal conductivity, and high electrical conductivity) allow for a wide range of technological uses. A disadvantage of CNTs is the tendency to aggregate those results in reduction of their surface area. But there are several ways to stop this right now, such as stabilizing the CNTs with a surfactant, polymerizing monomers in the presence of CNTs *in situ*, or combining several of these techniques (Cohen et al. 2020). Other variations, like reduced graphene oxide (rGO), have been used in sensor design even though neat graphene is more expensive and therefore less practical. After being made from graphite, graphene oxide is chemically (Kıranşan et al. 2018), thermally (Palaznik et al. 2019), or electrochemically (Shao et al. 2010) reduced, which lowers the amount of oxygen in the material.

According to one research, an electrochemical sensor based on rGO that had been tuned to detect various metal ions simultaneously showed great sensitivity, selectivity, and repeatability (Göde et al. 2017). In another study, a novel quick electrochemical pH-based nanosensor was synthesized that implemented a one-pot dye-free production to address the drawbacks mentioned above. Inverse emulsion polymerization of aniline was produced in an atmosphere that included various grades of CNTs or graphene, polycaprolactone as a structural reinforcement, and DBSA (dodecylbenzene sulfonic acid) as a dopant. Casting the resulting dispersion led to the formation of the nanocomposite film. This method included simultaneous measurements of conductivity with a two-probe device and pH with a pH meter in order to evaluate the pH sensitivity of the nanosensor. The results demonstrated that, of the two chemical residues used, aminophenol had better sensitivity. Due to their pH range of 5.0 to 8.0, these phenolic compounds provided good markers for water quality monitoring. Due to the increased pollutant load caused by the soil's acidification, wastewater typically has a pH between 6.0 and 7.5 (Mucci et al. 1995).

Future prospect of nanosensors in agroecosystem

Applications of nanotechnology in agriculture and food production

The sector of food and agriculture has a significant effect on a nation's economic development. Agriculture practices such as cultivating crops, raising animals and raising seafood are all included in the food business, as well as food processing, packaging, production control, and distribution. Agriculture includes farming, forestry, dairying, fruit-growing, raising chickens, beekeeping, and growing mushrooms. The Food and Agriculture Organization (FAO) estimates that between 20 and 45 percent of plant, meat, and fish products are lost or wasted, which equates to around 286 million tons of grain goods in industrialized nations. To assure food safety and commercial viability, product quality must be monitored at all phases of food production (Srivastava et al. 2017). With the help of this technology, store owners may recognize foods that have expired and receive a reminder to place a fresh purchase order. By 2010, it was expected that the global market for wireless sensors would reach $7 billion. By combining biotechnology and nanotechnology, sensors will become more sensitive, enabling a quicker reaction to environmental changes. Nanosensors made of carbon nanotubes (CNTs) or nanocantilevers, for instance, are so tiny that they can capture and quantify specific proteins or even tiny molecules. Electrical or chemical triggers can be created using nanoparticles or nanosurfaces, a warning when a pollutant like bacteria is present. Other nanosensors operate by activating an enzymatic process or by employing probes known as dendrimers, which are branching molecules that are nanoengineered and attach to particular chemicals and proteins. In the end, precision farming will enable increased production in agriculture by supplying correct information and assisting farmers in making better decisions. Nanotechnology applications are also being created to increase agricultural yield and soil fertility. Magnetic nanoparticles may clean up contaminated soil, and nanosensors could keep an eye on crop and animal health. The development of "lab on a chip" technologies may potentially have a big influence on underdeveloped countries.

Nanotechnology and food systems

The scope of nanotechnology for boosting food security should include the entirety of agricultural production and utilization systems since food systems include food availability, access, and utilization. Additionally, in an economy that is rapidly going global, increasing access to food and its consumption in rural areas will be largely determined by an increase in rural incomes. Value addition throughout the various links in the agricultural production-utilization cycle has been identified as the main driver of rising rural incomes (Barik et al. 2008). Farm inputs, farming production techniques, post-harvest handling, and finally markets and customers are all connected in this way. The application of nanotechnology must be expanded across all links of the agricultural value chain rather than being restricted to farm production levels in order to boost agricultural productivity, quality, consumer acceptance, and resource usage efficiency as illustrated in Figure 5. This will decrease farm costs, increase the value of output, boost rural incomes, and increase the amount of the natural resource base that sustains agricultural production systems, as reported by Kalpana Sastry. To promote both conventional and biotechnological breakthroughs, it is crucial to view nanotechnology as an enabling technology. Before integrating and developing new technologies, like nanotechnology, in agricultural food systems, it is also important to take into account their

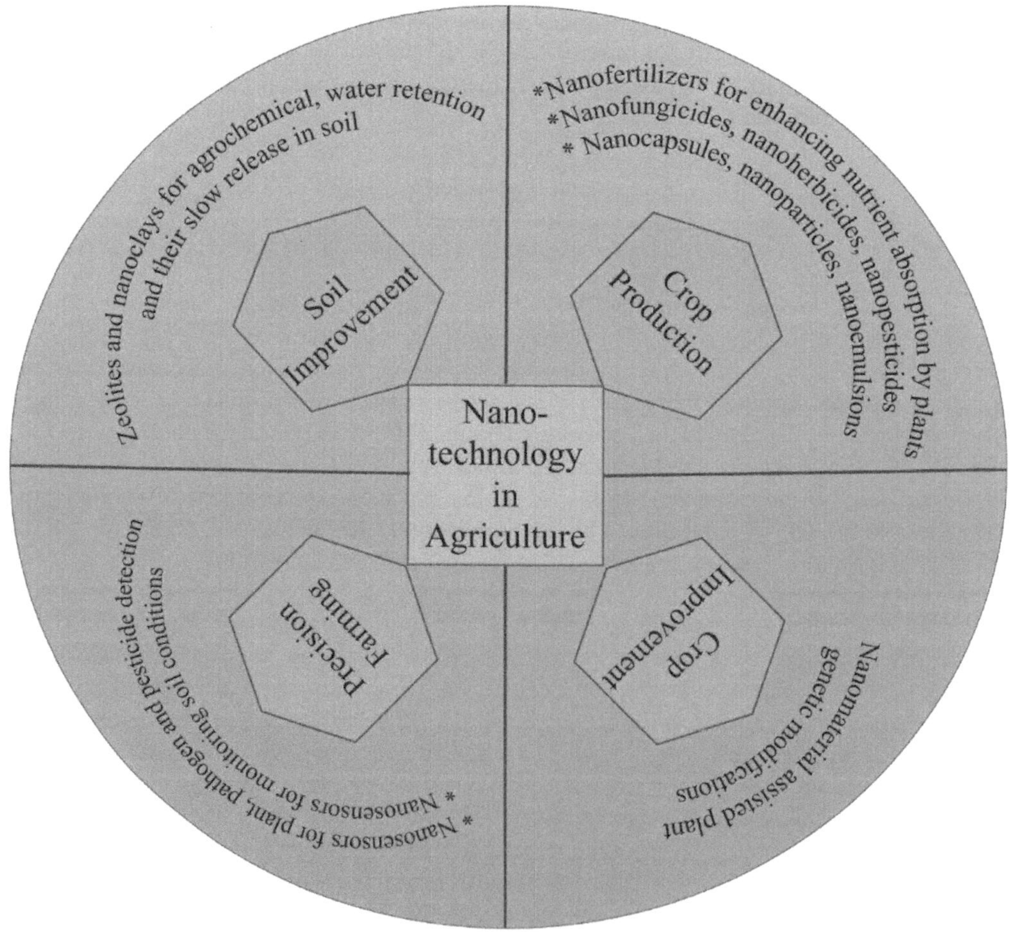

Figure 5. Application of nanotechnology in agroecosystem.

many socioeconomic and environmental implications. Since products based on agribiotechnology have been available in the market for the past 20 years, concerns about biosafety and consumer acceptability have developed (Abobatta 2018, Salerno et al. 2008, Sastry et al. 2010).

Nanotechnology application in food packaging

The necessity of nanosensors in food production is most obvious in the domain of food transport and packaging. Since food packaging prevents sensory experience, customers must rely on expiration dates supplied by manufacturers, which are predicated on many idealized assumptions about how the food is handled or kept. The quality of food may degrade if the storage conditions are prohibited for any length of time, which may not be apparent to the customer until the food item is opened or even consumed (Berekaa 2015). One of the six "Key Enabling Technologies" that the European Commission has identified as having the potential to considerably improve several industrial sectors is nanotechnology, although compared to other industries, such as drug delivery and pharmaceuticals, the use of nanotechnology in the food and agriculture sectors is still in its infancy. The latest concerns of food security, sustainability, and climate have increased researcher's interest in examining the role and significance of nanotechnology in the process of agrarian sector advancement (Parisi et al. 2015). Several uses of agriculture and food systems may be anticipated in the upcoming years, and nanotechnology has the potential to revolutionize the global food system (Rhouati et al. 2018). Novel agricultural and food safety systems, disease treatment delivery methods, tools for molecular and cellular biology, sensors for detecting plant pathogens and pesticides, intelligent packaging materials, environmental protection and education of the general public and future workforce are just a few examples of the significant impact that nanotechnology could have on the agriculture sector (Nsibande et al. 2016). The whole agri-food industry might be revolutionized by inexpensive sensors, cloud computing and intelligent software working together.

It is crucial to draw attention to the numerous micro-nano biosystems that the European Commission (2015) created as part of projects and utilized to implement smart agri-food systems. By improving the quality, quantity, sustainability, and cost-effectiveness of agricultural production, technology development, in particular the Internet of Things (IoT), this substantially helps in realization of smart agriculture as a growing reality where more and more gadgets are linked to people and other devices via the internet. The Internet of Nano Things (IoNT) is a new paradigm made possible by the creative use of nanotechnology in IoT. Due to their tiny size, nanosensors can gather data from a wide range of sources. Non-biological nanosensors like carbon nanotubes can perceive and transmit information, functioning as wireless nanoantennas (Sahani and Sharma 2021). Through their distinctive chemical and electro-optical properties, nanosensors can reduce the drawbacks of food packaging. The presence of gases, scents, chemical pollutants, diseases, and even changes in the environment may all be detected by nanosensors. By ensuring that consumers buy pleasant, fresh goods and lowering the incidence of foodborne diseases, nanosensors enhance food safety. There are signs that the large surface-to-volume ratios, plasmon resonance, distinctive surface functionalization properties, and photoactivity of NMs can be used to improve agri-food systems, even though these opportunities are typically less thoroughly investigated than applications for energy (Zhang et al. 2013) or water treatment (Qu et al. 2013). To achieve this, one can either

- Provide an increased surface area to facilitate surface reactions such as molecular adsorption.

- Produce electronic, electrochemical or optical outcome to enhance sensitive recognition.

- Offer extremely porous and surface functionalized materials for agrochemicals.

- Enable special photocatalytic and redox activity to enhance antimicrobial activity.

For example, high geographical and temporal resolution data on soil quality, crop performance, food spoilage/safety may be obtained via nanosensors. Additionally, they can make it possible to detect infections, toxins and contaminants in soil, water, air, food or plant tissues with extreme sensitivity. The use of sophisticated food packaging and nano-based antimicrobial disinfection technologies can reduce food loss and waste. With the end objective of reaching near-atom usage efficiency, there are opportunities to build multifunctional delivery platforms for agrochemicals that offer targeted distribution. Similar capabilities are available for animal vaccinations and veterinary medications, which improve effectiveness and reduce drug excretion into the environment and waste stream. Additionally, nanotechnology may enhance adsorption, fouling, and magnetic separation processes, allowing for greater selectivity and efficiency in nutrient and energy recovery, and water treatment. The following are promising prospects of nano-enabled technology to increase the sustainability of agri-food systems:

- Sensors for pathogen or toxin detection, chemical monitoring, and evaluation of chemical, physical, or biological characteristics and processes.
- Pathogen-controlling technologies to improve food safety and reduce food loss.
- Materials for scheduled and targeted distribution of agrochemical; sorbent and membrane technologies for resource recovery and spread-out water treatment.
- Materials for the targeted and timed distribution of agrochemicals.
- By enabling highly distributed, real-time detection of essential performance indicators for livestock and crops at the high spatial and temporal resolution required for precision agriculture approaches, nano-enabled sensors and probes are capable of significantly increasing agricultural productivity per unit of energy, water, or agrochemical input (Sonmez et al. 2007).
- Nanosensors monitor the concentration of an analyte and call for quick and reversible binding, meaning if the concentration decreases, the signal decreases.

Compared to conventional sensors, nanosensors and nanoprobes have several advantages. They usually make use of the distinct electrical and optical properties of NMs, high ratios of surface-to-volume, and easily functionalized surfaces in order to access and recognize the presence of particular chemicals or pathogens in environmental food samples with a level of precision and sensitivity higher than currently possible. Due to the lab-on-chip technology and its ability to miniaturize (Farahi et al. 2012), nano-enabled sensing may also be carried out with smaller sample quantities. Numerous agriculturally relevant analytes are the subject of extensive research and development for nanosensors in literature. For instance, it has been claimed that nanosensors can detect residues of organophosphates, organochlorines, and carbamates in water or vegetables, as well as carbofuran, 2,4-dichlorophenoxyacetic acid, methyl parathion, triazophos and dimethoate. Added indicators in water and agri-food systems include DNA, pesticides, fertilizers, and proteins (Khot et al. 2012, Vinayaka et al. 2009). Small chemical molecules, colorant additives, and food adulterants have been detected using nanoprobes while pollutants, glucose, antioxidants, and taste compounds have been measured using nanosensors.

Precision farming

Precision farming is now a reality because of computers, GPS and remote sensing tools. The main elements of a precise farming vision include monitoring and measuring environmental variables, making smart decisions at the right time, and taking targeted action based on the information obtained to maximize output while making the best use of available resources (i.e., herbicides, pesticides, fertilizers, etc.) (Liang et al. 2013). It is possible to fine-tune the use of seeds, fertilizer, chemicals, and water to lessen production costs and maybe increase productivity, all of which are advantageous to the farmer. Precision farming can also assist in decreasing agricultural waste and keep environmental degradation to a minimum, according to the Erosion, Technology, and Concentration (ETC) group. Future precision agricultural techniques will be greatly impacted by the tiny sensors and monitoring systems made possible by technology, even if they have not yet been completely applied. The greater usage of autonomous sensors connected to a GPS for real-time observing will be one of the main functions of nanotechnology-enabled gadgets. These nanosensors might be dispersed around the field to track crop development and soil quality. In some areas of Australia and the USA, wireless sensors are already in use. For instance, the IT firm Accenture assisted one of the California wineries, Pickberry, in Sonoma County with the installation of Wi-Fi equipment. The initial expense of installing such a system is justifiable since it makes it possible to cultivate the best grapes, which in turn generate superior wines that fetch a higher price. Of course, the vineyards are not the only places where such wireless networks are used; according to a Forbes Magazine piece, Honeywell, a worldwide technology R&D business, is using tiny nanosensors to keep an eye on Minnesota's grocery shops. By combining biotechnology and nanotechnology, sensors will become more sensitive, enabling a quicker reaction to environmental changes. For instance, nanosurfaces can be designed to induce a chemical or electrical reaction. Nanosensors using carbon nanotubes 12 or nanocantilevers 13 are tiny enough to capture and quantify individual proteins or even small molecules. Other nanosensors function by starting an enzymatic process or by employing probes made of dendrimers, which are nanoengineered branching molecules that attach to certain chemicals and proteins. In the end, precision farming will enable increased production in agriculture by supplying correct information and assisting farmers in making better decisions (Manjunatha et al. 2016).

Applications of nanomaterials in cleaning up the environment

Nanotechnology and nanomaterial are often used in cleaning up the environment because of their highly anticipated properties, such as nanosized, simplicity in injection of a lot of surfaces, enormous adsorption potential, and unique surface coatings for in-place and outdoor applications. The traditional techniques of cleaning up the environment required pumping wastewater and moving polluted soil for treatment; regrettably, this resulted in polluted waste that ultimately needed to be disposed of; as a result, it is a laborious procedure. Treatment of soils and wastewater may be done on-site using *in situ* nanoremediation methods without the need for pumping or transportation (Otto et al. 2008). Practically speaking, however, NMs are only useful close to their place of injection for environmental cleanup (Tratnyek et al. 2006). Numerous NMs including nanozeolites, metal oxides, carbon fibers and nanotubes, enzymes and composites of several nanomaterials, have been investigated for successful cleanup. Generally speaking, four types of NMs are utilized in environmental remediation: NMs based on carbon, NMs based on silica, NMs containing metals and their oxides and NMs based on polymers.

Controlled environment agriculture (CEA)

The use of sensors and automated delivery systems in agriculture will assist in tackling viruses and crop diseases. Insecticides and herbicides will soon be more effective due to nanostructured catalysis, allowing for the use of smaller amounts. Nanotechnology will indirectly save the environment through the use of renewable energy sources, filters, and catalysts to reduce pollution and eliminate existing toxic substances. Controlled environment agriculture (CEA) is a type of farming that is commonly practiced in the USA, Europe, and Japan. It effectively uses contemporary technology for crop management. Advanced and intense hydroponically based agriculture is known as CEA. To optimize manufacturing methods, plants are cultivated in a controlled atmosphere. The computerized system keeps an eye on and controls certain agricultural settings. The current state of CEA technology makes it a great starting point for the application of nanotechnology in agriculture. Nanotechnological tools for CEA that offer "scouting" skills could significantly enhance the grower's ability to access the best time of harvest for the crop, health of the crop, and food security issues, such as microbial or chemical contamination. This is due to the presence of multiple monitoring and control systems (Abobatta 2018, Manjunatha et al. 2016).

Conclusion

The incorporation of nanosensors in the soil-water agroecosystem is a revolutionary development with enormous potential for the future of agriculture, particularly in the context of precision agriculture. Nanosensors have proven they can revolutionize how we track and manage soil and water resources, improving agricultural operations for efficiency and sustainability. An important advantage of nanosensors is their capacity to deliver high-resolution, real-time data on important parameters like pH, moisture content, nutrient levels, and pollutants in both soil and water. With the help of this data granularity, farmers and other stakeholders can make wise decisions about irrigation, fertilization, and crop management, ultimately improving crop yields and resource efficiency. Additionally, nanosensors support accurate nutrient control, lower nutrient waste, and lessen environmental effects in sustainable farming operations. However, it is crucial to recognize and consider any potential difficulties and concerns related to the usage of nanosensors, including their price, deployment approaches, standardization of procedures, potential environmental effects, and public acceptance. Realizing the full potential of nanosensors while assuring their safe and ethical use requires ethical research and development, as well as interdisciplinary cooperation between scientists, engineers, agricultural experts, and policy makers. Overall, the incorporation of nanosensors into the soil-water agroecosystem marks a significant step toward increasing food security, environmental stewardship, and agricultural sustainability. Harnessing the full potential of nanosensors for the benefit of agriculture and society at large will depend critically on ongoing research, innovation, and appropriate application.

References

Abobatta, W.F. 2018. Nanotechnology application in agriculture. Acta Sci. Agric., 2(6).

Ahmad, R., Tripathy, N., Ahn, M.-S., Bhat, K.S., Mahmoudi, T., Wang, Y. et al. 2017. Highly efficient non-enzymatic glucose sensor based on CuO modified vertically-grown ZnO nanorods on electrode. Sci. Rep., 7(1): 5715.

Akoto, O., Oppong-Otoo, J., and Osei-Fosu, P. 2015. Carcinogenic and non-carcinogenic risk of organochlorine pesticide residues in processed cereal-based complementary foods for infants and young children in Ghana. Chemosphere, 132: 193–199.

Aragay, G., Pons, J., and Merkoçi, A. 2011. Recent trends in macro-, micro-, and nanomaterial-based tools and strategies for heavy-metal detection. Chem. Rev., 111(5): 3433–3458.

Ballen, S.C., Ostrowski, G.M., Steffens, J., and Steffens, C. 2021. Graphene oxide/urease nanobiosensor applied for cadmium detection in river water. IEEE Sens. J., 21(8): 9626–9633.

Barik, T., Sahu, B., and Swain, V. 2008. Nanosilica—from medicine to pest control. Parasitol. Res., 103: 253–258.

Berekaa, M. 2015. Review article nanotechnology in food industry. Int. J. Curr. Microbiol. Appl. Sci., 4(5): 345–357.

Biparva, P., Gorji, S., and Hedayati, E. 2020. Promoted reaction microextraction for determining pesticide residues in environmental water samples using gas chromatography-mass spectrometry. J. Chromatogr. A, 1612: 460639.

Bushdid, C., Magnasco, M.O., Vosshall, L.B., and Keller, A. 2014. Humans can discriminate more than 1 trillion olfactory stimuli. Science, 343(6177): 1370–1372.

Cesarino, I., Moraes, F.C., Lanza, M.R., and Machado, S.A. 2012. Electrochemical detection of carbamate pesticides in fruit and vegetables with a biosensor based on acetylcholinesterase immobilised on a composite of polyaniline–carbon nanotubes. Food Chem, 135(3): 873–879.

Chajanovsky, I., Cohen, S., Shtenberg, G., and Suckeveriene, R.Y. 2021. Development and characterization of integrated nano-sensors for organic residues and pH field detection. Sens, 21(17): 5842.

Chen, J., Andler, S.M., Goddard, J.M., Nugen, S.R., and Rotello, V.M. 2017. Integrating recognition elements with nanomaterials for bacteria sensing. Chem. Soc. Rev., 46(5): 1272–1283.

Christopher, F.C., Kumar, P.S., Christopher, F.J., Joshiba, G.J., and Madhesh, P. 2020. Recent advancements in rapid analysis of pesticides using nano biosensors: a present and future perspective. J. Clean. Prod., 269: 122356.

Cochran, F.V., Brunsell, N.A., and Suyker, A.E. 2016. A thermodynamic approach for assessing agroecosystem sustainability. Ecol. Indic., 67: 204–214.

Cohen, S., Zelikman, E., and Suckeveriene, R.Y. 2020. Ultrasonically induced polymerization and polymer grafting in the presence of carbonaceous nanoparticles. Processes, 8(12): 1680.

DeRosa, M.C., Monreal, C., Schnitzer, M., Walsh, R., and Sultan, Y. 2010. Nanotechnology in fertilizers. Nat. Nanotechnol., 5(2): 91–91.

Dhouib, I.B., Annabi, A., Jallouli, M., Marzouki, S., Gharbi, N., Elfazaa, S. et al. 2016. Carbamates pesticides induced immunotoxicity and carcinogenicity in human: A review. J. Appl. Biomed., 14(2): 85–90.

Du, D., Ye, X., Zhang, J., Zeng, Y., Tu, H., Zhang, A. et al. 2008. Stripping voltammetric analysis of organophosphate pesticides based on solid-phase extraction at zirconia nanoparticles modified electrode. Electrochem. Commun., 10(5): 686–690.

Emslie, K.R., Whaites, L., Griffiths, K.R., and Murby, E.J. 2007. Sampling plan and test protocol for the semiquantitative detection of genetically modified canola (*Brassica napus*) seed in bulk canola seed. J. Agric. Food Chem., 55(11): 4414–4421.

Farahi, R.H., Passian, A., Tetard, L., and Thundat, T. 2012. Critical issues in sensor science to aid food and water safety. ACS Nano., 6(6): 4548–4556.

Farré, M., Gajda-Schrantz, K., Kantiani, L., and Barceló, D. 2009. Ecotoxicity and analysis of nanomaterials in the aquatic environment. Anal. Bioanal. Chem., 393: 81–95.

Feregrino-Perez, A.A., Magaña-López, E., Guzmán, C., and Esquivel, K. 2018. A general overview of the benefits and possible negative effects of the nanotechnology in horticulture. Sci. Hortic., 238: 126–137.

Gascon, J., Oubiña, A., and Barceló, D. 1997. Detection of endocrine-disrupting pesticides by enzyme-linked immunosorbent assay (ELISA): application to atrazine. TrAC, Trends Anal. Chem., 16(10): 554–562.

Göde, C., Yola, M.L., Yılmaz, A., Atar, N., and Wang, S. 2017. A novel electrochemical sensor based on calixarene functionalized reduced graphene oxide: Application to simultaneous determination of Fe (III), Cd (II) and Pb (II) ions. J. Colloid Interface Sci., 508: 525–531.

Godfray, H.C.J., Beddington, J.R., Crute, I.R., Haddad, L., Lawrence, D., Muir, J.F. et al. 2010. Food security: the challenge of feeding 9 billion people. Science, 327(5967): 812–818.

Guan, J., Yang, J., Zhang, Y., Zhang, X., Deng, H., Xu, J. et al. 2021. Employing a fluorescent and colorimetric picolyl-functionalized rhodamine for the detection of glyphosate pesticide. Talanta, 224: 121834.

Guerra, F.D., Attia, M.F., Whitehead, D.C., and Alexis, F. 2018. Nanotechnology for environmental remediation: materials and applications. Mol., 23(7): 1760.

Gupta, N., Sharma, S., Mir, I.A., and Kumar, D. 2006. Advances in sensors based on conducting polymers. J. Sci. Ind. Res., 65(7): 549–557.

Han, D., Chu, Y., Yang, L., Liu, Y., and Lv, Z. 2005. Reversed micelle polymerization: a new route for the synthesis of DBSA–polyaniline nanoparticles. Colloids Surf. A: Physicochem. Eng., 259(1-3): 179–187.

Hellwig, M., and Henle, T. 2014. Baking, ageing, diabetes: a short history of the Maillard reaction. Angew. Chem. Int. Ed., 53(39): 10316–10329.

Hu, B., Shao, S., Ni, H., Fu, Z., Hu, L., Zhou, Y. et al. 2020. Current status, spatial features, health risks, and potential driving factors of soil heavy metal pollution in China at province level. Environ. Pollut. 266: 114961.

Huang, X., Aguilar, Z.P., Xu, H., Lai, W., and Xiong, Y. 2016. Membrane-based lateral flow immunochromatographic strip with nanoparticles as reporters for detection: A review. Biosens. Bioelectron, 75: 166–180.

Husen, A., and Siddiqi, K.S. 2014. Phytosynthesis of nanoparticles: concept, controversy and application. Nanoscale Res. Lett., 9(229): 1–24.

Ikram, F., Qayoom, A., Aslam, Z., and Shah, M.R. 2019. Epicatechin coated silver nanoparticles as highly selective nanosensor for the detection of Pb^{2+} in environmental samples. J. Mol. Liq., 277: 649–655.

Initiative, N.S. 2012. Nanotechnology for sensors and sensors for nanotechnology: improving and protecting health, safety, and the environment. Nanotechnol. Signat. Initiat, 1–11.

Jackson, T., Mansfield, K., Saafi, M., Colman, T., and Romine, P. 2008. Measuring soil temperature and moisture using wireless MEMS sensors. Measurement, 41(4): 381–390.

Jeevanandam, J., Kaliyaperumal, A., Sundararam, M., and Danquah, M.K. 2020. Nanomaterials as toxic gas sensors and biosensors. Nanosensor Technologies for Environmental Monitoring. Springer, 389–430.

Juárez-Maldonado, A., Ortega-Ortíz, H., Morales-Díaz, A.B., González-Morales, S., Morelos-Moreno, Á., Cabrera-De la Fuente, M. et al. 2019. Nanoparticles and nanomaterials as plant biostimulants. Int. J. Mol. Sci., 20(1): 162.

Kamat, P.V., and Meisel, D. 2003. Nanoscience opportunities in environmental remediation. C R Chim., 6(8-10): 999–1007.

Karn, B., Kuiken, T., and Otto, M. 2009. Nanotechnology and *in situ* remediation: a review of the benefits and potential risks. Environ. Health Perspect., 117(12): 1813–1831.

Khairy, M., Ayoub, H.A., and Banks, C.E. 2018. Non-enzymatic electrochemical platform for parathion pesticide sensing based on nanometer-sized nickel oxide modified screen-printed electrodes. Food Chem., 255: 104–111.

Khot, L.R., Sankaran, S., Maja, J.M., Ehsani, R., and Schuster, E.W. 2012. Applications of nanomaterials in agricultural production and crop protection: a review. Crop Prot., 35: 64–70.

Kim, S.J., Ko, S.H., Kang, K.H., and Han, J. 2010. Direct seawater desalination by ion concentration polarization. Nat. Nanotechnol., 5(4): 297–301.

Kim, Y.S., Chung, J., Song, M.Y., Jurng, J., and Kim, B.C. 2014. Aptamer cocktails: enhancement of sensing signals compared to single use of aptamers for detection of bacteria. Biosens. Bioelectron., 54: 195–198.

Kıranşan, K.D., Aksoy, M., and Topçu, E. 2018. Flexible and freestanding catalase-Fe$_3$O$_4$/reduced graphene oxide paper: Enzymatic hydrogen peroxide sensor applications. Mater. Res. Bull., 106: 57–65.

Kumar, A., Pandey, A.K., Singh, S.S., Shanker, R., and Dhawan, A. 2011. Engineered ZnO and TiO$_2$ nanoparticles induce oxidative stress and DNA damage leading to reduced viability of *Escherichia coli*. Free Radic Biol. Med., 51(10): 1872–1881.

Kuswandi, B., Restyana, A., Abdullah, A., Heng, L.Y., and Ahmad, M. 2012. A novel colorimetric food package label for fish spoilage based on polyaniline film. Food Control, 25(1): 184–189.

Lee, J.Y., Yang, J.-S., Han, M., and Choi, J. 2010. Comparison of the microbiological and chemical characterization of harvested rainwater and reservoir water as alternative water resources. Sci. Total Environ., 408(4): 896–905.

Lei, Y.-M., Xiao, B.-Q., Liang, W.-B., Chai, Y.-Q., Yuan, R., and Zhuo, Y. 2018. A robust, magnetic, and self-accelerated electrochemiluminescent nanosensor for ultrasensitive detection of copper ion. Biosens. Bioelectron., 109: 109–115.

Levin, S.A. 1998. Ecosystems and the biosphere as complex adaptive systems. Ecosyst., 1: 431–436.

Lewinski, N., Colvin, V., and Drezek, R. 2008. Cytotoxicity of nanoparticles. Small, 4(1): 26–49.

Li, B.-B., Clements, W.R., Yu, X.-C., Shi, K., Gong, Q., and Xiao, Y.-F. 2014. Single nanoparticle detection using split-mode microcavity Raman lasers. Proc. Natl. Acad. Sci., 111(41): 14657–14662.

Li, C., Dong, Y., Yang, J., Li, Y., and Huang, C. 2014. Modified nano-graphite/Fe$_3$O$_4$ composite as efficient adsorbent for the removal of methyl violet from aqueous solution. J. Mol. Liq., 196: 348–356.

Li, H., Yan, X., Qiao, S., Lu, G., and Su, X. 2018. Yellow-emissive carbon dot-based optical sensing platforms: cell imaging and analytical applications for biocatalytic reactions. ACS Appl. Mater. Interfaces, 10(9): 7737–7744.

Li, H., Yu, C., Chen, R., Li, J., and Li, J. 2012. Novel ionic liquid-type Gemini surfactants: Synthesis, surface property and antimicrobial activity. Colloids Surf A: Physicochem. Eng., 395: 116–124.

Li, M., Gou, H., Al-Ogaidi, I., and Wu, N. 2013. Nanostructured sensors for detection of heavy metals: a review. ACS Sustain. Chem. Eng., 1(7): 685–857.

Li, Y., Cu, Y.T.H., and Luo, D. 2005. Multiplexed detection of pathogen DNA with DNA-based fluorescence nanobarcodes. Nat. Biotechnol., 23(7): 885–889.

Liang, M., Fan, K., Pan, Y., Jiang, H., Wang, F., Yang, D. et al. 2013. Fe$_3$O$_4$ magnetic nanoparticle peroxidase mimetic-based colorimetric assay for the rapid detection of organophosphorus pesticide and nerve agent. Anal. Chem., 85(1): 308–312.

Liu, X., Lei, Z., Liu, F., Liu, D., and Wang, Z. 2014. Fabricating three-dimensional carbohydrate hydrogel microarray for lectin-mediated bacterium capturing. Biosens. Bioelectron., 58: 92–100.

Liu, Y., Song, Z., Gao, L., and Li, J. 2017. An optical pH sensor based on Diazocine. Chemistry Select., 2(26): 7956–7960.

López, M.G., Fussell, R.J., Stead, S.L., Roberts, D., McCullagh, M., and Rao, R. 2014. Evaluation and validation of an accurate mass screening method for the analysis of pesticides in fruits and vegetables using liquid chromatography–quadrupole-time of flight–mass spectrometry with automated detection. J. Chromatogr. A, 1373: 40–50.

Madianos, L., Skotadis, E., Tsekenis, G., Patsiouras, L., Tsigkourakos, M., and Tsoukalas, D. 2018. Impedimetric nanoparticle aptasensor for selective and label free pesticide detection. Microelectron. Eng., 189: 39–45.

Majdinasab, M., Yaqub, M., Rahim, A., Catanante, G., Hayat, A., and Marty, J.L. 2017. An overview on recent progress in electrochemical biosensors for antimicrobial drug residues in animal-derived food. Sens., 17(9): 1947.

Manjunatha, S., Biradar, D., and Aladakatti, Y.R. 2016. Nanotechnology and its applications in agriculture: A review. J. farm. Sci., 29(1): 1–13.

Mesci-Haftaci, S., Ankarali, H., Caglar, M., and Yavuzcan, A. 2014. Reliability of colposcopy in Turkey: correlation with Pap smear and 1-year follow up. Asian Pac. J. Cancer Prev., 15(17): 7317–7320.

Milik, J., and Pasela, R. 2018. Analysis of concentration trends and origins of heavy metal loads in stormwater runoff in selected cities: A review. E3S Web Conf., 44(00111): 8.

Mishra, V.K., and Kumar, A. 2009. Impact of metal nanoparticles on the plant growth promoting rhizobacteria. Dig. J. Nanomater. Biostruct., 4(3): 587–592.

Molden, D. 2013. Water for food water for life: A comprehensive assessment of water management in agriculture. International Water Management Institute.

Mossoba, M.M., Al-Khaldi, S.F., Schoen, B., and Yakes, B.J. 2010. Nanoparticle probes and mid-infrared chemical imaging for DNA microarray detection. Appl. Spectrosc., 64(11): 1191–1198.

Mucci, A., Montgomery, S., Lucotte, M., Plourde, Y., Pichet, P., and Tra, H.V. 1995. Mercury remobilization from flooded soils in a hydroelectric reservoir of northern Quebec, La Grande-2: results of a soil resuspension experiment. Can. J. Fish. Aquat. Sci., 52(11): 2507–2517.

Neuberger, T., Schöpf, B., Hofmann, H., Hofmann, M., and Von Rechenberg, B. 2005. Superparamagnetic nanoparticles for biomedical applications: Possibilities and limitations of a new drug delivery system. J. Magn. Magn. Mater., 293(1): 483–496.

Nguyen, H.L., Nguyen, H.N., Nguyen, H.H., Luu, M.Q., and Nguyen, M.H. 2014. Nanoparticles: synthesis and applications in life science and environmental technology. Adv. Nat. Sci.: Nanosci. Nanotechnol., 6(1): 015008.

Nsibande, S., and Forbes, P. 2016. Fluorescence detection of pesticides using quantum dot materials–a review. Anal. Chim. Acta, 945: 9–22.

O'Brien, N., and Cummins, E. 2008. Recent developments in nanotechnology and risk assessment strategies for addressing public and environmental health concerns. Hum. Ecol. Risk Assess, 14(3): 568–592.

O'Connell, P.J., Gormally, C., Pravda, M., and Guilbault, G.G. 2001. Development of an amperometric L-ascorbic acid (Vitamin C) sensor based on electropolymerised aniline for pharmaceutical and food analysis. Anal. Chim. Acta, 431(2): 239–247.

Ogilvie, A., Riaux, J., Massuel, S., Mulligan, M., Belaud, G., Le Goulven, P. et al. 2019. Socio-hydrological drivers of agricultural water use in small reservoirs. Agric. Water Manag., 218: 17–29.

Omanović-Mikličanina, E., and Maksimović, M. 2016. Nanosensors applications in agriculture and food industry. Bull. Chem. Technol. Bosnia Herzegovina, 47: 59–70.

Otto, M., Floyd, M., and Bajpai, S. 2008. Nanotechnology for site remediation. Remediation. Wiley Interscience, 19(1): 99–108.

Palaznik, O., Nedorezova, P., Pol'Shchikov, S., Klyamkina, A., Shevchenko, V., Krasheninnikov, V. et al. 2019. Production by *in situ* polymerization and properties of composite materials based on polypropylene and hybrid carbon nanofillers. Polym. Sci. Ser. B., 61: 200–214.

Parisi, C., Vigani, M., and Rodríguez-Cerezo, E. 2015. Agricultural nanotechnologies: what are the current possibilities? Nano Today, 10(2): 124–127.

Patel, V., Wilkes, S., Askey, S., Brewerton, H., Corcoran, S., Crombie, J. et al. 2011. www. rsc. org/analyst. Analyst, 136(2918): 2969–2974.

Paul, E. 2007. Soil microbiology, ecology, and biochemistry in perspective: Soil microbiology, ecology and biochemistry. Elsevier, 3–24.

Peng, J., and Loew, A. 2017. Recent advances in soil moisture estimation from remote sensing. Water, 9(7): 530.

Peng, J., Loew, A., Merlin, O., and Verhoest, N.E. 2017. A review of spatial downscaling of satellite remotely sensed soil moisture. Rev. Geophys., 55(2): 341–366.

Pham, T.B., Bui, H., and Do, T.C. 2020. Surface-enhanced Raman spectroscopy based on Silver nano-dendrites on microsphere end-shape optical fibre for pesticide residue detection. Optik, 219: 165172.

Phillips, R.L., Miranda, O.R., You, C.C., Rotello, V.M., and Bunz, U.H. 2008. Rapid and efficient identification of bacteria using gold-nanoparticle–poly (para-phenyleneethynylene) constructs. Angew. Chem. Int. Ed., 47(14): 2590–2594.

Philp, R., Leung, F., and Bradley, C. 2003. A comparison of the metal content of some benthic species from coastal waters of the Florida Panhandle using high-resolution inductively coupled plasma mass spectrometry (ICP-MS) analysis. Arch. Environ. Contam. Toxicol., 44: 0218–0223.

Picó, Y., and Andreu, V. 2014. Nanosensors and other techniques for detecting nanoparticles in the environment: Nanosensors for Chemical and Biological Applications. Woodhead Publ., 295–338.

Prasad, R., Kumar, M., and Kumar, V. 2017. Nanotechnology: an agricultural paradigm. Springer.

Prashar, P., and Shah, S. 2016. Impact of fertilizers and pesticides on soil microflora in agriculture: Sustainable Agriculture Reviews. Springer, 19: 331–361.

Qu, F., Zhou, X., Xu, J., Li, H., and Xie, G. 2009. Luminescence switching of CdTe quantum dots in presence of p-sulfonatocalix [4] arene to detect pesticides in aqueous solution. Talanta, 78(4-5): 1359–1363.

Qu, X., Alvarez, P.J., and Li, Q. 2013. Applications of nanotechnology in water and wastewater treatment. Water Res., 47(12): 3931–3946.

Qu, Y., Min, H., Wei, Y., Xiao, F., Shi, G., Li, X. et al. 2008. Au–TiO_2/Chit modified sensor for electrochemical detection of trace organophosphates insecticides. Talanta, 76(4): 758–762.

Raja, M.A., and Husen, A. 2020. Role of nanomaterials in soil and water quality management: Nanomaterials for agriculture and forestry applications. Elsevier, 491–503.

Rao, P.S., Subrahmanya, S., and Sathyanarayana, D. 2002. Inverse emulsion polymerization: a new route for the synthesis of conducting polyaniline. Synth. Met., 128(3): 311–316.

Rhouati, A., Majdinasab, M., and Hayat, A. 2018. A perspective on non-enzymatic electrochemical nanosensors for direct detection of pesticides. Curr. Opin. Electrochem., 11: 12–18.

Rowland, S., and Rook, J. 1961. Analytical methods. Int. J. Dairy Technol., 14(3): 112–114.

Rowley-Neale, S.J., Randviir, E.P., Dena, A.S.A., and Banks, C.E. 2018. An overview of recent applications of reduced graphene oxide as a basis of electroanalytical sensing platforms. Appl. Mater. Today, 10: 218–226.

Sadiq, N., Shafique, M., Akbar, M., Shakoor, M., Mujahid, A., Hussain, T. et al. 2023. An ion-imprinted polymer-receptor-based electrochemical sensor for the sensitive and selective detection of cadmium. ChemistrySelect, 8(10): e202204824.

Sahani, S., and Sharma, Y.C. 2021. Advancements in applications of nanotechnology in global food industry. Food Chem., 342: 128318.

Salerno, M., Landoni, P., and Verganti, R. 2008. Designing foresight studies for Nanoscience and Nanotechnology (NST) future developments. Technol. Forecast. Soc. Change, 75(8): 1202–1223.

Saliba, A.M., De Assis, M.-C., Nishi, R., Raymond, B., Marques, E.d.A., Lopes, U.G. et al. 2006. Implications of oxidative stress in the cytotoxicity of *Pseudomonas aeruginosa* ExoU. Microbes Infect., 8(2): 450–459.

Sastry, K., Rashmi, H., and Rao, N. 2010. Nanotechnology patents as R&D indicators for disease management strategies in agriculture. J. Intellect. Prop. Rights, 15: 197–205.

Satapathi, S., Kumar, V., Chini, M.K., Bera, R., Halder, K.K., and Patra, A. 2018. Highly sensitive detection and removal of mercury ion using a multimodal nanosensor. Nano-Struct. Nano-Objects, 16: 120–126.

Savk, A., Özdil, B., Demirkan, B., Nas, M.S., Calimli, M.H., Alma, M.H. et al. 2019. Multiwalled carbon nanotube-based nanosensor for ultrasensitive detection of uric acid, dopamine, and ascorbic acid. Mater. Sci. Eng. C, 99: 248–254.

Schets, F., Italiaander, R., Van Den Berg, H., and de Roda Husman, A. 2010. Rainwater harvesting: quality assessment and utilization in The Netherlands. J Water Health, 8(2): 224–235.

Sela, L., and Sobel, N. 2010. Human olfaction: a constant state of change-blindness. Exp. Brain Res., 205: 13–29.

Shao, Y., Wang, J., Engelhard, M., Wang, C., and Lin, Y. 2010. Facile and controllable electrochemical reduction of graphene oxide and its applications. J. Mater. Chem., 20(4): 743–748.

Sharma, A., Shukla, A., Attri, K., Kumar, M., Kumar, P., Suttee, A. et al. 2020. Global trends in pesticides: A looming threat and viable alternatives. Ecotoxicol. Environ. Saf., 201: 110812.

Simon, K.A., Mosadegh, B., Minn, K.T., Lockett, M.R., Mohammady, M.R., Boucher, D.M. et al. 2016. Metabolic response of lung cancer cells to radiation in a paper-based 3D cell culture system. Biomater., 95: 47–59.

Soares, B.G., Leyva, M.E., Barra, G.M., and Khastgir, D. 2006. Dielectric behavior of polyaniline synthesized by different techniques. Eur. Polym. J., 42(3): 676–686.

Sonmez, I., Kaplan, M., and Sonmez, S. 2007. Investigation of seasonal changes in nitrate contents of soils and irrigation waters in greenhouses Located in Antalya-Demn region. Asian J. Chem., 19(7): 5639.

Soylemez, S., Kesika, M., and Toppare, L. 2018. Biosensing devices: conjugated polymer based scaffolds. Encycl. Polym. Appl., 360–386.

Srivastava, A.K., Dev, A., and Karmakar, S. 2017. Nanosensors for food and agriculture: Nanoscience in Food and Agriculture, Springer 5: 41–79.

Sun, J., Ge, J., Liu, W., Wang, X., Fan, Z., Zhao, W. et al. 2012. A facile assay for direct colorimetric visualization of lipopolysaccharides at low nanomolar level. Nano Res., 5: 486–493.

Sun, Y., Fang, L., Wan, Y., and Gu, Z. 2018. Pathogenic detection and phenotype using magnetic nanoparticle-urease nanosensor. Sens. Actuators B: Chem., 259: 428–432.

Talari, F.F., Bozorg, A., Faridbod, F., and Vossoughi, M. 2021. A novel sensitive aptamer-based nanosensor using rGQDs and MWCNTs for rapid detection of diazinon pesticide. J. Environ. Chem. Eng., 9(1): 104878.

Tokalioglu, S., Kartal, S., and Elci, L. 2000. Speciation and determination of heavy metals in lake waters by atomic absorption spectrometry after sorption on amberlite XAD-16 resin. Anal. Sci, 16(11): 1169–1174.

Tothill, I.E. 2001. Biosensors developments and potential applications in the agricultural diagnosis sector. Comput. Electron. Agric., 30(1-3): 205–218.

Tratnyek, P.G., and Johnson, R.L. 2006. Nanotechnologies for environmental cleanup. Nano Today, 1(2): 44–48.

Ullah, N., Mansha, M., Khan, I., and Qurashi, A. 2018. Nanomaterial-based optical chemical sensors for the detection of heavy metals in water: Recent advances and challenges. TrAC, Trends Anal. Chem., 100: 155–166.

Umabharathi, P.S., and Karpagam, S. 2023. Real scenario of metal ion sensor: is conjugated polymer helpful to detect hazardous metal ion. Rev. Inorg. Chem., 43(3): 385–414.

Umasankareswari, T., Dharshana, V., Saravanadevi, K., Dorothy, R., Sasilatha, T., Nguyen, T.A. et al. 2022. Soil moisture nanosensors: Nanosensors for Smart Agriculture. Elsevier, 185–216.

Vaseashta, A., Vaclavikova, M., Vaseashta, S., Gallios, G., Roy, P., and Pummakarnchana, O. 2007. Nanostructures in environmental pollution detection, monitoring, and remediation. Sci. Technol. Adv. Mater, 8(1-2): 47.

Vashist, S.K., Zheng, D., Al-Rubeaan, K., Luong, J.H., and Sheu, F.-S. 2011. Advances in carbon nanotube based electrochemical sensors for bioanalytical applications. Biotechnol. Adv., 29(2): 169–188.

Verdian, A. 2018. Apta-nanosensors for detection and quantitative determination of acetamiprid–A pesticide residue in food and environment. Talanta, 176: 456–464.

Verma, M.S., Chen, P.Z., Jones, L., and Gu, F.X. 2014. "Chemical nose" for the visual identification of emerging ocular pathogens using gold nanostars. Biosens. Bioelectron., 61: 386–390.

Verma, M.S., Wei, S.-C., Rogowski, J.L., Tsuji, J.M., Chen, P.Z., Lin, C.-W. et al. 2016. Interactions between bacterial surface and nanoparticles govern the performance of "chemical nose" biosensors. Biosens. Bioelectron., 83: 115–125.

Vinayaka, A., Basheer, S., and Thakur, M. 2009. Bioconjugation of CdTe quantum dot for the detection of 2, 4-dichlorophenoxyacetic acid by competitive fluoroimmunoassay based biosensor. Biosens. Bioelectron., 24(6): 1615–1620.

Wang, B.-X., Wang, G.-Z., Sang, T., and Wang, L.-L. 2017. Six-band terahertz metamaterial absorber based on the combination of multiple-order responses of metallic patches in a dual-layer stacked resonance structure. Sci. Rep., 7(1): 41373.

Wang, J., Jiang, C., Wang, X., Wang, L., Chen, A., Hu, J. et al. 2016. Fabrication of an "ion-imprinting" dual-emission quantum dot nanohybrid for selective fluorescence turn-on and ratiometric detection of cadmium ions. Analyst., 141(20): 5886–5892.

Wang, J., and Musameh, M. 2005. Carbon-nanotubes doped polypyrrole glucose biosensor. Anal. Chim. Acta, 539(1-2): 209–213.

Wang, M., Huang, J., Wang, M., Zhang, D., and Chen, J. 2014. Electrochemical nonenzymatic sensor based on CoO decorated reduced graphene oxide for the simultaneous determination of carbofuran and carbaryl in fruits and vegetables. Food Chem., 151: 191–197.

Welbaum, G.E., Sturz, A.V., Dong, Z., and Nowak, J. 2004. Managing soil microorganisms to improve productivity of agro-ecosystems. Crit. Rev. Plant Sci., 23(2): 175–193.

Wencel, D., Abel, T., and McDonagh, C. 2014. Optical chemical pH sensors. Anal. Chem., 86(1): 15–29.

Xia, N., Feng, F., Liu, C., Li, R., Xiang, W., Shi, H. et al. 2019. The detection of mercury ion using DNA as sensors based on fluorescence resonance energy transfer. Talanta, 192: 500–507.

Xie, H., Bei, F., Hou, J., and Ai, S. 2018. A highly sensitive dual-signaling assay via inner filter effect between g-C$_3$N$_4$ and gold nanoparticles for organophosphorus pesticides. Sens. Actuators B: Chem., 255: 2232–2239.

Xie, Y., Yu, Y., Lu, L., Ma, X., Gong, L., Huang, X. et al. 2018. CuO nanoparticles decorated 3D graphene nanocomposite as non-enzymatic electrochemical sensing platform for malathion detection. J. Electroanal. Chem., 812: 82–89.

Yadav, A., Yadav, K., Ahmad, R., and Abd-Elsalam, K.A. 2023. Emerging frontiers in nanotechnology for precision agriculture: advancements, hurdles and prospects. Agrochemicals, 2(2): 220–256.

Yan, X., Li, H., and Su, X. 2018. Review of optical sensors for pesticides. TrAC, Trends Anal. Chem., 103: 1–20.

Yan, X., Shi, H., and Wang, M. 2012. Development of an enzyme-linked immunosorbent assay for the simultaneous determination of parathion and imidacloprid. Anal. Methods, 4(12): 4053–4057.

Yang, C., Ding, Y., and Qian, J. 2018. Design of magnetic-fluorescent based nanosensor for highly sensitive determination and removal of Hg^{2+}. Ceram. Int., 44(8): 9746–9752.

Yetim, N.K., Özkan, E.H., Özcan, C., and Sarı, N. 2020. Preparation of AChE immobilized microspheres containing thiophene and furan for the determination of pesticides by the HPLC-DAD method. J. Mol. Struct., 1222: 128931.

Yong, K.-T., Roy, I., Swihart, M.T., and Prasad, P.N. 2009. Multifunctional nanoparticles as biocompatible targeted probes for human cancer diagnosis and therapy. J. Mater. Chem., 19(27): 4655–4672.

Yosef, B.A., and Asmamaw, D.K. 2015. Rainwater harvesting: An option for dry land agriculture in arid and semi-arid Ethiopia. Int. J. Water Res. Environ. Eng., 7(2): 17–28.

Zhang, L., Peng, D., Liang, R.-P., and Qiu, J.-D. 2018. Graphene-based optical nanosensors for detection of heavy metal ions. TrAC, Trends Anal. Chem., 102: 280–289.

Zhang, M., Chen, W., Chen, X., Zhang, Y., Lin, X., Wu, Z. et al. 2013. Multiplex immunoassays of plant viruses based on functionalized upconversion nanoparticles coupled with immunomagnetic separation. J. Nanomater, 2013: 122–122.

Zhang, Q., Uchaker, E., Candelaria, S.L., and Cao, G. 2013. Nanomaterials for energy conversion and storage. Chem. Soc. Rev., 42(7): 3127–3171.

Zhou, C., Xu, L., Song, J., Xing, R., Xu, S., Liu, D. et al. 2014. Ultrasensitive non-enzymatic glucose sensor based on three-dimensional network of ZnO-CuO hierarchical nanocomposites by electrospinning. Sci. Rep., 4(1): 7382.

Zhou, L., Zhang, X., Ma, L., Gao, J., and Jiang, Y. 2017. Acetylcholinesterase/chitosan-transition metal carbides nanocomposites-based biosensor for the organophosphate pesticides detection. Biochem. Eng. J., 128: 243–249.

Zhou, X., Hui, E., Yu, X.-L., Lin, Z., Pu, L.-K., Tu, Z. et al. 2015. Development of a rapid immunochromatographic lateral flow device capable of differentiating phytase expressed from recombinant *aspergillus niger* phy A2 and genetically modified corn. J. Agric. Food Chem., 63(17): 4320–4326.

Zhou, Y., Ding, L., Wu, Y., Huang, X., Lai, W., and Xiong, Y. 2019. Emerging strategies to develop sensitive AuNP-based ICTS nanosensors. TrAC, Trends Anal. Chem., 112: 147–160.

4

Nanomaterials in Water Desalination

Rahat Javaid[1],* *and* *Umair Yaqub Qazi*[2],*

Introduction

Water is an essential feedstock for various industries, including petrochemical, electronics, food and beverage, pharmaceutical, and agricultural sectors. According to an estimation, by 2025, the annual industrial water consumption will increase 50% than in 1995 (Teow and Mohammad 2019). Due to the rapid growth in world's population, climate change, industrialization, urbanization, and stricter health-based water quality standards, the world faces enormous challenges to fulfill the increasing needs for clean water as freshwater supplies are depleted (Qu et al. 2013). Clean water sources are scarce everywhere, but they are particularly scarce in arid and dry regions. Even though 70% of the Earth's surface is occupied by water, freshwater is only 3%, and a very small fraction of this freshwater is readily accessible because most of the freshwater is either present as deep underground reservoirs or as frozen glaciers. On Earth, the most abundant source of water is seawater, but its high salinity renders it unsuitable for domestic and industrial applications. Conventional water treatment techniques are not efficient in reducing salinity. Desalination is a practical solution that can bridge the enormous gap between the present capacities and the rising demand for clean water. Seawater desalination is the process of removing minerals and salts to produce clean water (Misdan et al. 2012), but this process still consumes a lot of energy, and further study is required to make it more effective and sustainable in the long run.

Membrane-based and thermal procedures are the widely used desalination methods. Multiple effect distillation, multistage flash distillation, and mechanical vapor compression are examples of thermal desalination techniques. These procedures rely on condensation and evaporation, in which water is heated till its evaporation. This technique leaves salt behind as the vapor is condensed to form freshwater (Cho et al. 2011). In the past, thermal techniques produced a significant portion of drinkable water, specifically in regions where surplus heat from power plants is used for seawater desalination (Mezher et al. 2011). There are many advantages of thermal processes,

[1] Department of Chemical Engineering, University of South Carolina, Columbia, SC, USA.

[2] Department of Chemistry, Benedict College, Columbia, SC, USA.

* Corresponding authors: RQAZI@mailbox.sc.edu; umair.qazi@benedict.edu

including relatively simple and easy operations (Daer et al. 2015) with no requirements for pre-treatment procedures; therefore, their operational expenses are relatively lower (Daer et al. 2015). Despite the benefits mentioned above, thermal techniques face several challenges, such as scale formation on heat exchanger tubes from precipitation of calcium and magnesium salts, corrosion because they operate in an aggressive environment with seawater and non-condensable gasses, high energy needs, and substantial material and land requirements (Nair and Kumar 2013). Because of these disadvantages of thermal techniques, developing new procedures is inevitable. Thermal procedures can be replaced by membrane processes like electrodialysis and reverse osmosis (RO). RO has surpassed conventional thermal techniques as the most popular desalination method worldwide (Lee et al. 2011). Since RO achieves separation without phase change, its energy needs are lower than those of thermal processes. Moreover, RO systems are smaller requiring less land for construction with simpler expansion and maintenance (Nair and Kumar 2013). The available RO technique uses semipermeable membranes (0.0001–0.001 μm pore diameter), which enables retaining most of the molecules except for water. These membranes are often synthesized from polymers such as cellulose and acetate or ceramics (Nair and Kumar 2013). Forward osmosis (FO) uses semipermeable membrane along with osmosis processes to operate without external pressure. However, the designing and fabrication of effective membranes for FO is challenging.

RO and FO techniques currently use commercial membranes made from polymers, including polyamide (PA), cellulose triacetate, and thin film composite (TFC). The membrane distillation (MD) process uses membranes composed of polytetrafluoroethylene (PTFE) and polyvinylidene-fluoride (PVDF) (Daer et al. 2015). Capacitive deionization (CDI) is a new membranes-based separation technology that mostly relies on carbon-based membranes for separating solutes via electrosorption and regeneration methods. In all membrane-based techniques, membranes play a pivotal role in maintaining efficient water permeation and salt rejection. Recent advancement in material science in terms of synthesis, characterization, and modification enhanced the potential applicability of nanomaterials for various applications, including medical applications, environmental and energy related applications, electronics, water treatment, and desalination processes (Javaid et al. 2010, 2021, Yaqub Qazi and Javaid 2016, Javaid and Yaqub Qazi 2019, Javaid and Nanba 2022). This chapter summarizes research utilizing various nanomaterials such as zeolites, graphene, carbon nanotubes (CNTs), aquaporin, and metal oxides in desalination.

Carbon-based nanomaterials

Carbon Nanotubes (CNTs)

CNTs are carbon allotropes having a thickness of one atom rolled in a cylindrical hollow long structure with a diameter of 1 nm and several centimeters in length (Teow and Mohammad 2019). Generally, CNTs align into "ropes", held together via π-stacking, which are van der Waals forces. CNTs have hexagonal lattices with free-moving electrical charges. CNTs are known for their efficient adsorptive characteristics for both biological and chemical contaminants like phenols (Dehghani et al. 2016), heavy metals (Gouda and Al Ghannam 2016, Ihsanullah et al. 2016), natural organic matter (Yang and Xing 2009), and organic chemicals (Yang et al. 2006). Because of higher strength, low weight, and excellent adsorptive and electrical characteristics, CNTs have the potential for various applications in desalination. Incorporating CNTs into membranes has improved permeability, hydrophilicity, solute rejection, tensile strength, fouling resistance, and electrical conductivity (Khalid et al. 2015). Dumee et al. (Dumée et al. 2013) developed CNTs-based robust and stable support with potential applications in both FO and RO. Zhao et al. (Zhao et al. 2017) incorporated CNTs between the PVDF membrane and layer of PA (Figure 1). According to the results, CNTs as the intermediate/sublayer resulted in efficient separation for the FO membrane with two times higher water flux than that attained without CNTs, along with alleviation of concentration polarization problems usually faced in the FO technique.

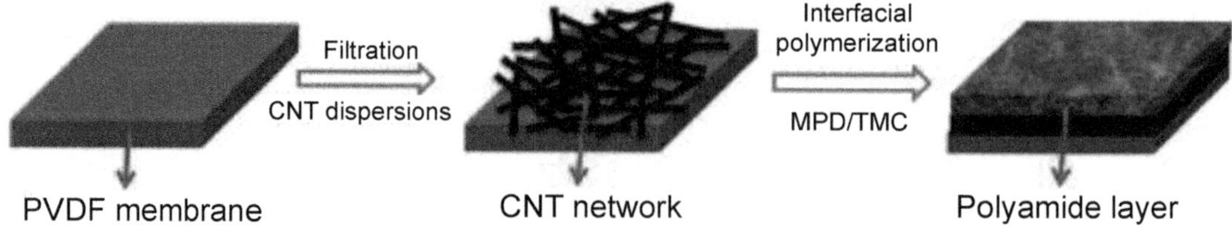

Figure 1. Schematic presentation of multilayer FO membrane. Reprinted from (Zhao et al. 2017).

Bhadra et al. (Bhadra et al. 2013) prepared carboxylated CNTs and incorporated these into the membrane to enhance the desalination efficacy. Carboxylation significantly improved the polarity, enhancing interactions between CNTs and water vapors during the MD process. Therefore, higher efficiency than the conventional membrane was achieved for desalination with 19.2 kg/m^2 h of permeate flux and 99% salt reduction. Farahbakhsh et al. (Farahbaksh et al. 2017) incorporated multi-walled carbon nanotubes (MWCNTs) into PA for the RO process. They compared the efficiency of membranes with and without oxidation of MWCNTs. According to the results, both membranes, i.e. with oxidized MWCNTs and non-oxidized MWCNTs, showed improved efficiency for saline solution flux. On the other hand, the membrane with oxidized MWCNTs showed a higher flux of 28.9 L/m^2 h, whereas the salt rejection was almost the same ($> 96\%$) for both membranes indicating a stable RO process. Likewise, Zarrabi et al. (Zarrabi et al. 2016) modified TFC/nanofiltration (TFC/NF) membrane with NH_2-modified MWCNT (NH_2-MWCNT) to improve hydrophilicity. These NH_2-MWCNT incorporated membranes improved the desalination efficiency with 95.72% rejection for Na_2SO_4 and 36.17% for NaCl. Moreover, the NH_2-MWCNT-modified membrane presented higher anti-fouling strength than the unmodified membrane.

To achieve various beneficial characteristics of a single membrane, researchers modified the membranes by incorporating two or more different nanomaterials, which rendered expected functionalities in terms of surface charge, hydrophilicity, and surface morphology. Wan Azelee et al. (Wan Azelee et al. 2017) developed a membrane by incorporating MWCNT and titania nanotubes (MWCNT-TNT). The incorporation of MWCNT-TNT provided the membrane with additional water channels that resulted in 57% higher water permeability than the non-modified membrane while maintaining a high rejection of 97.97% for NaCl and 98.07% for Na_2SO_4, attributing to the increased hydrophilicity and negative charge on the surface of the membrane. Likewise, Falath et al. (Falath et al. 2016) incorporated MWCNTs and Pluronic F-127 into poly(vinyl alcohol)/ diglycidyl ether (PVA/DGEBA) membranes for the RO process. This modification resulted in higher mechanical strength, hydrophilicity, salt rejection, surface roughness, and resistance against biofouling and chlorine, giving an excellent performance for the RO process. Zendehnam et al. (Zendehnam et al. 2014) formed a PVC-incorporated cation exchange membrane by incorporating MWCNT with copper (MWCNT-co-Cu). It was found that the optimization of Cu addition is necessary as the efficiency decreased in the beginning by adding additives, whereas 4 wt% was considered the optimized value giving the modified electrochemical characteristics. Many other researchers also emphasized the importance of optimization for the addition of nanomaterials as nanomaterials' concentration highly influences the membranes' efficiency (Chai et al. 2017, Ho et al. 2017, Teow et al. 2017). The surface charge density varied with changes in nanomaterial concentration. Higher surface charge density creates more conductive sights in the membrane, resulting in more flow channels giving easy passage for counter ions, ultimately increasing Donnan exclusion responsible for higher membrane potential (Hosseini et al. 2010). Like biological membranes, Li et al. (Li et al. 2016) tried to achieve higher selectivity by adding four polar charged species, i.e. NH_3^+, $CONH_2$, OH, COO^-, to CNTs. Modifying the RO membranes could attain a 13 times higher flux along with 100% water desalination without changing water conductance. Mishra and Ramaprabhu (Mishra and Ramaprabhu 2012) achieved higher efficiency for removing 73% calcium, 67% magnesium, and 70% sodium with only 1 V DC power supply using supercapacitors based on MWCNTs.

CNTs have been applied for CDI electrodes. Feng et al. (Feng et al. 2019) deposited TiO_2 nanoparticles on CNTs to attain hydrophilicity using the atomic layer deposition (ALD) method. They controlled the TiO_2 loading amount by monitoring the ALD cycles. 60 cycles were optimized to achieve enough TiO_2 loading for reasonable hydrophilicity by dropping the water contact angle to 37, which was attributed to the availability of hydroxyl groups on TiO_2 surface. Yang et al. (Yang et al. 2012) deposited manganese oxide (MnO_2) on CNTs for application as an electrode in membrane-based CDI (MCDI). They used sodium polystyrene sulfonate in a layer-by-layer deposition process to achieve better dispersion and growth of MnO_2 on CNTs. The MnO_2/CNTs resulted in 96.8% efficiency for salt removal with higher stability. A lot of research has been conducted to apply CNTs in desalination, but their recycling and reusability are still an issue of concern, especially for industrial-scale applications. Moreover, higher manufacturing costs refrain from using CNTs on a larger scale.

Graphene

Graphene, another allotrope of carbon atoms, made up of a 2D sheet with a hexagonal lattice, was found as relatively low-cost with superior characteristics for the desalination process (Qazi and Javaid 2023). Graphene has a theoretical surface area of 2600 m^2/g, double than that of powdered activated carbon (AC) (Stoller et al. 2008). Three bonds of the carbon atom are in the plane, whereas one bond is with an out-of-plane orientation. Carbon atoms are located in a sp2 hybridization, making it the most robust material. Graphene modified membranes are extensively studied for the desalination process. Figure 2 shows a schematic presentation of nanoporous graphene allowing water molecules to pass through while hindering the passage of salt ions (Teow and Mohammad 2019).

As pristine graphene has high hydrophobic characteristics, it cannot be efficiently used for desalination. Therefore, to decrease hydrophobicity, oxidation of graphene-to-graphene oxide (GO) is recommended, which introduces hydrophilic oxygen-containing groups like carbonyl, carboxyl groups, hydroxyl, etc., giving GO an asymmetric structure with excellent water transfer performance. Bhadra et al. (Bhadra et al. 2016) fabricated a distillation membrane with GO coating on the PTFE membrane. The coating of GO significantly enhanced the water flux to 97 kg/m^2. This effect was attributed to oxygen-containing polar functional groups, which enhanced the nanocapillary effect and selective adsorption. Due to nanopores' enhanced selectivity and shielding effect, water molecules could pass easily and efficiently even when the amount of salt was very high as 34,000 ppm. This research resulted in the advancement in the preparation of low-cost, stable membranes for desalination. Zahirifa et al. (Zahirifar et al. 2018) fabricated a membrane with two layers for the MD process by incorporating octadecyl amine functionalized GO (GO-ODA), which exhibited improved rejection of NaCl up to 98.3% and water flux of 16.7 kg/m^2h. This higher efficiency was suggested because of GO-ODA layer providing interconnected high surface area nanochannels accelerating water flow. Moreover, the GO-ODA layer provided low thermal conductivity and high hydrophobicity which reduced thermal diffusion, polarization, and pore wetting. Leong et al. (Leong et al. 2019) incorporated GO with PVA to fabricate a modified membrane for CDI. They found an optimized ratio of GO:10PVA with a surface area of 1067 m^2/g and 173 F/g capacitance. Due to the higher pore volume and surface area, enhanced adsorption of salt to 36.1 mg/g was attained in a 750 mg/L salt solution. The incorporation of GO with a cross-linker could increase spacing between layers. This spacing can be controlled as it is easy to monitor the cross-linking effect, which provides desired pore volume, pore size, and surface area, ultimately influencing the membrane's desalination efficiency. Qian et al. (Qian et al. 2018) incorporated GO with 1,4-cyclohexyl diamine (CDA). The incorporation of CDA influenced the interlayer spacing and improved hydrophilicity and water flux to 20.1 kg/m^2h with 99.9% rejection for salt ions. Likewise, to develop a defect-free dense layer, Feng et al. (Feng et al. 2016) applied another cross linking agent, i.e., 1,4-phenylene diisocyanate (PDI). Incorporating GO into PDI increased the interlayer spacing which reduced the

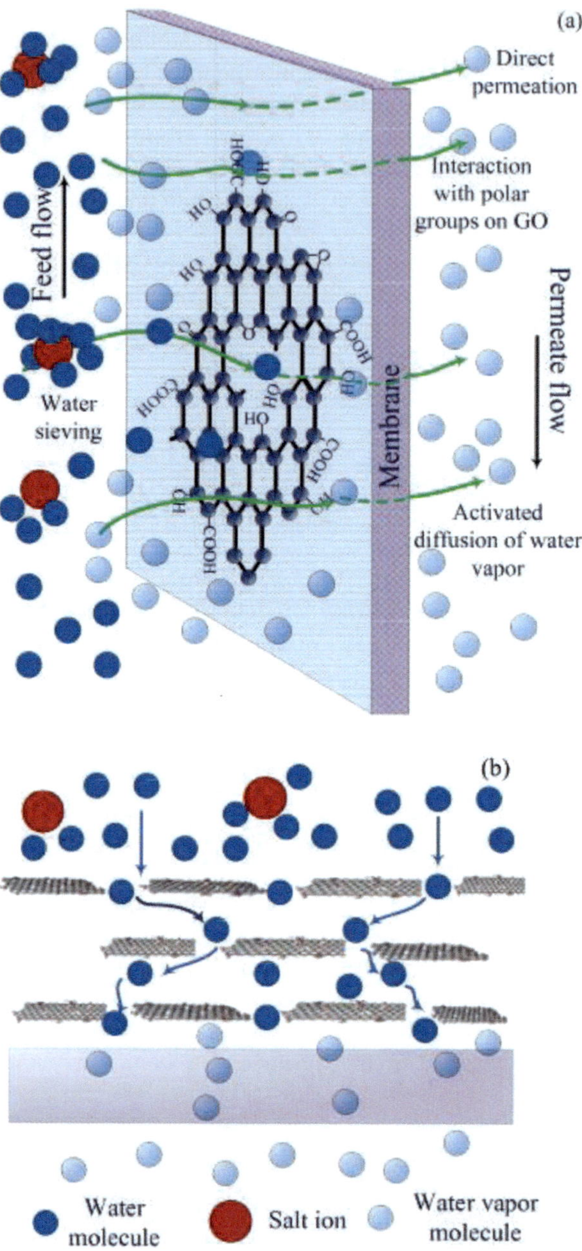

Figure 2. Schematic presentation of nanoporous graphene allowing water molecules to pass through while hindering the passage of salt ions. Reprinted from (Bhadra et al. 2016).

resistance for mass transfer. Hence, with a film thickness of 18 mm, an improved water flux of 11.4 kg/m^2 h was obtained at 90°C with 99.9% salt rejection. The above studies indicated the pivotal role of cross-linking agents for GO-based membranes.

Reduced graphene oxide (RGO) can be prepared by partially removing oxygen-containing groups. RGO has excellent thermal stability, mechanical properties, hydrophobicity, and high surface area (Qazi et al. 2022). Zhu et al. (Zhu et al. 2017) modified the membrane by adding RGO and poly(sodium-p-styrene sulfonate) on PAN films. They also immobilized halloysite nanotubes (HNTs) to facilitate an additional route to water, increasing the permeability rate of water to 8.8 L/m^2 h bar. In addition, the modified membrane showed good anti-fouling strength.

Li et al. (Li et al. 2018) applied a photocatalytic reduction to synthesize hybrid films of RGO/g-C_3N_4. Incorporating C_3N_4 partially removed the oxygen functional groups in GO resulting in RGO with reduced interlayer spacing. The degree of reduction of GO could be monitored to achieve efficient desalination performance. Similarly, Abdel-Karim et al. (Abdel-Karim et al. 2019) prepared RGO-based PVDF membranes for the MD process with various compositions as 36%, 58%, 65%, and 69%. The addition of RGO significantly increased the distillation efficiency of the composite membrane. Abdel-Karim et al. (Abdel-Karim et al. 2019) found an optimized RGO composition of 58% for the highest desalination efficiency with a salt rejection of > 99.99% and 7.0 L/m² h water flux, almost 1.7 times more efficient than the PVDF without RGO. Bharath et al. (Bharath et al. 2020) incorporated Mn_3O_4 with RGO to prepare electrodes. Incorporation of Mn_3O_4 with RGO introduced gradual pores (0.268–0.385 cm³/g pore volume) and a high surface area of 160 m²/g resulting in increased capacitance of 437 F/g. An adsorption capacity of 34.5 mg/g was attained with 1000 ppm NaCl under 1.2 V. Moreover, a higher efficiency of 98.5% could be acquired after 20 cycles.

Introducing nanopores in GO can improve the water flux, anti-fouling strength, and selectivity. Chen et al. (Chen et al. 2021) prepared a very thin GO film with controlled nanopores by incorporating with polymer on a ceramic substrate. This nanoporous-modified membrane showed good molecular sieve efficiency with an improved water flux of 472.3 ± 14.2 L/m² h and NaCl rejection efficiency as high as 99.99%. Hosseini et al. (Hosseini et al. 2018) used molecular dynamics simulation to fabricate nanoporous GO membranes. The nanoporous membranes with pore size 0.29–0.45 nm showed higher salt rejection of > 89% at pressures below 50 MPa. The modified membranes showed 2–5 times higher water flux than available RO membranes. Hosseini et al. (Hosseini et al. 2019) compared the desalination efficiency of fluorinated nanoporous GO (f-NPGO) and fluorinated nanoporous graphene (f-NPG). f-NPGO showed a significantly higher water permeability than that of f-NPG because of the presence of hydrophilic groups on f-NPGO surface. Moreover, f-NPGO efficiently rejected ions, especially chlorides giving a minimum salt rejection of 94.31%. Dianbudiyanto and Liu (Dianbudiyanto and Liu 2019) fabricated a hierarchical three-dimensional porous graphene (3DG) electrode using microwaves and hydrothermal-assisted methods. 3DG introduced a porous structure, large surface area, and high conductivity to the electrode reducing the resistance for charge transfer. Hence, 3DG showed an excellent capacitance and electrosorption capacity of 190 F/g and 21.58 mg/g, respectively. Though graphene-based membranes are easily modified according to their applications and are highly efficient for desalination, large-scale applicability is still impossible because of their high manufacturing cost.

Zeolites

Zeolites are crystalline aluminosilicates with regular intra-crystalline channels and cavities, allowing easy passage of the molecules and ions. Due to a regular porosity pattern, zeolites have specific surface selectivity, adsorption, ion exchange, and sieving properties, making them excellent materials for desalination (Dahe et al. 2012, Kim et al. 2013, Javaid et al. 2015, Ikhlaq et al. 2023). Zeolites have been used in various material science-related applications (Ikhlaq et al. 2021, 2022, Qazi et al. 2022), specifically in membrane-based technologies, because of their mechanical, thermal, and chemical strength and higher efficacy than conventional technologies (Xue et al. 2014, Rozhkovskaya et al. 2021). Recently, Lahnafi et al. (Lahnafi et al. 2022) developed nano-composite membranes based on zeolite Y (m-FAU) and zeolite A (m-LTA). The results showed that the membrane based on zeolite-Y was more efficient than the zeolite A-based modified membrane. Wajima et al. (Wajima et al. 2010) made changes to the natural zeolites (clinoptilolite) by heating at 60°C for 12 h, followed by treatment with silver nitrate ($AgNO_3$). They applied modified zeolites for NaCl removal from seawater resulting in 4.3% NaCl removal in single adsorption. In another study (Wibowo et al. 2017), natural zeolites from Indonesia were modified by heating at 225°C for 3 h and applied to reduce the salinity of seawater. The modification of clinoptilolite reduced salinity by up to 4.8%, which is comparable efficacy achieved by sorbent materials. Hence, the literature survey shows that natural zeolites can be potentially applicable as low-cost, efficient sorbent materials for desalination. Dong et al. (Dong et al. 2015) introduced NaY zeolite (0.15 wt%) on the TFC membrane for RO. They reported an increased water flux from 23.3 gal/ft^2 day to 43.7 gal/ft^2 day and 98.8% salt rejection for NaCl solution (16,000 ppm). Fathizadeh et al. (Fathizadeh et al. 2011) prepared PA TFC-based membrane by incorporating NaX zeolite (0.2%) and attained 1.8 times higher flux than that obtained with a non-modified PA membrane. Incorporating NaX zeolite into the PA membrane improved the surface characteristics by enhancing roughness, solid–liquid interface, and contact angle. Liu et al. used MFI zeolite membranes (Liu et al. 2008) and attained higher rejection efficiency (96.5%) for electrolytes with 0.33 L/m^2 h flux. In contrast, the rejection efficiency for non-electrolytes depended on the molecular dynamic size giving a separation efficiency of 99.5% for toluene and 17% for ethanol (Liu et al. 2008). Kazemimoghadam (Kazemimoghadam 2010) applied the membrane of HS zeolite to RO and evaluated the influence of experimental conditions on the efficiency of the membrane. An increase in flux was noticed within an experimental duration of 30–60 min on increasing feed rate and temperature. At optimized conditions (60°C temperature and 3 L/min feed rate), a permeate flux of 4 L/m^2 h was attained.

Much research shows applications of zeolites in ultrafiltration (UF). Liu et al. (Liu et al. 2014) modified polysulfone (PSf)-based UF membranes by adding 4A zeolite. Incorporating 4A zeolite improved the hydrophilicity, anti-fouling strength, and surface roughness with efficient nano-scale water flow and better mechanical as well as thermal stability. The modified membranes depicted enhanced water flux with rejection of protein. Han et al. (Han et al. 2009) incorporated NaA zeolite into composite UF membranes prepared from poly phthalazinone ether sulfone ketone (PPESK). Due to thermal, chemical, and mechanical stability, PPESK is commonly applied to fabricate F membranes (Han et al. 2009). PPESK has hydrophobic characteristics. The incorporation of NaA into PPESK resulted in higher anti-fouling strength, hydrophilicity, and water permeability. The NaA-modified PPESK membranes were claimed for higher efficiencies than other commercially available UF membranes. He et al. (He et al. 2014) incorporated zeolite (MCM-41) into polyvinylidene difluoride (PVDF) membranes. The incorporation of zeolite enhanced the permeability from 91.2 × 10^3 to 118.9 × 10^3 L/m^2 h bar, along with an improvement in mechanical strength from 22.5 MPa to 71.75 MPa (He et al. 2014).

Zeolite-based membranes have been applied for pervaporation (PV) (Wee et al. 2008), an energy-efficient solution for membrane-based separation mainly used for those separations that are challenging to attain by conventional methods. Kunnakorn et al. (Kunnakorn et al. 2011) applied NaA zeolite-based membranes fabricated on alumina support to separate water–ethanol mixture. The PV results confirmed high stability with an efficient separation resulting in 99.5% ethanol within a duration of 130 h along with a 1.0 kg/m^2 h water flux. Swenson et al. (Swenson et al. 2012) incorporated clinoptilolites for pervaporation desalination of water samples with different salinity. They observed higher than 97.5% rejection for K$^+$ and Na$^+$ and 98.52% and 99.99% rejection for Ca^{2+} and Mg^{2+}, respectively. Cho et al. (Cho et al. 2011) applied NaA zeolite membrane to synthetic seawater and achieved a salt rejection over 99.9% with 1.9 kg/m^2 water flux at 69°C.

Hosseini et al. (Hosseini et al. 2014) used zeolites for ion membranes such as PVC and applied them to the electrodialysis. The results presented efficient electrochemical properties with higher selectivity above 88% and lower electrical resistance of less than 6.5 Ω cm^2. This outcome is very important for techniques using electro-membranes like electrodialysis for water treatment. Garofalo et al. (Garofalo et al. 2014) first fabricated a 10 cm long MFI zeolites membrane and applied it for Vacuum Membrane Distillation (VMD) for desalination. After achieving improved efficiency by modifying the membrane, Garofalo et al. (Garofalo et al. 2016) successfully scaled it up to a 30 cm long membrane and applied for VMD, attaining the higher flux (13.7 kg/m^2 h) and 94.6% rejection for 75 g/L brine solution. Significant research on zeolite synthesis, modification, and characterization led to the advancement in its applications in the desalination process with high water flux and defect-free long-term stability, whereas it is still challenging to overcome the cost of the membranes. Therefore, it is desired to develop new techniques to prepare low-cost, efficient, and stable zeolite-based membranes, especially for large-scale industrial applications.

Aquaporin (AQP)

Most separation membranes applied in desalination are made of polymeric films with controlled and modified surfaces explicitly designed for their applications. Most synthetic membranes comprise polymer sheets with various-sized membrane pores; water must be forced through the membrane with pressure. The pore-forming proteins like AQP mimic the biological functions of lipid layers (Tang et al. 2013, Teow and Mohammad 2019, Porter et al. 2020). AQP comprises six transmembrane α-helices, which are embedded in the cell membrane. The halves resemble one another and appear to repeat a sequence of nucleotides, with the carboxyl and amino ends towards the inside of the cell (Gonen and Walz 2006, Fu and Lu 2007). When conditions are ideal, AQPs create channels, only water molecules can pass while rejecting ionic and polar species. Therefore, AQP can provide the ideal characteristics for developing low-energy water desalination processes.

Bowen et al. (Bowen 2006) proposed fabricating biomimicry membranes with improved permeability and better selectivity, whereas Kumar et al. (Kumar et al. 2007) suggested the incorporation of AQP into the desalination membrane. Aquaporin-based biomimetic membranes (ABMs) can achieve relatively higher membrane permeability at 60 L/m^2 h, two times more efficient than that attained by the commercial RO membrane, along with a selectivity of around 100%. Planar ABM was first fabricated by Zhong et al. (Zhong et al. 2012) using a cellulose acetate membrane functionalized with methacrylate groups. UV-irradiated polymerization and side-rupturing of triblock copolymer (ABA) vesicles were done to produce a surface layer for nanofiltration (NF). The improved NaCl rejection of more than 30% and water permeability of 34 L/m^2 h bar were attained by the NF membrane with an ABA:AQP ratio of 50:1. The salt rejection attained in Zhong et al.'s work (Zhong et al. 2012) was equivalent to several efficient NF membranes (Homayoonfal et al. 2010, Sun et al. 2010), but the flux showed a three-fold increase. In addition, the synthesized ABM had high mechanical strength and could tolerate pressures of up to 5 bar. Li et al. (Li et al. 2015) coated the inner surface of a hollow fiber membrane with AQP-incorporated proteoliposomes, and then a non-gas-aided interfacial polymerization method created a polyamide layer on top. According to morphological analysis, the injected proteoliposomes were completely immersed in the polyamide layer while maintaining their original spherical shape. Li et al. (Li et al. 2015) fabricated and tested hollow fiber membranes with various AQP densities. The higher covering density of AQPs improved the water flux to 40 L/m^2 h at 5 bars, along with above 97.5% rejection to a 500 ppm NaCl solution higher than a commercial membrane. However, due to the extremely specific fabrication techniques, producing and stabilizing ABMs for large-scale industrial applications is challenging. Zhao et al. (Zhao et al. 2012) introduced a new approach to ABM fabrication by the interfacial polymerization method. In this method, the APQ-incorporated polysulfone was immersed in the aqueous solution of m-phenylene-diamine, which was then treated with a solution of tri-mesoyl chloride to prepare a rejection layer (Figure 3). The prepared ABM showed higher mechanical stability while maintaining flux and rejection characteristics. Additionally, compared to commercial RO membrane (BW30), permeability of Zhao et al.'s (Zhao et al. 2012) prepared ABM was almost 40% higher. Despite the growing interest in ABMs, producing them as robust and defect-free for large-scale industrial applications is challenging.

Metal-oxides

Nearly 50% of the world's desalination systems for producing drinkable water use the filter membrane technique to desalinate salty seawater. It is a commonly used method for treating water because it is simple, economical, lowers the need for chemical-based additives, has fewer phase changes and higher productivity with easy scalability, and demonstrates effective removal capabilities.

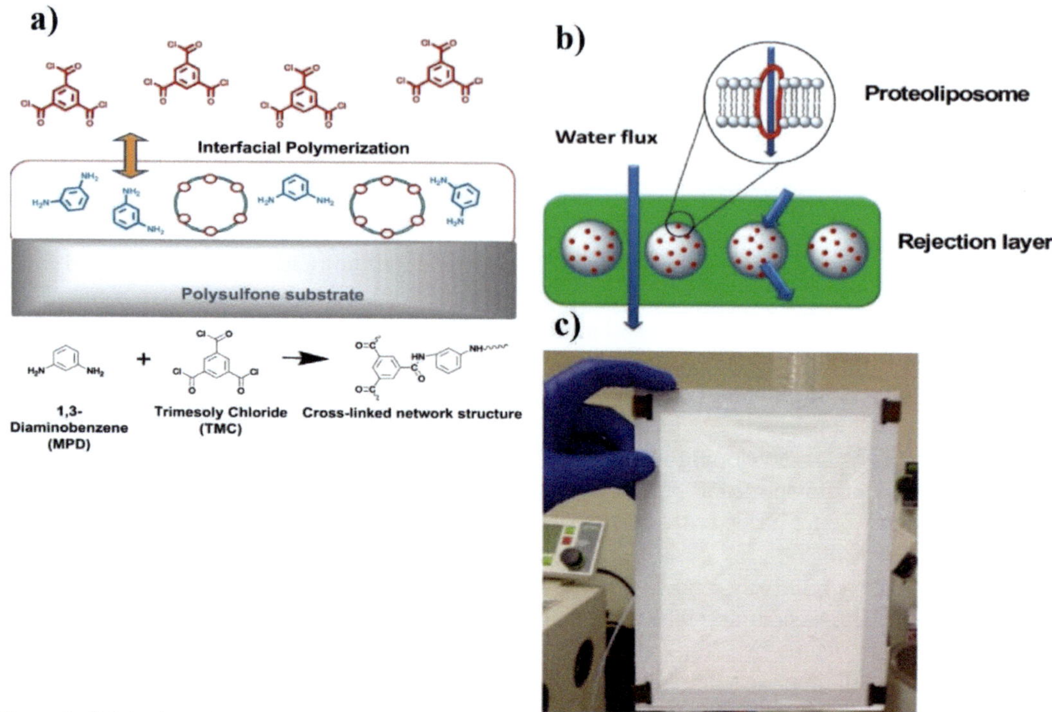

Figure 3. Fabrication of aquaporin-based biomimetic membrane (ABM): (a) schematic presentation of ABM synthesis, (b) conceptual model, (c) synthesized ABM. Reprinted from (Zhao et al. 2012).

Hydrophilic nanomaterials like CNTs, GO, silica (SiO_2), titanium dioxide (TiO_2), zeolites, and alumina (Al_2O_3) have been used successfully in the past to increase the effectiveness of TFC membranes significantly. The section summarizes the different nanomaterials used in reverse osmosis membranes.

Biofouling of membranes badly affects the efficacy of membrane handling. Silver (Ag) is well known for its anti-bacterial and antifungal characteristics. In the past few years, a lot of research is carried out on the immobilization of Ag nanomaterials on the surface of polyamide-TFC membranes to provide long-lasting antimicrobial capabilities (Yin et al. 2013). Ben-Sasson et al. (Ben-Sasson et al. 2016) incorporated Ag nanoparticles on the surface of the TFC membrane to enhance the antimicrobial strength. Research studies suggested that placing Ag nanomaterials at the interface of membrane-feed could result in better antimicrobial efficiencies by providing an instant interaction between the microbes and Ag nanoparticles (Gunawan et al. 2011). To avoid excessive leaching and release of Ag and Ag+ to the environment causing hazardous effects, Wijnhoven et al. (Wijnhoven et al. 2009) applied biochemical Ag immobilization on the surface of the membrane. Yin et al. (Yin et al. 2013) proficiently immobilized Ag nanoparticles (~ 15 nm diameter) on the outer side of the polyamide-TFC membrane by covalent linkage with the cysteamine. In contrast to the TFC membrane, Ag immobilized membrane showed comparatively greater water flux. Like Ag nanomaterials, copper (Cu) nanomaterials have also received significant attention in membrane filter technology for their bioactive properties. Cu nanomaterials are economically less expensive and abundantly available. Cu proved efficient anti-biofouling agent in the water treatment processes. Cu nanomaterials have also been applied to membrane technologies (Zareei and Hosseini 2019). Ben-Sasson et al. (Ben-Sasson et al. 2014) immobilized Cu nanomaterials with polyamide to cover the membrane's surface embedded with a pristine thin film.

TiO_2 is an excellent coating material because of its nontoxicity, photocatalytic activity, stability, and less expensiveness (Madaeni and Ghaemi 2007, Lee et al. 2016, Sirohi et al. 2023).

Shafiq et al. (Shafiq et al. 2018) coated CA/PEG membranes used for MD and RO with various TiO_2 coatings. The modified membrane with 15 wt% TiO_2 was the optimal concentration for CA/PEG membranes giving a higher salt rejection of 95.4%. Madaeni and Ghaemi (Madaeni and Ghaemi 2007) developed TiO_2 modified membrane for RO using ultraviolet (UV) irradiation. Incorporating TiO_2 into TFC increased the hydrophilic property of the membranes, which significantly influenced the water flux along and biofouling resistance of the membranes. Madaeni and Ghaemi (Madaeni and Ghaemi 2007) observed an increased self-cleaning and maximum flux of 50 L/m^2 h and 60 L/m^2 h after immobilizing 0.001 wt% and 0.003 wt% TiO_2, respectively. A further increase in TiO_2 loading to 0.005 wt% decreased the efficiency to 28 L/m^2 h because an excessive amount of TiO_2 particles blocked the membrane's pores. Due to hydrophilic and photocatalytic characteristics, TiO_2 nanoparticles were immobilized onto virgin membranes to optimize RO and MD processing. Shafiq et al. (Shafiq et al. 2018) used higher TiO_2 loading of 15 wt% to achieve enhanced salt rejection of 95.4% because they did not apply UV radiations to motivate hydrophilic and photocatalytic properties of TiO_2. In contrast, Madaeni and Ghaemi (Madaeni and Ghaemi 2007) attained a higher flux of 60 L/m^2 h with the only implication of 0.003 wt% TiO_2 loading as they applied UV radiation to optimize the hydrophilic and photocatalytic characteristics of the TiO_2-immobilized membranes. TiO_2-incorporated membranes under UV irradiation showed better self-cleaning properties (Emami et al. 2018, Stan et al. 2019). Kwak et al. (Kwak et al. 2001) incorporated pristine polyamide (PA) membranes with TiO_2-modified TFC to enhance anti-fouling strength. The enhanced anti-fouling characteristic was attributed to the photocatalytic activity of TiO_2. Safarpour et al. (Safarpour et al. 2015) developed a TFC membrane for RO by interfacial polymerization of trimesoyl chloride monomers and m-phenylenediamine followed by immobilization of TiO_2/reduced graphene oxide (TiO_2/rGO) on PA. Safarpour et al. (Safarpour et al. 2015) observed enhanced hydrophilicity of their developed membranes due to the presence of negatively charged hydrophilic functional groups such as hydroxyl, carboxyl, carbonyl, and epoxide groups on the membrane's surface. Ren et al. (Ren et al. 2017) developed a highly hydrophobic PVDF electrospun nanofiber membrane (ENM) immobilized with TiO_2 nanoparticles by hydrothermal technique followed by fluorosilanization to enhance its efficiency for direct contact membrane distillation (DCMD). This modified membrane showed a higher flux and salt rejection of 73.4 L/m^2 h and 99.99%, respectively, for 3.5 wt% NaCl solution.

SiO_2 has also been applied for the development of modified TFC membranes. Yin et al. (Yin et al. 2012) developed a TFC membrane for RO incorporating SiO_2 nanoparticles using an interfacial polymerization technique. Enhanced hydrophilicity increased the diffusion of water across the membrane, as shown in Figure 4.

Sabir et al. (Sabir et al. 2016) used a thermally induced phase inversion separation process to develop a polymer matrix (PM) membrane immobilized with SiO_2 nanoparticles for RO. The modified PM membrane showed an enhanced salt rejection with an improved flux of 2.39 L/m^2 h and better thermal stability. Ahmad et al. (Ahmad et al. 2015) studied the influence of SiO_2 nanoparticles in RO using CA/PEG membranes. The researchers noticed an improved mechanical and thermal stability for the SiO_2 immobilized CA/PEG membrane with enhanced flux and salt rejection of 2.46 L/m^2 h and 11.41%, respectively. Moreover, the modified membrane was highly resistant to fouling during RO. Pang and Zhang (Pang and Zhang 2018) developed a hydrophobic fluorinated SiO_2 nanoparticle-based TFC membrane to treat samples with a relatively higher salt content of 2000 ppm. The immobilization of SiO_2 enhanced the salt rejection to 98.6%. Lin et al. (Lin et al. 2014) modified a glass fiber membrane by immobilizing SiO_2 nanoparticles which was further treated for fluorination and coating of polymer to fabricate an omniphobic membrane for the treatment of sodium dodecyl sulfate (SDS) solution via DCMD. The SiO_2 nanoparticles incorporated modified amphiphobic membranes, which showed higher efficiency for various feed solutions.

Among various types of metal oxide-based nanomaterials used for membrane modification, zinc oxide (ZnO) nanoparticles can be excellent options as they have low cost; high thermal, physical, mechanical, and chemical stability; high surface area; and excellent antimicrobial and anticorrosive

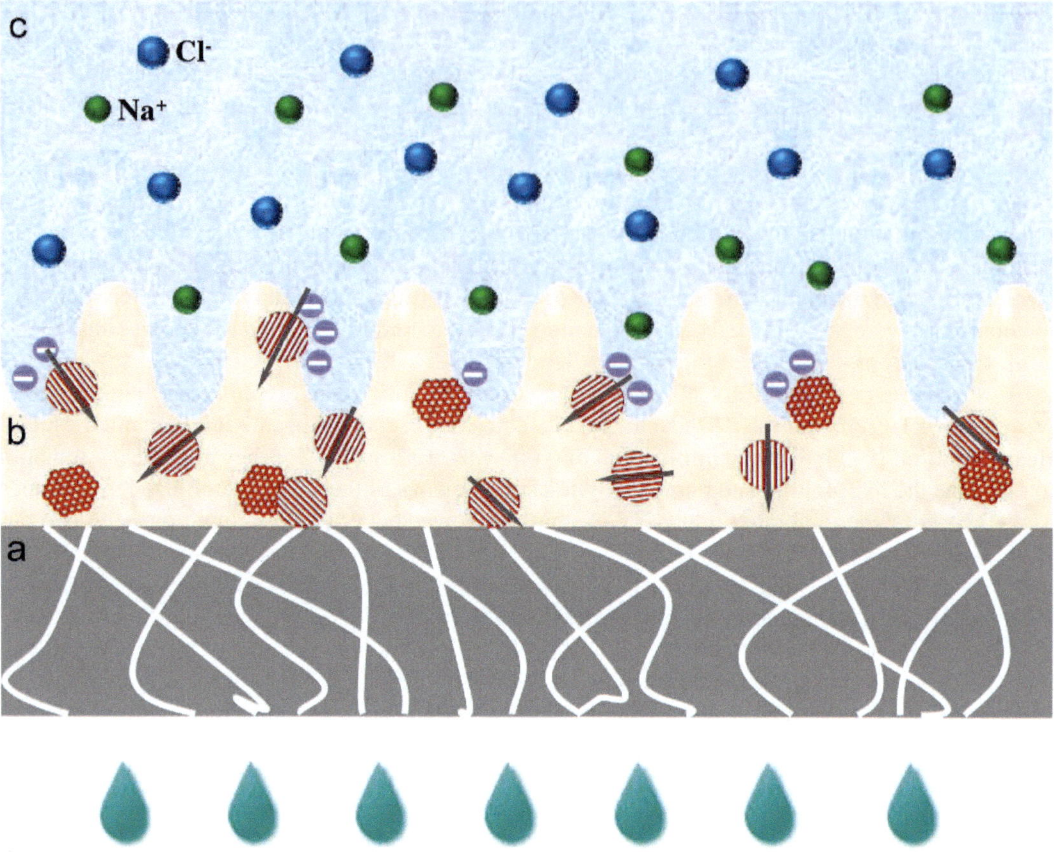

Figure 4. A schematic presentation of the hypothesized mechanism of MCM-41 (SiO$_2$) nanoparticle-enhanced TFC membrane. (a) presents porous PSU support, (b) thin film of PA with MCM-41 SiO$_2$ (red particles), (c) feed solution of NaCl. Reprinted from (Yin et al. 2012).

characteristics (Mahmoudi et al. 2018, Nasrollahi, Aber et al. 2018, Nasrollahi, Vatanpour et al. 2018). The immobilization of ZnO can enhance hydrophilicity by increasing permeability and resistance against fouling (Nasrollahi et al. 2018). Anitha et al. (Anitha et al. 2012) immobilized ZnO nanoparticles to the CA membrane to improve anti-bacterial resistance against biofouling. Huang et al. (Huang et al. 2017) fabricated ZnO immobilized PTFE membrane for self-cleaning during vacuum membrane distillation (VMD). The ZnO immobilized modified membrane with a higher surface area (22.6 m^2/g) showed higher thermal and chemical stability and an efficient salt rejection of 99.9%. Chen et al. (Chen et al. 2018) applied ZnO nanoparticles to modify a glass fiber membrane to attain an omniphobic membrane for DCMD. The modified omniphobic membrane showed high resistance against wetting with 30 L/m^2 h flux and 99.99% salt rejection. Hence, the literature survey indicates that immobilizing metal oxide nanoparticles can fabricate improved membranes. Due to the enhanced interaction of water to the membrane surface, these modified membranes provide better water permeability and salt rejection. Still, it is required to establish efficient technologies to modify membranes with metal nanoparticles to attain stable antimicrobial activities while maintaining higher water flux and ionic rejection.

Conclusion

This chapter reviewed various nanomaterials applied in desalination technology. Nanotechnology introduced revolutionary and novel water desalination techniques to overcome the freshwater shortage. Due to their selective ionic molecular sieving abilities, mechanical stability, and ultrafast permeation, nanomaterials have been used in a variety of ways to improve desalination efficiency, including ion exchange, CDI, distillation, filtration, adsorption, electrodialysis, and pervaporation. Incorporating nanomaterials in desalination is a key factor in preventing future water shortages. The application of nanomaterials in desalination has brought a new horizon for developing improved membranes with higher efficiency and stability. Due to the high costs of synthesizing nanomaterials, their use in commercial applications has been restricted. In addition, accidental leakage of nanomaterials can pose risks to the environment and public health. These associated concerns of nanomaterials must be thoroughly studied, and appraised, and appropriate management must be strengthened. Further developments and conclusive results may be seen in the future, particularly in economic and environmental sustainability.

References

Abdel-Karim, A., Luque-Alled, J.M., Leaper, S., Alberto, M., Fan, X., Vijayaraghavan, A. et al. 2019. PVDF membranes containing reduced graphene oxide: Effect of degree of reduction on membrane distillation performance. Desalination, 452 (November 2018): 196–207.

Ahmad, A., Waheed, S., Khan, S.M., e-Gul, S., Shafiq, M., Farooq, M. et al. 2015. Effect of silica on the properties of cellulose acetate/polyethylene glycol membranes for reverse osmosis. Desalination, 355: 1–10.

Anitha, S., Brabu, B., Thiruvadigal, D.J., Gopalakrishnan, C., and Natarajan, T.S. 2012. Optical, bactericidal and water repellent properties of electrospun nano-composite membranes of cellulose acetate and ZnO. Carbohydrate Polymers, 87(2): 1065–1072.

Ben-Sasson, M., Lu, X., Nejati, S., Jaramillo, H., and Elimelech, M. 2016. *In situ* surface functionalization of reverse osmosis membranes with biocidal copper nanoparticles. Desalination, 388: 1–8.

Ben-Sasson, M., Zodrow, K.R., Genggeng, Q., Kang, Y., Giannelis, E.P., and Elimelech, M. 2014. Surface functionalization of thin-film composite membranes with copper nanoparticles for antimicrobial surface properties. Environmental Science and Technology, 48(1): 384–393.

Bhadra, M., Roy, S., and Mitra, S. 2013. Enhanced desalination using carboxylated carbon nanotube immobilized membranes. Separation and Purification Technology, 120: 373–377.

Bhadra, M., Roy, S., and Mitra, S. 2016. Desalination across a graphene oxide membrane via direct contact membrane distillation. Desalination, 378: 37–43.

Bharath, G., Arora, N., Hai, A., Banat, F., Savariraj, D., Taher, H. et al. 2020. Synthesis of hierarchical Mn3O4 nanowires on reduced graphene oxide nanoarchitecture as effective pseudocapacitive electrodes for capacitive desalination application. Electrochimica Acta, 337: 135668.

Bowen, W.R. 2006. Biomimetic separations - learning from the early development of biological membranes. Desalination, 199(1–3): 225–227.

Chai, P.V., Mahmoudi, E., Teow, Y.H., and Mohammad, A.W. 2017. Preparation of novel polysulfone-Fe3O4/GO mixed-matrix membrane for humic acid rejection. Journal of Water Process Engineering, 15: 83–88.

Chen, L.H., Huang, A., Chen, Y.R., Chen, C.H., Hsu, C.C., Tsai, F.Y. et al. 2018. Omniphobic membranes for direct contact membrane distillation: Effective deposition of zinc oxide nanoparticles. Desalination, 428 (November 2017): 255–263.

Chen, X., Zhu, Y.B., Yu, H., Liu, J.Z., Easton, C.D., Wang, Z. et al. 2021. Ultrafast water evaporation through graphene membranes with subnanometer pores for desalination. Journal of Membrane Science, 621 (November 2020): 118934.

Cho, C.H., Oh, K.Y., Kim, S.K., Yeo, J.G., and Sharma, P. 2011. Pervaporative seawater desalination using NaA zeolite membrane: Mechanisms of high water flux and high salt rejection. Journal of Membrane Science, 371 (1-2): 226–238.

Daer, S., Kharraz, J., Giwa, A., and Hasan, S.W. 2015. Recent applications of nanomaterials in water desalination: A critical review and future opportunities. Desalination, 367: 37–48.

Dahe, G.J., Teotia, R.S., and Bellare, J.R. 2012. The role of zeolite nanoparticles additive on morphology, mechanical properties and performance of polysulfone hollow fiber membranes. Chemical Engineering Journal, 197: 398–406.

Dehghani, M.H., Mostofi, M., Alimohammadi, M., McKay, G., Yetilmezsoy, K., Albadarin, A.B. et al. 2016. High-performance removal of toxic phenol by single-walled and multi-walled carbon nanotubes: Kinetics, adsorption, mechanism and optimization studies. Journal of Industrial and Engineering Chemistry, 35: 63–74.

Dianbudiyanto, W., and Liu, S.H. 2019. Outstanding performance of capacitive deionization by a hierarchically porous 3D architectural graphene. Desalination, 468 (March): 114069.

Dong, H., Zhao, L., Zhang, L., Chen, H., Gao, C., and Winston Ho, W.S. 2015. High-flux reverse osmosis membranes incorporated with NaY zeolite nanoparticles for brackish water desalination. Journal of Membrane Science, 476: 73–383.

Dumée, L., Lee, J., Sears, K., Tardy, B., Duke, M., and Gray, S. 2013. Fabrication of thin film composite poly(amide)-carbon-nanotube supported membranes for enhanced performance in osmotically driven desalination systems. Journal of Membrane Science, 427: 422–430.

Emami, F., Shekarriz, S., Shariatinia, Z., and Moridi Mahdieh, Z. 2018. Self-cleaning properties of nylon 6 fabrics treated with corona and TiO_2 nanoparticles under both ultraviolet and daylight irradiations. Fibers and Polymers, 19(5): 1014–1023.

Falath, W., Sabir, A., and Jacob, K.I. 2016. Highly improved reverse osmosis performance of novel PVA/DGEBA cross-linked membranes by incorporation of Pluronic F-127 and MWCNTs for water desalination. Desalination, 397: 53–66.

Farahbaksh, J., Delnavaz, M., and Vatanpour, V. 2017. Investigation of raw and oxidized multiwalled carbon nanotubes in fabrication of reverse osmosis polyamide membranes for improvement in desalination and antifouling properties. Desalination, 410: 1–9.

Fathizadeh, M., Aroujalian, A., and Raisi, A. 2011. Effect of added NaX nano-zeolite into polyamide as a top thin layer of membrane on water flux and salt rejection in a reverse osmosis process. Journal of Membrane Science, 375(1-2): 88–95.

Feng, B., Xu, K., and Huang, A. 2016. Covalent synthesis of three-dimensional graphene oxide framework (GOF) membrane for seawater desalination. Desalination, 394: 123–130.

Feng, J., Xiong, S., and Wang, Y. 2019. Atomic layer deposition of TiO_2 on carbon-nanotube membranes for enhanced capacitive deionization. Separation and Purification Technology, 213 (December 2018): 70–77.

Fu, D. and Lu, M. 2007. The structural basis of water permeation and proton exclusion in aquaporins (Review). Molecular Membrane Biology, 24(5-6): 366–374.

Garofalo, A., Carnevale, M.C., Donato, L., Drioli, E., Alharbi, O., Aljlil, S.A. et al. 2016. Scale-up of MFI zeolite membranes for desalination by vacuum membrane distillation. Desalination, 397: 205–212.

Garofalo, A., Donato, L., Drioli, E., Criscuoli, A., Carnevale, M.C., Alharbi, O. et al. 2014. Supported MFI zeolite membranes by cross flow filtration for water treatment. Separation and Purification Technology, 137: 28–35.

Gonen, T., and Walz, T. 2006. The structure of aquaporins. Quarterly Reviews of Biophysics, 39(4): 361–396.

Gouda, A.A., and Al Ghannam, S.M. 2016. Impregnated multiwalled carbon nanotubes as efficient sorbent for the solid phase extraction of trace amounts of heavy metal ions in food and water samples. Food Chemistry, 202: 409–416.

Gunawan, P., Guan, C., Song, X., Zhang, Q., Leong, S.S.J., Tang, C. et al. 2011. Hollow fiber membrane decorated with Ag/MWNTs : Toward ffective Water. *ACS Nano*, 5(12): 10033–10040.

Han, R., Zhang, S., Liu, C., Wang, Y., and Jian, X. 2009. Effect of NaA zeolite particle addition on poly(phthalazinone ether sulfone ketone) composite ultrafiltration (UF) membrane performance. Journal of Membrane Science, 345(1-2): 5–12.

He, T., Zhou, W., Bahi, A., Yang, H., and Ko, F. 2014. High permeability of ultrafiltration membranes based on electrospun PVDF modified by nanosized zeolite hybrid membrane scaffolds under low pressure. Chemical Engineering Journal, 252: 327–336.

Ho, K.C., Teow, Y.H., Ang, W.L., and Mohammad, A.W. 2017. Novel GO/OMWCNTs mixed-matrix membrane with enhanced antifouling property for palm oil mill effluent treatment. Separation and Purification Technology, 177: 337–349.

Homayoonfal, M., Akbari, A., and Mehrnia, M.R. 2010. Preparation of polysulfone nanofiltration membranes by UV-assisted grafting polymerization for water softening. Desalination, 263(1–3): 217–225.

Hosseini, M., Azamat, J., and Erfan-Niya, H. 2018. Improving the performance of water desalination through ultra-permeable functionalized nanoporous graphene oxide membrane. Applied Surface Science, 427: 1000–1008.

Hosseini, M., Azamat, J., and Erfan-Niya, H. 2019. Water desalination through fluorine-functionalized nanoporous graphene oxide membranes. Materials Chemistry and Physics, 223 (September 2018): 277–286.

Hosseini, S.M., Madaeni, S.S., and Khodabakhshi, A.R. 2010. Preparation and characterization of PC/SBR heterogeneous cation exchange membrane filled with carbon nano-tubes. Journal of Membrane Science, 362(1-2): 550–559.

Hosseini, S.M., Rafiei, S., Hamidi, A.R., Moghadassi, A.R., and Madaeni, S.S. 2014. Preparation and electrochemical characterization of mixed matrix heterogeneous cation exchange membranes filled with zeolite nanoparticles: Ionic transport property in desalination. Desalination, 351: 138–144.

Huang, Q.L., Huang, Y., Xiao, C.F., You, Y.W., and Zhang, C.X. 2017. Electrospun ultrafine fibrous PTFE-supported ZnO porous membrane with self-cleaning function for vacuum membrane distillation. Journal of Membrane Science, 534 (399): 73–82.

Ihsanullah, Abbas, A., Al-Amer, A.M., Laoui, T., Al-Marri, M.J., Nasser, M.S., Khraisheh, M. et al. 2016. Heavy metal removal from aqueous solution by advanced carbon nanotubes: Critical review of adsorption applications. Separation and Purification Technology, 157: 141–161.

Ikhlaq, A., Fatima, R., Yaqub Qazi, U., Javaid, R., Akram, A., Shamsah, S.I. et al. 2021. Combined iron-loaded zeolites and ozone-based process for the purification of drinking water in a novel hybrid reactor: removal of faecal coliforms and arsenic. Catalysts 2021, 11(3): 373.

Ikhlaq, A., Hussain, A., Gilani, S.R., Qazi, U.Y., Akram, A., Al-Sodani, K.A.A. et al. 2023. Synergistically ozone and Fe-zeolite based catalytic purification of milk from heavy metals and pathogens. International Journal of Environmental Science and Technology.

Ikhlaq, A., Javaid, R., Akram, A., Yaqub Qazi, U., Erfan, J., Madkour, M. et al. 2022. Application of attapulgite clay-based Fe-zeolite 5A in uv-assisted catalytic ozonation for the removal of ciprofloxacin. Journal of Chemistry, 2022(5): 1–10.

Javaid, R., Kawanami, H., Chatterjee, M., Ishizaka, T., Suzuki, A., and Suzuki, T.M. 2010. Fabrication of microtubular reactors coated with thin catalytic layer (M=Pd, Pd-Cu, Pt, Rh, Au). Catalysis Communications, 11(14): 1160–1164.

Javaid, R., and Nanba, T. 2022. Stability of Cs/Ru/MgO catalyst for ammonia synthesis as a hydrogen and energy carrier. Energies, 15(10): 3506.

Javaid, R., Urata, K., Furukawa, S., and Komatsu, T. 2015. Factors affecting coke formation on H-ZSM-5 in naphtha cracking. *Applied Catalysis A: General*, 491: 100–105.

Javaid, R., and Yaqub Qazi, U. 2019. Catalytic Oxidation process for the degradation of synthetic dyes: an overview. International Journal of Environmental Research and Public Health, 16(11): 2066.

Javaid, R., Yaqub Qazi, U., Ikhlaq, A., Zahid, M., and Alazmi, A. 2021. Subcritical and supercritical water oxidation for dye decomposition. Journal of Environmental Management, 290: 112605.

Kazemimoghadam, M. 2010. New nanopore zeolite membranes for water treatment. Desalination, 251(1–3): 176–180.

Khalid, A., Al-Juhani, A.A., Al-Hamouz, O.C., Laoui, T., Khan, Z., and Atieh, M.A. 2015. Preparation and properties of nanocomposite polysulfone/multi-walled carbon nanotubes membranes for desalination. Desalination, 367: 134–144.

Kim, S.G., Hyeon, D.H., Chun, J.H., Chun, B.H., and Kim, S.H. 2013. Nanocomposite poly(arylene ether sulfone) reverse osmosis membrane containing functional zeolite nanoparticles for seawater desalination. Journal of Membrane Science, 443: 10–18.

Kumar, M., Grzelakowski, M., Zilles, J., Clark, M., and Meier, W. 2007. Highly permeable polymeric membranes based on the incorporation of the functional water channel protein Aquaporin Z. Proceedings of the National Academy of Sciences of the United States of America, 104(52): 20719–20724.

Kunnakorn, D., Rirksomboon, T., Aungkavattana, P., Kuanchertchoo, N., Atong, D., Kulprathipanja, S. et al. 2011. Performance of sodium A zeolite membranes synthesized via microwave and autoclave techniques for water-ethanol separation: Recycle-continuous pervaporation process. Desalination, 269 (1–3): 78–83.

Kwak, Seung-Yeop, Kim, Sung Ho, and Kim, S.S. 2001. Hybrid Organic/Inorganic Reverse Osmosis (RO) Membrane for preparation and characterization of TiO_2 nanoparticle self-assembled aromatic polyamide membrane. Environ. Sci. Technol., 35(11): 2388–2394.

Lahnafi, A., Elgamouz, A., Tijani, N., Jaber, L., and Kawde, A.N. 2022. Hydrothermal synthesis and electrochemical characterization of novel zeolite membranes supported on flat porous clay-based microfiltration system and its application of heavy metals removal of synthetic wastewaters. Microporous and Mesoporous Materials, 334 (October 2021): 111778.

Lee, E.J., An, A.K., He, T., Woo, Y.C., and Shon, H.K. 2016. Electrospun nanofiber membranes incorporating fluorosilane-coated TiO_2 nanocomposite for direct contact membrane distillation. Journal of Membrane Science, 520: 145–154.

Lee, K.P., Arnot, T.C., and Mattia, D. 2011. A review of reverse osmosis membrane materials for desalination-Development to date and future potential. Journal of Membrane Science, 370(1-2): 1–22.

Leong, Z.Y., Lu, G., and Yang, H.Y. 2019. Three-dimensional graphene oxide and polyvinyl alcohol composites as structured activated carbons for capacitive desalination. Desalination, 451 (May 2017): 172–181.

Li, Q., Yang, D., Shi, J., Xu, X., Yan, S., and Liu, Q. 2016. Biomimetic modification of large diameter carbon nanotubes and the desalination behavior of its reverse osmosis membrane. Desalination, 379: 164–171.

Li, X., Chou, S., Wang, R., Shi, L., Fang, W., Chaitra, G. et al. 2015. Nature gives the best solution for desalination: Aquaporin-based hollow fiber composite membrane with superior performance. Journal of Membrane Science, 494: 68–77.

Li, Z., Xing, Y., Fan, X., Lin, L., Meng, A., and Li, Q. 2018. rGO/protonated g-C3N4 hybrid membranes fabricated by photocatalytic reduction for the enhanced water desalination. Desalination, 443 (June): 130–136.

Lin, S., Nejati, S., Boo, C., Hu, Y., Osuji, C.O., and Elimelech, M. 2014. Omniphobic Membrane for Robust Membrane Distillation. Environmental Science & Technology Letters, 1(11): 443–447.

Liu, F., Ma, B.R., Zhou, D., Xiang, Y.H., and Xue, L.X. 2014. Breaking through tradeoff of Polysulfone ultrafiltration membranes by zeolite 4A. Microporous and Mesoporous Materials, 186: 113–120.

Liu, N., Li, L., McPherson, B., and Lee, R. 2008. Removal of organics from produced water by reverse osmosis using MFI-type zeolite membranes. Journal of Membrane Science, 325(1): 357–361.

Madaeni, S.S., and Ghaemi, N. 2007. Characterization of self-cleaning RO membranes coated with TiO_2 particles under UV irradiation. Journal of Membrane Science, 303(1-2): 221–233.

Mahmoudi, E., Ng, L.Y., Mohammad, A.W., Ba-Abbad, M.M., and Razzaz, Z. 2018. Enhancement of polysulfone membrane with integrated ZnO nanoparticles for the clarification of sweetwater. International Journal of Environmental Science and Technology, 15(3): 561–570.

Mezher, T., Fath, H., Abbas, Z., and Khaled, A. 2011. Techno-economic assessment and environmental impacts of desalination technologies. Desalination, 266(1–3): 263–273.

Misdan, N., Lau, W.J., and Ismail, A.F. 2012. Seawater Reverse Osmosis (SWRO) desalination by thin-film composite membrane-Current development, challenges and future prospects. Desalination, 287: 228–237.

Mishra, A.K., and Ramaprabhu, S. 2012. The role of functionalised multiwalled carbon nanotubes based supercapacitor for arsenic removal and desalination of sea water. Journal of Experimental Nanoscience, 7(1): 85–97.

Nair, M., and Kumar, D. 2013. Water desalination and challenges: The Middle East perspective: a review. Desalination and water treatment, 51(10–12): 2030–2040.

Nasrollahi, N., Aber, S., Vatanpour, V., and Mahmoodi, N.M. 2018. The effect of amine functionalization of CuO and ZnO nanoparticles used as additives on the morphology and the permeation properties of polyethersulfone ultrafiltration nanocomposite membranes. Composites Part B: Engineering, 154 (March): 388–409.

Nasrollahi, N., Vatanpour, V., Aber, S., and Mahmoodi, N.M. 2018. Preparation and characterization of a novel polyethersulfone (PES) ultrafiltration membrane modified with a CuO/ZnO nanocomposite to improve permeability and antifouling properties. Separation and Purification Technology, 192 (June 2017): 369–382.

Pang, R., and Zhang, K. 2018. Fabrication of hydrophobic fluorinated silica-polyamide thin film nanocomposite reverse osmosis membranes with dramatically improved salt rejection. Journal of Colloid and Interface Science, 510: 127–132.

Porter, C.J., Werber, J.R., Zhong, M., Wilson, C.J., and Elimelech, M. 2020. Pathways and challenges for biomimetic desalination membranes with sub-nanometer channels. ACS Nano, 14(9): 10894–10916.

Qazi, U.Y., Iftikhar, R., Ikhlaq, A., Riaz, I., Jaleel, R., Nusrat, R. et al. 2022. Application of Fe-RGO for the removal of dyes by catalytic ozonation process. Environmental Science and Pollution Research.

Qazi, U.Y., and Javaid, R. 2023. Graphene utilization for efficient energy storage and potential applications: challenges and future implementations. Energies, 16: 2927.

Qazi, U.Y., Javaid, R., Ikhlaq, A., Khoja, A.H., and Saleem, F. 2022. A comprehensive review on zeolite chemistry for catalytic conversion of biomass/waste into green fuels. Molecules, 27(23).

Qian, Y., Zhou, C., and Huang, A. 2018. Cross-linking modification with diamine monomers to enhance desalination performance of graphene oxide membranes. Carbon, 136: 28–37.

Qu, X., Brame, J., Li, Q., and Alvarez, P.J.J. 2013. Nanotechnology for a safe and sustainable water supply: Enabling integrated water treatment and reuse. Accounts of Chemical Research, 46(3): 834–843.

Ren, L., Tao, J., Chen, H., Bian, Y., Yang, X., Chen, G. et al. 2017. Myeloid differentiation protein 2-dependent mechanisms in retinal ischemia-reperfusion injury. Toxicology and Applied Pharmacology, 317: 1–11.

Rozhkovskaya, A., Rajapakse, J., and Millar, G.J. 2021. Process engineering approach to conversion of alum sludge and waste glass into zeolite LTA for water softening. Journal of Water Process Engineering, 43(April): 102177.

Sabir, A., Shafiq, M., Islam, A., Jabeen, F., Shafeeq, A., Ahmad, A. et al. 2016. Conjugation of silica nanoparticles with cellulose acetate/polyethylene glycol 300 membrane for reverse osmosis using MgSO4 solution. Carbohydrate Polymers, 136: 551–559.

Safarpour, M., Khataee, A., and Vatanpour, V. 2015. Thin film nanocomposite reverse osmosis membrane modified by reduced graphene oxide/TiO2 with improved desalination performance. Journal of Membrane Science, 489: 43–54.

Shafiq, M., Sabir, A., Islam, A., Khan, S.M., Gull, N., Hussain, S.N. et al. 2018. Cellulaose acetate based thin film nanocomposite reverse osmosis membrane incorporated with TiO_2 nanoparticles for improved performance. Carbohydrate Polymers, 186 (January): 367–376.

Sirohi, R., Kumar, Y., Madhavan, A., Sagar, N.A., Sindhu, R., Bharatiraja, B. et al. 2023. Engineered nanomaterials for water desalination: Trends and challenges. Environmental Technology & Innovation, 30: 103108.

Stan, M.S., Badea, M.A., Pircalabioru, G.G., Chifiriuc, M.C., Diamandescu, L., Dumitrescu, I. et al. 2019. Designing cotton fibers impregnated with photocatalytic graphene oxide/Fe, N-doped TiO_2 particles as prospective industrial self-cleaning and biocompatible textiles. Materials Science and Engineering C, 94(August 2018): 318–332.

Stoller, M.D., Park, S., Zhu, Y., An, J., and Ruoff, R.S. 2008. Graphene-based ultracapacitors. Nano Letters, 8: 3498–3502.

Sun, S.P., Wang, K.Y., Peng, N., Hatton, T.A., and Chung, T.S. 2010. Novel polyamide-imide/cellulose acetate dual-layer hollow fiber membranes for nanofiltration. Journal of Membrane Science, 363 (1-2): 232–242.

Swenson, P., Tanchuk, B., Gupta, A., An, W., and Kuznicki, S.M. 2012. Pervaporative desalination of water using natural zeolite membranes. Desalination, 285: 68–72.

Tang, C.Y., Zhao, Y., Wang, R., Hélix-Nielsen, C., and Fane, A.G. 2013. Desalination by biomimetic aquaporin membranes: Review of status and prospects. Desalination, 308: 34–40.

Teow, Y.H., and Mohammad, A.W. 2019. New generation nanomaterials for water desalination: A review. Desalination, 451(May 2017): 2–17.

Teow, Y.H., Ooi, B.S., and Ahmad, A.L. 2017. Study on PVDF-TiO_2 mixed-matrix membrane behaviour towards humic acid adsorption. Journal of Water Process Engineering, 15: 99–106.

Wajima, T., Shimizu, T., Yamato, T., and Ikegami, Y. 2010. Removal of NaCl from seawater using natural zeolite. Toxicological & Environmental Chemistry, 92(1): 21–26.

Wan Azelee, I., Goh, P.S., Lau, W.J., Ismail, A.F., Rezaei-DashtArzhandi, M., Wong, K.C. et al. 2017. Enhanced desalination of polyamide thin film nanocomposite incorporated with acid treated multiwalled carbon nanotube-titania nanotube hybrid. Desalination, 409: 163–170.

Wee, S.L., Tye, C.T., and Bhatia, S. 2008. Membrane separation process-pervaporation through zeolite membrane. Separation and Purification Technology, 63(3): 500–516.

Wibowo, E., Sutisna, Rokhmat, M., Murniati, R., Khairurrijal, and Abdullah, M. 2017. Utilization of natural zeolite as sorbent material for seawater desalination. Procedia Engineering, 170: 8–13.

Wijnhoven, S.W.P., Peijnenburg, W.J.G.M., Herberts, C.A., Hagens, W.I., Oomen, A.G., Heugens, E.H.W. et al. 2009. Nano-silver – a review of available data and knowledge gaps in human and environmental risk assessment. Nanotoxicology, 3(2): 109–138.

Xue, Z., Li, Z., Ma, J., Bai, X., Kang, Y., Hao, W. et al. 2014. Effective removal of Mg2+ and Ca2+ ions by mesoporous LTA zeolite. Desalination, 341(1): 10–18.

Yang, J., Zou, L., and Song, H. 2012. Preparing MnO 2/PSS/CNTs composite electrodes by layer-by-layer deposition of MnO 2 in the membrane capacitive deionisation. Desalination, 286: 108–114.

Yang, K., and Xing, B. 2009. Adsorption of fulvic acid by carbon nanotubes from water. Environmental Pollution, 157(4): 1095–1100.

Yang, K., Zhu, L., and Xing, B. 2006. Adsorption of polycyclic aromatic hydrocarbons by carbon nanomaterials. Environmental Science and Technology, 40(6): 1855–1861.

Yaqub Qazi, U., and Javaid, R. 2016. A Review on metal nanostructures: preparation methods and their potential applications. Advances in Nanoparticles, 05(01): 27–43.

Yin, J., Kim, E.-S., Yang, J., and Deng, B. 2012. Fabrication of a novel thin-film nanocomposite (TFN) membrane containing MCM-41 silica nanoparticles (NPs) for water purification. Journal of Membrane Science, 423-424: 238–246.

Yin, J., Yang, Y., Hu, Z., and Deng, B. 2013. Attachment of silver nanoparticles (AgNPs) onto thin-film composite (TFC) membranes through covalent bonding to reduce membrane biofouling. Journal of Membrane Science, 441: 73–82.

Zahirifar, J., Moosavian, S.M.A., Hadi, A., Khadiv-Parsi, P., and Karimi-Sabet, J. 2018. Fabrication of a novel octadecylamine functionalized graphene oxide/PVDF dual-layer flat sheet membrane for desalination via air gap membrane distillation. Desalination, 428 (May 2017): 227–239.

Zareei, F., and Hosseini, S.M. 2019. A new type of polyethersulfone based composite nanofiltration membrane decorated by cobalt ferrite-copper oxide nanoparticles with enhanced performance and antifouling property. Separation and Purification Technology, 226(May): 48–58.

Zarrabi, H., Yekavalangi, M.E., Vatanpour, V., Shockravi, A., and Safarpour, M. 2016. Improvement in desalination performance of thin film nanocomposite nanofiltration membrane using amine-functionalized multiwalled carbon nanotube. Desalination, 394: 83–90.

Zendehnam, A., Mokhtari, S., Hosseini, S.M., and Rabieyan, M. 2014. Fabrication of novel heterogeneous cation exchange membrane by use of synthesized carbon nanotubes-co-copper nanolayer composite nanoparticles: Characterization, performance in desalination. Desalination, 347: 86–93.

Zhao, X., Li, J., and Liu, C. 2017. A novel TFC-type FO membrane with inserted sublayer of carbon nanotube networks exhibiting the improved separation performance. Desalination, 413: 176–183.

Zhao, Y., Qiu, C., Li, X., Vararattanavech, A., Shen, W., Torres, J. et al. 2012. Synthesis of robust and high-performance aquaporin-based biomimetic membranes by interfacial polymerization-membrane preparation and RO performance characterization. Journal of Membrane Science, 423-424, 422–428.

Zhong, P.S., Chung, T.S., Jeyaseelan, K., and Armugam, A. 2012. Aquaporin-embedded biomimetic membranes for nanofiltration. Journal of Membrane Science, 407-408: 27–33.

Zhu, L., Wang, H., Bai, J., Liu, J., and Zhang, Y. 2017. A porous graphene composite membrane intercalated by halloysite nanotubes for efficient dye desalination. Desalination, 420 (March): 145–157.

5

Nanotoxicity Effects of Metal Oxide Nanoparticles on Aquatic and Soil Ecosystems

Josef Jampílek[1,2,*] and *Katarína Kráľová*[3]

Introduction

The robust development of nanotechnology, mainly since the beginning of the century, has led to an increase in the production and use of metal-based nanoparticles (NPs), including metal oxide NPs (MONPs) in various sectors of the economy. In agriculture, MONPs are used, for example, as nanopriming agents and nanofertilizers (Jampílek and Kráľová 2017, 2019a, Kráľová and Jampílek 2021a, 2022a), eventually for mitigation of adverse impact of heavy metal-induced stress in plants (Jampílek and Kráľová 2020, Kráľová and Jampílek 2021a). MONPs are also used as important technological and biomedical materials in theranostics, immunotherapy, regenerative medicine, etc. (Nikolova and Chavali 2020) and they are also applied in electronics, automotive, and construction industry, where they are used in sensors, biosensing materials in electronics and optoelectronics, in integrated circuits, batteries, solar cells and antennas. Global MONPs market was valuated as \$0.9 billion in 2020 and is predicted to reach \$1.8 billion by 2030, growing at compound annual growth rate (CAGR) of 7.3% from 2021 to 2030 (Allied Market Research 2022). As a result, the amount of MONP contaminating aquatic and terrestrial ecosystems is also gradually increasing.

Unlike bulk material and microparticles (MPs) of the same chemical composition, NPs have different physical and chemical properties due to their small size, with higher biological activity of NPs being associated with their greater surface area per mass compared to larger particles of the same chemical composition (Sukhanova et al. 2018, Jampílek et al. 2019, Kráľová et al. 2019, Saxena et al. 2020, Zein et al. 2020, Jampílek and Kráľová 2021, 2022, Mitchell et al. 2021, Bommakanti et al. 2022, Yuan et al. 2022). The toxic effect of metal NPs and MONPs can be caused either by

[1] Department of Analytical Chemistry, Faculty of Natural Sciences, Comenius University, Ilkovičova 6, 842 15 Bratislava, Slovakia.

[2] Department of Chemical Biology, Faculty of Science, Palacky University Olomouc, Šlechtitelů 27, 783 71 Olomouc, Czech Republic.

[3] Institute of Chemistry, Faculty of Natural Sciences, Comenius University, Ilkovičova 6, 842 15 Bratislava, Slovakia.

* Corresponding author: josef.jampilek@gmail.com

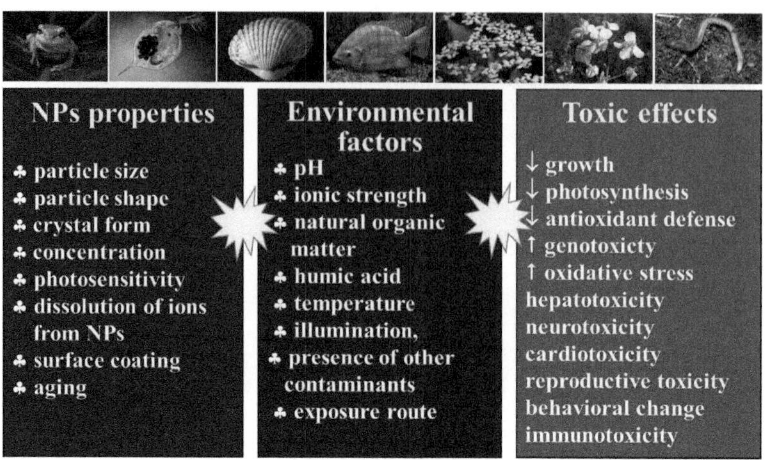

Figure 1. Properties of metal-based NPs, environmental factors affecting their toxicity and toxic effect of such on aquatic and terrestrial fauna and flora.

metal ions dissolved from the NPs or through the nanospecific effects of the NPs (Thwala et al. 2016, Peng et al. 2017, Mahana et al. 2021). The properties of metal-based NPs affecting their toxic impact on living organisms are particle size and shape, crystalline form, concentration, photosensitivity, dissolution of metal ions from NPs, surface coating and aging of NPs, while main environmental factors influencing the extent of toxic effects are pH, ionic strength, temperature, illumination, presence of natural organic matter, humic acids or other contaminants and exposure route (Figure 1). Indeed, the sensitivity of individual living organisms to definite NPs can differ each from other.

Under environmental conditions, MONPs undergo transformations that strongly influence their fate, transport, persistence, bioavailability, and toxicity (Amde et al. 2017), even though environmental transformation can be suppressed by surface functionalization of NPs affecting aggregation of NPs, their mobility and dissolution, and thereby the economic potential of NPs as well (Shelving et al. 2018, Zhang et al. 2018a). Coating of NPs with natural organic matter, humic acids or natural organic compounds such as chitosan (CS), alginate (ALG) or citric acid generally reduces their toxicity (Peng et al. 2015, Lei et al. 2018, Lopez-Moreno et al. 2018, Jonas et al. 2019). The fate of NPs in soils and toxicity impact on soil organisms, along with soil factors controlling toxicity of NPs, was discussed by Xu et al. (2022).

Metal ions, metal NPs and MONPs usually exhibit hormetic effects on photosynthesizing organisms causing growth stimulation and improve their performance at low concentrations but show dose-dependent inhibition at higher concentrations (Masarovičová et al. 2014, Kráľová and Jampílek 2021b,c, Kráľová et al. 2021). Even though some metal oxides of essential metals (e.g., Cu, Zn, Fe or Mg) are used as nutrients necessary for growth and development of living organisms, their excess in environmental matrices, whether in bulk or nanoscale form, shows adverse impact on aquatic and terrestrial flora manifested in impaired growth, inhibited photosynthetic electron transport mainly in photosystem (PS) II, reduced levels of photosynthetic pigments (chlorophyll (Chl) and carotenoids) and non-enzymatic antioxidants (e.g., glutathione (GSH), phenols, and flavonoids), suppression of the levels and activities of antioxidant enzymes (e.g., superoxide dismutase (SOD), catalase (CAT), peroxidase (POD), S-glutathione transferase (GST)), modified expression of genes and even in genotoxic effects (Masarovičová et al. 2014, Kráľová et al. 2019, 2021, Kráľová and Jampílek 2021b,c). To MONPs causing cellular toxicity and genotoxicity belong, for example, nanosized particles of Al_2O_3, CuO, Fe_2O_3, ZnO or TiO_2 (Karahalil 2021). Rajput et al. (2019) discussed structural and ultrastructural changes in NPs-exposed plants focusing on the phytotoxic impact of NPs on germination and morphometric parameters of plants as well as on cellular damages due to lipid peroxidation, protein damage, membrane destruction, suppressed transpiration rate and photosynthesis, etc., resulting in adverse impact on plant growth. Ecotoxicological impact of NPs on algae was overviewed by several researchers (Wang et al. 2019, Nguyen et al. 2020, Giri et al. 2021).

However, metal-based MPs including MONPs also adversely affect soil biota, including microbial communities needed for some biological processes and the nutrient cycling and can impair soil health and fertility (Kwak and An 2016, Parada et al. 2019, Khanna et al. 2021). For example, CuO NPs in soil modified denitrification function genes and bacterial community composition, and suppressed denitrification (Zhao et al. 2020). Changes in the relative numbers of some microbes related to N and P cycling in soil were observed at treatment with CeO_2 NPs (Li et al. 2022). Important inhibitory effects on microbial growth, their diversity in the soil and their activity after exposure to CuO NPs were also observed by Samarajeewa et al. (2021) and Salas-Leiva et al. (2021). Adverse impact on total bacteria population in alkali soil was exhibited by ZnO NPs (You et al. 2018). The negative impact of chronic contamination with TiO_2 NPs on soil microbiological functioning exceeded that of single exposure because at repeated treatment with TiO_2 NPs transport of them to deep layers of soil and groundwater was restricted (Simonin et al. 2016). However, NPs also exhibit ecotoxicological impact on organisms from soil micro- and mesofauna (Kiss et al. 2021), CuO NPs were reported to reduce population of *Eisenia fetida* (Scott-Fordsmand et al. 2022) and NPs including CuO NPs and ZnO NPs exhibited reproductive toxic effect to *Caenorhabditis elegans*, which depended on the size, charge, surface modification and shape of NPs (Yao et al. 2022). Negative impact of MONPs on *Meloidogyne incognita* second-stage juveniles (Tauseef et al. 2021a,b, Elarabi et al. 2022, Khan et al. 2022) or *Folsomia candida* (Fischer et al. 2021, 2022) were reported as well.

Various species of aquatic fauna, such as crustacean (Dogra et al. 2016, Borase et al. 2021, Li et al. 2021a, Lai et al. 2021), snails (Ramskov et al. 2015, Ibrahim et al. 2022, Caixeta et al. 2021), bivalves (Chelomin et al. 2017, Moezzi et al. 2018, Scola et al. 2021, Zhou et al. 2021) amphibians (Motta et al. 2023, do Amaral et al. 2022, Glavas et al. 2022) or fishes (Naeemi et al. 2020, Abdel-Latif et al. 2021, Shahzad et al. 2022, Rastgar et al. 2022) exposed to metal ions, metal NPs and MONPs, which generate oxidative stress, were adversely affected. Upgrowth and expansion were negatively influenced with these toxic NPs, with observed higher toxicity in certain species and life stages, and mortality in sensitive species; some MONPs have caused hepatotoxicity, neurotoxicity, cardiotoxicity, reproductive toxicity, genotoxicity, immunotoxicity, and behavioral changes (Figure 1). Similar to aquatic and terrestrial photosynthesizing organisms, oxidative stress was considered to be the main mechanism of action in species of aquatic fauna.

Environmental contamination with MONPs might pose a risk also for human health, because of their trophic transfer particularly in aquatic food webs, where aquatic organisms such as algae compose the basic trophic levels for many other organisms, including terrestrial ones (Gupta et al. 2017, Peng et al. 2017, Mahana et al. 2021, Pereira et al. 2021). Nanoscale materials can reach up to three trophic levels of the food chain, and up to two trophic levels the CeO_2 NPs showed a biomagnification factor > 1 (Gupta et al. 2017).

This chapter provides comprehensive overview of the toxic effects and relevant mechanisms of action of MONPs (CuO, ZnO Fe_2O_3, Fe_3O_4, NiO, MgO, Al_2O_3 NPs, CeO_2 NPs and TiO_2 NPs) on freshwater and marine algae and cyanobacteria, aquatic plants and aquatic animals (mainly crustaceans, snails, bivalves, amphibians or fishes), on the number and composition of soil microbial communities and the toxic impact on soil fauna, while adverse impact on terrestrial plants is briefly discussed as well.

Impact of metal oxide nanoparticles on aquatic ecosystem

Impact of metal oxides on aquatic photosynthesizing organisms

Algae

Algae represent photosynthesizing organisms; they are the primary producers living in fresh, salt, and brackish water that create around 50% of all oxygen on Earth (Chapman 2013, Thomas 2019). Higher concentrations of bulk metal-based materials as well as their nanoforms inhibit photosynthetic processes in algae via damaging function of PS II, reduction of the levels of photosynthetic pigments, and inducing oxidative stress, when excess amounts of reactive oxygen species (ROS) cause lipid peroxidation and destruction of membranes and impair activity of antioxidant enzymes, ultimately resulting in algal growth inhibition (Masarovičová et al. 2014, Jampílek and Kráľová 2019b, Kráľová et al. 2019, Kráľová and Jampílek 2021c). Impact of metal NPs on marine and freshwater algae was comprehensively overviewed by Kráľová and Jampílek (2021b). Metal-based NPs can be internalized by algal cells and are toxic to algae causing oxidative damage, although they are mostly less toxic than the corresponding metal ions. Proposed interactions of nanocolloids (NCs) such as MONPs with photosynthesizing aquatic organisms at the cellular level are shown in Figure 2. The NPs, which were internalized by cells via direct penetration or dissolution, can induce oxidative stress, mitochondrial dysfunction and DNA damage, and adhesive NPs in the membrane can adsorb contaminants in the aquatic environment and alter permeability of the membrane (Ouyang et al. 2022).

Nevertheless, the trophic transfer of metal-based NPs in aquatic food webs poses a risk for aquatic organisms (Mahana et al. 2021). Saxena and Harish (2018) discussed the effects of MONPs on algal physiology, focusing primarily on the toxic impact of TiO_2 NPs, ZnO NPs, CuO NPs, SiO_2 NPs, CeO_2 NPs Al_2O_3 NPs and iron oxide NPs reflected in the induction of oxidative stress as well as on main factors affecting toxicity of NPs, including their size and shape, pH of medium, and exposure time as well as photocatalytic activity of NPs. Low concentrations of metal-based NPs promoted microalgal growth and photosynthesis, while their high concentrations damaged photosynthetic proteins, which resulted in impaired algal growth. The toxic impact of MONPs on microalgae decreased in following order: ZnO > TiO_2 > CuO > Fe_2O_3 (Lau et al. 2022).

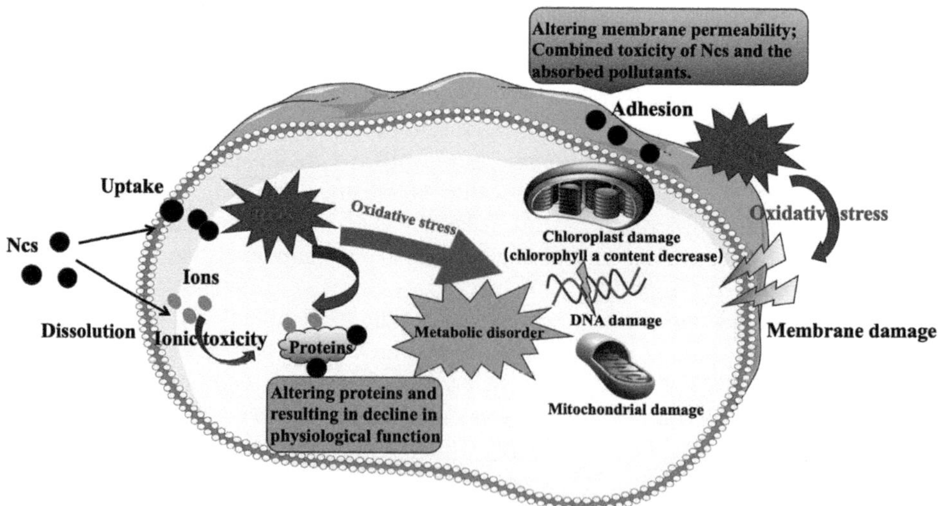

Figure 2. Proposed interactions of nanocolloids (NCs) such as MONPs with photosynthesizing aquatic organisms at the cellular level. Reprinted from Ouyang, S., Li, Y., Zheng, T., Wu, K., Wang, X., and Zhou, Q., 2022. Ecotoxicity of natural nanocolloids in aquatic environment. Water, 14: 2971. This article is an open access article distributed under the terms and conditions of the Creative Commons Attribution (CC BY) license (https://creativecommons.org/licenses/by/4.0/).

Environmental matrix (such as lake water) attenuated adverse impact of CuO NPs effects on microalga *Chlamydomonas reinhardtii* during 24 h exposure (Moos et al. 2015). CuO NPs were able to cross the cell wall of *C. reinhardtii* and be taken up into the algal cell. Their toxic impact on algae was manifested by algal growth inhibition (72-h IC_{50}: 150.45 ± 1.17 mg/L), reduced levels of carotenoids, increased ROS generation and enhanced lipid peroxidation along with changes in the activity of enzymatic antioxidants (Melegari et al. 2013). The 75-h IC_{50} value of CuO NPs related to growth inhibition *Chlorella pyrenoidosa* was reported as 45.7 mg/L by Zhao et al. (2016). At exposure to CuO NPs, the NPs were attached to the surface of algal cells, and the layer composed of extracellular polymeric substances, which was situated on the surface of algal cells to ensure protection of algae from NPs toxicity, showed approximately 4-fold thickening. However, despite such protection, the CuO NPs were internalized by cells and stored in algal vacuoles, whereby the internalized CuO NPs were transformed to Cu_2O NPs (approx. 5 nm). At exposure of algae to CuO NPs, serious membrane damage, ROS overproduction and mitochondrial depolarization was observed. CuO NPs inhibited growth of *Tetradesmus obliquus* algae (96-h IC_{50}: 169.2 mg/L; however, at co-exposure of CuO NPs with multi-walled carbon nanotubes, heteroaggregation of these contaminants reduced the contact of NPs with cell membranes resulting in lower physical damage and reduced ROS production (Fang et al. 2022). The toxic effect of MONPs on *Nannochloropsis oculata* green alga living predominantly in the marine environment decreased in following order: CuO NPs > ZnO NPs > Fe_2O_3 NPs. Treatment with more toxic CuO NPs and ZnO NPs resulted in the shrinkage of cell wall and accumulation of NPs on the surface of algal cell wall (Fazelian and Yousefzadi 2022). On investigation of a short-term (72-h) toxic impact Fe_3O_4 NPs and CuO NPs on freshwater *C. reinhardtii* and marine *Chlamydomonas euryale* algal species, it was found that ethylenediaminetetraacetic acid (EDTA) increased solubility of NPs, while salinity decreased their solubility. CuO NPs showing higher solubility were more toxic than Fe_3O_4 NPs in freshwater conditions. Formation of larger NPs agglomerates at a salinity of 32 g/L reduced their toxicity to *C. euryale*. On the other hand, EDTA addition altered the nutrient uptake of algal cells and exhibited toxic impact on algae, whereby *C. reinhardtii* were more sensitive to EDTA treatment. Co-treatment of NPs (particularly CuO NPs) with EDTA reduced toxic impact of NPs on algae as well (Canuel et al. 2021).

ZnO NPs at doses < 10 mg/L exhibited low toxicity to *Pseudokirchneriella subcapitata* microalgae; however, at higher doses, considerable toxic effects along with changes in morphology of algal cells were observed (Pereira et al. 2020). ZnO NPs applied at doses 10–200 mg/L reduced growth of *Scenedesmus opoliensis* algae and their Chl*a* and carbohydrate content in a concentration-dependent manner, while protein and lipid contents gradually increased with increasing ZnO NPs concentration. Moreover, at higher doses of ZnO NPs, morphological changes in the microalgal cells were observed (Elrefaey et al. 2021). ZnO NPs affected the growth of *Gymnodinium* algae only slightly but ROS generation induced by NPs resulted in activation of intracellular antioxidant defensive system, which also includes the activation of SOD and adenosine triphosphatase (ATPase). The NPs were allocated to the cell surface and made the cells shrink (Zhu et al. 2022a). ZnO NPs applied at a dose 1 mg/L entrapping the cell surface and causing oxidative stress to freshwater green microalgae *Coelastrella terrestris* were found to be more toxic to algae than the bulk ZnO particles, which was manifested by increased liberation of lactate dehydrogenase (LDH), lipid peroxidation, and CAT activity. The toxicity increased with increasing ZnO NPs concentration and prolongation of the exposure, whereby at higher ZnO NPs concentrations, organelle damage, cell wall breakage, and cytoplasm shrinkage were observed (Saxena and Harish 2019). Aggregation of ZnO NPs was strongly reduced in the presence of naturally-derived cellulose nanocrystals leading to improved bioavailability and increased toxicity of ZnO NPs to *Eremosphaera viridis*, whereas the extracellular ROS can be generated via adhering ZnO NPs on the surface of the algal cell, and dissolution of ZnO NPs on the cell wall can be stimulated. This results in strong ROS generation causing damage in membrane lipids, reduced activity of antioxidant enzymes and inhibition of lipid synthesis; simultaneously, levels of Zn^{2+} ions inside the algal cell increased via intracellular transportation (Yin et al. 2022). Although ZnO nanorods (NR) inhibited the growth of freshwater microalga *Desmodesmus subspicatus* as well as of marine microalga *Tetraselmus* sp., higher sensitivity of freshwater alga to these ZnO NRs was reflected in pronounced changes in the CAT activity at a dose 10 mg/L of ZnO NRs after 72 h compared to control, in contrast to *Tetraselmis* sp., in which CAT activity did not change. The toxic effect on *D. subspicatus* was associated with oxidative stress generated by ZnO NRs (de Almeida et al. 2021).

Comparison of the phytotoxic effect of ZnO and CoO NPs on microalgae showed that while to ZnO NPs toxicity contributed, in comparable extent, ionic dissolution and NPs-induced ROS generation, in particle-induced ROS generation and the damage of algal membrane by CoO NPs partially soluble CoO NPs played the decisive role (Sharan and Nara 2019). The 25-d IC_{50} values of ZnO NPs and Fe_2O_3 NPs on freshwater *C. vulgaris* were estimated as 0.258 mg/L and 12.99 mg/L, respectively, confirming their higher toxicity compared to respective bulk-forms (1.255 mg/L and 17.88 mg/L, respectively). Both NPs caused pronounced decrease of Ch levels along with an increase in proline levels and SOD and CAT activities, and increased LDH level indicated membrane disintegration. The NPs were deposited on the surface of algal cells and caused disintegration of algal cell wall (Saxena et al. 2021).

The toxicity of iron oxides showing comparable particle size of 20–30 nm to *C. pyrenoidosa* decreased in following order: Fe_3O_4 NPs > α-Fe_2O_3 NPs > γ-Fe_2O_3 NPs. The toxic impact of iron oxides on algae was mainly due to NPs-induced ROS generation, and dissolved ions contributed to the toxicity only slightly, in contrast to heteroagglomeration of NPs with algal cells and physical interactions (Lei et al. 2016). Treatment of *C. vulgaris* with 50 and 100 mg/L α-Fe_2O_3 NPs reduced algal biomass at exponential growth phase by 41.2% and 83.7% and increased total lipid contents by 39.7% and 25.5%, respectively, along with reduction in fatty acids, and morphological deformations were observed as well; however, α-Fe_2O_3 NPs did not serve as Fe source to promote algal growth (Bibi et al. 2021). Monitoring of *C. terrestris* treated with 50 mg/L Fe_2O_3 NPs for 25 days showed that the NPs were absorbed by algal cells achieving accumulation factor of 2.984, but stronger reduction of algal growth was observed compared to that observed with bulk Fe_2O_3 particles (33.33% vs. 27.77%); oxidative stress induced by Fe_2O_3 NPs was found to be responsible for the algal growth inhibition (Saxena et al. 2020). Superparamagnetic iron oxide NPs (SPION) induced ROS generation resulting in lipid peroxidation, mitochondrial dysfunction and photosynthesis inhibition in *C. reinhardtii*. SPION were internalized by algae and were sequestered in autophagic vacuoles (Hurtado-Gallego et al. 2020).

TiO_2 NPs, ZnO NPs and Fe_2O_3 NPs restricted the motility in marine microalgae, *Platymonas subcordiformis*, greatly reduced the levels of intracellular ATP and photosynthetic pigments, and diminished the activities of enzymes catalyzing glycolysis. Pronounced reduction in the cell viability of *P. subcordiformis* was associated with impaired photosynthesis and overproduction of ROS, leading to lipid peroxidation and cell damage (Du et al. 2021).

Ocean acidification resulted in rising of the toxic impact of TiO_2 NPs on the marine microalga *C. vulgaris* resulting in more effective inhibition of algal growth due to synergistically increased oxidative stress generated by TiO_2 NPs at lower pH. Moreover, lower aggregation of TiO_2 NPs in acidified seawater was accompanied with increased internalization of TiO_2 NPs in algal cells leading to higher toxicity (Xia et al. 2018). TiO_2 NPs adversely affected the growth of *C. vulgare* algae (96-h IC_{50}: 9.1 mg/L), caused cellular and sub-cellular injuries and deteriorated photosynthetic function of chloroplasts. TiO_2 NPs generated oxidative stress in algae, inhibited photosynthesis via damaging PS II reaction center, reduced ATP and glucose production and up-regulated genes encoding Chl*a* and PS II, thereby disrupting material and energy metabolisms in the photosynthesis (Middepogu et al. 2018). The hydrodynamic size and the dose of TiO_2 NPs played crucial role in the acute toxicity to *C. pyrenoidosa*, although the photocatalytic properties of TiO_2 NPs may improve photosynthesis at doses < 40 μg/mL (AL-Ammari et al. 2021). Combined exposure of *C. pyrenoidosa* to TiO_2 NPs and UVB radiation inhibited growth and photosynthesis in algae via oxidative damage, NP shading and reduced production of extracellular polymeric substances resulting in impaired protection of algal cells, which contributed to the enhanced internalization of TiO_2 NPs (Zhu et al. 2022b). Increasing concentrations of TiO_2 NPs damaged *C. reinhardtii* cells, caused a decline in the number of chloroplasts and damage to some other organelles and to plasmolysis, whereby TiO_2 NPs were situated inside cell wall and membrane; TiO_2 NPs surrounded the surface of algal cells, inhibited photosynthesis and caused lipid peroxidation (Chen et al. 2012). Fate and behavior of TiO_2 NPs in seawater and their negative impacts on marine microalgae, including mechanisms of action, was comprehensively discussed by Thiagarajan et al. (2021). The toxicity of TiO_2 NPs to marine microalga *Nitzschia closterium* greatly enhanced with decreasing NPs size, which was reflected in respective 96-h IC_{50} values of 88.78, 118.80 and 179.05 mg/L for 21 nm, 60 nm and 400 nm TiO_2 NPs. At higher concentration of 5 mg/L TiO_2 NPs, the reduction of antioxidant enzymes along with increases in malondialdehyde (MDA) and ROS levels and damaged membrane integrity were observed; TiO_2 NPs were internalized in *N. closterium* cells and generated enhanced ROS levels inside algal cells (Xia et al. 2015). TiO_2 NPs inhibited the growth of marine microalgae *Chaetoceros muelleri*, whereby the estimated 3-d and 10-d IC_{50} values were 10.08 and 5.01 mg/L, respectively, and pronounced reduction of Chl*a* and Chl*c* levels was observed as well. An exposure of *C. muelleri* to 400 mg/L TiO_2 NPs lasting 10 days reduced the protein and lipid contents up to 13.02% and 47.6%, respectively (Bameri et al. 2022). Toxic impact of TiO_2 NPs on marine algae, *Dunaliella salina* was associated with high ROS production and was reflected in 48-h IC_{50} of 4.21 mg/L (Bhuvaneshwari et al. 2018).

Growth inhibition of the marine dinoflagellate *Karenia brevis* (72-h IC_{50}: 10.69 mg/L) and *Skeletonema costatum* diatom (72-h IC_{50}: 7.37 mg/L) by nanoanatase particles of 5–10 nm was accompanied with enhanced ROS generation in chloroplasts, damage of cell membrane, increased MDA levels as well as with destroyed balance between oxidation and antioxidation due to changed levels of antioxidant enzymes (Li et al. 2015).

TiO_2 NPs adversely affected algae and bacteria in periphytic biofilms causing growth inhibition of algae without imparting the primary production of periphytic biofilms and causing significant changes among bacteria, predominantly including *Alphaproteobacteria*, *Gammaproteobacteria*, *Cytophagia*, *Flavobacteriia*, *Sphingobacteriia*, *Synechococcophycideae* and *Oscillatoriophycideae*. Enhanced production of extracellular polymeric substances in periphytic biofilms helped to prevent TiO_2 NPs toxicity. The negative effect of TiO_2 NPs on periphytic biofilms was observed already at a dose 1 mg/L (Hou et al. 2019).

NiO NPs showing only 6.42% of soluble fraction (free Ni^{2+} ions) for 100 mg/L was toxic to *Chlorella vulgaris* algae (96-h IC_{50}: 13.7 mg/L), inhibited cell division, impaired Chl biosynthesis and photosynthetic electron transport in algae and caused oxidative stress via ROS generation. After crossing biological membranes, the NiO NPs accumulated inside algal cells (Oukarroum et al. 2017). Toxic impact of NiO NPs on marine algae *C. vulgaris* reflected in algal growth inhibition and reduced Chl levels due to shading effect caused by aggregation NiO NPs in seawater and oxidative stress was reported by Gong et al. (2019).

Cyanobacteria

Cyanobacteria (called also as blue-green algae) are photosynthesizing organisms, which can produce toxins, thereby negatively affecting the health of aquatic organisms and humans (Falconer and Humpage 2005, Paerl and Otten 2013, Rastogi et al. 2015, Tang et al. 2018). Under adequate light availability, high temperature and high concentrations of nutrients, some cyanobacteria living in rivers, lakes and sea can form harmful blooms, which limit growth and survival of other aquatic organisms (Paerl and Otten 2013). On the other hand, growth of cyanobacteria is also considerably suppressed in the presence of metal-based NPs (Dedman et al. 2021, Kráľová and Jampílek 2021b, Xu et al. 2021, Alghanmi and Al-Khazali 2022).

Green synthesized CuO NPs with mean size of 551 nm and zeta potential of –26.6 mV showing various shapes (spherical, rod or irregular) and applied at doses 0–50 mg/L exhibited dose-dependent inhibition of the growth of *Microcystis aeruginosa* (by 89.7% at 50 mg/L CuO NPs) and reduction of photosynthetic pigments. The enhanced ROS level generated by CuO NPs caused prompt oxidative stress to the cyanobacteria resulting in the damage of membrane integrity and alteration of mitochondrial membrane potential ($\Delta\psi_m$), and ultimately in strong mitochondrial injury in *M. aeruginosa* cells exposed to CuO NPs (Sankar et al. 2014). CuO NPs were toxic to *Synechocystis* sp. and showed inhibition of cyanobacterial biomass production (72-h IC_{50}: 0.003 mg/L) estimated in OECD TG 201 medium (OECD 2011), whereby for the toxic effect the dissolved fraction from CuO NPs showing strong adverse impact on photosynthesis was mostly responsible. On the other hand, in nutrient-adjusted natural water the toxic impact of CuO NPs was reduced due to binging of Cu ions to natural organic matter resulting in lower bioavailability. In contrast, TiO_2 NPs did not decrease the biomass production of *Synechocystis* sp. up to 100 mg/L (Joonas et al. 2019). Cyanobacterium *Wollea salina* cultured in BG11 medium in the presence of 0.1185 mg/L CuO NPs reduced growth, and decreased concentrations of Chl*a*, carotenoids, saccharides, proteins and total lipids compared to control, and morphological changes reflected in swelling of *W. salina* cells were observed as well (Alghanmi and Al-Khazali 2022).

Whereas treatment of *M. aeruginosa* with 1 mg/L ZnO NPs for 28 days resulted in considerably greater growth inhibition compared to treatment with 0.71 mg/L Zn^{2+} ions (47% vs. 15%), at exposure to higher doses of Zn compounds (10 and 50 mg/L) the ionic Zn^{2+} form was more toxic. Already after one week of exposure to 10 and 50 mg/L ZnO NPs, an increase in MDA levels (83% and 53%) and SOD activity (106% and 61%) was observed, even though the corresponding concentrations of Zn^{2+} ions practically did not affect these parameters. Consequently, it can be considered that at early phase of exposure, the NPs caused lipid peroxidation, but with prolongation of the exposure, the antioxidant response of *M. aeruginosa* was manifested (Du et al. 2019). ZnO NPs at a dose > 1.0 mg/L caused damage of *M. aeruginosa* cells and the release of intracellular organic substances into the water environment, which promoted the formation of microcystin-LR (MC-LR), although for effective stimulation of MC-LR production, several fold higher ZnO NPs concentrations were necessary (Tang et al. 2018).

Magnetic Fe_3O_4 NPs showed hormetic effect on growth of *M. aeruginosa*, at a dose of 182 mg/L; they improved inhibition of cyanobacterial growth by allelopathic *p*-hydroxybenzoic acid via enhanced generation of •OH radicals, resulting in a 50% decrease of the IC_{50} value, destroyed cellular integrity and caused strong reduction of the total protein content. Moreover, cyanobacterial growth suppression can be enhanced by addition of Fe^{3+} ions and with increased light intensity. Hence, magnetic Fe_3O_4 NPs have potential to be used as cyanobactericide for harmful cyanobacterial bloom management (Zuo et al. 2021). A review paper discussing environmental behavior of iron-based NPs and their impact on cyanobacteria, biomass harvesting of cyanobacteria and removal of metabolites was presented by Yang et al. (2022). Cyanobacteria blooms are the most common type of harmful algal blooms in surface water and for inactivation of such toxic cyanobacteria species, NPs can be used. For example, γ-Fe_2O_3/TiO_2 NPs inactivated microcystin/nodularin- and saxitoxin-producing cyanobacteria species from a lake in southern Illinois, whereby such NPs showing magnetic properties can also be recycled and reused. It should be noted that inactivation efficiency can be enhanced under visible light, suggesting potential of γ-Fe_2O_3/TiO_2 NPs application under sunlight in removing harmful cyanobacterial blooms from surface water (Madany et al. 2021).

Exposure of *Synechocystis* sp. cells to TiO_2 NPs showing a size of 10 nm led to considerable reduction of cyanobacteria growth, PS II quantum yields, reduced phycocyanin and allophycocyanin levels but enhanced ROS levels and increased SOD activity. TiO_2 NPs also affected the phycobilisomes and up-regulation of D1 and D2 protein genes (*psbA* and *psbD*), ferredoxin gene (*petF*) and F-type ATPase genes (e.g., *atpB*); down-regulation of *psbM* and *psb28-2* in PS II was observed as well (Xu et al. 2021). Investigation of the impact of sublethal concentrations of TiO_2 NPs on the nitrogen metabolism of nitrogen-fixing cyanobacteria *Anabaena* PCC 7120 showed that under illumination, most of the genes linked to cellular nitrogen status were upregulated when doses 6–60 mg/L were used, while gene downregulation was detected primarily at the end of dark cycle when low cell metabolism and ATP levels were observed. The formation of intracellular metabolites (amino acids), both from the glutamine synthetase-glutamine oxoglutarate aminotransferase (GS-GOGAT) nitrogen assimilation pathway and the intracellular nitrogen storage pathway, depended on the dose of applied TiO_2 NPs, suggesting activation or repression of amino acid production or loss. Consequently, cyanobacterium responded to stress induced by TiO_2 NPs via increasing synthesis of proteins participating in detoxification (Cherchi et al. 2021).

A dose of 50 mg/L CeO_2 NPs inhibited growth of *M. aeruginosa* (FACHB-942) by $11.48 \pm 5.76\%$, impaired photosynthesis and caused ROS generation after 72 h exposure, whereby the toxic effects were due to the adsorption of CeO_2 NPs on the surface of cyanobacterium cells, generating ROS and the intracellular Ce content. This high dose of CeO_2 NPs also boosted the production of intracellular microcystins, which after releasing into aquatic environment can be harmful to aquatic ecosystems (Wu et al. 2022). At exposure of marine cyanobacterium *Prochlorococcus* to 100 µg/L CeO_2 NPs under environmentally relevant conditions for 72 h, the cell density was decreased by up to 68.8%. However, under simulated natural conditions, the cyanobacterial populations treated with CeO_2 NPs for 10 days were found to be recovered. On the other hand, as the cultivation was performed under optimal conditions, *Prochlorococcus* cultures responded more sensitively to extended exposure. Treatment with high CeO_2 NPs concentration such as 100 mg/L in natural oligotrophic seawater and nutrient enriched media reduced cell density up to 95.7% and 82%, respectively, which can be due to massive aggregation of NPs at entering into natural seawater. It was stated that for decline of *Prochlorococcus* cells, the co-aggregation and co-sedimentation with CeO_2 NPs is primarily responsible (Dedman et al. 2021).

Aquatic plants

Macrophytes, i.e., macroscopic water plants growing in or near water, can colonize lakes, wetlands, rivers, or marine environments; they are divided in three groups—floating, emergent, and submerged macrophytes—and can be used for decontamination of aquatic environment from organic and inorganic polluters (Wani et al. 2017, Ebrahimbabai et al. 2020, Qu et al. 2020, Kráľová and Jampílek 2022b). Nevertheless, some excess of metal-based material, whether in bulk or nanoform, have a negative effect on growth and development of aquatic plants (Yue et al. 2018, Ceschin et al. 2021, Krayem et al. 2021, Dumont et al. 2022).

A floating plant *Lemna minor*, exposed to CuO NPs (150 µg/L) for 7 days, showed a decrease in the frond number, frond surface area and dry weights of whole plants by 32%, 47% and 33%, and the toxic impact of CuO NPs was also manifested by strong damage of the epidermis fronds, damaged cell integrity in roots, and shrinkage of both chloroplast and starch grains in the frond cells. After treatment with 1 mg/L CuO NPs, root Cu concentration increased threefold compared to shoot Cu concentration, and H_2O_2 and •OH levels increased (by 56 and 57%, respectively), and chloroplasts can be considered as the site of ROS production (Yue et al. 2018). In a floating plant *Eichhornia crassipes* (water hyacinth), treated with 50 mg/L of CuO NPs for 8 days, remarkable inhibition of both roots and shoots, severe damage of root caps and meristematic zone of plants with damaged cell arrangement as well as eliminated extension zone of root tips was observed. CuO NPs were able to translocate from both roots and submerged leaves, and they were detected in roots, submerged leaves, and emerged leaves. Moreover, transformation of CuO NPs was confirmed by the presence of Cu_2S and other Cu species in these tissues. During the contact of CuO NPs with leaf, stomatal closure was caused by the formation of H_2O_2 and enhanced Ca levels in leaf guard cells were observed (Zhao et al. 2017). Toxic effect of CuO NPs applied at doses 5 and 70 mg/L on *Myriophyllum spicatum* for 10 days was associated with progressive leaching of Cu^{2+} ions from CuO NPs (Dumont et al. 2022). *Schoenoplectus tabernaemontani* exposed to CuO NPs in hydroponic mesocosms using doses of 0.5 mg/L and 50 mg/L accumulated in the roots 14 µg Cu/g and 4,057 µg Cu/g, respectively, whereby a root uptake percentage (of the total initial NP mass) achieved 40.6–68.4%. On the other hand, limited translocation of CuO NPs from the roots to the shoots was observed (Zhang et al. 2014). CuO NPs were able to enter the roots of *Spirodela polyrrhiza* and showed adverse impact on plant growth. In treated plants, considerable changes in antioxidant enzyme activity were observed: CAT, POD and SOD activity enhanced, suggesting involvement of the plant's defense system in ROS scavenging. Due to phytotoxicity of CuO NPs, reduction in relative frond number and Chl content was observed (Khataee et al. 2017). Humic acids (HA) and to a lesser extent also fulvic acids can reduce the toxicity of both CuO NPs and Cu^{2+} ions to *Landoltia punctata* plants, probably via formation of complexes between dissolved organic matter and Cu, resulting in diminished Cu availability (Rippner et al. 2018). Cetyltrimethylammonium bromide (CTAB) released from CTAB-modified Cu_2O NPs inhibited the growth of *L. punctata* already at subtoxic Cu concentrations, whereby HA and EDTA can reduce the extent of this inhibition, likely via electrostatic and hydrophobic interactions of these organic ligands with CTAB (Rippner et al. 2020).

M. spicatum plants treated with 0.8 and 2 ppm ZnO NPs for 1, 4, and 7 days in tap water showed time-dependent removal of NPs with no further progress after 4 days, achieving 29.5–70.3% removal of ZnO NPs; in pond water slightly higher removal was observed (Ergonul et al. 2021).

By combining Zn stable isotopes and microanalysis, Caldelas et al. (2020) investigated uptake and toxicity of ZnO NPs (0–1000 mg/L) in *Phragmites australis*. ZnO NPs exhibited adverse impact on plant growth, Chl content, photosynthetic efficacy and transpiration, and resulted in Zn precipitation outside the plasma membranes of root cells. ZnO NPs with a size < 50 nm were found to release more Zn^{2+} resulting in stronger toxic impact, increased Zn precipitation and accumulation in the roots and reduced Zn isotopic fractionation during Zn uptake. Zn^{2+} was the main form transported to the shoots. Whereas reduced growth of the submerged aquatic plant *Hydrilla verticillata* treated with 1000 mg/L ZnO NPs was observed during the early stages of the experiment, considerable growth reduction of the emerged aquatic plant *P. australis* began after a few weeks. Higher adverse impact of ZnO NPs reflected in reduced Chl content, modified activity of antioxidant enzymes and higher accumulated Zn amounts was observed in *P. australis*, which absorbed more Zn than *H. verticillata*, suggesting that nutrient and water uptake whether by leaves or roots affect the NPs-induced toxicity in aquatic plants (Song and Lee 2016). *Salvinia natans* plants cultivated in culture medium in the presence of 50 mg/L ZnO NPs (25 nm) for 7 days noticeably increased SOD and CAT activities and suppressed the levels of photosynthetic pigments without considerably affecting plant growth; after 8 h exposure, the majority of ZnO NPs precipitated at the bottom and the rest of the ZnO NPs were subjected to dissolution and aggregation on the roots (Hu et al. 2014). Comparison of phytotoxicity of Zn^{2+} ions, bulk ZnO as well as undoped and Er-doped ZnO NPs on the growth of *L. minor* showed that it decreased in the following order: Zn^{2+} >> ZnO NPs >> Er-doped NPs >> bulk ZnO. A strong increase in SOD activity with increasing concentration of studied Zn compounds was observed as well (Torbati et al. 2017).

Fe_3O_4 NPs applied at a dose 200 mg/L considerably inhibited the growth of roots and leaves of *E. crassipes* after a 21-day exposure and strongly damaged the root tips, which thinned out, while the root epidermis peeled off; a damaged root tip elongation zone, reduced Chl content and CAT activity, along with increasing MDA levels were also observed. Fe_3O_4 NPs were detected in the root epidermis, intercellular space and protoplasts and in the leaf cytoplasm and chloroplasts. In the leaves, besides Fe_3O_4 NPs, which were observed in original form, FeS formed by reduction via reactions with plant components during translocation was detected as well (Ding et al. 2019). Exposure to negatively charged Fe_3O_4 NPs with galate biocaping for 15 days reduced relative growth rate and relative frond number and noticeably decreased POD activity in *Azolla filiculoides* aquatic plants, while activity of SOD after initial increase showed a decrease; the treatment was also accompanied with a remarkable increase of total phenol and flavonoid levels, suggesting activation of the enzymatic and non-enzymatic antioxidant defense systems of plant against ZnO NPs induced stress (Jafarirad et al. 2019).

Aquatic plant *S. natans* exposed to 20 mg/L of TiO_2 NPs accumulated in roots and shoots 14.9 mg Ti/g and 2.2 mg Ti/g, respectively. The bioconcentration factors (BCF) showed a decrease with increasing concentration of TiO_2 NPs in the culture medium. Ti enrichment in plants resulted in an increase of nutrients' (P, S, and Ca) levels in the roots and leaves, resulting in beneficial impact on the growth and development of this aquatic plant, although Fe levels were decreased. Consequently, it can be stated that TiO_2 NPs can modify the bioaccumulation of metal ions in the plant (Shan et al. 2021). Treatment with increasing concentrations of TiO_2 NPs (anatase; mean crystalline size of 8 nm) pronouncedly reduced growth characteristics and POD activity in *S. polyrrhiza* aquatic plants, while activity of SOD considerably increased, suggesting activation of antioxidant defense system to scavenge ROS. On the other hand, decreased POD activity was likely associated with enhanced damage of this enzyme or plant defense system caused by ROS (Movafeghi et al. 2018). Exposure of macrophyte *H. verticillata* to TiO_2 NPs at concentrations 0.1 mg/L and 10 mg/L for 24–168 h showed that plants were able to adapt to generated oxidation stress via elevated activities of CAT and glutathione reductase (GR) and GSH-dependent pathways counteracting ROS formation. However, an increase in H_2O_2 level was observed only at treatment with 10 mg/L TiO_2 NPs, whereby normalization in antioxidative stress response of treated plants was achieved after 168 h of exposure (Spengler et al. 2017). *Vallisneria natans* plants exposed to TiO_2 NPs (anatase; 5 and 20 nm) applied at doses of 5 and 20 mg/L for 30 days caused damage to plant leaf cells, ruptured epiphytic diatoms membranes and enhanced the ratio of free-living microbes. Moreover, treatment with TiO_2 NPs led to pronouncedly greater decrease of total nitrogen, compared to application of bulk TiO_2 (600–1000 nm) due to the TiO_2 NPs-damaged *V. natans* plants, diminished diversity of epiphytic bacterial community and abundance of nitrogen cycle bacteria. The reduced relative bacterial abundance, including phylum *Cyanobacteria*, *Planctomycetes*, and *Verrucomicrobia*, depended on the size of TiO_2 NPs; this decrease was considerably higher at treatment with TiO_2 NPs compared to bulk TiO_2 and control (Alklaf et al. 2022).

In contrast to both bulk TiO_2 and TiO_2 NPs, which did not exhibit serious phytotoxic impact on *L. minor* plants, both bulk CuO and CuO NPs strongly reduced the number of fronds and colonies as well as the length of roots and fronds, decreased photochemical efficiency by up to 35%, and caused more than 2.4-fold increase in the activities of guaiacol peroxidase, ascorbate peroxidase (APX) and GR increased by more than 240%. In addition, in treated plants increased occurrence of necrosis and bleaching in the colonies of *L. minor* was observed (Koce 2017).

Impact of metal oxide nanoparticles on aquatic animals

Crustaceans

Crustaceans are arthropods and include animals such as decapods, seed shrimp, amphipods, etc. Most crustaceans are free-living aquatic animals, but shrimp and prawns are also made by fishing or farming for human food. Moreover, these aquatic invertebrates are significant in the food chains of aquatic ecosystems and can be used to judge the extent of harmful contaminants in aquatic environments (Abdel-Salam and Hamdi 2014, Bertrand et al. 2018, Susanto 2021). *Daphnia magna* is a crustacean characterized with small body sizes, able to generate rapidly big populations, its genome is completely mapped, and therefore using of a multitude of omics techniques can be utilized for the study of its responses to stress induced by NPs as well as for risk prediction (Liu et al. 2022, Martinez et al. 2022). Gutierrez et al. (2021) in a review article summarized the findings related to the impact of metal-based and metalloid NPs on freshwater microcrustaceans and discussed the dominant factors affecting toxicity of these NPs along with ecotoxicological effects.

In acute toxicity test using *D. magna* model organism, 48-h EC_{50} values of 22 mg/L and 223.6 mg/L were determined for CuO NPs and CuO MPs, respectively. On the other hand, in chronic toxicity tests, both types of CuO particles showed pronounced adverse impact on the growth and reproduction parameters of the *D. magna* compared to control along with morphological changes, such as lack of apical spine development and malformed carapaces (Rossetto et al. 2014). The average Cu body burden was 2.8–42-fold higher in daphnids treated with 0.05 mg Cu/L and

1 mg Cu/L of CuO NPs than in daphnids treated with equal or equitoxic concentrations of $CuSO_4$ (0.025 mg Cu/L and 0.05 mg Cu/L). In natural freshwater, higher Cu burden was observed after exposure to CuO NPs than in artificial water. The total Cu body burden in daphnids treated with CuO NPs strongly decreased after depuration in the presence of *Raphidocelis subcapitata* algae lasting 24-h (Muna et al. 2017).

At exposure of *D. magna* to 0.3 mg Zn/L using ZnO NPs, these were effectively internalized in the midgut epithelial cells between 48 h and 9 d, and translocated to other tissues. After exposure of *D. magna* to ZnO NPs for 21 days, swelling of mitochondria was found, but animals previously exposed to ZnO NPs at low concentrations were able to completely recover full reproduction potential, in contrast to crustacean treated with $ZnSO_4$. In ZnO NPs, cytotoxicity of the soluble form plays a crucial role, whereby ZnO NPs can locally increase Zn levels inside the cells, including the ovary (Bacchetta et al. 2017). While LC_{50} of $ZnSO_4$ for *D. magna* was estimated as 2.15 mg/L, those of 20, 40 and 300 nm ZnO NPs were 1.68, 1.71 mg/L and 6.35 mg/L; the toxic impact of ZnO NPs was reduced in the presence of surfactant. Zn accumulated along the digestive system regardless of the particle size and any depuration of ZnO NPs were observed 24-h post exposure (Santos-Rasera et al. 2022). The 90 nm ZnO NPs penetrated and accumulated in *D. magna* more easily than smaller ZnO NPs of 30 and 50 nm and the largest tested ZnO NPs also showed considerably higher MDA levels per unit Zn accumulation in *D. magna* compared to smaller ones, suggesting higher oxidative damage to treated crustaceans, which was due to smaller hydrodynamic diameter of 90 nm ZnO NPs (Li et al. 2021a). ZnO NPs of 500 nm caused higher mortality of freshwater crustacean *Moina macrocopa* than ZnO NPs of 250 nm (48-h LC_{50}: 0.0092 ± 0.0012 mg/L vs. 0.0337 ± 0.0133 mg/L). ZnO NPs inhibited acetylcholine esterase (AChE) and digestive enzymes (trypsin, amylase), increased CAT and GST levels as well as activity of β-galactosidase, while reduced SOD activity. The enzyme activities were likely disrupted via accumulation of ZnO NPs in the digestive tract of *M. macrocopa* (Borase et al. 2021). ZnO NPs showed higher toxic impact on *Hyalella azteca*, an epibenthic crustacean, compared to Zn^{2+} ions, and even though the dissolution of ZnO NPs contributed to the toxicity, the researchers hypothesized that ZnO NPs can provide an improved exposure route for Zn^{2+} uptake into *H. azteca* (Poynton et al. 2013). Bare ZnO NPs and hydrophobic ZnO NPs exhibited lower toxicity to marine copepod, *Tigriopus japonicus*, belonging to small crustaceans than hydrophilic ZnO NPs. On the other hand, except least toxic hydrophobic ZnO NPs, the other tested particles exhibited chronic toxicity similar to that of Zn^{2+} because of their dissolution into Zn^{2+} ions, which was found already at low studied concentrations (Lai et al. 2021). Exposure to ZnO NPs biosynthesized using *Sargassum wightii* extract or ingestion of these ZnO NPs improved immune parameters (count of hemocytes, phenoloxidase and SOD activity) of the green tiger shrimp, *Penaeus semisulcatus* (Ishwarya et al. 2018). ZnO NPs applied at doses 10–100 mg/L to freshwater decapod crustacean *Macrobrachium rosenbergii* during 90 days exhibited a dose-dependent effect on the release of crustacean hyperglycemic hormone (CHH) from the X-organ into the hemolymph and pronouncedly affected SOD and CAT activities as well. Whereas at exposure of prawns to 50 mg/L ZnO NPs there was spawning and larvae died promptly after hatching, at a dose of 100 mg/L ZnO NPs spawning of brood-stock was not estimated; ZnO NPs also affected viability rate and dry weight of eggs, brood-stock inter-spawn period and egg clutch somatic index reproductive variables (Tavabe et al. 2020).

The 96-h LC_{50} of Fe_2O_3 NPs and Co_2O_3 NPs against *D. magna* were 163.21 and 121.04 mg/L, respectively, but higher adverse effects on ecosystems and aquatic organisms in long run can be expected. Fe_2O_3 NPs were found to be less toxic to *D. magna* compared to Co_2O_3 NPs (Farsi et al. 2021). Low toxic impact of Fe_3O_4 NPs on *D. magna* at short-term exposure (96-h LC_{50}: 654.65 mg/L) was reported by Shariati et al. (2020). Although both Fe_3O_4 NPs (27.2 ± 9.8 nm) and bulk magnetite particles (144.2 ± 67.7 nm) exhibited only minor toxicity to *D. magna* (EC_{50} > 100 ppm), under exposure to subtoxic concentrations of 10 and 100 ppm, the number of neonates hatched from *D. magna* ephippia was reduced, and short-term (48-h) treatment of neonate daphnids with 10 and 100 ppm Fe_3O_4 NPs or bulk magnetite can have a considerable impact on the long-term survival and reproductive potential of daphnids, resulting in disruption of the stability of *D. magna* populations (Blinova et al. 2017). At exposure of *Carcinus aestuarii* crabs to three studied metal oxides using a dose of 1 mg/L for 14 days, the distribution of studied MONPs in individual tissues decreased as follows: hepatopancreas > hemolymph $\geq$ gill > muscle (for CuO NPs); gill > hepatopancreas > muscle > hemolymph (for α-Fe_2O_3 NPs); and gill > muscle $\geq$ hemolymph > hepatopancreas (for α-Al_2O_3 NPs). The oxidative damage caused by the CuO NPs was only minor but α-Fe_2O_3 NPs and α-Al_2O_3 NPs caused lipid peroxidation, and affected activity of antioxidant enzymes. In addition, the tested MONPs inhibited Na/K-ATPase activity in the gill and depleted hemolymph and carcass ion concentrations, resulting in osmoregulatory and ionoregulatory toxicity (Gurkan et al. 2018).

Even though both the hydrophilic and hydrophobic TiO_2 NPs did not show acute toxicity to *D. magna*, at their co-exposure with Cu^{2+} ions mortality rates of 40–50% were observed. The researchers assumed that increased toxicity of Cu^{2+} in the presence of hydrophobic TiO_2 NPs increased bioaccumulation of Cu and Ti, subsequently causing strong oxidative stress, while in the presence of hydrophilic TiO_2 NPs the intensified Cu^{2+} toxicity can be due to increased intestinal membrane damage (Liu et al. 2019). The toxicity of undissociated TiO_2 NPs inducing molting and accelerating energy metabolism in *Daphnia pulex* was less than that of dissociated ZnO NPs, which inhibited the molting via released Zn^{2+} ions. Undissociated TiO_2 NPs affected the motility of *D. pulex* and induced their molting after adsorption to the body surface; gene expressions of molting (*eip*) and energy metabolism (*scot* and *idh*) were increased, and improved energy metabolism probably participates in the increased toxicity of these NPs to *D. pulex*. It should be noted that molting and gene expressions of *eip*, *scot* and *idh* were inhibited by dissociated ZnO NPs, whereby their toxicity was associated mainly with the released Zn^{2+} ions (Wang et al. 2021).

Using TiO_2 NPs, remarkable difference was reported between the phototoxicity ratios of aquatic species belonging to order Cladocera and all other aquatic species, whereby order Cladocera very sensitively responded to photoactive TiO_2 NPs, producing ROS under natural sunlight irradiation and causing oxidative damage, cell injury, and death of organisms. Whereas the illumination increased TiO_2 NPs toxicity to non-Cladocera 20-fold, toxicity to Cladocera showed even an 1867-fold increase (Jovanovic 2015). de Lucca et al. (2018) used the microalga *R. subcapitata* exposed to TiO_2 NPs for 96-h as food for females of freshwater cladoceran species *Ceriodaphnia silvestrii* and found that doses 1 and 10 mg/L of TiO_2 NPs exhibited pronounced toxic impact on body length and total number of neonates as well as eggs produced by this crustacean. As the most sensitive parameter responding to dietary exposure of *C. silvestrii* to TiO_2 NPs, the survival was detected because it showed a decrease already at treatment with 0.01 mg/L of TiO_2 NPs.

Brine shrimp *Artemia salina* larvae (also called nauplii) treated with TiO_2 NPs for 24-h in artificial seawater on microplates showed only low percentages of immobilization and cysts, suggesting no effect of TiO_2 NPs on their hatching (Pecoraro et al. 2021a). The 48-h LC_{50} values of TiO_2 NPs for *A. salina* was found as 4.21 mg/L. At direct exposure of TiO_2 NPs to *A. salina* higher ROS generation and SOD and CAT activities were observed compared to feeding of *Artemia* with *D. salina* algae containing accumulated TiO_2 NPs, although biomagnification and trophic transfer from algae to *Artemia* via feeding was not detected. Toxicity of TiO_2 NPs to *Artemia* was also reflected in the morphological and internal damages in the crustacean species (Bhuvaneshwari et al. 2018).

CeO_2 NPs applied at doses < 10 mg/L did not show toxicity to *D. magna* but a dose of 10 mg/L caused 100% mortality by day 7 of exposure. The toxicity of CeO_2 NPs is likely associated with reduced feeding and physical interference with the daphnids' carapace leading to diminished swimming ability (Gaiser et al. 2011). At investigation of the uptake and release of CeO_2 NPs by/from *D. pulex* over a molting stage, it was found that CeO_2 NPs were mainly administered through food, and the NPs were situated in the gut, in direct contact with the peritrophic membrane and on the cuticle. However, 40–100% of the uptaken CeO_2 NPs was not released after the depuration process (24-h) using *Chlorella pseudomonas*. As the key mechanism affecting the released CeO_2 NPs, the shedding of the chitinous exoskeleton was considered (Auffan et al. 2013). A 10-day exposure to 12.5 mg/L of CeO_2 NPs did not affect survival of *Corophium volutator*, an amphipod crustacean grown in marine sediments, but pronouncedly enhanced single-strand DNA breaks, lipid peroxidation and SOD activity. CeO_2 NPs were found to produce more ROS in both deionized and saline water compared to bulk CeO_2. Whereas in both types of medium the bulk CeO_2 was composed of Ce^{4+}, the CeO_2 NPs were mostly Ce^{3+} in saline waters and Ce^{4+} in deionized water, suggesting that redox cycling of CeO_2 NPs between Ce^{3+} and Ce^{4+} increases in saline waters and causes sublethal oxidative damage in tissues of *C. volutator* (Dogra et al. 2016).

Bivalves

Bivalve molluscs are living in marine and freshwater ecosystems, which extensively filter feeders coupling the water column and benthos. Due to their ability to accumulate nutrients as well as environmental contaminants in soft tissue and shells, they are suitable for environmental monitoring (Vaughn and Hoellein 2018, Clements and Comeau 2019). Cultivated bivalves remove 49,000 tons of N and 6,000 tons of P, worth a potential \$1.20 billion (Van der Schatte et al. 2018). Although metal NPs usually occur in the environment at levels, which do not cause noticeable toxicity, their sublethal doses can cause immunotoxicity with adverse health effects. In bivalve species, cell-mediated immunity was found to be an important target for NPs *in vitro*. Under *in vivo* exposure of bivalve species to NPs, the NP agglomerates/aggregates taken up by the gills and subsequently directed to the digestive gland are then intracellularly taken up where they induce lysosomal perturbations and oxidative stress (Canesi et al. 2012). Rastgar et al. (2022) discussed immunotoxic potential of metal NPs on bivalves and fish.

In the marine eastern mussel *Mytilus trossulus* exposed to CuO NPs, much higher Cu absorption in the digestive gland was observed as compared with the gills, while pronouncedly enhanced DNA damage was estimated only in mussel gill cells (Chelomin et al. 2017). CuO NPs applied at a dose 10 μg/L to the mussel *Mytilus galloprovincialis* for 14 days exhibited genotoxic effect reflected in DNA damage in hemolymph cells (Gomes et al. 2013). Marine bivalve *Scrobicularia plana* exposed to sediment-associated CuO NPs accumulated less Cu than bivalves treated with soluble $CuCl_2$. Under exposure to CuO NPs, more powerful detoxification was observed associated with speedy mobilization of CuO NPs to metal-containing granules and delayed induction of metallothionein (MT)-like proteins. Even though high concentrations of either CuO NPs or $CuCl_2$ caused powerful acute toxic effect resulting in the mortality of all treated bivalves, using CuO NPs led to the delayed outcome in exposed *S. plana*. CuO NPs bioaccumulated in the enzymatic and mitochondrial metabolically available fractions, generated oxidative stress causing toxicity to bivalves, whereas soluble Cu associated to the mitochondrial fraction impaired ATP synthesis capacity at mitochondrial level (Scola et al. 2021). In the swan mussel *Anodonta cygnea* exposed to CuO NPs (0.25–25.0 μg/L) for 12 days, changes in the length and form of gill lamellae as well as changes in inter-lamellar spaces, epithelial hyperplasia, atrophy and tissue rupture was observed. CuO NPs accumulated in exposed mussels until day 4 and with prolongation of the exposure, no further changes in accumulated amounts were observed, whereby at treatment with 25.0 μg/L CuO NPs the level of accumulated NPs was considerably lower compared to treatment with lower tested concentrations. Hence, CuO NPs exhibited detrimental effect on the filtration activity of *A. cygnea* (Moezzi et al. 2018).

CuO NPs and Fe_3O_4 NPs considerably stimulated bioaccumulation and disrupted the As distribution in green mussel *Perna viridis* due to powerful adsorption of As onto NPs. Metal oxide NPs increased the toxicity of As by disturbing osmoregulation in mussels, which was accompanied with reduction of activity of Na/K-ATPase and average weight loss of *P. viridis* mussels. Moreover, tested MONPs decreased the biotransformation effect of more toxic inorganic As to less toxic organic As, thereby inhibiting As detoxifying process of mussels. Under exposure to CuO NPs and Fe_3O_4 NPs overproduction of ROS increased the levels of SOD and lipid peroxidation, while repressed GST activity and declined GSH content in mussel was detected (Zhou et al. 2021).

Dissolved Zn caused higher metabolic damage to the blue mussel *Mytilus edulis* than ZnO NPs through suppressing the mitochondrial ATP synthesis capacity, binding to anaerobic metabolism and modification of the free amino acid profiles, whereby the biological impact of both ZnO NPs and dissolved Zn gradually diminished with increasing salinity stress (Noor et al. 2021). On treatment of *M. galloprovincialis* with 50 and 100 mg/L of AuNPs-decorated ZnO NPs for two weeks, increased Cu, Fe, Mn, Zn and Au levels were detected in mussel tissues, but pronounced increase was observed only for Zn. Following the exposure to AuNPs-decorated ZnO NPs, the H_2O_2 level, GSH)/glutathione disulfide (GSSG) ratio, SOD, CAT and AChE activities in the gills and the digestive glands of the mussel were greatly modified (Sellami et al. 2021). ALG/ZnO NPs nanocomposites (10–15 nm) applied at doses 12–50 mg/L exhibited oxidation stress in *Coelatora aegyptica*, enhanced MDA and NO levels but reduced CAT and GSH levels and caused histopathological alterations in gills and mantle of this freshwater bivalve, whereby the observed physiological and histological changes were more pronounced compared to treatment with ZnO NPs (5–10 nm) (Omar et al. 2022). ZnO NPs showing sizes 28–88 nm in ultrapure water and 1–2 μm in seawater were toxic to Pacific oysters *Crassostrea gigas*. Due to effective dissociation from ZnO NPs, the Zn^{2+} ions were released in seawater, and can be considered as a reason of ZnO NPs toxicity (96-h LC_{50} of about 30 mg/L was estimated for both soluble Zn and ZnO NPs). Exposure to 4 mg/L ZnO NPs for 6–48 h resulted in NPs accumulation in gills (24 and 48 h) and digestive glands (48 h). Electron-dense vesicles near the cell membrane and loss of mitochondrial cristae were observed after 6 h exposure, swollen mitochondria and increased loss of mitochondrial cristae occurred after 24-h exposure, while mitochondria with disrupted membranes and an increased number of cytosolic vesicles displaying electron-dense material were detected after 48 h exposure. Gills incorporated Zn prior (24 h) to digestive gland (48 h) resulting in earlier mitochondrial disruption and oxidative stress, whereby the majority of the biochemical alterations detected in gills was not observed in digestive gland (Trevisan et al. 2014).

Environmental transformation of TiO_2 NPs in the aquatic systems and their toxic impact on bivalve molluscs was comprehensively overviewed by Abdel-Latif et al. (2020). Li et al. (2021b) performed comprehensive analysis of toxic impacts of nanosized TiO_2 in bivalves, focusing on the toxic impact of TiO_2 NPs on the antioxidant system and cell physiology of bivalves, corresponding mechanisms of action and environmental factors affecting the toxicological effects. TiO_2 NPs applied at a concentration that should be relevant for the environment (1 mg/L) showed a strong toxic effect against marine scallop *Chlamys farreri* due to strong oxidative stress. The treatment was accompanied with increased activities of antioxidant enzymes SOD and CAT as well as MDA contents, indicating lipid peroxidation of polyunsaturated fatty acids in cellular membranes. Neurotoxic effect of TiO_2 NPs was manifested by increased AChE activities, and dysplastic and necrosis in the gill and digestive gland of *C. farreri* was observed as well (Xia et al. 2017). In hemocytes of three marine bivalve species, *Crenomytilus grayanus*, *Modiolus modiolus*, and *Arca boucardi*, exposed to TiO_2 NPs using a wide concentration range from 1 to 1000 mg/L, cell membrane depolarization and mortality of cells indicating an early toxic response was observed. While treatment with TiO_2 NPs resulted in considerable mortality of *C. grayanus* and *M. modiolus* hemocytes, relatively low cytotoxic impact of TiO_2 NPs on *A. boucardi* hemocytes was observed only at a dose 1000 mg/L after 6 h exposure (Pikula et al. 2020a). Exposure of freshwater mussel *Unio ravoisieri* to TiO_2 NPs (10, 100 and 1000 μg/L) led to a significant increase of CAT activity and MDA and H_2O_2 levels in gills and digestive glands in a concentration-dependent manner, suggesting disruption of the antioxidant system. Based on AChE activity as a disturbance threshold for the cholinergic system, concentration < 1 mg TiO_2/L was estimated (Smii et al. 2021). Exposure of Pacific oyster, *C. gigas*, to TiO_2 NPs at doses > 10 μg/L pronouncedly reduced total counts of hemocytes and phagocytosis activity, while doses 10 and 15 μg/L considerably reduced the effect of all antioxidant enzymes and glutathione peroxidase (GPX) activity and increased the MDA levels, suggesting that TiO_2 NPs exhibit negative effect on the immunity and antioxidant defense in *C. gigas* (Khoei and Rezaei 2022).

Comparison of the toxic impact of naked, ALG- and CS-coated CeO_2 NPs on the freshwater bivalve *Dreissena polymorpha* for two weeks using a dose 100 µg/L showed that CS coating suppressed NP aggregation in water. All investigated CeO_2 NPs considerably diminished ROS levels and reduced SOD, GPX and GST activities, the greatest impact being observed with ALG-coated CeO_2 NPs, which also modulated amino acid metabolism (Della Torre et al. 2021). CeO_2 NPs applied at a dose of 100 µg/L CeO_2 NPs for 96 h selectively modulated different physiological processes in the marine mussel *M. galloprovincialis*. CeO_2 NPs exhibited specific immunomodulatory and antioxidant effects at different levels of biological organization without Ce accumulation in the tissue but caused a small reduction in normal embryo development (Auguste et al. 2019). Filter-feeding activity of the freshwater bivalve *D. polymorpha* resulted in considerable removal of CeO_2 NPs from water environment, whereby the mussels bioaccumulated threefold higher amount of citrate-coated CeO_2 NPs compared to bare CeO_2 NPs, and exposure to both forms of CeO_2 NPs for 3 weeks did not exhibit serious toxic impact on the mussels (Garaud et al. 2016). Exposure to CeO_2 NPs resulted in the reduction of the size of the lysosomal system, CAT effect and lipoperoxidation in *D. polymorpha* mussel digestive glands and had adverse effect on haemolymph ion concentrations (Garaud et al. 2015). In bivalve *Corbicula flutninea* exposed to CeO_2 NPs (10 µg/L and 100 µg/L) for 6 days, the observed DNA damage was pronouncedly higher compared to control already at second day of the treatment and it increased with time. Even though after 6 days of exposure, the total antioxidant capacity, CAT, GST, caspase-3 and LDH activity showed an increase, only the induction of caspase pathway and DNA damages were found to be significant in treated *C. flutninea* (Koehle-Divo et al. 2018).

M. galloprovincialis mussels treated with γ-Al_2O_3 NPs of 5 nm at doses 5–40 mg/L for 96-h had histopathological findings in gill and digestive gland, whereby at exposure to 20 and 40 mg/L γ-Al_2O_3 NPs in digestive gland, very high proportion of atrophic phase tubules was observed. Although CAT activities in gills and digestive gland were practically not affected, significant differences in SOD activities in digestive gland at doses from 20 mg/L γ-Al_2O_3 NP and in GPX activities in gills at doses from 40 mg/L γ-Al_2O_3 NPs were observed compared to control (Gürkan and Gürkan 2021). The use of *Mytilus* hemocytes for screening the potential impacts of MONPs on the cells of aquatic invertebrates was recommended already by Ciacci et al. (2012). Investigation of *Unio tigridis* freshwater mussels fed with *C. vulgaris*, which were treated with Al_2O_3, CuO, and TiO_2 NPs (1–9 mg/L) for 14 days, showed that the MONPs were taken via the lysosomes or endosoms and accumulated in gills, digestive gland and muscle. The highest metal accumulation (76.51 mg Al/g d.w., 111.63 mg Cu/g d.w., and 113.83 mg Ti/g d.w.) was found in the gills, the lowest one in the muscles (Canli et al. 2022). Canli and Canli (2021) treated the *U. tigridis* freshwater mussels with Al_2O_3 and CuO NPs (30–270 µg/L) and found that no ouabain-sensitive ATPase activity in the gill up to 10 mM ouabain concentrations was observed. While the Na-ATPase and Ca-ATPase *in vitro* were not pronouncedly modified by the tested NPs, treatment with the same doses of ionic metals greatly decreased the enzyme activities.

Snails

Freshwater gastropods living in various aquatic environments are sensitive bioindicator of environmental metal pollution and can serve as a model organism for risk assessment of heavy metal toxicity in freshwater ecosystems (Habib et al. 2016, Oliveira-Filho et al. 2017, Amorim et al. 2019). Caixeta et al. (2020) summarized the findings related to the bioaccumulation, reproductive and transgenerational toxicity, embryotoxicity, genotoxicity and potential molluscicidal activity of nanomaterials in 21 snail species; the toxicity depended on the physical and chemical properties of nanomaterials, their environmental transformation and on experimental design, and was associated with generated ROS causing oxidative damage to DNA, lipids and proteins.

CuO NPs exhibited molluscicidal activity against *Biomphalaria alexandrina* reflected in LC_{50}/LC_{90} values of 40/64.3 mg/L, and were able to reduce the growth and reproductive rates as well as egg viability already at sublethal doses. Under exposure to 15.6–27.18 mg/L CuO NPs, the head-foot and mantle of treated snails showed morphological alterations in the ultrastructure. In addition, CuO NPs were also toxic to miracidiae and cercariae of *Schistosoma mansoni* (Ibrahim et al. 2022). Whereas exposure to sediment mixed with CuO NPs showing spherical, rod or platelet shape (207 µg Cu/g dry weight sediment) for 14 days practically did not affect the mortality of sediment-dwelling gastropod, *Potamopyrgus antipodarum*, exposure to spheres and platelets of CuO NPs mixed with sediment reduced the growth of snails. *P. antipodarum* snails strongly accumulated Cu at exposure to three tested CuO NPs forms and retained it in their tissues practically without elimination, suggesting a risk of Cu toxicity (Ramskov et al. 2015).

Investigation of the chronic toxicity of gluconic acid-functionalized iron oxide NPs (GLA-IONPs) and FeCl$_3$ applied at doses 1.0–15.6 mg/L to freshwater snail *Biomphalaria glabrata* for 28 days showed effective Fe bioaccumulation in the soft tissue portion of GLA-IONPs-treated snails along with increased behavioral impairments, which were more pronounced compared to those observed with Fe^{3+} ions and the control. Moreover, chronic exposure to GLA-IONPs reduced the fecundity and fertility and caused reproductive toxicity in the *B. glabrata* (Caixeta et al. 2021). Lauric acid (LA) bilayer-functionalized iron oxide NPs (LA-IONPs) exhibited developmental toxicity in *B. glabrata* and adversely affected embryos and newly-hatched snails, although their toxicity was lower than that of LA. LA-IONPs, iron ions and LA stimulated death rate, concentration-dependent inhibition of hatching and morphological changes in snail embryos; at treatment with LA-IONPs or Fe ions, embryos also bioaccumulated Fe. The protective role of ovigerous masses during the early developmental stage was manifested by considerable toxicity found in newly-hatched snails compared to embryos (Pena et al. 2022).

The 96-h EC$_{50}$ value of ZnO NPs on freshwater snail *Pila virens* was estimated as 180.49 µg/L. Treatment of *P. virens* with sublethal concentration of ZnO NPs (135 µg/L) for 24 and 96 h, respectively, resulted in ROS generation, protein carbonyl and lipid peroxidation and reduced activities of GST and GPX activity compared to control. Using an *in silico* approach, strong binding of ZnO NPs to the target protein at its active site resulting in the loss of GST activity was shown (Thummala et al. 2022). Freshwater snail *B. alexandrina* treated with 7 and 35 µg ZnO NPs/mL, respectively, for three successive weeks resulted in considerable increase of MDA and NO levels and reduction of GSH and GST levels in hemolymph and soft tissues of tested snails; pronounced drop in overall protein and albumin concentrations along with enhanced levels of total lipids and cholesterol and increased effects of aspartate aminotransferase (AST), alanine aminotransferase (ALT), and alkaline phosphatase (ALP) in hemolymph and soft tissues of exposed tested snails was also found. The estimated LC$_{50}$ value of ZnO NPs against *B. alexandrina* was 145 µg/mL (Fahmy et al. 2014). Remarkable decrease in GSH, GST and GPX along with gradual increase in MDA level and CAT was also observed in digestive gland of *L. luteola* exposed to 32 µg/mL ZnO NPs for 24 and 96 h, and genotoxic impact of ZnO NPs in digestive gland cells due to created oxidative stress was estimated as well (Ali et al. 2012).

Bi$_2$O$_3$ NPs exhibited toxic impact on *Lymnaea luteola* snails (96-h LC$_{50}$: 72.6 µg/mL). Exposure to sublethal concentrations of Bi$_2$O$_3$ NPs for 1, 3 and 7 days resulted in pronounced increase of MDA, SOD and GST levels along with GSH reduction in pancreatic gland tissues of tested snails, and meaningful percentage of apoptotic/necrotic heamocytes in hemolymph and DNA damage in pancreatic cells of tested snails was observed as well (Al-Abdan et al. 2021).

Amphibians

Amphibians are cold-blooded vertebrates possessing permeable skins, which are able to exploit both aquatic and terrestrial habitats, show high sensitivity to inorganic and organic environmental contaminates and are suitable to be used as pollution indicators (Zhang et al. 2018b, Sievers et al. 2019, Tabat et al. 2022).

Exposure of agile frog, *Rana dalmatina* embryos to both CuO NPs (40–80 nm; 0.01–0.3 mg/L) and $CuSO_4$ for 16 days considerably reduced carbohydrate content and LDH activity, while treatment with CuO NPs showed stronger adverse impact on size and weight of tadpoles than $CuSO_4$. However, it should also be mentioned that the adverse effect of $CuSO_4$ on the lipid content was higher compared to CuO NPs. The researcher stated higher susceptibility of *R. dalmatiana* to the adverse impact of both Cu forms compared to *Xenopus laevis* (Glavas et al. 2022).

Polypedates maculatus tadpoles treated with ZnO NPs (1, 10 and 50 mg/L) effectively bioaccumulated Zn in the blood, liver, kidney and bones compared to control animals. Treatment with ZnO NPs considerably modified the concentrations of mean corpuscular hemoglobin, mean corpuscular volume and hemoglobin, and pronouncedly reduced SOD and CAT levels compared to control group, suggesting toxicity of ZnO NPs applied at sublethal doses to amphibians (Murthy et al. 2022). In the tadpoles of *Lithobates catesbeianus* treated with ZnO NPs or $ZnCl_2$ at a dose 10 mg/L for 7 days, morphological and behavioral changes were observed: the animals did not have oral plate structures, it was smaller but elongated and changes in the location and coiling of the intestine as well as malformation of the chondrocranium were observed as well, suggesting retardations in the metamorphosis and reduced overall condition of the tadpoles; reduced mobility of amphibians due to treatment with 10 mg/L ZnO NPs or $ZnCl_2$ was found as well (Motta et al. 2023).

In the tadpoles *Dendropsophus minutus* exposed to 0.1–10 mg/L TiO_2 NPs or dissolved TiO_2 for 7 days, decrease of total size, body length, and width, and in the height of the musculature of the tail of the tadpoles as well as reduced mobility and a considerable DNA damage were observed (do Amaral et al. 2022).

The impact of citrate-coated spherical CeO_2 NPs (2–5 nm) and larger uncoated CeO_2 nanoplates (20–60 nm) on the amphibian larvae *X. laevis* and *Pleurodeles waltl* was investigated by Bour et al. (2015). Treatment of *Xenopus* larvae with both types of CeO_2 NPs at doses 1 and 10 mg/L for 16 days considerably inhibited their growth, while pronounced growth inhibition of *Pleurodeles* was observed only when treated with 10 mg/L of uncoated CeO_2 nanoplates. Moreover, treatment with 10 mg/L of uncoated CeO_2 nanoplates resulted in 35% mortality of *Xenopus* larvae but no death of *Pleurodeles* was detected after 12-d treatment. On the other hand, dose-dependent genotoxic effect was induced in *Pleurodeles* exposed to CeO_2 nanoplates, but no genotoxicity was observed in *Xenopus*. Consequently, it can be stated that different properties such as size, shape or coating of studied CeO_2 NPs affected their different responses in tested amphibians. In another study, Bour et al. (2016) investigated toxicity of CeO_2 NPs in environmentally relevant conditions using mesocosms. While 35.3% mortality and a mean Ce concentration of 13.5 ± 3.9 mg/kg were reported for *Pleurodeles*, a direct or dietary exposure of *Pleurodeles* to CeO_2 NPs was not toxic to animals at 100 mg Ce/kg. Hence, the toxicity of CeO_2 NPs to *Pleurodeles* in mesocosm is likely indirect, caused by interaction of microorganisms with CeO_2 NPs, or dissolution of NPs in mesocosm.

Sea urchins

Sea urchins are echinoderms inhabiting marine benthic (sea bed) habitats. TiO_2 NPs showed only low inhibition of sea urchin *Strongylocentrotus intermedius* egg fertilization (EC_{50}: 626.6 mg/L), while the EC_{50} values related to embryo mortality decreased with increasing exposure time from 192.9 mg/L (2 h) to 32.3 mg/L (48-h) (Pikula et al. 2020b). There was a remarkable toxic impact of some MONPs on sea urchin *Paracentrotus lividus* embryos but no adverse effect on sperm fertilization capability was reported previously (Gambardella et al. 2013, 2015, 2016).

Freshwater shredders and rotifers

Freshwater shredders are aquatic macroinvertebrates involved in nutrient cycling in aquatic ecosystems and are responsible for processing coarse particulate organic matter. In freshwater shredder *Allogamus ligonifer* exposed to 75 mg/L CuO NPs via contaminated tap water or pre-contaminated food, significant inhibition of leaf consumption rate (up to 47%) and shredder growth rate (up to 46%) was observed, whereby Cu accumulated in the larval body. Leached ionic Cu from CuO NPs adsorbed or accumulated in *A. ligonifer* likely affected its feeding behavior and growth (Pradhan et al. 2012). Under exposure of *A. ligonifer* to CuO NPs at concentrations between LC_{10} and LC_{30} for 96-h, the activity of GST increased, while that of CAT decreased, and oxidation stress induced by CuO NPs resulted even in the death of shredders. In addition, strong inhibition of AChE was observed even at concentrations $<LC_{10}$, indicating neuronal stress in *A. ligonifer* induced by CuO NPs (Pradhan et al. 2016).

Rotifers (wheel animalcules) are microscopic aquatic invertebrates mostly living in freshwater and functioning as grazers, suspension feeders and predators within the zooplankton community. Under exposure of the freshwater planktonic rotifer *Brachionus calyciflorus* to TiO_2 NPs at 25°C, the 24-h LC_{50} of 117.14 mg/L decreased with prolongation of the exposure to 60.11 mg/L (48-h LC_{50}) and increasing temperature resulted in higher toxicity of TiO_2 NPs to rotifers. Using a dose of 200 µg/L TiO_2 NPs, the swimming linear speed of rotifers considerably increased compared to control. GSH content and CAT activity in *B. calyciflorus* treated with TiO_2 NPs at doses > 40 µg/L were pronouncedly reduced, suggesting a toxic impact of TiO_2 NPs on the freshwater rotifer. On the other hand, observed MDA consent was comparable to the control (Dong et al. 2020).

Aquatic insect

Treatment of the water insect *Hydropsyche excellent* Dufour, 1841 with TiO_2 NPs greatly increased abdominal contractions, increased lipid peroxidation and reduced CAT activity, and 56% mortality was observed, following pulse exposure to TiO_2 NPs for 2 weeks (Torres-Garcia et al. 2020).

Fishes

Fish are aquatic, craniate, vertebrate animals with respirator organ gills able to extract dissolved O_2 from water and excrete CO_2, which allow fish to respire underwater. According to FishBase (2022), there are 34 900 fish species, of which ca. 40% inhabit permanently freshwater systems (Tedesco et al. 2017). Fishes are crucial biological indicators of environmental contamination and water quality (López-López and Sedoño-Diaz 2015, Ali et al. 2020), able to accumulate toxic metals in their tissues (Javed and Usmani 2016, Adegbola et al. 2021, El Bahgy et al. 2021). The review paper of Aziz and Abdullah (2022) focused on the toxicity of MONPs in freshwater fish; factors affecting nanotoxicity, including the size and shape of NPs, their functionalization, chemical composition, solubility, and pH of the system were comprehensively discussed, along with mechanisms of their toxicity associated with ROS overproduction, the liberation of metal ions and the negative effect of oxidative stress on cell membranes and DNA.

A comprehensive overview of the toxic impact of MONPs on zebrafish (*Danio rerio*) during both adulthood and growth stages, highlighting the role of oxidative stress, was presented by d'Amora et al. (2022). Zebrafish is widely used for studying the toxic impact of NPs, including MONPs, on fishes, and because zebrafish genome shares about 70% similarity with human genome (Howe et al. 2013, Kettleborough et al. 2013), the zebrafish model is also suitable for studying cellular mechanisms of various clinical disease entities (Hsu et al. 2007). For example, Au@Cu_2O NPs with a gold core, which were internalized in zebrafish embryonic cells, were quickly disintegrated to Cu ions that subsequently underwent lysosome-mediated exocytosis. The uptake rate of 130 nm NPs was smaller than that of 200 nm ones, but the activation of exocytosis by smaller NPs was faster due to their faster dissolution on cells. The prompt toxicity of Cu_2O NPs was associated with the fast release of Cu ions, and the cell deaths arose predominantly from necrosis. Even though the turnover of intracellular Cu at treatment with sublethal doses was hundred times faster compared to the basal values, only twofold increase in labile Cu^+ concentration was observed due to the buffering ability of GSH (Wang and Wang 2022). Neurotoxicity of various NPs to zebrafish was discussed by Zhao et al. (2022a).

Exposure of *Oreochromis niloticus* to 50 mg/L CuO NPs (68.92 ± 3.49 nm) for 25 days resulted in considerable upregulation of the transcription of pro-inflammatory cytokines (tumor necrosis factor alpha (TNF-α), interleukin-1β, interleukin-12, and interleukin-8), heat shock protein 70, apoptosis-related gene (caspase-3), and SOD, CAT and GPX genes in liver and gills of treated fish compared to control; histopathological injuries in the hepatopancreatic tissues, posterior kidneys, and gills in treated fish were observed as well (Abdel-Latif et al. 2021). Firat et al. (2022) reported that in *O. niloticus* treated with 0.05 mg/L CuO NPs for 3 weeks, enhanced plasma ALP, AST, ALT, LDH, cortisol, glucose, creatinine, blood urea nitrogen, and tissue MDA levels were observed. The treatment also reduced plasma total protein, and tissue SOD, CAT, GST, GR and GSH levels, and the toxicity of CuO NPs was comparable with that of $CuSO_4$. Similarly, in *O. niloticus* fed with a normal food and treated with 15 mg/L of CuO NPs, pronounced enhancement in SOD and CAT activities, serum total protein, glucose, AST, ALT, creatinine, and uric acid levels were observed, while hematological indices showed remarkable decrease. A great enhancement in the percentage of poikilocytosis and nuclear abnormalities of red blood cells, along with histopathological modifications of the brain, liver, intestine, and kidney, were also found with ZnO NP treatment. However, a diet supplemented with *Spirulina platensis* allowed to mitigate adverse effect of CuO NPs on *O. niloticus* (Soliman et al. 2021). In common carp (*Cyprinus carpio*) larvae treated with 0.1–1 mg/L CuO NPs for 7 days, pronouncedly enhanced activities of antioxidant enzymes were observed, along with considerably enhanced expression of Bax gene and reduced expression of Bcl-2 gene compared to control. The enzymes' activities reached a peak at a dose of 0.5 mg/L of CuO NPs, while cell apoptosis and bioaccumulation of NPs in *C. carpio* larvae culminated at exposure to 1 mg/L of CuO NPs (Naeemi et al. 2020). CuO NPs exhibited a beneficial impact on rainbow trout (*Oncorhynchus mykiss* W.) spermatozoa velocity, linearity, and motility duration, particularly up to 24-h of storage; however, with increasing concentrations of CuO NPs in an incubation medium, the amount of motile sperm was reduced (Garncarek et al. 2022). In *Labeo rohita* exposed to CuO NPs (0.5–1.5 µg/L) for 45 days, the accumulation of Cu in fish organs decreased as follows: gill > kidney > liver > heart, and strong reduction in the GST in these organs due to enhanced MDA content was estimated. Cu from CuO NPs accumulated in soft tissues to a greater extent and caused perturbation of antioxidant defenses (Riaz et al. 2020). Exposure of *D. rerio* to 100 µg/L and 200 µg/L of CuO NPs nanorods for 24 h increased metabolic rate by 294% and 321%, reduced specific ammonia excretion by 34% and 83% as well as swimming capacity by 34% and 55%, respectively, compared to the control. On the other hand, exposure to nanoscale CuO with spherical shape did not exhibit greater changes suggesting that different morphological structures do not have the same effect on fish metabolism and can cause different toxic effects (Eiras et al. 2022).

Fertilized zebrafish eggs exposed to ZnO NPs caused hatching delay, but did not cause larval mortality or malformation. However, application of high concentrations of ZnO NPs (5–10 mg/L) modified larval activity, mean velocity, and maximum velocity (Chen et al. 2014). Kidney, liver, gill, and intestine damage, osmoregulatory changes and immune disorder in *O. niloticus* exposed to ZnO NPs for 7 and 14 days were observed, whereby respiratory burst and potential killing activity of small ZnO NPs (10–30 nm) pronouncedly increased compared to the control group, in contrast to application of large ZnO NPs of 100 nm, which caused a decrease of these parameters compared to control. ZnO NPs inhibited the Na/K-ATPase activity and increased serum Ca^{2+} and Cl^- levels, particularly in gill (Kaya et al. 2016). *O. niloticus* exposed to sublethal concentrations of ZnO NPs up to 4 weeks accumulated Zn in fish tissues to a higher extent than at treatment with bulk ZnO particles, whereby Zn accumulation in gill tissue exceeded that observed at exposure to bulk ZnO particles. However, the accumulated Zn levels were considerably reduced by supplementation with E and C vitamins, mainly at low concentrations of ZnO particles, Moreover, exposure of fish to both types of ZnO particles resulted in a minor increase of moisture, and ash content increased in contrast to protein and fat, which were reduced, suggesting ameliorated chemical composition in treated *O. niloticus* fish (Mohamed et al. 2022). Exposure of *O. niloticus* to 760 µg/L and 76,000 µg/L ZnO NPs for 72 h did not affect animals' locomotor abilities and anxiety-predictive behavior but the accumulated Zn levels in body tissues correlated with changes in eating behavior and deficits in antipredatory defensive behavior (de Campos et al. 2019). When *Oreochromis mossambicus* was exposed to 1.5 mg/L ZnO NPs, Zn accumulation in the liver achieved 3.0643 mg/kg, and NPs accumulated in the soft tissues caused respiratory problems such as oxidative stress, and genotoxicity (Shahzad et al. 2019).

Comparison of the effect of biogenic ZnO NPs (hexagonal shape) and chemical ZnO NPs with spherical to cubic in shape on juvenile *C. carpio* showed that chemical NPs adversely affected CAT, lactoperoxidase (LPO), GST, and GSH activities and caused higher tissue damage than the eco-friendly green synthesized NPs causing less toxic effects to aquatic organisms (Rasool et al. 2022). On investigation of immunological responses of *C. carpio* skin mucus to commercial and green synthesized ZnO NPs of 58 nm, it was found that biogenic ZnO NPs exhibited lower acute toxicity at 96 h exposure compared to commercially available ones (LC_{50}: 78.9 mg/L vs. 59.95 mg/L) and also induced lower immunosuppressive effects on key parameters of fish skin mucus than the commercially available ZnO NPs (Rashidian et al. 2021). In *C. carpio* exposed to biogenic ZnO NPs (0.382–1.146 mg/L), the hematological parameter was modified and generation of ROS resulted in enhanced activities of SOD, CAT, GPX, GST and diminished GSH activity in the fish. In addition, treatment with ZnO NPs was associated with lamellar fusion, aneurism, cytoplasmic vacuolation, nuclear alteration, necrotic muscle fiber and pyknotic nuclei in the gills, liver and muscles of treated fish, along with pronounced upregulation of the overlapping expressions of SOD1, CAT, GPx1a, GST-α, CYP1A, and Nrf-2 genes. At exposure to the highest tested ZnO NPs concentration, Zn bioaccumulation in *C. carpio* organs decreased in following order: gill (35.03 ± 2.50 µg/g) > liver (5.33 ± 0.73 µg/g) > muscle (2.30 ± 0.20 µg/g) (Rajkumar et al. 2022). Mahboub et al. (2022) described the protective action of *Allium hirtifolium* extract supplemented to the diet of *C. carpio* treated with foodborne ZnO NPs via triggering antioxidant responses both at tissue and molecular levels and reducing oxidative stress generated by ZnO NPs. ZnO NPs (20–30 nm; zeta potential of +26.0 mV) exhibited negative impact on the testis function in common carp: downregulation of the expression of several transcription factors and few steroidogenic enzyme genes and slow progression of spermatogenesis in the treated testis was observed, while CAT, SOD, and GST in testis pronouncedly increased compared to control. Addition of ZnO NPs to TM3 Leydig cell culture resulted in loss of adhesion; clumping with reduced viability and a considerable rising in the apoptotic cells and DNA damage due to strong oxidative stress was observed as well (Deepa et al. 2019). In a commercial carp feed with a diet containing 500 mg/kg of ZnO NPs for six weeks, the abundances of 32 proteins in the treated intestinal folds and 28 proteins in the muscular parts were significantly changed, and downregulation of pathways attributed to protein synthesis in both parts of the treated intestine was observed. Intestinal folds exposed to ZnO NPs responded with apoptosis, while in the muscular parts enhanced levels of protein associated with cancerous cell survival were detected. ZnO NPs were found to affect the protein abundances associated with cell motility, immune system response, oxidative stress response, as well as cell metabolism (Chupani et al. 2018).

Rutilus rutilus caspicus (Caspian roach) exposure to ZnO NPs showed increased lesions in gill, liver and kidney, whereby the observed alterations correlated with accumulated Zn concentration in the target organs (Khosravi-Katuli et al. 2018). In goldfish (*Carassius auratus*) exposed to ZnO NPs under elevated CO_2 levels (600 ± 10 µL/L), higher Zn accumulation in liver, brain and muscle by 43.3%, 86.4% and 22.5%, respectively, was detected compared to exposure at ambient CO_2 levels (400 ± 10 µL/L) along with stronger oxidative damage due to increased ROS generation, which was reflected in increased MDA and lower GSH content in liver and brain (Yin et al. 2017). In juvenile *C. auratus* exposed to CuO NPs, ZnO NPs and CeO_2 NPs (alone and in mixtures; 20–320 mg/L) for 4 days, strong inhibition of Na/K-ATPase in gill, and SOD and CAT activity in liver at doses ≥ 160 mg/L was observed (with the exception of CeO_2 NPs). The response of above mentioned integrated biomarkers calculated by combining multiple biomarkers into a single value increased as follows: CeO_2 NPs $\approx$ ZnO NPs/CeO_2 NPs $\approx$ CuO NPs/CeO_2 NPs < CuO NPs/ZnO NPs/CeO_2 NPs < ZnO NPs < CuO NPs < CuO NPs/ZnO NPs, suggesting synergistic effect for mixtures of CuO NPs and ZnO NPs, additive effect for the ternary mixture and antagonistic effect for the binary mixtures containing CeO_2 NPs (Xia et al. 2013). Co-treatment of *C. auratus* with ZnO NPs and Al_2O_3 NPs or a single treatment with these NPs at sublethal concentrations increased CAT, SOD, and GST activity as well as lipid peroxidation in the gills and livers of fish. Gill hyperplasia and liver degeneration were observed in exposed *C. auratus*, gills being impacted to greater extent. Combined treatment with ZnO NPs and Al_2O_3 NPs showed somewhat greater toxicity compared to individual MONPs (Benavides et al. 2016).

The 96-h LC_{50} value ZnO NPs green synthesized using *Satureja hortensis* estimated for *O. mykiss* showing mean weight and length of 20 ± 3 g and 15 ± 2 cm respectively, was estimated as 25.50 mg/L, suggesting low toxicity of these NPs. However, the lifetime of fishes was reduced with increasing the concentration of ZnO (1–100 mg/L) (Taherian et al. 2020). Comparison of gill damages and Zn accumulation in gills of *O. mykiss* exposed to 500 µg/L ZnO NPs and 500 µg/L Zn^{2+} following 14 days showed that despite accumulation capability of Zn^{2+} compared to ZnO NPs, stronger damage in the gill tissue, including shortening and fusion of secondary lamellae, surface epithelium hypertrophy, and hyperplasia of the primary lamellae, was observed using ZnO NPs (Mansouri et al. 2018).

In *O. mykiss* fed with *Melanopsis praemorsa* molluscs, which were previously fed with *Elodea canadensis* with absorbed Fe_3O_4 NPs, the NPs were absorbed through microvilli of the epithelial cells of the tunica mucosa in the intestine. They were then passed through the vessels of the lamina propria of the tunica mucosa, then via blood circulation, they achieved the sinusoids of the liver and were subsequently trapped from the endothelium of the sinusoid to the cytoplasm of liver hepatocytes and to mitochondria and lysosome (Agayeva et al. 2020). In *O. mykiss* treated with 0.013 mL/L magnetic Fe_3O_4 NPs for 48 h and 96 h, a decrease of AChE activity and brain-derived neurotrophic factor in the brain tissue over time was observed, along with reduced activities of SOD, CAT, GPX, and GSH levels and increased lipid peroxidation. The treatment with Fe_3O_4 NPs was also accompanied with increased TNF-α, IL-6, 8-OHdG and caspase-3 concentrations and decreased Nrf-2 concentrations (Ucar et al. 2022). In *O. mykiss* treated with increasing concentrations (up to 25 mg/L) of rod-like α-Fe_2O_3 (20–110 nm; zeta potential: -23.5 ± 1.3) and mostly spherical γ-Fe_2O_3 NPs (25–130 nm; zeta potential: -19.3 ± 1.1) generating oxidative stress for 10 days, followed by 10-day recovery without NPs, enhancement of melanomacrophage aggregations, epithelial tissue deformations, cytoplasmic vacuolizations, fatty changes, necrosis, pyknosis, hyperplasia, hypertrophy, lamellar fusions, and capillary dilatations was observed, whereby γ-Fe_2O_3 NPs accumulated in kidney, liver and gill of rainbow trout more than α-Fe_2O_3 NPs; the highest accumulation of studied NPs was observed in liver (Gurkan et al. 2021). *O. niloticus* exposed to aqueous suspensions of α-Fe_2O_3 NPs and γ-Fe_2O_3 NPs for 60 days showed the largest accumulation of NPs in spleen, followed by intestine, kidney, liver, gills, brain and muscle tissue, whereby Fe levels in all organs were higher in fishes treated with γ-Fe_2O_3 NPs. Treatment with both Fe_2O_3 NPs resulted in the suppression of lysozyme activity suggesting an immunosuppressive impact of these NPs. On the other hand, exposure to α-Fe_2O_3 NPs greatly increased myeloperoxidase levels but the effect of γ-Fe_2O_3 NPs was only minor, suggesting impact of morphological differences of Fe_2O_3 NPs on the uptake and immunotoxic effects on *O. niloticus* on long-term exposure (Ates et al. 2016). In *O. niloticus* exposed to 10 mg/L Fe_2O_3 NPs and Al_2O_3 NPs, the accumulation of Fe in fish organs decreased as follows: liver > kidney > gills > skin > muscles, while for Al accumulation following order was observed: liver > gills > kidney > skin > muscle; accumulated amounts of Fe in these organs exceeded the amount determined for Al. Treatment with both NPs resulted in reduction of red blood cells count, hemoglobin content, hematoxicity values, and mean corpuscular hemoglobin concentration, along with rising of mean corpuscular volume and mean corpuscular hemoglobin; histopathological changes in gills, liver, and kidneys were observed as well. Addition of rice husk had beneficial impact on above mentioned parameters due to effective absorption of both MONPs (Abdel-Khalek et al. 2020).

Dietary NiO NPs (100 and 500 mg/kg feed) for 60 days caused necrosis of acinar cells, edema of connective tissue and cellular shrinkage in exposed rainbow trout; pancreas tissue necrosis, nuclear degeneration and vascular dilation was observed in fish group treated with higher NiO NPs concentration. However, silymarin supplementation in the feed was able to partially prevent the adverse impact of NiO NPs on enzymatic activity and helped the fish maintain alimentary volume (Nazdar et al. 2018). Investigating the *in vitro* cytotoxic impact of NiO NPs on HEPA-E1 hepatocytes of *Anguilla japonica* exposed to NiO NPs (0.5–5 $\mu g/cm^2$) for 4 h or 24 h showed decreased cell viability and considerable production of $O_2^{\cdot-}$, suggesting increased oxidative stress resulting in increased Ca^{2+} levels and a pro-inflammatory response in HEPA-E1, mitochondrial membrane potential reduction and Ca^{2+} overload in mitochondria, resulting in apoptosis. The toxic actions of NiO NPs can also be harmful for lipid metabolism and reproduction of eels (Germande et al. 2022). At exposure of *Heteropneustes fossilis* to NiO NPs (12–48 mg/L) for 2 weeks, changes in hematological (counts of red and white blood cells) and biochemical parameters (cholesterol, triglycerides, glucose, total protein, albumin, globulin, bilirubin and creatinine levels as well as in effects of ALT, AST, ALP and LDH) were found. Ni capturing and modification in cortisol concentration in the blood of *H. fossilis* were estimated as well; in red blood cells of the fish treated with NiO NPs, membrane and nuclear disintegration, micronucleus, deformed and vacuolated cells, and enucleation were also detected (Samim et al. 2021). Altered activities of antioxidant enzymes, metabolic enzymes and functional groups of biomolecules, along with structural damages in gills, disturbed expressions of metallothionein and CYP1A protein and increased Ni accumulation in gills and liver tissues of *Heteropneustes fossilis* treated with NiO NPs was also reported (Samim et al. 2022).

MgO NPs adversely affected early developmental and larval stages of zebrafish, induced cellular apoptosis (96-h LC_{50} on zebrafish survival: 428 mg/L) and intracellular ROS generation, and reduced hatching rates of the embryos (48-h EC_{50}: 175 mg/L), whereby the treated embryos also showed different types of malformations (Ghobadian et al. 2015).

Treatment with Nd_2O_3 NPs damaged the development of transgenic zebrafish embryo at doses > 200 $\mu g/mL$, the mortality and malformation rate stepwise increased with increasing Nd_2O_3 concentration and at 120 h post fertilization, LD_{50} of 203.4 $\mu g/mL$ was evaluated. In treated embryos, profound arrhythmia and decreased heart rate was found, along with remarkable cerebrovascular disappearance at doses 100 and 200 $\mu g/mL$ (Chen et al. 2020a).

ZrO_2 NPs of 15–20 nm applied at doses $\geq$ 0.5–1 $\mu g/mL$ 24–96 h post fertilization exhibited acute developmental toxicity to treated zebrafish embryos, caused mortality, hatching delay, and malformation, and induced axis bent, tail bent, spinal cord curvature, yolk-sac, and pericardial edema (Karthiga et al. 2019).

The 96-h LC_{50} of Al^{3+} ions and Al_2O_3 NPs for zebrafish larvae were 3.26 $\pm$ 0.38 and 130.19 $\pm$ 5.59 mg/L, respectively, and genotoxic impact of both Al form on larval cells showed a dose-dependent increase, whereby Al^{3+} ions caused considerably higher damage to DNA than Al_2O_3 NPs. Moreover, exposure of larvae to Al^{3+} and Al_2O_3 NPs pronouncedly induced the stress-associated MT2 gene (Boran and Saffak 2020). Treatment of zebrafish embryos with Al_2O_3 NPs (13 nm) resulted in Al accumulation in brain and progressively damaged neurodevelopmental behaviors and latent learning as well as memory performance. Al_2O_3 NPs induced oxidative stress and disruption of dopaminergic transmission in zebrafish brain tissues considerably reduced the number of neural cells in the telencephalon tissue; exposure of embryos to Al_2O_3 NPs resulted in a dose-dependent and time-dependent damage on learning and memory performance in adult zebrafish (Chen et al. 2020b).

O. niloticus exposed to 2.6 ppm and 5.2 ppm Al_2O_3 for 3 and 7 days showed considerably reduced activities of SOD, CAT, and GPX, even though GST enzyme effect was pronouncedly enhanced in 7 days. The treatment was accompanied with remarkable reduction of GSH but great increase of 8-hydroxy-2'-deoxyguanosine and thiobarbituric acid reactive substances (TBARS) levels with increasing Al_2O_3 NPs concentration and exposure time was observed and the HSP70 levels were reduced as well. The 96-h LC_{50} value of Al_2O_3 NPs for *O. niloticus* was found to be 52.4 ppm (Temiz and Kargin 2022). Acute 96 h exposure of *O. mossambicus* to 180 ppm of Al_2O_3 NPs caused serious damage and adversely altered histoarchitecture in the brain, gill, intestine, kidney, and muscle tissues of exposed fishes (Murali et al. 2018). Treatment of *O. niloticus* with 1 and 10 mg/L Al_2O_3 or TiO_2 NPs for 2 and 20 days resulted in NPs accumulation in fish tissues with the highest level in gill tissue. Increased salinity (10 ppt) considerably enhanced NPs accumulation in the tissues in acute exposure of 2 days and pronouncedly reduced it in chronic exposure lasting 3 weeks; salinity increase also greatly reduced ATP-ase effects in the gill of the control fish (up to 54%) as well as in *O. niloticus* treated with MONPs (up to 83%); reduction of ATP-ase activities using NPs alone in the no saline medium was up to 80%, suggesting risks of salt inputs and NP discharges into freshwater systems (Pepe et al. 2022). Canli and Canli (2020) measured activities of Na/K-ATPase, Mg-ATPase, Ca-ATPase, and AChE in the brain and muscle of *O. niloticus* treated with 1–25 mg/L of Al_2O_3 NPs, CuO NPs and TiO_2 NPs. Whereas treatment with CuO NPs and TiO_2 NPs reduced AChE activity in the brain, Al_2O_3 NPs practically did not affect it. However, great decrease of Ca-ATPase effect in the muscle was observed at treatment with Al_2O_3 NPs and CuO NPs, but not with TiO_2 NPs. Overall ATPase and Mg-ATPase effects were not considerably affected in the brain, but treatment with CuO NPs led to a reduction of Na/K-ATPase activity. Some NPs were preserved in the tissues also after the depuration lasting 14 days.

Early-life stage of zebrafish treated with TiO_2 NPs at doses 0.01–1.0 mg/L considerably reduced the body length and weight, decreased the swimming speed and clockwise rotation times of the larvae, showed negative impact on motor neuron axon length in Tg (*hb9-GFP*) zebrafish, decreased central nervous system neurogenesis in Tg (*HuC-GFP*) zebrafish, and downregulated the expression levels of neurodevelopment genes (Gu et al. 2021). Treatment of zebrafish embryos with low TiO_2 NPs doses (5–40 µg/L) greatly decreased spatial recognition memory and levels of norepinephrine, dopamine, and 5-hydroxytryptamine and strongly increased NO levels; overproliferation of glial cells and neuron apoptosis was observed as well. Exposure to TiO_2 NPs also altered expressions of memory behavior-related genes (Sheng et al. 2016). Although treatment with TiO_2 NPs (0.1–10 µg/L) exhibited only low toxicity to zebrafish embryos, co-exposure with Cd (10 µg/L) for 68 h resulted in toxic impact manifested as cardiotoxicity, morphological and morphometric changes (Mamboungou et al. 2022). ROS formation and cell death of hypothalamus was observed in brain of zebrafish larvae containing accumulated TiO_2 NPs and treatment with TiO_2 NPs also resulted in increased expression of the *pink1*, *parkin*, α-syn and *uchl1* gene and loss of dopaminergic neurons in zebrafish *in vitro*. Thus, neurotoxicity induced by TiO_2 NPs *in vivo* and *in vitro* presents a considerable risk factor for the development of Parkinson's disease (Hu et al. 2017). After 48 h, green prepared TiO_2 NPs using *Echinophora cinerea* extract were evaluated as a less hazardous material to zebrafish (45-h LC_{50}: 300 mg/L) (Jafari et al. 2022). However, Rocco et al. (2015) described that treatment of *D. rerio* with 10 µg/L of TiO_2 NPs for 14 and 21 days resulted in pronounced genotoxic effect. In *D. rerio* group fed with food containing TiO_2 NPs for one or two weeks, vacuolization and disruption of the apical cytoplasm of epithelial cells that covered the intestinal villi were observed; kidney morphology was not modified, the lumen of the proximal tubule was capitate, and the cells of the distal tubule were vacuolated; no morphological modifications were observed in the liver. Nevertheless, reduced SOD effect signaled liver changes occurring at the molecular level. On the other hand, during experiment neither mortalities nor body weight losses in TiO_2 NPs-treated fishes were detected (Cunha and Brito-Gitirana 2020).

At exposure of *O. mossambicus* to TiO_2 NPs (0.5–1.5 mg TiO_2 NPs/L), accumulation of NPs in soft tissues was accompanied with oxidative stress. Using a dose of 1.5 mg/L higher NPs accumulation in gills was observed compared to muscles and liver tissues. Whereas CAT, GSH and lipid peroxidation levels were pronouncedly higher in the gills, SOD levels were considerably higher in the liver. In addition, at high TiO_2 NPs dose thickening and fusion in lamellae, rupturing of gill filaments, and hyperplasia of gills were observed as well, along with increased necrosis and apoptosis in the liver; generation of sinusoid spaces and condensed nuclear bodies and increased values of olive tail movement and % tail DNA suggested pathological impact and ecotoxicity of TiO_2 NPs to *O. mossambicus* (Shahzad et al. 2022). Wang et al. (2016) investigated benthic trophic transfer of TiO_2 NPs from clamworm (*Perinereis aibuhitensis*) to juvenile turbot (*Scophthalmus maximus*). During dietary exposure, the TiO_2 NPs accumulation in *S. maximus* enhanced with enhancing Ti levels in clamworms but biomagnification of NPs was not found. For both direct and dietary treatment with TiO_2 NPs, their content in turbot organs decreased in following order: gill > intestine > stomach > skin > liver > muscle, although higher Ti accumulation was observed under direct exposure of turbot to TiO_2 NPs compared to dietary exposure. Moreover, both types of exposure resulted in lower protein and higher lipid contents, suggesting reduced nutrition quality of the fish. In addition, at day 20 of dietary exposure, reduced growth of *S. maximus* along with abnormal symptoms of liver and spleen were estimated as well. Treatment of *C. auratus* and *C. carpio* with Cr-, Fe- and Ni-doped TiO_2 NPs applied at a dose 10 mg/L for 7 days induced oxidative stress responses in gills and histopathological changes in both fish species, including hyperplasia, fusion, and aneurism in the gills as well as degeneration, integration of villi, necrosis and erosion of the intestine, whereby magnetic saturation degree of the metal-doped TiO_2 NPs increased their toxicity (Pirsaheb et al. 2019). In *Clarias gariepinus* exposed to 1–10 mg/L TiO_2 NPs over 7 days, serum glucose and albumin levels increased while serum total protein, cholesterol and triglyceride levels decreased compared to control. Treatment with 10 mg/L TiO_2 reduced serum Na^+, K^+ and Cl^- levels and decreased cholinesterase levels, while activities of AST, ALT, ALP and LDH increased compared to control (Tuncsoy et al. 2021). Carmo et al. (2019) reported that TiO_2 NPs have potential to affect immune system and enhancement energy expenditure, thereby decreasing the capability of Neotropical detritivorous fish, *Prochilodus lineatus*, to avoid predator and to resist pathogens. Acute exposure of fishes to TiO_2 NPs resulted in Ti accumulation in the liver, muscle, and brain and reduced muscular AChE effect, suggesting neurotoxic potential of these NPs.

TiO_2 NPs used at a dose 10 mg/L reduced the velocities of rainbow trout sperm cells and increased SOD activity and total GSH in sperms (Ozgur et al. 2018). Rainbow trout exhibited higher uptake of TiO_2 and CeO_2 NPs coated with polyethylene glycol or citrate compared to that of uncoated MONPs. Whereas Ti was eliminated from the tissues during the first day of depuration, remaining Ce levels in stomach, intestine and liver were detected, even though differences in depuration and redistribution of tested CeO_2 NPs were observed. All tested MONPs did not show bioaccumulation potential in the fish (Connolly et al. 2022).

Acute exposure of *O. mykiss* for 96 h to 25 mg/L CeO_2 NPs adversely affected biochemical parameters, exhibited genotoxic effect and caused histopathological damages in gills, liver and kidney of treated fishes, which was reflected in significantly increased genetic damage index and pathological indices of the organs (Correia et al. 2020). Similar histopathological changes in gills and liver of *O. mykiss* were also observed after chronic (4 weeks) exposure of fishes to 0.001–0.1 µg/L CeO_2 NPs (Correia et al. 2019). In freshwater fish, *Catostomus commersonii*, exposed to 1.0 mg/L CeO_2 NPs for 25 h, oxidative, cardiorespiratory or osmoregulatory stress was not observed; mild toxicity of CeO_2 to tissues outside of the cardiorespiratory system was reflected only in 6-fold increase of plasma cortisol levels (Rundle et al. 2016). Exposure of zebrafish larvae to commercial, as-prepared and modified CeO_2 NPs (> 1 mg/L) for 96 h led to an increased expression of ethoxyresorufin *O*-deethylase (EROD) compared to control, although metallothionein and HSP70 expression was not pronouncedly affected (Pecoraro et al. 2021b).

At the end of this long section, an overview of the LC_{50} values of some MONPs for selected aquatic animals is presented in Table 1.

Table 1. LC_{50} values of some metal oxide NPs for selected aquatic animals.

Animal	NPs	Test period [h]	LC_{50} [mg/L]	References
Pila virens	ZnO	96	180	Thummala et al. (2022)
Biomphalaria alexandrina	ZnO	24	145	Fahmy et al. (2014)
Biomphalaria alexandrina	CuO	24	40	Ibrahim et al. (2022)
Lymnaea luteola	Bi_2O_3	96	72.6	Al-Abdan et al. (2021)
Crassostrea gigas	ZnO	96	30	Trevisan et al. (2014)
Brachionus calyciflorus	TiO_2	24 48	117 60.1	Dong et al. (2020)
Daphnia magna	Fe_3O_4	96	655	Shariati et al. (2020)
Daphnia magna	ZnO 20 nm 40 nm 300 nm	48	 1.68 1.71 6.35	Santos-Rasera et al. (2022)
Daphnia magna	CuO	48	22	Rossetto et al. (2014)
Moina macrocopa	ZnO 500 nm 250 nm	48	 0.0092 0.0337	Borase et al. (2021)
Daphnia magna	Fe_2O_3	96	163	Farsi et al. (2021)
Daphnia magna	Co_2O_3	96	121	Farsi et al. (2021)
Artemia salina	TiO_2	48	4.21	Bhuvaneshwari et al. (2018)
Moina macrocopa	ZnO < 100 nm	48	2.71	Nam et al. (2017)
Danio rerio	TiO_2 (green)	45	300	Jafari et al. (2022)
Cyprinus carpio	ZnO 58 nm	96	78.9	Rashidian et al. (2021)
Danio rerio	MgO	96	428	Ghobadian et al. (2015)
Oncorhynchus mykiss	ZnO	96	25.5	Taherian et al. (2020)
Rutilus rutilus caspicus	ZnO	24 48 72 96	78 61 53 48	Khosravi-Katuli et al. (2018)

Impact of metal oxide nanoparticles on soil ecosystem

Impact on microbial communities

Metal and metal oxides belong to nanomaterials showing most toxic impact on soil biota, because they present a risk to soil health and fertility, and may adversely affect microbial communities indispensable for some biological processes and the nutrient cycling (Parada et al. 2019). The multifaceted roles that play microorganisms in the geochemical cycling of nutrients are shown in Figure 3. Negative effects of NPs on soil health, enzymes and microbial proliferation in the soil were summarized by Khanna et al. (2021), while Kwak and An (2016) comprehensively discussed the impact of NPs on terrestrial environments along with adverse impact of NPs on terrestrial species, including soil microorganisms, plants, and earthworms.

The presence of MONPs in soil affects soil bacteria depending on their concentration. Low concentrations of metal oxides can stimulate the survival of bacteria, while at high concentrations the relative amount of bacteria shows a decrease. The reactivity depends on both the type of soil and the size of NPs. NPs with sizes about 20 nm are much more reactive and have negative impact

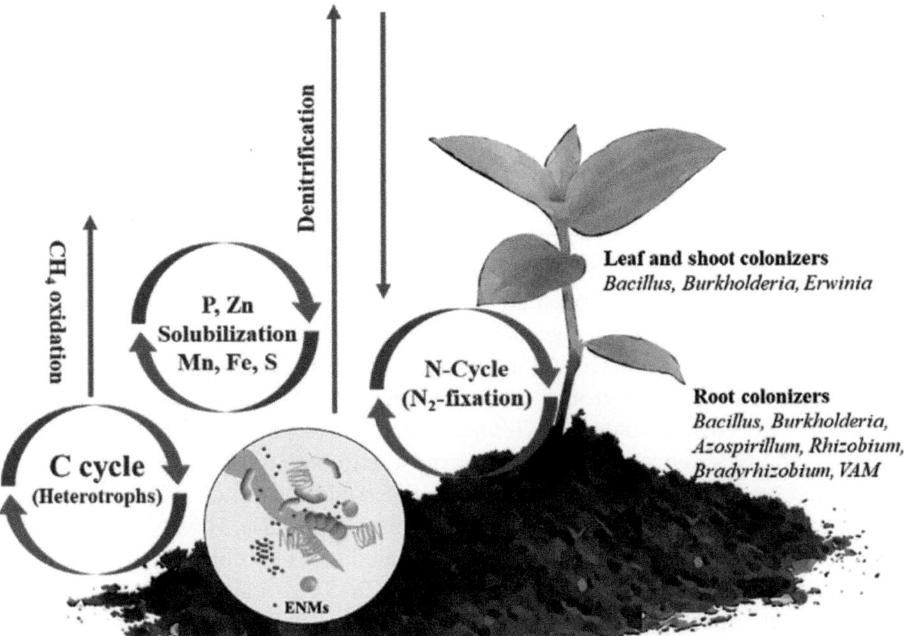

Figure 3. Versatile roles played by microorganisms in the geochemical cycling of nutrients. Reprinted from Khan, S.T., Adil, S,F., Shaik, M.R., Alkhathlan, H.Z., Khan, M., and Khan, M., 2021b. Engineered nanomaterials in soil: Their impact on soil microbiome and plant health. Plants, 11: 109. This article is an open access article distributed under the terms and conditions of the Creative Commons Attribution (CC BY) license (https://creativecommons.org/licenses/by/4.0/).

on the survival of microorganisms. Plant seeds treated with metal or MONPs (up to 100 nm) before planting and applied up to 1000 g/ton do not affect the soil metal content (Churilov et al. 2021).

Municipal biosolids, which were amended with CuO NPs (< 50 nm) and mixed into a sandy loam soil stimulated nitrification and β-glucosidase in soil at lower CuO NPs levels (≤ 265 mg Cu/kg soil), while no considerable inhibitory activities on the soil microbial growth or diversity were observed at 627 mg Cu/soil, except the dehydrogenase and phosphatase enzyme activities (Samarajeewa et al. 2021). Exposure to CuO NPs (10, 100, and 1000 mg/kg soil) for 45 days in microcosms showed only minor impact on diversity and metabolic profiles of the prokaryotic community in pecan tree (*Carya illinoinensis*) rhizospheric soil, whereby in the prokaryotic microbial community of the soils predominantly Thaumarchaeota, Proteobacteria, Actinobacteria, Bacteroidetes, Acidobacteria and Firmicutes (> 80%) were present. However, treatment with 1000 mg CuO NPs/kg soil pronouncedly increased the abundance of Caulobacterales, Burkholderiales, Xanthomonadales and Clostridiales and CuO NPs did not greatly affect the metabolic profile of soil microbial communities mainly linked to N, S and halogen cycles, suggesting that they can be used to increase soil Cu levels in Cu poor soils (Salas-Leiva et al. 2021). Treatment with CuO NPs for 3 weeks considerably enhanced the alpha-diversity of the gut microbiota of the non-target soil invertebrate *Enchytraeus crypticus*, caused pronounced diminution of the relative amount of Planctomycetes (by 17.46%), and greatly increased the relative abundance of Bacteroidetes, Firmicutes and Acidobacteria; simultaneously, at *E. crypticus* exposure to CuO NPs, the antibiotic resistance genes (ARGs) in its gut decreased by 44.90% compared to control (Ma et al. 2020).

Denitrification is the conversion of soil NO_3^- to N_2O and N_2 gases by soil bacteria that use NO_3^- as a terminal electron acceptor in the absence of oxygen. Exposure to 500 mg CuO NPs/kg soil pronouncedly suppressed denitrification, resulting in an 11-fold enhancement in NO_3^- picking up, whereby N_2O emission rates were reduced by 10.2–24.1%; simultaneously, the effects of nitrate reductase and nitric oxide reductase decreased by 21.1–42.1% and 10.3–163%. Remarkable inhibition of electron transport system activity by CuO NPs was observed as well and modification of functional denitrification genes and composition of bacterial communities also affected the denitrification process (Zhao et al. 2020). CuO NPs influenced nitrate reduction of *Pseudomonas tolaasii* Y-11and inhibited the ammonium transformation of strain Y-11 via releasing Cu^{2+} ions; using 20 mg/L CuO NPs, the removal efficiency of NO_3^- and NH_4^+ decreased by one order. In the presence of exogenous Fe^{2+} aggregation of CuO NPs resulted in lower contact of NPs with bacteria, thereby deaccelerating the harm of CuO NPs to the strain Y-11 (Yang et al. 2020).

Soil acidity has crucial effect on dissolution of the MONPs. The potentially bioavailable fraction of Zn from ZnO NPs showed an increase when soil pH declined to 4 and achieved 42.1–45 mg Zn/kg, i.e., similar concentration observed at treatment with $ZnCl_2$, On the other hand, for CuO NPs lower Cu content (24.7 mg/kg) was determined compared to $CuCl_2$ (36.5 mg/kg) application. However, in the alkaline soil, the level of exchangeable fraction of both metals was below 5.0 mg/kg (Josko et al. 2021). CuO and ZnO NPs raised soil microbial communities manifested by enhanced number of extractable bacterial or fungal groups. Sphingomonadales showed an increase with increasing concentrations of ZnO NPs, in contrast to *Actinobacteria bacterium* IMCC26256, which decreased with increasing CuO NPs concentrations. At co-exposure to both types of NPs, only a sum of their individual impacts on soil bacterial community was observed. In contrast to ZnO NPs, elevated soil organic carbon attenuated the activities of CuO NPs on soil bacteria. However, it should be noted that the essential metal ions released from NPs stimulated soil microbial community richness and diversity; nevertheless, at concentrations > 213.08 mg Cu /kg soil and 162.73 mg Zn/kg soil this beneficial effect was suppressed (Liu et al. 2021). Exposure of Pb-polluted soil to > 10 mg ZnO NPs/kg soil for two months reduced concentration of diethylenetriaminepentaacetic acid-extractable Pb by 10.6–21.3%. Moreover, pH of soil exposed to ZnO NPs for one week decreased from 6.18 to 6.08, also affecting β-glucosidase activity, and microbial energy demand for the conversion of carbon sources into biomass declined as well. On the contrary, it should be noted that the long-term exposure to ZnO NPs showed beneficial impact on the microbial diversity and richness of some metal-tolerant bacteria (Huang et al. 2022). Treatment with 10 mg/kg ZnO NPs and bulk ZnO increased the biomass and net photosynthetic rate in *Lactuca sativa* plants and at doses ≥ 10 mg/kg, the level of accumulated Zn in plant tissues was higher at treatment with ZnO NPs. A dose of 10 mg ZnO NPs per kg soil also considerably modified the soil bacterial community structure, while alpha diversity was only slightly affected; *Cyanobacteria* and other phyla individually showed comparable or different answers to ZnO NPs and bulk ZnO (Xu et al. 2018).

ZnO NPs more effectively controlled the root-knot nematode compared to bulk ZnO or oxamyl (used as a chemical control). Whereas in an *in vitro* study, treatment with ZnO NPs alone caused 72.86% mortality of second stage juveniles 2 (J2s) of *M. incognita*, at combined treatment using ZnO NPs and oxamyl, the mortality achieved even 98.91%; under *in vivo* conditions, remarkable reduction of J2s community in soil and galls number in roots (by 82.77 and 81.87%, respectively) was observed under direct exposure, and ZnO NPs were distributed and accumulated on the body of *M. incognita* JS2 causing toxicity to root-knot nematode (El-Ansary et al. 2022).

Elevated CO_2 levels considerably ameliorated the bioavailability of ZnO NPs and mitigated their negative impact on *Oryza sativa* growth but reduced protein as well as K and P levels in grains; it also attenuated adverse impacts of ZnO NPs on the soil bacterial species and altered organization of the microbial community (Du et al. 2022). On the other hand, at elevated CO_2 concentration (570 μmoL/mol.), the TiO_2 NPs doses of 50 mg/kg and 200 mg/kg, respectively, suppressed *O. sativa* biomass/grain yield by 17.9%/20.8% and 22.1%/44.1%, respectively, and treatment with 200 mg TiO_2 NPs/kg pronouncedly increased the levels of Ca, Mg, Mn, P, Zn, and Ti, while reduced fat and total sugar amounts in grains. Besides affecting the nutritional quality of crops exposed to TiO_2 NPs, the elevated CO_2 levels also modified the impact of TiO_2 NPs on the functional composition of soil microbial communities (Du et al. 2017). Investigation of soil amended with ZnO NPs or Fe_2O_3 NPs (10 and 40 mg/kg) and subjected to heat stress (48°C, 24 h) showed that alkaline phosphatase and fluorescein diacetate hydrolyzing activities were higher at exposure to ZnO NPs compared to Fe_2O_3 NPs, while the presence of ZnO NPs did not affect the activity of dehydrogenase.

Microbial count showed improved resistance values under Fe_2O_3 NPs treatment compared to that of ZnO NPs, except *Pseudomonas*, and the recovery indices for the microbial counts observed using Fe_2O_3 NPs were 0.34 for *Actinobacteria*, 0.38 for fungi, 0.33 for *Pseudomonas* and 0.28 for *Azotobacter*, suggesting that microbial indicators in soil can recover with time resulting in reconstitution and resilience of soil systems (Kumar et al. 2019). In addition, FeO NPs at doses 1 and 10 mg/kg had good impacts on soil microbial metabolic activity and on soil nitrification potential at doses 0.1 and 1 mg/kg soil. The presence of FeO NPs was able to modify the C and N cycles of the agricultural soil via affecting soil microbial metabolism (He et al. 2016). On the other hand, in soil amended with Fe_3O_4 and SnO_2 NPs using doses 10 and 100 mg Fe and Sn per kilogram of dry soil, both microbial biomass C and N were practically not affected after one week, although the microbial C/N ratio increased, likely due to the prevailing microbial communities such as ectomycorrhizae. Increased metabolic quotient (qCO_2) in NPs amended soil suggested microbial stress or changing of the bacterial/fungal biomass ratio (Antisari et al. 2013). Comparison of the effects of ZnO NPs, TiO_2 NPs, CeO_2 NPs and Fe_3O_4 NPs on black soil and saline-alkali soil showed that ZnO NPs affected soil enzymatic activities more than other tested NPs. Pronounced reduction in total bacteria population was estimated only in saline-alkali soil treated with 0.5–2.0 mg/g ZnO NPs, where particularly *Bacilli, Alphaproteobacteria*, and *Gammaproteobacteria* were affected. Soil type was considered as the key component responsible for the toxic impact of the tested NPs on the composition and size of the bacterial community (You et al. 2018).

Exposure to 150 nm, 200 nm and 900 nm ZnO NPs influenced the soil microbial metabolic potential, while treatment with 43 nm ZnO NPs affected the IC_{50} value. Investigation of the dynamic behavior of ZnO NPs using a time weighted average (TWA) approach showed that the effect of initial concentrations was higher compared to that expressed as TWA (Zhai et al. 2017). Pristine and sulfidized ZnO NPs showed similar effect on soil microbes, caused remarkable changes in *Nitrospirae* and *Actinobacteria*, which are the two most dominant phylum associated with the reduction of NH_4^+–N and dissolved organic carbon, suggesting potential long-term effects of transformed ZnO NPs on soil carbon and nitrogen cycling (Chen et al. 2021). The addition of ZnO NPs (200–1000 mg/kg) to soil samples from solid-waste land enhanced enzymatic effects associated with soil C cycle and modified bacterial community structure at the phylum and genus levels, and a significant reduction in the abundance of hydrocarbon-degrading taxa was observed. ZnO NPs negatively affected the phenoloxidase activity of *Bacteroidetes* degrading hydrocarbons, and *Acidobacteria* was able to serve as an indicator of soil pollution with ZnO NPs. On the other hand, growth of *Phytolacca americana* plants cultivated on ZnO NPs-treated soil was not significantly inhibited (Shi et al. 2021). Green synthesized ZnO NPs applied at doses 100 or 1000 mg/kg soil decreased microbial biomass carbon and increased cumulative C mineralization (i.e., soil respiration), whereby addition of rice-straw derived biochar at 1 or 5% did not mitigate the ephemeral toxicity found after 24 d of incubation (Shemawar et al. 2021).

In soil amended with 1 mg/g ZnO NPs or CeO_2 NPs, impaired thermogenic metabolism, decreased numbers of soil *Azotobacter*, P-solubilizing and K-solubilizing bacteria and suppressed enzymatic effects were observed. It should be also noted that TiO_2 NPs caused inconsiderable reduction in the plenty of functional bacteria and enzymatic effect, while SiO_2 NPs showed rather small stimulation of the soil microbial activity (Chai et al. 2015). Treatment with 50 and 500 mg CeO_2 NPs/kg soil reduced the effects of acid protease and acid phosphatase involved in soil N and P cycles, and this inhibitory impact was suppressed in the presence of earthworms that caused considerable change in the plenty of some microbes related with the soil N and P cycles (*Flavobacterium, Pedobacter, Streptomyces, Bacillus, Bacteroidota, Actinobacteria*, and *Firmicutes*). A change in metabolites involved in microbial N and P metabolism was observed as well. Thus, both CeO_2 NPs and earthworms modified soil bacteria and soil metabolite profiles; whereas greater changes in soil bacteria and metabolites were observed with co-exposure to CeO_2 NPs and earthworms (Li et al. 2022).

Besides affecting physiology and enzymatic effect of soil microorganisms, the CeO_2 NPs contaminating the soil can also be taken over by plants resulting in their trapping in plant tissues and subsequent interference with crucial metabolic processes of plants (Prakash et al. 2021). In soil amended with 1 mg CeO_2 NPs/kg (3.5 or 31 nm; bare or citrate-coated NPs), in which *Brassica napus* plants were cultivated, the crucial parameter influencing the effect of CeO_2 NPs on root microbes was particle size, whereby citrate coating reduced the impact of CeO_2 NPs on microbial enzymatic activities along with supporting the variability in the structure of bacterial community near the plant root. Some CeO_2 NPs supported taxa were closest relatives to hydrocarbon-degrading bacteria and hindered taxa showed disease-suppressive activity against plant pathogens (Hamidat et al. 2016).

Combined exposure of soil to CeO_2 or Cr_2O_3 NPs and elevated CO_2 showed beneficial impact on dehydrogenase, acid phosphatase and urease activities as well as microbial biomass carbon and resulted in higher bacterial α-diversity, suggesting that higher levels of CO_2 attenuated adverse impacts of NPs on soil microorganisms. Such co-exposure affected α- and β-diversity and functional profile of bacterial communities and increased CO_2 concentration stimulated the resilience of NP-resistant bacterial populations, particularly *Alphaproteobacteria*, *Gammaproteobacteria* and *Bacteroidia*, showing fast carbon use capability. Attenuation of toxic impact of NPs by elevated CO_2 might be associated with considerable reduction of available metals disassociated from NPs and available carbon level, along with stimulation of rapid carbon-metabolizing microbes (Luo et al. 2020). Elevated CO_2 greatly attenuated the toxicity of Cr_2O_3 NPs, increased community α-diversity in a sandy loam soil, and considerably reduced community divergences mainly in clay loam soil. Elevated CO_2 supported the growth of oligotrophic (*Acidobacteriaceae*, *Bryobacteraceae*) rather than the copiotrophic bacteria (*Sphingomonadaceae*, *Caulobacteraceae*, *Bacteroidaceae*), which can improve the community recovery and enhance available carbon utilization efficiency. The extent of the attenuation of Cr_2O_3 NPs toxicity by elevated CO_2 depended on soil type and organic matter content, resilience of the resistant bacterial taxa, and microbial network complexity in distinct soils (Luo et al. 2021). In the presence of 87.8 mg Co_3O_4 NPs/kg soil arylsulphatase, acid phosphatase and soil basal respiration were greatly affected; enzymes' activity related to carbon and nitrogen cycling and basal respiration was stimulated but the microbial biomass carbon was not pronouncedly affected (Bouguerra et al. 2022).

Using Li_2O NPs at a dose of 474 µg Li/g soil resulted in 3.45-fold higher CO_2 release compared to control and reduction of β-glucosidase effect, along with increased urease activity. Temporary inhibition of β-glucosidase activity in soil was also observed at treatment with MoO_3 NPs and NiO NPs. On treatment with tested MONPs, increasing metal concentrations affected all three domains (Archaea, Bacteria, and Eukarya) of soil microbial communities, and pronounced changes compared to control were observed particularly in Archaeal communities (Avila-Arias et al. 2019).

TiO_2 NPs incorporated into soil up to a dose of 20 mg TiO_2 NPs/kg soil increased soil enzyme activities and microbial biomass, and while with increasing TiO_2 NPs concentration up to 100 mg TiO_2 NPs/kg a significant reduction in biomass was observed, soil respiration and metabolic quotient showed increase, indicating a sublethal stress on the microbial community. Enhancement of biomass of total phospholipid fatty acid, Gram-positive, Gram-negative bacteria, fungi, actinomyctetes and anaerobes was observed up to a dose of 80 mg TiO_2 NPs/kg soil, but strongly decreased at a dose of 100 mg TiO_2 NPs/kg soil (Bhattacharjya et al. 2021). After short-term (2 weeks) treatment of soil with TiO_2 NPs, considerable impact on enzyme activity, bacterial community structure and composition, and community functioning was observed in the clay soils with high organic matter but not in sandy soils, whereby alterations were estimated in taxa belonging to Acidobacteria and Verrucomicrobia, and functional pathways concerning carbohydrates impairment. With increasing exposure time, the bacterial community recovered after two months, indicating that bacterial community was able to adapt and overcome the impact of TiO_2 NPs (Zhai et al. 2021). In a previous study, it was also found that despite considerable alterations in the taxonomic composition over time and TiO_2 NPs exposure concentrations, no remarkable change in the community functional profile and activity after 2 months exposure was observed (Zhai et al. 2019). In examining the effect of TiO_2 NPs (0.05–500 mg/kg dry soil) on the activity and plenty of ammonia-oxidizing archaea (AOA) and bacteria (AOB) as well as nitrite-oxidizing bacteria (*Nitrobacter* and *Nitrospira*), it was found that after 3 months of exposure, AOA abundance decreased independently of the TiO_2 NPs concentration used by 40%, with *Nitrospira* being less sensitive to TiO_2 NPs treatment compared to AOA. AOB and *Nitrobacter* abundances were reduced at intermediate TiO_2 NPs doses. TiO_2 NPs affected soil nitrification by both indirect impacts on the plenty of nitrifier groups and by direct impacts on nitrification effect; direct impacts of TiO_2 NPs on nitrifier abundances were found, but only on ammonia-oxidizers and not nitrite-oxidizer (Simonin et al. 2017).

Vigna unguiculata plants inoculated with arbuscular mycorrhizal fungi (AMF) and grown in Cd-polluted soil amended with TiO_2 NPs (200 mg/kg) showed stimulated germination and Chl contents and reduced uptake of Cd along with decreasing its translocation to shoots more than single inoculation with AMF; moreover, TiO_2 NPs reduced lipid peroxidation in plants more than sole inoculation with AMF, suggesting their beneficial impact on stress tolerance. Hence, TiO_2 NPs promoted the activity of AMF via mitigating Cd stress in pre-flowering *V. unguiculata* plants (Ogunkunle et al. 2020).

Since under repeated exposures to TiO_2 NPs, carriage of NPs to deep soil layers and groundwater is prohibited, chronic contamination more negatively affect soil microbiological functioning than a single exposure (Simonin et al. 2016). In flooded paddy soil, exposure to CuO NPs caused higher reduction in soil microbial biomass and the overall phospholipid fatty acids and enzyme effects, including urease, phosphatase and dehydrogenase, than TiO_2 NPs. CuO NPs also decreased the composition and variety of the paddy soil microbes. Although both TiO_2 NPs and CuO NPs can increase perturbations on the microbes in flooded paddy soil, higher bioavailability of CuO NPs is likely the reason for their stronger toxicity to microbes (Xu et al. 2015).

Impact on soil fauna

Ecotoxicological impact of ZnO NPs on investigated organisms from soil micro- and mesofauna was discussed by Kiss et al. (2021). For the study of soil–NPs–organism interactions, *C. elegans*, a free-living transparent nematode, is frequently used (Cox et al. 2021). Transgenerational and multigenerational toxicity of various environmental contaminants in *C. elegans* and the underlying mechanisms were discussed by Zhao et al. (2022b). Toxic impact of TiO_2 NPs investigated in 5 different soils using *C. elegans* was affected predominantly by pH and dissolved organic matter. Although at pH 5 enhanced adhesion of positively charged TiO_2 NPs resulted in contact nanotoxicity, humic acid was able prevent TiO_2 NPs' adhesion onto the epidermis due to its negative charge and overstimulated stress response pathway, thereby reducing nanotoxicity (Hou et al. 2022). Administration of TiO_2 NPs at doses up to 1.0 mg/mL did not affect the viability and lifespan of *C. elegans* but greatly attenuated its pharyngeal function, reproduction, and development (Sitia et al. 2022). In addition, reproductive toxicity of CuO NPs and ZnO NPs in *C. elegans* was critically discussed by Yao et al. (2022).

Earthworms (*E. fetida*) incubated in sandy loam and silt loam with ZnO NPs exhibited auto-regulation of their Zn content, in contrast to earthworms exposed to CuO NPs, in which the bioaccumulated Cu correlated with the extractable Cu concentrations. In general, accumulation of Cu and Zn was higher in juvenile earthworms compared to adults. Higher pH value of silt loam was responsible for higher bioavailability of bioavailable metals, but aging suppressed the variabilities among the effects of NPs and ionic metal compounds on *E. fetida* (Josko et al. 2021). Investigation of full life cycle of *E. fetida* exposed to CuO nanomaterials and $CuCl_2$ showed that the survival during the first 28 days post-hatching was particularly susceptible to the treatment, and concentration-dependent reduction in population growth was found using both tested Cu forms, whereby at higher concentrations prompt decline in growth rate was observed (Scott-Fordsmand et al. 2022). Nanosized and bulk rare earth oxides La_2O_3 and Yb_2O_3 used in soil at a dose of 100 mg/kg increased earthworm mortality by 33–35% and 13–15%, and decreased reproduction by 10–32% and 10–12%, respectively, while exposure to higher doses of 500–1000 mg/kg caused abnormalities in internal organelles, including mitochondria, Golgi apparatus and chloragosomes. La_2O_3 NPs and Yb_2O_3 NPs reduced earthworm digestive and cast enzymes more than their bulk counterparts. On the other hand, earthworms reduced toxicity of rare earth oxides to microbial biomass carbon and soil enzymes (Adeel et al. 2021).

Nematodes can infest agricultural crops, resulting in massive decline of yields, and MONPs can effectively kill these harmful organisms along with having favorable effect on plant growth infested with plant-parasitic nematodes (Khan et al. 2021a). CuO NPs exhibited inhibitory effects on *M. incognita* second-stage juveniles (J2S), both *in vitro* and *in vivo* and treated nematodes showed distorted, crinkled cuticle compared to control (Tauseef et al. 2021a). A dose of 200 ppm of CuO NPs showed strong inhibition of hatching from egg masses in *M. incognita* J2s after six days of exposure and also caused the highest mortality of this root-knot nematode (Khan et al. 2022). Cu-doped ZnO NPs coated with diethylene glycol showed EC_{50} of 2.60 µg/mL against *Meloidogyne javanica*. Foliar application of these NPs resulted in decrease of galling and population of *M. javanica* ranging from 39.32% to 32.29%, and treated *L. sativa* plants were characterized with increased antioxidant activity (Tryfon et al. 2022). At exposure of *M. incognita* J2s to ZnO NPs for 24-h, LC_{50}/LC_{90} values of 63.56/208.5 ppm were observed, while using 100, 200, and 300 ppm ZnO NPs resulted in nematode mortality rate of 58, 78, and 91%, respectively. Moreover, at treatment with ZnO NPs downregulation of parasitism genes (*Xyl-1*, *16D10*, and *msp-20* genes), neuropeptidergic gene *Ace-2*, and expansion-like proteins *MAP-1* were estimated as well, while the oxidative stress response gene *GSTS-1* was upregulated (Elarabi et al. 2022). *M. incognita* J2s treated with MgO NPs had indentations, roughness and distortions in the cuticular surface compared to control nematodes. The application of 100 ppm MgO NPs in root dip reduced nematode fecundity, diminished the number of galls and reduced their size; improved growth of treated *Vigna unguiculata* plants along with an increase in the level of photosynthetic pigments, protein in seed as well as N contents in root and shoots (Tauseef et al. 2021b).

Exposure to CuO NPs perturbed the collembolan gut microbial community structure, decreased the gut microbial diversity, induced decrease in both variety and plenty of associated antibiotic resistance genes (ARGs), and changed the metabolism of collembolans, which was manifested by modified C and N stable isotope constitution and was associated with changes in Cu bioaccumulation and the gut microbiota (Ding et al. 2020). Toxicity of low CuO NPs doses to springtails was modified by clay types added to artificial test soils. At field-realistic amount of 3 mg/kg CuO NPs in soil containing 30% montmorillonite, the reproduction of *F. candida* was reduced by about 40%; however, such reduction was not observed at a dose 32 mg/kg CuO NPs. The toxicity observed in CuO NPs–montmorillonite soil could be associated with reduced CAT activity resulting in increased ROS generation. Similar toxic impact was not observed at exposure of *F. candida* to CuO NPs in soil spiked with 30% kaolin (Fischer et al. 2022). Exposure of loamy soils to CuO NPs at a dose 5 mg Cu/kg soil for 28 days reduced reproduction of springtails up to 61%, whereby at higher CuO NPs this reduction was lower. Treatment with the low CuO NPs dose also reduced the growth of *F. candida* by 28% and inhibited Cu elimination from springtails. The size of CuO NPs taken up by springtails and interactions of CuO NPs with clay were found to be decisive for NPs toxicity (Fischer et al. 2021).

Impact on terrestrial plants

When plants are cultivated in soil, rhizospheric interactions between plant roots, environmental pollutants, soil, microorganisms, NPs and/or nanomaterials are exerted (El-Ramady et al. 2022, see Figure 4). Bioavailable metal ions or metal-based NPs are taken up from soil solution by the roots, can accumulate in plant organs, and depending on the applied doses and the chemical properties of the NPs can exhibit beneficial or adverse impact on plants, whereby green synthesized NPs being less toxic than those, which are prepared via chemical routes (Masarovičová et al. 2014, Kráľová et al. 2019, 2021, Jiang et al. 2022).

Nanosized particles of essential metals or their oxides serving as nutrients when applied at low concentrations to soil or via foliar treatment have beneficial impact on plants (see Figure 5), resulting in enhanced photosynthetic rate, improved plant defense, regulation of photosynthetic pigment accumulation, ROS inhibition, osmolyte and phytohormone regulations, maintenance of osmotic homeostasis, improved crop yield and increased water use efficiency and growth-related traits, along with maintenance of nutrient balance (Masarovičová et al. 2014, Kráľová et al. 2021, Kumari et al. 2022). However, exposure to high doses of such NPs results in decreased photosynthetic processes, decreased concentrations of photosynthetic pigments, reduced plant growth, decline in the contents of non-enzyme antioxidants, and impaired activity of antioxidant enzymes (e.g., SOD, CAT, GPX); due to enhanced oxidative stress induced by these NPs, accompanied with ROS overproduction, interruption of redox homeostasis, lipid peroxidation, impaired mitochondrial function, and

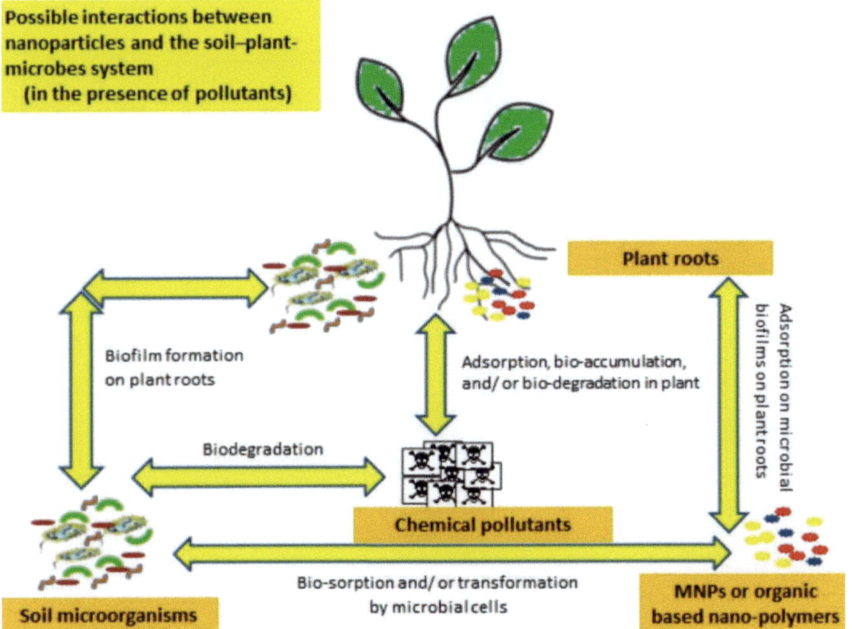

Figure 4. Soil rhizospheric interactions between plant roots, environmental pollutants, soil micro-organisms NPs and or nanomaterials (NMs), and their interactions in the soil solution. Reprinted from El-Ramady, H., Brevik, E.C., Fawzy, Z.F., Elsakhawy, T., Omara, A.E.D., Amer, M. et al. 2022. Nano-restoration for sustaining soil fertility: A pictorial and diagrammatic review article. Plants, 11: 2392. This article is an open access article distributed under the terms and conditions of the Creative Commons Attribution (CC BY) license (https://creativecommons.org/licenses/by/4.0/).

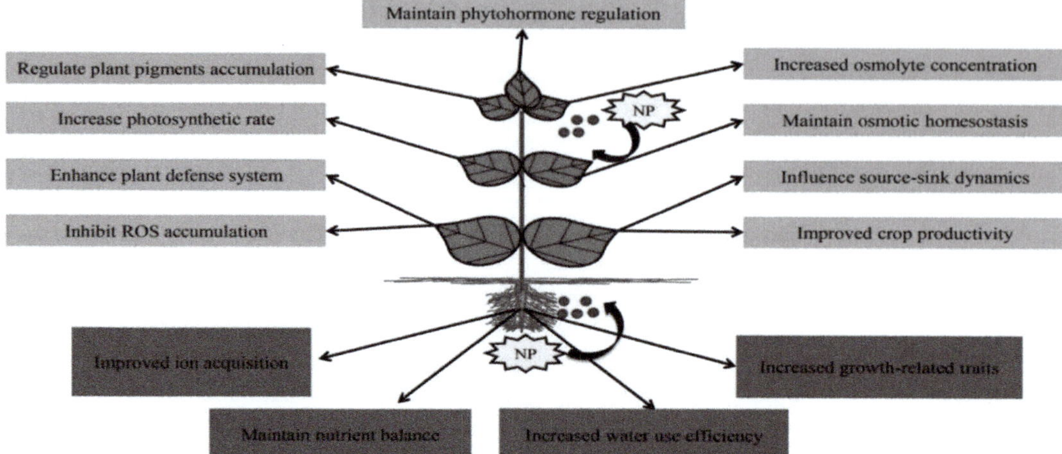

Figure 5. Physiological and biochemical responses of plants to low doses of metals or metal oxide NPs applied via foliar spraying or amended to soil. Reprinted from Kumari, S., Khanna, R.R., Nazir, F., Albaqami, M., Chhillar, H., Wahid, I. et al. 2022. Bio-synthesized nanoparticles in developing plant abiotic stress resilience: A new boon for sustainable approach. Int. J. Mol. Sci., 23: 4452. This article is an open access article distributed under the terms and conditions of the Creative Commons Attribution (CC BY) license (https://creativecommons.org/licenses/by/4.0/).

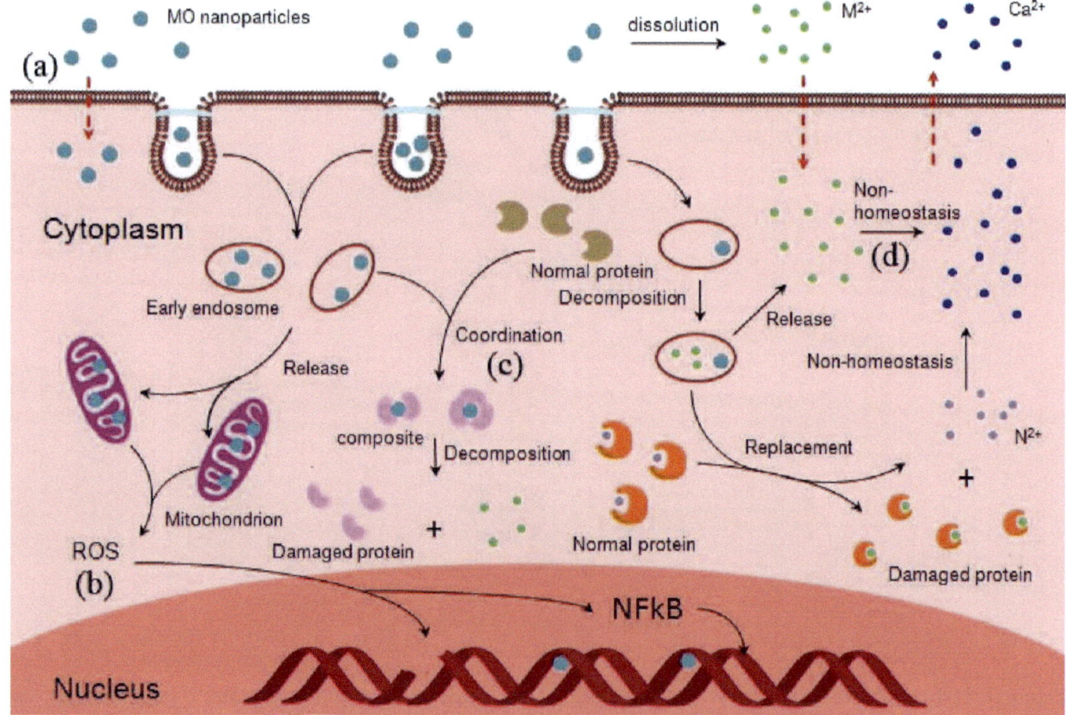

Figure 6. Different pathways inducing cellular toxicity by CuO and ZnO NPs. (a) Potential mechanisms of CuO and ZnO NPs' entry into cells; (b) ROS effect of intracellular CuO and ZnO NPs; (c) coordination effect of Cu^{2+} and Zn^{2+} released from NPs in cell; (d) non-homeostasis effect disrupted by Cu^{2+} and Zn^{2+}. Reprinted from Chang, Y.N., Zhang, M.Y., Xia, L., Zhang, J., and Xing, G.M. 2012. The toxic effects and mechanisms of CuO and ZnO nanoparticles. Materials, 5: 2850–2871. This article is an open access article distributed under the terms and conditions of the Creative Commons Attribution (CC BY) license (https://creativecommons.org/licenses/by/4.0/).

membrane lesion also occur (Masarovičová et al. 2014, Hossain et al. 2015, Li et al. 2020, Kráľová et al. 2021). Moreover, high doses of MONPs exhibit genotoxic impacts in plants. For example, at CuO NPs treatment, these were manifested as abnormalities of mitotic division on interphase, prophase, metaphase, anaphase, and telophase (Tasar 2022); on treatment with TiO_2 NPs, various chromosomal abnormalities like anaphase bridges, telophase bridges, laggard chromosomes and binucleate cells in *Allium cepa* roots were observed (Roy et al. 2021), while higher doses of Al_2O_3 NPs caused raise in chromosome aberrations, micronuclei and DNA strand breaks (De et al. 2016).

Chang et al. (2015) presented a schematic summary of the various pathways inducing cellular toxicity of CuO and ZnO NPs (Figure 6).Whereas the small NPs can diffuse directly across the membrane, other NPs can enter cells via "endocytosis", Cu^{2+} and Zn^{2+} dissolved from NPs can be transferred into cell via ion channels. In the cell, the NPs interact with mitochondria and redox active proteins and promote generation of ROS, which can cause DNA strand breaks, and induce gene expression. Chelate formation between Cu^{2+} and Zn^{2+} and entrapment of the metal ions in metalloproteins can inactivate the functional protein, and increased local concentration of Cu^{2+} and Zn^{2+} ions released from CuO and ZnO NPs disrupts cellular metal cation homeostasis leading to toxicity. Negative effects of higher doses of MONPs on staple crops such as wheat (Amooaghaie et al. 2017, Chen et al. 2019, Dias et al. 2019), rice (Yang et al. 2015, Da Costa and Sharma 2016, Wang et al. 2022a), maize (Yang et al. 2015, Ahmed et al. 2022, Roy et al. 2022) and soybean (Li et al. 2018, 2020) as well as other crops consumed in relatively large quantities such as tomatoes (Amooaghaie et al. 2017, Wang et al. 2022b), lettuce (Hayes et al. 2020, Xiong et al. 2021, Zhao et al. 2021), common bean (Salehi et al. 2021), or cucumber (Ahmed et al. 2021) can adversely affect ensuring a sufficient safe food of high nutritional quality for the growing human population. Considering the above results, to avoid adverse impact of MONPs on plants, it is desirable to prevent their entry into soil in excessive amounts, and in agricultural practice it is necessary to avoid overuse of metal oxides applied as agrochemicals whether in nanoform or in bulk form.

Conclusions

Intensive anthropogenic activities contribute significantly to the increasing contamination of environmental matrices. Although low concentrations of nanoparticles based on essential metals serving as nutrients have been found to have a beneficial effect on the growth and development of many organisms, in the last 20 years, when the large use of nanomaterials in industry and agriculture significantly increased, metal oxide NPs and their nanocomposites, which reached soil and aquatic ecosystems in greater quantities, showed an adverse impact on non-target organisms. Metal-based NPs that enter the food chain through soil, water, and food represent a hazard to macrophytes and aquatic animals and may ultimately threaten even human health. It is also important to mention that even government agendas, rules and regulations are not uniform, but with their often confused arrangement, they allow anyone to both sell and apply any preparation, because money-oriented values and the promotion of economic interests become the only mantra. It is desirable to decrease the excess amount of metal-based NPs in environmental matrices. Vegetated buffers, grass strips or wood species that can tolerate and effectively accumulate higher metal amounts and are appropriate for phytoremediation of metal-contaminated areas can be used to decrease the contamination of the aquatic ecosystem by metal oxide-based contaminants and to prevent the release of metal oxide NPs applied as nanofertilizers, nanoprimers or fungicides from crop fields. Appropriate metal-accumulating aquatic plants can also be used to decontaminate the aquatic environment from excess metal contaminants. Moreover, the use of effective adsorbents such as biochar can reduce the concentrations of bioavailable metals in soil. In addition, there is a need to increase public awareness of the safe use of "artificial" nanosized additives in agriculture based on rational facts and to improve regulations and standards at the governmental and international levels for their safe and environmentally friendly applications to prevent further pollution of soil and aquatic ecosystems.

Acknowledgement

This work was supported by project APVV-22-0133.

References

Abdel-Khalek, A.A., Badran, S.R., and Marie, M.A.S. 2020. The efficient role of rice husk in reducing the toxicity of iron and aluminum oxides nanoparticles in *Oreochromis niloticus*: Hematological, bioaccumulation, and histological endpoints. Water Air Soil Pollut., 231: 53.

Abdel-Latif, H.M.R., Dawood, M.A.O., Menanteau-Ledouble, S., and El-Matbouli, M. 2020. Environmental transformation of n-TiO$_2$ in the aquatic systems and their ecotoxicity in bivalve mollusks: A systematic review. Ecotoxicol. Environ. Saf., 200: 110776.

Abdel-Latif, H.M.R., Dawood, M.A.O., Mahmoud, S.F., Shukry, M., Noreldin, A.E., Ghetas, H.A. et al. 2021. Copper oxide nanoparticles alter serum biochemical indices, induce histopathological alterations, and modulate transcription of cytokines, HSP70, and oxidative stress genes in *Oreochromis niloticus*. Animals, 11: 652.

Abdel-Salam, H.A., and Hamdi, S.A.H. 2014. Heavy metals monitoring using commercially important crustacean and mollusks collected from Egyptian and Saudi Arabia coasts. Anim. Vet. Sci., 2: 49–61.

Adeel, M., Shakoor, N., Ahmad, M.A., White, J.C., Jilani, G., and Rui, Y. 2021. Bioavailability and toxicity of nanoscale/bulk rare earth oxides in soil: Physiological and ultrastructural alterations in *Eisenia fetida*. Environ. Sci. Nano., 8: 1654–1666.

Adegbola, I,P., Aborisade, B.A., and Adetutu, A. 2021. Health risk assessment and heavy metal accumulation in fish species (*Clarias gariepinus* and *Sarotherodon melanotheron*) from industrially polluted Ogun and Eleyele Rivers, Nigeria. Toxicol. Rep., 8: 1445–1460.

Agayeva, N.J., Rzayev, F.H., Gasimov, E.K., Mamedov, C.A., Ahmadov, I.S., Sadigova, N.A. et al. 2020. Exposure of rainbow trout (*Oncorhynchus mykiss*) to magnetite (Fe$_3$O$_4$) nanoparticles in simplified food chain: Study on ultrastructural characterization. Saudi J. Biol. Sci., 27: 3258–3266.

Ahmed, B., Rizvi, A., Syed, A., Jailani, A., Elgorban, A.M., Khan, M.S. et al. 2021. Differential bioaccumulations and ecotoxicological impacts of metal-oxide nanoparticles, bulk materials, and metal-ions in cucumbers grown in sandy clay loam soil. Environ. Pollut., 289:117854.

Ahmed, B., Rizvi, A., Syed, A., Rajput, V.D., Elgorban, A.M., Al-Rejaie, S.S. et al. 2022. Understanding the phytotoxic impact of Al^{3+}, nano-size, and bulk Al$_2$O$_3$ on growth and physiology of maize (*Zea mays* L.) in aqueous and soil media. Chemosphere, 300: 134555.

Al-Abdan, M.A., Bin-Jumah, M.N., Ali, D., and Alarifi, S. 2021. Investigation of biological accumulation and eco-genotoxicity of bismuth oxide nanoparticle in fresh water snail *Lymnaea luteola*. J. King Saud. Univ. Sci., 33: 101355.

Al-Ammari, A., Zhang, L., Yang, J.Z., Wei, F., Chen, C.T., and Sun, D.P. 2021. Toxicity assessment of synthesized titanium dioxide nanoparticles in fresh water algae *Chlorella pyrenoidosa* and a zebrafish liver cell line. Ecotoxicol. Environ. Saf., 211: 111948.

Alghanmi, H.A., and Al-Khazali, Z.K.M. 2022. Impact of copper oxide nanoparticles on the growth and biochemical content of cyanobacterium *Wollea salina* Chatchawan, Kozlíková, Komárek & Kaštovský. Geomicrobiol. J., 39: 552–565.

Ali, S., Alarifi, S., Kumar, S., Ahamed, M., and Siddiqui, M.A. 2012. Oxidative stress and genotoxic effect of zinc oxide nanoparticles in freshwater snail *Lymnaea luteola* L. Aquat. Toxicol., 124–125: 83–90.

Ali, D., Almarzoug, M.H.A., Ali, H.A., Samdani, M.S., Hussain, S.A., and Alarifi, S. 2020. Fish as bio indicators to determine the effects of pollution in river by using the micronucleus and alkaline single cell gel electrophoresis assay. J. King Saud. Univ. Sci., 32: 2880–2885.

Alklaf, S.A., Zhang, S.H., Zhu, J.Z., Manirakiza, B., Addo, F.G., Guo, S.Z. et al. 2022. Impacts of nano-titanium dioxide toward *Vallisneria natans* and epiphytic microbes. J. Hazard. Mater., 436: 129066.

Allied Market Research, 2022. Metal Oxide Nanopartile Market. Global Opportunity Analysis and Industry Forcast, 2021-2030, January 2022, https://www.alliedmarketresearch.com/metal-oxide-nanoparticles-market-A15611.

Amde, M., Liu, J.F., Tan, Z.Q., and Bekana, D. 2017. Transformation and bioavailability of metal oxide nanoparticles in aquatic and terrestrial environments. A review. Environ. Pollut., 230: 250–267.

Amooaghaie, R., Norouzi, M., and Saeri, M. 2017. Impact of zinc and zinc oxide nanoparticles on the physiological and biochemical processes in tomato and wheat. Botany, 95: 441–455.

Amorim, J., Abreu, I., Rodrigues, P., Peixoto, D., Pinheiro, C., Saraiva, A. et al. 2019. *Lymnaea stagnalis* as a freshwater model invertebrate for ecotoxicological studies. Sci. Total Environ., 669: 11–28.

Antisari, L.V., Carbone, S., Gatti, A., Vianello, G., and Nannipieri, P. 2013. Toxicity of metal oxide (CeO_2, Fe_3O_4, SnO_2) engineered nanoparticles on soil microbial biomass and their distribution in soil. Soil Biol. Biochem., 60: 87–94.

Ates, M., Demir, V., Arslan, Z., Kaya, H., Yilmaz, S., and Camas, M. 2016. Chronic exposure of tilapia (*Oreochromis niloticus*) to iron oxide nanoparticles: Effects of particle morphology on accumulation, elimination, hematology and immune responses. Aquat Toxicol., 177: 22–32.

Auffan, M., Bertin, D., Chaurand, P., Pailles, C., Dominici, C., Rose, J. et al. 2013. Role of molting on the biodistribution of CeO_2 nanoparticles within *Daphnia pulex*. Water Res., 47: 3921–3930.

Auguste, M., Balbi, T., Montagna, M., Fabbri, R., Sendra, M., Blasco, J., Canesi, L., 2019. *In vivo* immunomodulatory and antioxidant properties of nanoceria ($nCeO_2$) in the marine mussel *Mytilus galloprovincialis*. Comp. Biochem. Physiol. C Toxicol. Pharmacol., 219: 95–102.

Avila-Arias, H., Nies, L.F., Gray, M.B., and Turco, R.F. 2019. Impacts of molybdenum-, nickel-, and lithium-oxide nanomaterials on soil activity and microbial community structure. Sci. Total Environ., 652: 202–211.

Aziz, S., and Abdullah, S. 2022. Toxicity of metal oxide nanoparticles in freshwater fish. *In*: Kaushik, P., and Tean, Z. (Eds.). Nanomaterials in the Battle Against Pathogens and Disease Vectors. CRC Press, eBook ISBN 9781003126256, 26 pages, https://www.taylorfrancis.com/chapters/edit/10.1201/9781003126256-10/toxicity-metal-oxide-nanoparticles-freshwater-fish-sana-aziz-sajid-abdullah.

Bacchetta, R., Santo, N., Marelli, M., Nosengo, G., and Tremolada, P. 2017. Chronic toxicity effects of $ZnSO_4$ and ZnO nanoparticles in *Daphnia magna*. Environ. Res. 152, 128–140.

Bameri, L., Sourinejad, I., Ghasemi, Z., and Fazelian, N. 2022. Toxicity of TiO_2 nanoparticles to the marine microalga *Chaetoceros muelleri* Lemmermann, 1898 under long-term exposure. Environ. Sci. Pollut. Res., 29: 30427–30440.

Benavides, M., Fernandez-Lodeiro, J., Coelho, P., Lodeiro, C., and Diniz, M.S. 2016. Single and combined effects of aluminum (Al_2O_3) and zinc (ZnO) oxide nanoparticles in a freshwater fish, *Carassius auratus*. Environ. Sci. Pollut. Res. Int., 23: 24578–24591.

Bertrand, L., Monferrán, M.V., Mouneyrac, C., and Amé, M.V. 2018. Native crustacean species as a bioindicator of freshwater ecosystem pollution: A multivariate and integrative study of multi-biomarker response in active river monitoring. Chemosphere, 206: 265–277.

Bhattacharjya, S., Adhikari, T., Sahu, A., and Patra, A.K. 2021. Ecotoxicological effect of TiO_2 nano particles on different soil enzymes and microbial community. Ecotoxicology, 30: 719–732.

Bhuvaneshwari, M., Thiagarajan, Nemade, P., Chandrasekaran, P.N., and Mukherjee, A. 2018. Toxicity and trophic transfer of P25 TiO_2 NPs from *Dunaliella salina* to *Artemia salina*: Effect of dietary and waterborne exposure. Environ. Res., 160: 39–46.

Bibi, M., Zhu, X.Y., Munir, M., and Angelidaki, I. 2021. Bioavailability and effect of α-Fe_2O_3 nanoparticles on growth, fatty acid composition and morphological indices of *Chlorella vulgaris*. Chemosphere, 282: 131044.

Blinova, I., Kanarbik, L., Irha, N., and Kahru, A. 2017. Ecotoxicity of nanosized magnetite to crustacean *Daphnia magna* and duckweed *Lemna minor*. Hydrobiologia, 798: 141–149.

Bommakanti, V., Banerjee, M., Shah, D., Manisha, K., Sri, K., and Banerjee, S. 2022. An overview of synthesis, characterization, applications and associated adverse effects of bioactive nanoparticles. Environ. Res., 214: 113919.

Boran, H., and Saffak, S. 2020. Transcriptome alterations and genotoxic influences in zebrafish larvae after exposure to dissolved aluminum and aluminum oxide nanoparticles. Toxicol. Mech. Methods, 30: 546–554.

Borase, H.P., Muley, A.B., Patil, S.V., and Singhal, R.S. 2021. Enzymatic response of *Moina macrocopa* to different sized zinc oxide particles: An aquatic metal toxicology study. Environ. Res., 194: 110609.

Bouguerra, S., Gavina, A., Natal-da-Luz, T., Sousa, J.P., Ksibi, M., and Pereira, R. 2022. The use of soil enzymes activity, microbial biomass, and basal respiration to assess the effects of cobalt oxide nanomaterial in soil microbiota. Appl. Soil. Ecol., 169: 104246.

Bour, A., Mouchet, F., Verneuil, L., Evariste, L., Silvestre, J., Pinelli, E. et al. 2015. Toxicity of CeO_2 nanoparticles at different trophic levels - Effects on diatoms, chironomids and amphibians. Chemosphere, 120: 230–236.

Bour, A., Mouchet, F., Cadarsi, S., Silvestre, J., Verneuil, L., Baque, D. et al. 2016. Toxicity of CeO_2 nanoparticles on a freshwater experimental trophic chain: A study in environmentally relevant conditions through the use of mesocosms. Nanotoxicology, 10: 245–255.

Caixeta, M.B., Araújo, P.S., Gonçalves, B.B., Silva, L.D., Grano-Maldonado, M.I., and Rocha, T.L. 2020. Toxicity of engineered nanomaterials to aquatic and land snails: A scientometric and systematic review. Chemosphere, 260: 127654.

Caixeta, M.B., Araújo, P.S., Rodrigues, C.C., Gonçalves, B.B., Araújo, O.A., Bevilaqua, G.B. et al. 2021. Risk assessment of iron oxide nanoparticles in an aquatic ecosystem: A case study on *Biomphalaria glabrata*. J. Hazard. Mater., 401: 123398.

Caldelas, C., Poitrasson, F., Viers, J., and Araus, J.L. 2020. Stable Zn isotopes reveal the uptake and toxicity of zinc oxide engineered nanomaterials in *Phragmites australis*. Environ. Sci. Nano., 7: 1927–1941.

Canesi, L., Ciacci, C., Fabbri, R., Marcomini, A., Pojana, G., and Gallo, G. 2012. Bivalve molluscs as a unique target group for nanoparticle toxicity. Mar. Environ. Res., 76: 16–21.

Canli, E.G., and Canli, M. 2020. Investigations of the nervous system biomarkers in the brain and muscle of freshwater fish (*Oreochromis niloticus*) following accumulation of nanoparticles in the tissues. Turk. J. Zool., 44: 90–103.

Canli, E.G., and Canli, M. 2021. Characterization of ATPases in the gill of freshwater mussel (*Unio tigridis*) and effects of ionic and nanoparticle forms of aluminium and copper. Comp. Biochem. Physiol. C Toxicol. Pharmacol., 247: 109059.

Canli, E.G., Celenk, A., and Canli, M. 2022. Accumulation and distribution of nanoparticles (Al_2O_3, CuO, and TiO_2) in tissues of freshwater mussel (*Unio tigridis*). Bull. Environ. Contam. Toxicol., 108: 702–707.

Canuel, E., Vaz, C., Matias, W.G., and Dewez, D. 2021. Interaction effect of EDTA, salinity, and oxide nanoparticles on alga *Chlamydomonas reinhardtii* and *Chlamydomonas euryale*. Plants, 10: 2118.

Carmo, T.L.L., Siqueira, P.R., Azevedo, V.C., Tavares, D., Pesenti, E.C., Cestari, M.M. et al. 2019. Overview of the toxic effects of titanium dioxide nanoparticles in blood, liver, muscles, and brain of a *Neotropical detritivorous*. Environ. Toxicol., 34: 457–468.

Ceschin, S., Bellini, A., and Scalici, M. 2021. Aquatic plants and ecotoxicological assessment in freshwater ecosystems: A review. Environ. Sci. Pollut. Res., 28: 4975–4988.

Chai, H.K., Yao, J., Sun, J.J., Zhang, C., Liu, W.J., Zhu, M.J. et al. 2015. The effect of metal oxide nanoparticles on functional bacteria and metabolic profiles in agricultural soil. Bull. Environ. Contam. Toxicol., 94: 490–495.

Chang, Y.N., Zhang, M.Y., Xia, L., Zhang, J., and Xing, G.M. 2012. The toxic effects and mechanisms of CuO and ZnO nanoparticles. Materials, 5: 2850–2871.

Chapman, R.L. 2013. Algae: the world's most important "plants"—an introduction. Mitig. Adapt. Strateg. Glob. Change, 18: 5–12.

Chelomin, V.P., Slobodskova, V., Zakhartsev, M., and Kukla, S. 2017. Genotoxic potential of copper oxide nanoparticles in the bivalve mollusk *Mytilus trossulus*. J. Ocean Univ. China, 16: 339–345.

Chen, L.Z., Zhou, L.N., Liu, Y.D., Deng, S.Q., Wu, H., and Wang, G.H. 2012. Toxicological effects of nanometer titanium dioxide (nano-TiO_2) on *Chlamydomonas reinhardtii*. Ecotoxicol. Environ. Saf., 84: 155–162.

Chen, T.H., Lin, C.C., and Meng, P.J. 2014. Zinc oxide nanoparticles alter hatching and larval locomotor activity in zebrafish (*Danio rerio*). J. Hazard. Mater., 277: 134–140.

Chen, Y.E., Wu, N., Mao, H.T., Zhou, J., Su, Y.Q., Zhang, Z.W. et al. 2019. Different toxicities of nanoscale titanium dioxide particles in the roots and leaves of wheat seedlings. RSC Adv., 9: 19243–19252.

Chen, Y., Zhu, W., Shu, F., Fan, Y., Yang, N., Wu, T. et al. 2020a. Nd_2O_3 nanoparticles induce toxicity and cardiac/cerebrovascular abnormality in zebrafish embryos via the apoptosis pathway. Int. J. Nanomedicine, 15: 387–400.

Chen, J., Fan, R., Wang, Y.H., Huang, T., Shang, N., He, K.H. et al. 2020b. Progressive impairment of learning and memory in adult zebrafish treated by Al_2O_3 nanoparticles when in embryos. Chemosphere, 254: 126608.

Chen, C., Unrine, J.M., Hu, Y.W., Guo, L.L., Tsyusko, O.V., Fan, Z. et al. 2021. Responses of soil bacteria and fungal communities to pristine and sulfidized zinc oxide nanoparticles relative to Zn ions. J. Hazard. Mater., 405: 124258.

Cherchi, C., Lin, Y.S., and Gu, A.Z. 2021. Nano-titanium dioxide exposure impacts nitrogen metabolism pathways in cyanobacteria. Environ. Eng. Sci., 38: 469–480.

Chupani, L., Niksirat, H., Luensmann, V., Haange, S.B., von Bergen, M., Jehmlich, N. et al. 2018. Insight into the modulation of intestinal proteome of juvenile common carp (*Cyprinus carpio* L.) after dietary exposure to ZnO nanoparticles. Sci. Total Environ., 613: 62–71.

Churilov, D.G., Churilov, G.I., Churilova, V.V., Polischuk, S.D., Borychev, S.N., and Byshov, N.V. 2021. Influence of nanoparticles on soil microflora. Int. J. Nanotechnol., 18: 788–802.

Ciacci, C., Canonico, B., Bilanicova, D., Fabbri, R., Cortese, K., Gallo, G. et al. 2012. Immunomodulation by different types of *N*-oxides in the hemocytes of marine bivalve *Mytilus galloprovincialis*. PloS One, 7: e36937.

Clements, J.C., and Comeau, L.A. 2019. Nitrogen removal potential of shellfish aquaculture harvests in eastern Canada: A comparison of culture methods. Aquac. Rep., 13: 100183.

Connolly, M., Hernandez-Moreno, D., Conde, E., Garnica, A., Navas, J.M., Torrent, F. et al. 2022. Influence of citrate and PEG coatings on the bioaccumulation of TiO_2 and CeO_2 nanoparticles following dietary exposure in rainbow trout. Environ. Sci. Eur., 34: 1.

Correia, A.T., Rebelo, D., Marques, J., and Nunes, B. 2019. Effects of the chronic exposure to cerium dioxide nanoparticles in *Oncorhynchus mykiss*: Assessment of oxidative stress, neurotoxicity and histological alterations. Environ. Toxicol. Pharmacol., 68: 27–36.

Correia, A.T., Rodrigues, S., Ferreira-Martins, D., Nunes, A.C., Ribeiro, M.I., and Antunes, S.C. 2020. Multi-biomarker approach to assess the acute effects of cerium dioxide nanoparticles in gills, liver and kidney of *Oncorhynchus mykiss*. Comp. Biochem. Physiol. C Toxicol. Pharmacol., 238: 108842.

Cox, A., and Sharma, N. 2021. *Caenorhabditis elegans*: A unique animal model to study soil–nanoparticles–organism interactions. pp.73–101. *In*: Sharma, N., and Sahi, S. (Eds.). Nanomaterial Biointeractions at the Cellular, Organismal and System Levels. Springer, Cham, Switzerland.

Cunha, R.L.D.D., and Brito-Gitirana, L. 2020. Effects of titanium dioxide nanoparticles on the intestine, liver, and kidney of *Danio rerio*. Ecotoxicol. Environ. Saf., 203: 111032.

Da Costa, M.V.J., and Sharma, P. K. 2016. Effect of copper oxide nanoparticles on growth, morphology, photosynthesis, and antioxidant response in *Oryza sativa*. Photosynthetica, 54: 110–119.

d'Amora, M., Schmidt, T.J.N., Konstantinidou, S., Raffa, De Angelis, F., and Tantussi, F. 2022. Effects of metal oxide nanoparticles in zebrafish. Oxid. Med. Cell. Longev., 2022: 3313016.

de Almeida, A.C.O., dos Santos, L.F., Vicentini, D.S., Matias, W.G., and Melegari, S.P. 2021. Evaluation of toxicity of zinc oxide nanorods on green microalgae of freshwater and marine ecosystems. pp. 85–90. *In*: Kannan, K. (Ed.). Environmental Chemistry and Ecotoxicology, vol. 3. KeAi Communications Co. Ltd.

De, A., Chakrabarti, M., Ghosh, I., and Mukherjee, A. 2016. Evaluation of genotoxicity and oxidative stress of aluminium oxide nanoparticles and its bulk form in *Allium cepa*. Nucleus-India, 59: 219–225.

de Campos, R.P., Chagas, T.Q., Alvarez, T.G.D., Mesak, C., Vieira, J.E.D., Paixao, C.F.C. et al. 2019. Analysis of ZnO nanoparticle-induced changes in *Oreochromis niloticus* behavior as toxicity endpoint. Sci. Total Environ., 682: 561–571.

Dedman, C.J., Rizk, M.M.I., Christie-Oleza, J.A., and Davies, G.L. 2021. Investigating the impact of cerium oxide nanoparticles upon the ecologically significant marine cyanobacterium *Prochlorococcus*. Front. Mar. Sci., 8: 116508.

Deepa, S., Murugananthkumar, R., Gupta, Y.R., Gowda, K.S.M., and Senthilkumaran, B. 2019. Effects of zinc oxide nanoparticles and zinc sulfate on the testis of common carp, *Cyprinus carpio*. Nanotoxicology, 13: 240–257.

Della Torre, C., Maggioni, D., Nigro, L., Farè, F., Hamza, H., Protano, G. et al. 2021. Alginate coating modifies the biological effects of cerium oxide nanoparticles to the freshwater bivalve *Dreissena polymorpha*. Sci. Total Environ., 773: 145612.

de Lucca, G.M., Freitas, E.C., and Melao, M.D.G. 2018. Effects of TiO_2 nanoparticles on the neotropical cladoceran *Ceriodaphnia silvestrii* by waterborne and dietary routes. Water Air Soil Pollut., 229: 307.

Dias, M.C., Santos, C., Pinto, G., Silva, A.M.S., and Silva, S. 2019. Titanium dioxide nanoparticles impaired both photochemical and non-photochemical phases of photosynthesis in wheat. Protoplasma, 256: 69–78.

Ding, Y.Y., Bai, X., Ye, Z.F., Ma, L.G., and Liang, L. 2019. Toxicological responses of Fe_3O_4 nanoparticles on *Eichhornia crassipes* and associated plant transportation. Sci. Total Environ., 671: 558–567.

Ding, J., Liu, J., Chang, X.B., Zhu, D., and Lassen, S.B. 2020. Exposure of CuO nanoparticles and their metal counterpart leads to change in the gut microbiota and resistome of collembolans. Chemosphere, 258: 127347.

do Amaral, D.F., Guerra, V., Almeida, K.L., Signorelli, L., Rocha, T.L., and Silva, D.D.M. 2022. Titanium dioxide nanoparticles as a risk factor for the health of Neotropical tadpoles: A case study of *Dendropsophus minutus* (Anura: Hylidae). Environ. Sci. Pollut. Res., 29: 50515–50529.

Dogra, Y., Arkill, K.P., Elgy, C., Stolpe, B., Lead, J., Valsami-Jones, E. et al. 2016. Cerium oxide nanoparticles induce oxidative stress in the sediment-dwelling amphipod *Corophium volutator*. Nanotoxicology, 10: 480–487.

Dong, L.L., Wang, H.X., Ding, T., Li, W., and Zhang, G. 2020. Effects of TiO_2 nanoparticles on the life-table parameters, antioxidant indices, and swimming speed of the freshwater rotifer *Brachionus calyciflorus*. J. Exp. Zool. A Ecol. Integr. Physiol., 333: 230–239.

Du, W.C., Gardea-Torresdey, J.L., Xie, Y.W., Yin, Y., Zhu, J.G., Zhang, X.W. et al. 2017. Elevated CO_2 levels modify TiO_2 nanoparticle effects on rice and soil microbial communities. Sci. Total Environ., 578: 408–416.

Du, J.J., Guo, R.L., Li, K., Ma, B.B., Chen, Y., and Lv, Y.N. 2019. Contributions of Zn ions to ZnO nanoparticle toxicity on *Microcystis aeruginosa* during chronic exposure. Bull. Environ. Contam. Toxicol., 103: 802–807.

Du, X.Y., Zhou, W.S., Zhang, W.X., Sun, S.G., Han, Y., Tang, Y. et al. 2021. Toxicities of three metal oxide nanoparticles to a marine microalga: Impacts on the motility and potential affecting mechanisms. Environ. Pollut., 290: 118027.

Du, W.C., Xu, M.L., Yin, Y., Sun, Y.Y., Wu, J.C., Zhu, J.G. et al. 2022. Elevated CO_2 levels alleviated toxicity of ZnO nanoparticles to rice and soil bacteria. Sci. Total Environ., 804: 149822.

Dumont, E.R., Elger, A., Azema, C., Michel, H.C., Surble, S., and Larue, C. 2022. Cutting-edge spectroscopy techniques highlight toxicity mechanisms of copper oxide nanoparticles in the aquatic plant *Myriophyllum spicatum*. Sci. Total Environ., 803: 150001.

Ebrahimbabaie, P., Meeinkuirt, W., and Pichtel, J. 2020. Phytoremediation of engineered nanoparticles using aquatic plants: Mechanisms and practical feasibility. J. Environ. Sci., 93: 151–163.

Eiras, M.I.O., Costa, L.S.D., and Barbieri, E. 2022. Copper II oxide nanoparticles (CuONPs) alter metabolic markers and swimming activity in zebra-fish (*Danio rerio*). Comp. Biochem. Physiol. C Toxicol. Pharmacol., 257: 109343.

El-Ansary, M.S.M., Hamouda, R.A., and Elshamy, M.M. 2022. Using biosynthesized zinc oxide nanoparticles as a pesticide to alleviate the toxicity on banana infested with parasitic-nematode. Waste Biomass Valorization. 13: 405–415.

Elarabi, N.I., Abdel-Rahman, A.A., Abdel-Haleem, H., and Abdel-Hakeem, M. 2022. Silver and zinc oxide nanoparticles disrupt essential parasitism, neuropeptidergic, and expansion-like proteins genes in *Meloidogyne incognita*. Exp. Parasitol., 243: 108402.

El Bahgy, H.E.K., Elabd, H., and Elkorashey, R.M. 2021. Heavy metals bioaccumulation in marine cultured fish and its probabilistic health hazard. Environ. Sci. Pollut. Res., 28: 41431–41438.

El-Ramady, H., Brevik, E.C., Fawzy, Z.F., Elsakhawy, T., Omara, A.E.D., Amer, M. et al. 2022. Nano-restoration for sustaining soil fertility: A pictorial and diagrammatic review article. Plants, 11: 2392.

Elrefaey, A., El-Gamal, A., Hamed, S.M., and El-Belely, E.F. 2021. Growth, primary metabolites, and cell morphogenesis of *Scenedesmus opoliensis* in response to zinc oxide nanoparticles stress. Egypt. J. Phycol., 2: 100–118.

Ergonul, M.B., Nassouhi, D., Celik, M., and Atasagun, S. 2021. A comparison of the removal efficiencies of *Myriophyllum spicatum* L. for zinc oxide nanoparticles (ZnO NP) in different media: A microcosm approach. Environ. Sci. Pollut. Res., 28: 8556–8568.

Fahmy, S.R., Abdel-Ghaffar, F., Bakry, F.A., and Sayed D.A. 2014. Ecotoxicological effect of sublethal exposure to zinc oxide nanoparticles on freshwater snail *Biomphalaria alexandrina*. Arch. Environ. Contam. Toxicol., 67: 192–202.

Falconer, I.R., and Humpage, A.R. 2005. Health risk assessment of cyanobacterial (blue-green algal) toxins in drinking water. Int. J. Environ. Res. Public Health, 2: 43–50.

Fang, R., Gong, J.L., Cao, W.C., Chen, Z.P., Huang, D.L., Ye, J. et al. 2022. The combined toxicity and mechanism of multi-walled carbon nanotubes and nano copper oxide toward freshwater algae: *Tetradesmus obliquus*. J. Environ. Sci., 112: 376–387.

Farsi, L., Sabzalipour, S., Khodadadi, M., Fard, N.J.H.F., and Jamali-Shini, F. 2021. The ecotoxicity of nanoparticles Co_2O_3 and Fe_2O_3 on *Daphnia magna* in freshwater. J. Water Chem. Technol., 43: 509–516.

Fazelian, N., and Yousefzadi, M. 2022. Influence of metal oxide nanoparticles on the cell wall structure of *Nannochloropsis oculata*. Aquat. Physiol. Biotechnol., 10: 55–69.

Fırat, Ö., Erol, R., and Fırat, Ö. 2022. Effects of individual and co-exposure of copper oxide nanoparticles and copper sulphate on Nile tilapia *Oreochromis niloticus*: Nanoparticles enhance pesticide biochemical toxicity. Acta Chim. Slov., 15: 81–90.

Fischer, J., Evlanova, A., Philippe, A., and Filser, J. 2021. Soil properties can evoke toxicity of copper oxide nanoparticles towards springtails at low concentrations. Environ. Pollut., 270: 116084.

Fischer, J., Talal, G.D.A., Schnee, L.S., Otomo, P.V., and Filser, J. 2022. Clay types modulate the toxicity of low concentrated copper oxide nanoparticles toward springtails in artificial test soils. Environ. Toxicol. Chem., 41: 2454–2465.

FishBase, 2022. https://www.fishbase.se/search.php.

Gaiser, B.K., Biswas, A., Rosenkranz, P., Jepson, M.A., Lead, J.R., Stone, V. et al. 2011. Effects of silver and cerium dioxide micro- and nano-sized particles on *Daphnia magna*. Environ. Monit. Assess., 13: 1227–1235.

Gambardella, C., Aluigi, M.G., Ferrando, S., Gallus, L., Ramoino, P., Gatti, A.M. et al. 2013. Developmental abnormalities and changes in cholinesterase activity in sea urchin embryos and larvae from sperm exposed to engineered nanoparticles. Aquat. Toxicol., 130: 77–85.

Gambardella, C., Ferrando, S., Morgana, S., Gallus, L., Ramoino, P., Ravera, S. et al. 2015. Exposure of *Paracentrotus lividus* male gametes to engineered nanoparticles affects skeletal bio-mineralization processes and larval plasticity. Aquat. Toxicol., 158: 181–191.

Gambardella, C., Ferrando, S., Gatti, A.M., Cataldi, E., Ramoino, P., Aluigi, M.G. et al. 2016. Review: Morphofunctional and biochemical markers of stress in sea urchin life stages exposed to engineered nanoparticles. Environ. Toxicol., 31: 1552–1562.

Garaud, M., Trapp, J., Devin, S., Cossu-Leguille, C., Pain-Devin, S., Felten, V. et al. 2015. Multibiomarker assessment of cerium dioxide nanoparticle ($nCeO_2$) sublethal effects on two freshwater invertebrates, *Dreissena polymorpha* and *Gammarus roeseli*. Aquat. Toxicol., 158: 63–74.

Garaud, M., Auffan, M., Devin, S., Felten, V., Pagnout, C., Pain-Devin, S. et al. 2016. Integrated assessment of ceria nanoparticle impacts on the freshwater bivalve *Dreissena polymorpha*. Nanotoxicology, 10: 935–944.

Garncarek, M., Dziewulska, K., and Kowalska-Góralska, M. 2022. The effect of copper and copper oxide nanoparticles on rainbow trout (*Oncorhynchus mykiss* W.) spermatozoa motility after incubation with contaminants. Int. J. Environ. Res. Public Health, 19: 8486.

Germande, O., Beaufils, F., Daffe, G., Gonzalez, P., Mornet, S., Bejko, M. et al. 2022. Cellular and molecular mechanisms of NiONPs toxicity on eel hepatocytes HEPA-E1: An illustration of the impact of Ni release from mining activity in New Caledonia. Chemosphere 303, Part, 2: 135158.

Ghobadian, M., Nabiuni, M., Parivar, K., Fathi, M., and Pazooki, J. 2015. Toxic effects of magnesium oxide nanoparticles on early developmental and larval stages of zebrafish (*Danio rerio*). Ecotoxicol. Environ. Saf., 122: 260–267.

Giri, S., Thiagarajan, V., Chandrasekaran, N., and Mukherjee, A. 2022. Ecotoxicity of nanomaterials to freshwater microalgae and fish. pp. 143–160. *In*: Guo, LH., and Mortimer, M. (Eds.). Advances in Toxicology and Risk Assessment of Nanomaterials and Emerging Contaminants. Springer, Singapore.

Glavas, O.J., Stjepanovic, N., and Hackenberger, B.K. 2022. Influence of nano and bulk copper on agile frog development. Ecotoxicology, 31: 357–365.

Gomes, T., Araujo, O., Pereira, R., Almeida, A.C., Cravo, A., and Bebianno, M.J. 2013. Genotoxicity of copper oxide and silver nanoparticles in the mussel *Mytilus galloprovincialis*. Mar. Environ. Res., 84: 51–59.

Gong, N., Shao, K.H., Che, C., and Sun, Y.Q. 2019. Stability of nickel oxide nanoparticles and its influence on toxicity to marine algae *Chlorella vulgaris*. Mar. Pollut. Bull., 149: 10532.

Gu, J., Guo, M., Huang, C., Wang, X., Zhu, Y., Wang, L. et al. 2021. Titanium dioxide nanoparticle affects motor behavior, neurodevelopment and axonal growth in zebrafish (*Danio rerio*) larvae. Sci. Total Environ., 754: 142315.

Gupta, G.S., Shanker, R., Dhawan, A., and Kumar, A. 2017. Impact of nanomaterials on the aquatic food chain. pp. 309–333 *In*: Ranjan, S, Dasgupta, N., and Lichtfouse, E. (Eds.). Nanoscience in Food and Agriculture 5. Sustainable Agriculture Reviews vol. 26. Springer Nature, Cham, Switzerland.

Gurkan, M. 2018. Effects of three different nanoparticles on bioaccumulation, oxidative stress, osmoregulatory, and immune responses of *Carcinus aestuarii*. Toxicol. Environ. Chem., 100: 693–716.

Gürkan, S.E, and Gürkan, M. 2021. Toxicity of gamma aluminium oxide nanoparticles in the Mediterranean mussel (*Mytilus galloprovincialis*): Histopathological alterations and antioxidant responses in the gill and digestive gland. Biomarkers, 26: 248–259.

Gurkan, M., Gurkan, S.E., Yilmaz, S., and Ates, M. 2021. Comparative toxicity of alpha and gamma iron oxide nanoparticles in rainbow trout: Histopathology, hematology, accumulation, and oxidative stress. Water Air Soil Pollut., 232: 37.

Gutierrez, M.F., Ale, A., Andrade, V., Bacchetta, C., Rossi, A., and Cazenave, J. 2021. Metallic, metal oxide, and metalloid nanoparticles toxic effects on freshwater microcrustaceans: An update and basis for the use of new test species. Water Environ. Res., 93: 2505–2526.

Habib, M.R., Mohamed, A.H., Osman, G.Y., Mossalem, H.S., Sharaf El-Din, A.T., and Croll, R.P. 2016. *Biomphalaria alexandrina* as a bioindicator of metal toxicity. Chemosphere, 157: 97–106.

Hamidat, M., Barakat, M., Ortet, P., Chanéac, C., Rose, J., Bottero, J.Y. et al. 2016. Design defines the effects of nanoceria at a low dose on soil microbiota and the potentiation of impacts by the canola plant. Environ. Sci. Technol., 50: 6892–6901.

Hayes, K.L., Mui, J., Song, B., Sani, E.S., Eisenman, S.W., Sheffield, J.B. et al. 2020. Effects, uptake, and translocation of aluminum oxide nanoparticles in lettuce: A comparison study to phytotoxic aluminum ions. Sci. Total Environ., 719: 137393.

He, S.Y., Feng, Y.Z., Ni, J., Sun, Y.F., Xue, L.H., Feng, Y.F. et al. 2016. Different responses of soil microbial metabolic activity to silver and iron oxide nanoparticles. Chemosphere, 147: 195–202.

Hossain, Z., Mustafa, G., and Komatsu, S. 2015. Plant responses to nanoparticle stress. Int. J. Mol. Sci., 16: 26644–26653.

Hou, J., Li, T.F., Miao, L.Z., You, G.X., Xu, Y., and Liu, S.Q. 2019. Effects of titanium dioxide nanoparticles on algal and bacterial communities in periphytic biofilms. Environ. Pollut., 251: 407–414.

Hou, J., Hu, C., Wang, Y.L., Zhang, J.Y., White, J.C., Yang, K. et al. 2022. Nano-bio interfacial interactions determined the contact toxicity of $nTiO_2$ to nematodes in various soils. Sci. Total Environ., 835: 155456.

Howe, K., Clark, M.D., Torroja, C.F., Torrance, J., Berthelot, C., Muffato, M. et al. 2013. The zebrafish reference genome sequence and its relationship to the human genome. Nature, 496: 498–503.

Hsu C.H., Wen Z.H., Lin C.S., and Chakraborty C. 2007. The zebrafish model: Use in studying cellular mechanisms for a spectrum of clinical disease entities. Curr. Neurovasc. Res., 4: 111–120.

Hu, C.W., Liu, X., Li, X.L., and Zhao, Y.J. 2014. Evaluation of growth and biochemical indicators of *Salvinia natans* exposed to zinc oxide nanoparticles and zinc accumulation in plants. Environ. Sci. Pollut. Res., 21: 732–739.

Hu, Q., Guo, F., Zhao, F., and Fu, Z. 2017. Effects of titanium dioxide nanoparticles exposure on parkinsonism in zebrafish larvae and pc12. Chemosphere, 173: 373–379.

Huang, H.Y., Chen, J.S., Liu, S.B., and Pu, S.Y. 2022. Impact of ZnO nanoparticles on soil lead bioavailability and microbial properties. Sci. Total Envion., 806: 150299.

Hurtado-Gallego, J., Pulido-Reyes, G., González-Pleiter, M., Salas, G., Leganés, F., Rosal, R. et al. 2020. Toxicity of superparamagnetic iron oxide nanoparticles to the microalga *Chlamydomonas reinhardtii*. Chemosphere, 238: 124562.

Ibrahim, A.M., Abdel-Ghaffar, F.A., Hassan, H.AM., and Fol, M.F. 2022. Assessment of molluscicidal and larvicidal activities of CuO nanoparticles on *Biomphalaria alexandrina* snails. Beni-Suef Univ. J. Basic Appl. Sci., 11: 84.

Ishwarya, R., Vaseeharan, B., Subbaiah, S., Nazar, A.K., Govindarajan, M., Alharbi, N.S. et al. 2018. *Sargassum wightii*-synthesized ZnO nanoparticles - from antibacterial and insecticidal activity to immunostimulatory effects on the green tiger shrimp *Penaeus semisulcatus*. J. Photochem. Photobiol. B, 183: 318–330.

Jafari, A., Rashidipour, M., Kamarehi, B., Alipour, S., and Ghaderpoori, M. 2022. Toxicity of green synthesized TiO_2 nanoparticles (TiO_2 NPs) on zebra fish. Environ. Res., 212, Part E: 113542.

Jafarirad, S., Ardehjani, P.H., and Movafeghi, A. 2019. Are the green synthesized nanoparticles safe for environment? A case study of aquatic plant *Azolla filiculoides* as an indicator exposed to magnetite nanoparticles fabricated using microwave hydrothermal treatment and plant extract. J. Environ. Sci. Health A Tox. Hazard. Subst. Environ. Eng., 54: 506–517.

Jampílek, J., and Kráľová, K. 2017. Nanomaterials for delivery of nutrients and growth-promoting compounds to plants. pp. 177–226. *In*: Prasad, R. (Ed.). Nanotechnology: An Agricultural Paradigm. Springer Nature, Singapore.

Jampílek, J., and Kráľová, K. 2019a. Beneficial effects of metal- and metalloid-based nanoparticles on crop production. pp. 161–219. *In*: Panpatte, D.G., and Jhala, Y.K. (Eds.). Nanotechnology for Agriculture. Springer Nature, Singapore.

Jampílek, J., and Kráľová. K. 2019b. Impact of nanoparticles on photosynthesizing organisms and their use in hybrid structures with some components of photosynthetic apparatus. pp. 255–332. *In*: Prasad, R. (Ed.). Plant Nanobionics-Nanotechnology in the Life Sciences, Vol. 1. Springer Nature, Cham, Switzerland.

Jampílek, J., Kos, J., and Kráľová, K. 2019. Potential of nanomaterial applications in dietary supplements and foods for special medical purposes. Nanomaterials, 9: 296.

Jampílek J., and Kráľová K. 2020. Nanoparticles for improving and augmenting plant functions. pp. 171–227. *In*: Jogaiah, S., Singh, H.B., Fraceto, L.F., and De Lima, R. (Eds.). Advances in Nano-Fertilizers and Nano-Pesticides Application for Crop Improvement. Woodhead Publishing & Elsevier, Kidlington, UK.

Jampílek, J., and Kráľová, K. 2021. Advances in drug delivery nanosystems using graphene-based materials and carbon nanotubes. Materials, 14: 1059.

Jampílek, J., and Kráľová, K. 2022. Advances in nanostructures for antimicrobial therapy. Materials, 15: 2388.

Javed, M., and Usmani, N. 2016. Accumulation of heavy metals and human health risk assessment via the consumption of freshwater fish *Mastacembelus armatus* inhabiting, thermal power plant effluent loaded canal. SpringerPlus, 5: 776.

Jiang, Y.Q., Zhou, P.F., Zhang, P., Adeel, M., Shakoor, N., Li, Y.B. et al. 2022. Green synthesis of metal-based nanoparticles for sustainable agriculture. Environ. Pollut., 309: 119755.

Joonas, E., Aruoja, V., Olli, K., and Kahru, A. 2019. Environmental safety data on CuO and TiO_2 nanoparticles for multiple algal species in natural water: Filling the data gaps for risk assessment. Sci. Total Environ., 647: 973–980.

Josko, I., Kusiak, M., and Oleszczuk, P. 2021. The chronic effects of CuO and ZnO nanoparticles on *Eisenia fetida* in relation to the bioavailability in aged soils. Chemosphere, 266: 128982.

Jovanovic, B. 2015. Review of titanium dioxide nanoparticle phototoxicity: Developing a phototoxicity ratio to correct the endpoint values of toxicity tests. Environ. Toxicol. Chem., 34: 1070–1077.

Karahalil, B. 2021. Nanomaterials causing cellular toxicity and genotoxicity. pp. 125–138. *In*: Kumar, V., Guleria, P., Ranjan, S., Dasgupta, N., and Lichtfouse, E. (Eds.). Nanotoxicology and Nanoecotoxicology Vol. 1. Environmental Chemistry for a Sustainable World, vol. 59. Springer, Cham, Switzerland.

Karthiga, P., Ponnanikajamideen, M., Rajendran, R.S., Annadurai, G., and Rajeshkumar, S. 2019. Characterization and toxicology evaluation of zirconium oxide nanoparticles on the embryonic development of zebrafish, *Danio rerio*. Biomed. Environ. Sci., 42: 104–111.

Kaya, H., Aydin, F., Gurkan, M., Yilmaz, S., Ates, M., Demir, V. et al. 2016. A comparative toxicity study between small and large size zinc oxide nanoparticles in tilapia (*Oreochromis niloticus*): Organ pathologies, osmoregulatory responses and immunological parameters. Chemosphere, 144: 571–582.

Kettleborough, R.N., Busch-Nentwich, E.M., Harvey, S.A., Dooley, C.M., de Bruijn, E., van Eeden, F. et al. 2013. A systematic genome-wide analysis of zebrafish protein-coding gene function. Nature, 496: 494–497.

Khan, F., Ansari, T., Shariq, M., and Siddiqui, M.A. 2021a. Nanotechnology: A new beginning to mitigate the effect of plant-parasitic nematodes. *In*: Singh, R.K., and Gopala (Eds.). Innovative Approaches in Diagnosis and Management of Crop Diseases. Apple Academic Press, Chapter 2, 25 pp.

Khan, S.T., Adil, S.F., Shaik, M.R., Alkhathlan, H.Z., Khan, M., and Khan, M. 2021b. Engineered nanomaterials in soil: Their impact on soil microbiome and plant health. Plants, 11: 109.

Khan, A., Mfarrej, M.F.B., Danish, M., Shariq, M., Khan, M.F., Ansari, M.S. et al. 2022. Synthesized copper oxide nanoparticles *via* the green route act as antagonists to pathogenic root-knot nematode, *Meloidogyne incognita*. Green Chem. Lett. Rev., 15: 491–507.

Khanna, K., Kohli, S.K., Handa, N., Kaur, H., Ohri, P., Bhardwaj, R. et al. 2021. Enthralling the impact of engineered nanoparticles on soil microbiome: A concentric approach towards environmental risks and cogitation. Ecotox. Environ. Saf., 222: 112459.

Khataee, A., Movafeghi, A., Mojaver, N., Vafaei, F., Tarrahi, R., and Dadpour, M.R. 2017. Toxicity of copper oxide nanoparticles on *Spirodela polyrrhiza*: Assessing physiological parameters. Res. Chem. Intermed., 43: 927–941.

Khoei, A.J., and Rezaei, K. 2022. Toxicity of titanium nano-oxide nanoparticles (TiO_2) on the pacific oyster, *Crassostrea gigas*: Immunity and antioxidant defence. Toxin Rev., 41: 237–246.

Khosravi-Katuli, K., Lofrano, G., Nezhad, H.P., Giorgio, A., Guida, M., Aliberti, F. et al. 2018. Effects of ZnO nanoparticles in the Caspian roach (*Rutilus rutilus caspicus*). Sci. Total Environ., 626: 30–41.

Kiss, L.V., Seres, A., Boros, G., Sárospataki, M., and Nagy, P.I. 2021. Ecotoxicological effects of zinc oxide nanoparticles on test organisms from soil micro- and mesofauna. pp. 569–588 *In*: Abd-Elsalam, K.A. (Ed.). Nanobiotechnology for Plant Protection, Elsevier, Amterdam, the Netherlands.

Koehle-Divo, V., Cossu-Leguille, C., Pain-Devin, S., Simonin, C., Bertrand, C., Sohm, B. et al. 2018. Genotoxicity and physiological effects of CeO_2 NPs on a freshwater bivalve (*Corbicula fluminea*). Aquat. Toxicol., 198: 141–148.

Kráľová, K., Masarovičová, E., and Jampílek, J. 2019. Plant responses to stress induced by toxic metals and their nanoforms. pp. 479–522. *In*: Pessarakli, M. (Ed.), Handbook of Plant and Crop Stress, 4th ed. CRC Press, Boca Raton, FL, USA.

Kráľová, K., and Jampílek, J. 2021a. Nanotechnology as effective tool for improved crop production under changing climatic conditions. pp. 463–512. *In*: Sarma, H., Joshi, S.J., Prasad, R., and Jampilek, J. (Eds.). Biobased Nanotechnology for Green Application. Springer Nature, Cham, Switzerland.

Kráľová, K., and Jampílek, J. 2021b. Impact of metal nanoparticles on marine and freshwater algae. pp. 889–921. *In*: Pessarakli, M. (Ed.). Handbook of Plant and Crop Physiology, 4th ed. CRC Press, Boca Raton, FL, USA.

Kráľová, K., and Jampílek, J. 2021c. Responses of medicinal and aromatic plants to engineered nanoparticles. Appl. Sci., 11: 1813.

Kráľová, K., Masarovičová, E., and Jampílek, J. 2021. Risks and benefits of metal-based nanoparticles for vascular plants. pp. 923–963. *In*: Pessarakli, M. (Ed.). Handbook of Plant and Crop Physiology, 4th ed. CRC Press, Boca Raton, FL, USA.

Kráľová, K., and Jampílek, J. 2022a. Metal- and metalloid-based nanofertilizers and nanopesticides for advanced agriculture. pp. 295–361. *In*: Fraceto, L.F., de Carvalho, H.W.P., de Lima, R., Ghoshal, S., and Santaella, C. (Eds.). Inorganic Nanopesticides and Nanofertilizers. Springer Nature, Cham, Switzerland.

Kráľová, K., and Jampílek, J. 2022b. Phytoremediation of environmental matrices contaminated with photosystem II-inhibiting herbicides. pp. 31–80. *In*: Siddiqui, S., Meghvansi, M.K., and Chaudhary, K.K. (Eds.). Pesticides Bioremediation. Springer Nature, Cham, Switzerland.

Krayem, M., ElKhatib, S., Hassan, Y., Deluchat, V., and Labrousse, P. 2021. In search for potential biomarkers of copper stress in aquatic plants. Aquat. Toxicol., 239: 105952.

Koce, J.D. 2017. Effects of exposure to nano and bulk sized TiO_2 and CuO in *Lemna minor*. Plant Physiol. Biochem., 119: 43–49.

Kumar, A., Rakshit, R., Bhowmik, A., Mandal, N., Das, A., and Adhikary, S. 2019. Nanoparticle-induced changes in resistance and resilience of sensitive microbial indicators towards heat stress in soil. Sustainability, 11: 862.

Kumari, S., Khanna, R.R., Nazir, F., Albaqami, M., Chhillar, H., Wahid, I. et al. 2022. Bio-synthesized nanoparticles in developing plant abiotic stress resilience: A new boon for sustainable approach. Int. J. Mol. Sci., 23: 4452.

Kwak, J.I., and An, Y.J. 2016. The current state of the art in research on engineered nanomaterials and terrestrial environments: Different-scale approaches. Environ. Res., 151: 368–382.

Lai, R.W.S., Kang, H.M., Zhou, G.J., Yung, M.M.N., He, Y.L., Ng, A.M.C. et al. 2021. Hydrophobic surface coating can reduce toxicity of zinc oxide nanoparticles to the marine copepod *Tigriopus japonicus*. Environ. Sci. Technol., 55: 6917–6925.

Lau, Z.L., Low, S.S., Ezeigwe, E.R., Chew, K.W., Chai, W.S., Bhatnagar, A. et al. 2022. A review on the diverse interactions between microalgae and nanomaterials: Growth variation, photosynthetic performance and toxicity. Bioresour. Technol., 351: 127048.

Lei, C., Zhang, L.Q., Yang, K., Zhu, L.H., and Lin, D.H. 2016. Toxicity of iron-based nanoparticles to green algae: Effects of particle size, crystal phase, oxidation state and environmental aging. Environ. Pollut., 218: 505–512.

Lei, C., Sun, Y.Q., Tsang, D.C.W., and Lin, D.H. 2018. Environmental transformations and ecological effects of iron-based nanoparticles. Environ. Pollut., 232: 10–30.

Li, F.G., Liang, Z., Zheng, X., Zhao, W., Wu, M., and Wang, Z.Y. 2015. Toxicity of nano-TiO_2 on algae and the site of reactive oxygen species production. Aquat. Toxicol., 158: 1–13.

Li, J.X., Song, Y.C., Wu, K.R., Tao, Q., Liang, Y.C., and Li, T.Q. 2018. Effects of Cr_2O_3 nanoparticles on the chlorophyll fluorescence and chloroplast ultrastructure of soybean (*Glycine max*). Environ. Sci. Pollut. Res., 25: 19446–19457.

Li, J.X., Mu, Q.L., Du, Y.L., Luo, J.P., Liu, Y.K., and Li, T.Q. 2020. Growth and photosynthetic inhibition of cerium oxide nanoparticles on soybean (*Glycine max*). Bull. Environ. Contam. Toxicol., 105: 119–126.

Li, N.J., Xu, Z., Yao, H.Y., Chen, J.W., and Li, X.H. 2021a. Impact of particle size of zinc oxide nanoparticles on its bioaccumulation and oxidative stress responses. Chin. Sci. Bull., 66: 3219–3226.

Li, Z.Q., Hu, M.H., Song, H.T., Lin, D.H., and Wang, Y.J. 2021b. Toxic effects of nano-TiO_2 in bivalves—A synthesis of meta-analysis and bibliometric analysis. J. Environ. Sci., 104: 188–203.

Li, W.X., Zhang, P.H., Qiu, H., Van Gestel, C.A.M., Peijnenburg, W.J.G.M., Cao, X.D. et al. 2022. Commonwealth of soil health: How do earthworms modify the soil microbial responses to CeO_2 nanoparticles? Environ. Sci. Technol., 56: 1138–1148.

Liu, S., Cui, M.M., Li, X.M., Thuyet, D.Q., and Fan, W.H. 2019. Effects of hydrophobicity of titanium dioxide nanoparticles and exposure scenarios on copper uptake and toxicity in *Daphnia magna*. Water Res., 154: 162–170.

Liu, Y., Li, Y., Pan, B., Zhang, X.Y., Zhang, H., Steinberg, C.E.W. et al. 2021. Application of low dosage of copper oxide and zinc oxide nanoparticles boosts bacterial and fungal communities in soil. Sci. Total Environ., 757: 143807.

Liu, Z.Q., Malinowski, C.R., and Sepulveda, M.S. 2022. Emerging trends in nanoparticle toxicity and the significance of using *Daphnia* as a model organism. Chemosphere, 291: 132941.

López-López, E., and Sedeño-Díaz, J.E. 2015. Biological indicators of water quality: The role of fish and macroinvertebrates as indicators of water quality. pp. 643–661. *In*: Armon, R., and Hänninen, O. (Eds.). Environmental Indicators. Springer, Dordrecht, the Netherlands.

Lopez-Moreno, M.L., Cedeno-Mattei, Y., Bailon-Ruiz , S.J., Vazquez-Nunez, E., Henandez-Viezcas, J.A., Perales-Pérez, O.J. et al. 2018. Environmental behavior of coated NMs: Physicochemical aspects and plant interactions. J. Hazard Mater., 347: 196–217.

Luo, J.P., Song, Y.C., Liang, J.B., Li, J.X., Islam, E., and Li, T.Q. 2020. Elevated CO_2 mitigates the negative effect of CeO_2 and Cr_2O_3 nanoparticles on soil bacterial communities by alteration of microbial carbon use. Environ. Pollut., 263, Part B: 114456.

Luo, J.P., Guo, X.Y., Liang, J.B., Song, Y.C., Liu, Y.K., Li, J.X. et al. 2021. The influence of elevated CO_2 on bacterial community structure and its co-occurrence network in soils polluted with Cr_2O_3 nanoparticles. Sci. Total Environ., 779: 146430.

Ma, J., Chen, Q.L., O'Connor, P., and Sheng, G.D. 2020. Does soil CuO nanoparticles pollution alter the gut microbiota and resistome of *Enchytraeus crypticus*? Environ. Pollut., 256: 113463.

Madany, P., Xia, C., Bhattacharjee, L., Khan, N., Li, R.P., and Liu, J. 2021. Antibacterial activity of $\gamma Fe_2O_3/TiO_2$ nanoparticles on toxic cyanobacteria from a lake in southern Illinois. Water Environ. Res., 93: 2807–2818.

Mahana, A., Guliy, O.I., and Mehta, S.K. 2021. Accumulation and cellular toxicity of engineered metallic nanoparticle in freshwater microalgae: Current status and future challenges. Ecotoxicol. Environ. Saf., 208: 111662.

Mahboub, H.H., Rashidian, G., Hoseinifar, S.H., Kamel, S., Zare, M., Ghafarifarsani, H. et al. 2022. Protective effects of *Allium hirtifolium* extract against foodborne toxicity of zinc oxide nanoparticles in common carp (*Cyprinus carpio*). Comp. Biochem. Physiol. C Toxicol. Pharmacol., 257: 109345.

Mamboungou, J., Canedo, A., Qualhato, G., Rocha, T.L., and Vieira, L.G. 2022. Environmental risk of titanium dioxide nanoparticle and cadmium mixture: Developmental toxicity assessment in zebrafish (*Danio rerio*). J. Nanopart. Res., 24: 186.

Mansouri, B., Johari, S.A., Azadi, N.A., and Sarkheil, M. 2018. Effects of waterborne ZnO nanoparticles and Zn^{2+} ions on the gills of rainbow trout (*Oncorhynchus mykiss*): Bioaccumulation, histopathological and ultrastructural changes. Turkish J. Fish. Aquat. Sci., 18: 739–746.

Martinez, D.S.T., Ellis, L.J.A., Da Silva, G.H., Petry, R., Medeiros, A.M.Z., Davoudi, H.H. et al. 2022. *Daphnia magna* and mixture toxicity with nanomaterials - Current status and perspectives in data-driven risk prediction. Nano Today, 43: 101430.

Masarovičová, E., Kráľová, K., and Zinjarde, S.S. 2014. Metal nanoparticles in plants. Formation and action. pp. 683–731. *In*: Pessarakli, M. (Ed.). Handbook of Plant and Crop Physiology, 3rd ed. CRC Press, Boca Raton, FL, USA.

Melegari, S.P., Perreault, F., Costa, R.H.R., Popovic, R., and Matias, W.G. 2013. Evaluation of toxicity and oxidative stress induced by copper oxide nanoparticles in the green alga *Chlamydomonas reinhardtii*. Aquat. Toxicol., 142-143: 431–440.

Middepogu, A., Hou, J., Gao, X.A., and Lin, D.H. 2018. Effect and mechanism of TiO_2 nanoparticles on the photosynthesis of *Chlorella pyrenoidosa*. Ecotoxicol. Environ. Saf., 161: 497–506.

Mitchell, M.J., Billingsley, M.M., Haley, R.M., Wechsler, M.E., Peppas, N.A., and Langer, R. 2021. Engineering precision nanoparticles for drug delivery. Nat. Rev. Drug Discov., 20: 101–124.

Moezzi, F., Hedayati, S.A., and Ghadermazi, A. 2018. Ecotoxicological impacts of exposure to copper oxide nanoparticles on the gill of the Swan mussel, *Anodonta cygnea* (Linnaeus, 1758). Molluscan Res., 38: 187–197.

Mohamed, A.S., Ghannam, H.E., and Soliman, H.A. 2022. The protective role of vitamins (E plus C) on Nile tilapia (*Oreochromis niloticus*) exposed to ZnO NPs and Zn ions: Bioaccumulation and proximate chemical composition. Ann. Anim. Sci., 22: 633–642.

Moos, N., Maillard, L., and Slaveykova, V.I. 2015. Dynamics of sub-lethal effects of nano-CuO on the microalga *Chlamydomonas reinhardtii* during short-term exposure. Aquat. Toxicol., 161: 267–275.

Motta, A.G.C., Guerra, V., do Amaral, D.F., Araújo, A.P.D., Vieira, L.G., Silva, DDE. et al. 2023. Assessment of multiple biomarkers in *Lithobates catesbeianus* (Anura: Ranidae) tadpoles exposed to zinc oxide nanoparticles and zinc chloride: Integrating morphological and behavioral approaches to ecotoxicology. Environ. Sci. Pollut. Res., 30: 13755–13772.

Movafeghi, A., Khataee, A., Abedi, M., Tarrahi, R., Dadpour, M., and Vafaei, F. 2018. Effects of TiO_2 nanoparticles on the aquatic plant *Spirodela polyrrhiza*: Evaluation of growth parameters, pigment contents and antioxidant enzyme activities. J. Envion. Sci., 64: 130–138.

Muna, M., Heinlaan, M., Blinova, I., Vija, H., and Kahru A. 2017. Evaluation of the effect of test medium on total Cu body burden of nano CuO-exposed *Daphnia magna*: A TXRF spectroscopy study. Environ. Pollut., 231: 1488–1496.

Murali, M., Athif, P., Suganthi, P., Bukhari, A.S., Mohamed, H.E.S., Basu, H. et al. 2018. Toxicological effect of Al_2O_3 nanoparticles on histoarchitecture of the freshwater fish *Oreochromis mossambicus*. Environ. Toxicol. Pharmacol., 59: 74–81.

Murthy, M.K., Mohanty, C.S., Swain, P., and Pattanayak, R. 2022. Assessment of toxicity in the freshwater tadpole *Polypedates maculatus* exposed to silver and zinc oxide nanoparticles: A multi-biomarker approach. Chemosphere, 293: 133511.

Nazdar, N., Imani, A., Noori, F., and Moghanlou, K.S. 2018. Effect of silymarin supplementation on nickel oxide nanoparticle toxicity to rainbow trout (*Oncorhynchus mykiss*) fingerlings: Pancreas tissue histopathology and alkaline protease activity. Iran. J. Sci. Technol. Trans. A: Sci., 42: 353–361.

Naeemi, A.S., Elmi, F., Vaezi, G., and Ghorbankhah, M. 2020. Copper oxide nanoparticles induce oxidative stress mediated apoptosis in carp (*Cyprinus carpio*) larva. Gene Rep., 19: 100676.

Nguyen, M.K., Moon, J.Y., and Lee, Y.C. 2020. Microalgal ecotoxicity of nanoparticles: An updated review. Ecotoxicol. Environ. Saf., 201: 110781.

Nikolova, M.P., and Chavali, M. 2020. Metal oxide nanoparticles as biomedical materials. Biomimetics, 5: 27.

Noor, M.N., Wu, F.L., Sokolov, E.P., Falfushynska, H., Timm, S., Haider, F. et al. 2021. Salinity-dependent effects of ZnO nanoparticles on bioenergetics and intermediate metabolite homeostasis in a euryhaline marine bivalve, *Mytilus edulis*. Sci. Total Environ., 774: 145195.

OECD. (2011). OECD Guidelines for the testing of chemicals. Test Guideline 201: Freshwater alga and cyanobacterial, growth inhibition test. Organisation for Economic Co-operation and Development, Paris, https://www.oecd.org/env/test-no-201-alga-growth-inhibition-test-9789264069923-en.htm, Accessed on: September 12, 2022.

Ogunkunle, C.O., El-Imam, A.M.A., Bassey, E., Vishwakarma, V., and Fatoba, P.O. 2020. Co-application of indigenous arbuscular mycorrhizal fungi and nano-TiO_2 reduced Cd uptake and oxidative stress in pre-flowering cowpea plants. Environ. Technol. Innov., 20: 101163.

Oliveira-Filho, E.C., Nakano, E., and Tallarico, L.D.F. 2017. Bioassays with freshwater snails *Biomphalaria* sp.: From control of hosts in public health to alternative tools in ecotoxicology. Invertebr. Reprod. Dev., 61: 49–57.

Omar, T.Y., Elshenawy, H.I.A., Abdelfattah, M.A., Shawoush, A.M.A., Mohamed, A.S., and Saad, D.Y. 2022. Biointerfrence between zinc oxide/alginate nanocomposites and freshwater bivalve. Biointerface Res. Appl. Chem., 13: 277.

Oukarroum, A., Zaidi, W., Samadani, M., and Dewez, D. 2017. Toxicity of nickel oxide nanoparticles on a freshwater green algal strain of *Chlorella vulgaris*. Biomed. Res. Int., 2017: 9528180.

Ouyang, S., Li, Y., Zheng, T., Wu, K., Wang, X., and Zhou, Q. 2022. Ecotoxicity of natural nanocolloids in aquatic environment. Water, 14: 2971.

Ozgur, M.E., Balcioglu, S., Ulu, A., Ozcan, I., Okumus, F., Koytepe, S. et al. 2018. The *in vitro* toxicity analysis of titanium dioxide (TiO_2) nanoparticles on kinematics and biochemical quality of rainbow trout sperm cells. Environ. Toxicol. Pharmacol., 62: 11–19.

Paerl, H.W., and Otten, T.G. 2013. Harmful cyanobacterial blooms: Causes, consequences, and controls. Microb. Ecol., 65: 995–1010.

Parada, J., Rubilar, O., Fernandez-Baldo, M.A., Bertolino, F.A., Duran, N., Seabra, A.B. et al. 2019. The nanotechnology among US: Are metal and metal oxides nanoparticles a nano or mega risk for soil microbial communities? Crit. Rev. Biotechnol., 39: 157–172.

Pecoraro, R., Scalisi, E.M., Messina, G., Fragala, G., Ignoto, S., Salvaggio, A. et al. 2021a. *Artemia salina*: A microcrustacean to assess engineered nanoparticles toxicity. Microsc. Res. Techn., 84: 531–536.

Pecoraro, R., Scalisi, E.M., Iaria, C., Capparucci, F., Rizza, M.T., Ignoto, S. et al. 2021b. Toxicological assessment of CeO_2 nanoparticles on early development of zebrafish. Toxicol. Res., 10: 570–578.

Pena, R.V., Machado, R.C., Caixeta, M.B., Araújo, P.S., Oliveira, E.C., Silva, S.M. et al. 2022. Lauric acid bilayer-functionalized iron oxide nanoparticles disrupt early development of freshwater snail *Biomphalaria glabrata* (Say, 1818). Acta Trop., 229:, 106362.

Peng, C., Zhang, H., Fang, H.X., Xu, C., Huang, H.M., Wang, Y. et al. 2015. Natural organic matter-induced alleviation of the phytotoxicity to rice (*Oryza sativa* L.) caused by copper oxide nanoparticles. Environ. Toxicol. Chem., 34: 1996–2003.

Peng, C., Zhang, W., Gao, H.P., Li, Y., Tong, X., Li, K.G. et al. 2017. Behavior and potential impacts of metal-based engineered nanoparticles in aquatic environments. Nanomaterials, 7: 21.

Pepe, N., Canli, E.G., and Canli, M. 2022. Salinity and/or nanoparticles (Al_2O_3, TiO_2) affect metal accumulation and ATPase activity in freshwater fish (*Oreochromis niloticus*). Environ. Toxicol. Pharmacol., 94: 103931.

Pereira, F.F., Paris, E.C., Bresolin, J.D., Mitsuyuki, M.C., Ferreira, M.D., and Correa, D.S. 2020. The effect of ZnO nanoparticles morphology on the toxicity towards microalgae *Pseudokirchneriella subcapitata*. J. Nanosci. Nanotechnol., 20: 48–63.

Pereira, F.F., Ferreira, M.D., Jonsson, C.M., de Jesus, K.R., de Castro, V.L.S., and Correa, D.S. 2021. Toxicity of engineered nanostructures in aquatic environments. pp. 71–202. *In*: Kumar, V., Guleria, P., Ranjan, S., Dasgupta, N., Lichtfouse, E. (Eds.). Nanotoxicology and Nanoecotoxicology Vol. 1. Environmental Chemistry for a Sustainable World, vol 59. Springer, Cham, Switzerland.

Pikula, K., Chaika, V., Zakharenko, A., Savelyeva, A., Kirsanova, I., Anisimova, A. et al. 2020a. Toxicity of carbon, silicon, and metal-based nanoparticles to the hemocytes of three marine bivalves. Animals, 10: 827.

Pikula, K., Zakharenko, A., Chaika, V., Em, I., Nikitina, A., Avtomonov, E. et al. 2020b. Toxicity of carbon, silicon, and metal-based nanoparticles to sea urchin *Strongylocentrotus intermedius*. Nanomaterials, 10: 1825.

Pirsaheb, M., Azadi, N.A., Miglietta, M.L., Sayadi, M.H., Blahova, J., Fathi, M. et al. 2019. Toxicological effects of transition metal-doped titanium dioxide nanoparticles on goldfish (*Carassius auratus*) and common carp (*Cyprinus carpio*). Chemosphere, 215: 904–915.

Poynton, H.C., Lazorchak, J.M., Impellitteri, C.A., Blalock, B., Smith, M.E., Struewing, K. et al. 2013. Toxicity and transcriptomic analysis in *Hyalella azteca* suggests increased exposure and susceptibility of epibenthic organisms to zinc oxide nanoparticles. Environ. Sci. Technol., 47: 9453–9460.

Pradhan, A., Seena, S., Pascoal, C., and Cassio, F. 2012. Copper oxide nanoparticles can induce toxicity to the freshwater shredder *Allogamus ligonifer*. Chemosphere, 9: 1142–1150.

Pradhan, A., Silva, C.O., Silva, C., Pascoal, C., and Cassio, F. 2016. Enzymatic biomarkers can portray nanoCuO-induced oxidative and neuronal stress in freshwater shredders. Aquat. Toxicol., 180: 227–235.

Prakash, V., Peralta-Videa, J., Tripathi, D.K., Ma, X.M., and Sharma, S. 2021. Recent insights into the impact, fate and transport of cerium oxide nanoparticles in the plant-soil continuum. Ecotox. Environ. Saf., 221: 112403.

Qu, M.J., Liu, G.L., Zhao, J.W., Li, H.D., Liu, W., Yan, Y.P. et al. 2020. Fate of atrazine and its relationship with environmental factors in distinctly different lake sediments associated with hydrophytes. Environ. Pollut., 256: 113371.

Rajkumar, K.S., Sivagaami, P., Ramkumar, A., Murugadas, A., Srinivasan, V., Arun, S. et al. 2022. Bio-functionalized zinc oxide nanoparticles: Potential toxicity impact on freshwater fish *Cyprinus carpio*. Chemosphere, 290: 133220.

Rajput, V.D., Minkina, T., Sushkova, S., Mandzhieva, S., Fedorenko, A., Lysenko, V. et al. 2019. Structural and ultrastructural changes in nanoparticle exposed plants. pp. 281–295. *In*: Pudake, R.N., Chauhan, N., and Kole, C. (Eds.). Nanoscience for Sustainable Agriculture. Springer Nature, Cham, Switzerland.

Ramskov, T., Croteau, M.L., Forbes, V.E., and Selck, H. 2015. Biokinetics of different-shaped copper oxide nanoparticles in the freshwater gastropod, *Potamopyrgus antipodarum*. Aquat. Toxicol., 163: 71–80.

Rashidian, G., Lazado, C.C., Mahboub, H.H., Mohammadi-Aloucheh, R., Prokić, M.D., Nada, H.S. et al. 2021. Chemically and green synthesized ZnO nanoparticles alter key immunological molecules in common carp (*Cyprinus carpio*) skin mucus. Int. J. Mol. Sci., 22: 3270.

Rasool, S., Faheem, M., Hanif, U., Bahadur, S., Taj, S., Liaqat, F. et al. 2022. Toxicological effects of the chemical and green ZnO NPs on *Cyprinus carpio* L. observed under light and scanning electron microscopy. Microsc. Res. Tech., 85: 848–860.

Rastgar, S., Ardeshir, R.A., Segner, H., Tyler, C.R., Peijnenburg, W.J.G.M., Wang, Z.J. et al. 2022. Immunotoxic effects of metal-based nanoparticles in fish and bivalves. Nanotoxicology, 16: 88–113.

Rastogi, R.P., Madamwar, D., and Incharoensakdi, I. 2015. Bloom dynamics of cyanobacteria and their toxins: Environmental health impacts and mitigation strategies. Front. Microbiol., 6: 254.

Riaz, A., Riaz, M.A., Shahzad, K., Ijaz, B., and Khan, M.S. 2020. Deposition trend of subchronic exposure of copper oxide nanoparticles (CuO-NPs) and its effect on the antioxidant system of *Labeo rohita*. Int. Nano Lett., 10: 279–285.

Rippner, D.A., Green, P.G., Young, T.M., and Parikh, S.J. 2018. Dissolved organic matter reduces CuO nanoparticle toxicity to duckweed in simulated natural systems. Environ. Pollut., 234: 692–698.

Rippner, D.A., Lien, J., Balla, H., Guo, T., Green, P.G., Young, T.M. et al. 2020. Surface modification induced cuprous oxide nanoparticle toxicity to duckweed at sub-toxic metal concentrations. Sci. Total Environ., 22: 137607.

Rocco, L., Santonastaso, M., Mottola, F., Costagliola, D., Suero, T., Pacifico, S. et al. 2015. Genotoxicity assessment of TiO_2 nanoparticles in the teleost *Danio rerio*. Ecotoxicol. Environ. Saf., 113: 223–230.

Rossetto, A.L.D.O.F., Melegari, S.P., Ouriques, L.C., and Matias, W.G. 2014. Comparative evaluation of acute and chronic toxicities of CuO nanoparticles and bulk using *Daphnia magna* and *Vibrio fischeri*. Sci. Total Environ., 490: 807–814.

Roy, B., Kadam, K., Krishnan, S.P., Natarajan, C., and Mukherjee, A. 2021. Assessing combined toxic effects of tetracycline and P25 titanium dioxide nanoparticles using *Allium cepa* bioassay. Front. Environ. Sci. Eng., 15: 6.

Roy, D., Adhikari, S., Adhikari, A., Ghosh, S., Azahar, I., Basuli, D. et al. 2022. Impact of CuO nanoparticles on maize: Comparison with CuO bulk particles with special reference to oxidative stress damages and antioxidant defense status. Chemosphere, 287:131911.

Rundle, A., Robertson, A.B., Blay, A.M., Butler, K.M.A., Callaghan, N.I., Dieni, C.A. et al. 2016. Cerium oxide nanoparticles exhibit minimal cardiac and cytotoxicity in the freshwater fish *Catostomus commersonii*. Comp. Biochem. Physiol. C Toxicol. Pharmacol., 181: 19–26.

Salas-Leiva, J., Salas-Leiva, D.E., Tovar-Ramirez, D., Herrera-Perez, G., Tarango-Rivero, S., Luna-V. et al. 2021. Copper oxide nanoparticles slightly affect diversity and metabolic profiles of the prokaryotic community in pecan tree (*Carya illinoinensis*) rhizospheric soil. Appl. Soil Ecol., 157: 103772.

Salehi, H., Diego, N.D., Rad, A.C., Benjamin, J.J., Trevisan, M., and Lucini, L. 2021. Exogenous application of ZnO nanoparticles and $ZnSO_4$ distinctly influence the metabolic response in *Phaseolus vulgaris* L. Sci. Total Environ., 778: 146331.

Samarajeewa, A.D., Velicogna, J.R., Schwertfeger, D.M., Princz, J.I., Subasinghe, R.M., Scroggins, R.P. et al. 2021. Ecotoxicological effects of copper oxide nanoparticles (nCuO) on the soil microbial community in a biosolids-amended soil. Sci Total Environ., 763: 143037.

Samim, A.R., and Vaseem, H. 2021. Assessment of the potential threat of nickel(II) oxide nanoparticles to fish *Heteropneustes fossilis* associated with the changes in haematological, biochemical and enzymological parameters. Environ. Sci. Pollut. Res., 28: 54630–54646.

Samim, A.R., Singh, V.K., and Vaseem, H. 2022. Assessment of hazardous impact of nickel oxide nanoparticles on biochemical and histological parameters of gills and liver tissues of *Heteropneustes fossilis*. J. Trace Elem. Med. Biol., 74: 127059.

Sankar, R., Prasath, B.B., Nandakumar, R., Santhanam, P., Shivashangari, K.S., and Ravikumar, V. 2014. Growth inhibition of bloom forming cyanobacterium *Microcystis aeruginosa* by green route fabricated copper oxide nanoparticles. Environ. Sci. Pollut. Res., 21: 14232–14240.

Santos-Rasera, J.R., Monteiro, R.T.R., and de Carvalho, H.W.P. 2022. Investigation of acute toxicity, accumulation, and depuration of ZnO nanoparticles in *Daphnia magna*. Sci. Total Environ., 821: 153307.

Saxena, P., and Harish. 2018. Nanoecotoxicological reports of engineered metal oxide nanoparticles on algae. Curr. Pollut. Rep., 4: 128–142.

Saxena, P., and Harish. 2019. Toxicity assessment of ZnO nanoparticles to freshwater microalgae *Coelastrella terrestris*. Environ. Sci. Pollut. Res., 26: 26991–27001.

Saxena, P., Sangela, V., and Harish. 2020. Toxicity evaluation of iron oxide nanoparticles and accumulation by microalgae *Coelastrella terrestris*. Environ. Sci., Pollut. Res., 27: 19650–19660.

Saxena, P., Saharan, V., Baroliya, P.K., Gour, V.S., Rai, M.K., and Harish, 2021. Mechanism of nanotoxicity in *Chlorella vulgaris* exposed to zinc and iron oxide. Toxicol. Rep., 8: 724–731.

Scola, S., Blasco, J., and Campana, O. 2021. "Nanosize effect" in the metal-handling strategy of the bivalve *Scrobicularia plana* exposed to CuO nanoparticles and copper ions in whole-sediment toxicity tests. Sci. Total Environ., 760: 143886.

Scott-Fordsmand, J.J., Irizar, A., and Amorim, M.J.B. 2022. Full life cycle test with *Eisenia fetida* – copper oxide NM toxicity assessment. Ecotox. Envion. Saf., 241: 113720.

Sellami, B., Bouzidi, I., Hedfi, A., Almalki, M., Rizk, R., Pacioglu, O. et al. 2021. Impacts of nanoparticles and phosphonates in the behavior and oxidative status of the mediterranean mussels (*Mytilus galloprovincialis*). Saudi J. Biol. Sci., 28: 6365–6374.

Shahzad, K., Khan, M.N., Jabeen, F., Chaudhry, A.S., Khan, M.K.A., Ara, C. et al. 2022. Study of some toxicological aspects of titanium dioxide nanoparticles through oxidative stress, genotoxicity, and histopathology in tilapia, *Oreochromis mossambicus*. Bio. Nano. Sci., 12: 1116–1124.

Shan, Q., Liu, Y., Zhang, X.L., Shao, J.F., Hei, D.Q., Ling, J.S. et al. 2020. EDXRF analysis of TiO_2 nanoparticles bioaccumulation in aquatic plant, *Salvinia natans*. Microchem. J., 155: 104784.

Sharan, A., and Nara, S. 2019. Phytotoxic properties of zinc and cobalt oxide nanoparticles in algaes. pp. 1–22. *In*: Tripathi, D.K., Ahmad, P., Sharma, S., Chauhan, D.K., and Dubey, N.K. (Eds.). Nanomaterials in Plants, Algae and Microorganisms. Concepts and Controversies, vol. 2, Elsevier, Amterdam, the Netherlands.

Shariati, F., Poordeljoo, T., and Zanjanchi, P. 2020. The acute toxicity of SiO_2 and Fe_3O_4 nano-particles on *Daphnia magna*. Silicon., 12: 2941–2946.

Shahzad, K., Khan, M.N., Jabeen, F., Kosour, N., Chaudhry, A.S., Sohail, M. et al. 2019. Toxicity of zinc oxide nanoparticles (ZnO-NPs) in tilapia (*Oreochromis mossambicus*): Tissue accumulation, oxidative stress, histopathology and genotoxicity. Int. J. Environ. Sci. Technol., 16: 1973–1984.

Shemawar, Mahmood, A., Hussain, S., Mahmood, F., Iqbal, M., Shahid, M., Ibrahim, M. et al. 2021. Toxicity of biogenic zinc oxide nanoparticles to soil organic matter cycling and their interaction with rice-straw derived biochar. Sci. Rep., 11: 8429.

Sheng, L., Wang, L., Su, M., Zhao, X., Hu, R., Yu, X. et al. 2016. Mechanism of TiO_2 nanoparticle-induced neurotoxicity in zebrafish (*Danio rerio*). Environ. Toxicol., 31: 163–175.

Shevlin, D., O'Brien, N., and Cummins, E. 2018. Silver engineered nanoparticles in freshwater systems – Likely fate and behaviour through natural attenuation processes. Sci. Total Environ., 621: 1033–1046.

Shi, Y., Xiao, Y.M., Li, Z.Q., Zhang, X.Y., Liu, T., Li, Y. et al. 2021. Microorganism structure variation in urban soil microenvironment upon ZnO nanoparticles contamination. Chemosphere, 273: 128565.

Sievers, M., Hale, R., Parris, K.M., Melvin, S.D., Lanctot, C.M., and Swearer, S.E. 2019. Contaminant-induced behavioural changes in amphibians: A meta-analysis. Sci. Total Environ., 693: 133570.

Simonin, M., Martins, J.M.F., Uzu, G., Vince, E., and Richaume, A. 2016. Combined study of titanium dioxide nanoparticle transport and toxicity on microbial nitrifying communities under single and repeated exposures in soil columns. Environ. Sci. Technol., 50: 10693–10699.

Simonin, M., Martins, J.M.F., Le Roux, X., Uzu, G., Calas, A., and Richaume, A. 2017. Toxicity of TiO_2 nanoparticles on soil nitrification at environmentally relevant concentrations: Lack of classical dose-response relationships. Nanotoxicology, 11: 247–255.

Sitia, G., Fiordaliso, F., Violatto, M.B., Alarcon, J.F., Talamini, L., Corbelli, A. et al. 2022. Food-grade titanium dioxide induces toxicity in the nematode *Caenorhabditis elegans* and acute hepatic and pulmonary responses in mice. Nanomaterials, 12: 1669.

Smii, H., Khazri, A., Ali, M.B., Mezni, A., Hedfi, A., Albogami, B. et al. 2021. Titanium dioxide nanoparticles are toxic for the freshwater mussel *Unio ravoisieri*: Evidence from a multimarker approach. Diversity, 13: 679.

Soliman, H.A.M., Hamed, M., and Sayed, A.E.H. 2021. Investigating the effects of copper sulfate and copper oxide nanoparticles in Nile tilapia (*Oreochromis niloticus*) using multiple biomarkers: The prophylactic role of *Spirulina*. Environ. Sci. Pollut. Res., 28: 30046–30057.

Song, U., and Lee, S. 2016. Phytotoxicity and accumulation of zinc oxide nanoparticles on the aquatic plants *Hydrilla verticillata* and *Phragmites australis*: Leaf-type-dependent responses. Environ. Sci. Pollut. Res., 23: 8539–8545.

Spengler, A., Wanninger, L., and Pflugmacher, S. 2017. Oxidative stress mediated toxicity of TiO_2 nanoparticles after a concentration and time dependent exposure of the aquatic macrophyte *Hydrilla verticillata*. Aquat. Toxicol., 190: 32–39.

Sukhanova, A., Bozrova, S., Sokolov, P., Berestovoy, M., Karaulov, A., and Nabiev, I. 2018. Dependence of nanoparticle toxicity on their physical and chemical properties. Nanoscale Res. Lett., 13: 44.

Susanto, N.G. 2021. Crustacea: The increasing economic importance of crustaceans to humans. *In*: Ranz, R.E.R. (Ed.). Arthropods, IntechOpen, Rijeka, Croatia, doi:10.5772/intechopen.9625.

Tabat, J.L., Adakole, J.A., Gadzama, I.M.K., and Yerima, R. 2022. Sublethal toxicity of cadmium chloride ($CdCl_2$) on the growth parameters of *Xenopus laevis* tadpoles (Daudin, 1802). Int. J. Fish. Aquat. Stud., 10: 23–28.

Taherian, S.M.R., Hosseini, S.A., Jafari, A., Etminan, A., and Birjandi, M. 2020. Acute toxicity of zinc oxide nanoparticles from *Satureja hortensis* on rainbow trout (*Oncorhynchus mykiss*). Turkish J. Fish. Aquat. Sci., 20: 481–489.

Tang, Y.L., Xin, H.J., Yang S., Guo, M.T., Malkoske, T., Yin, D.Q. et al. 2018. Environmental risks of ZnO nanoparticle exposure on *Microcystis aeruginosa*: Toxic effects and environmental feedback. Aquat. Toxicol., 204: 19–26.

Tasar, N. 2022. Mitotic effects of copper oxide nanoparticle on root development and root tip cells of *Phaseolus vulgaris* L. seeds. Microsc. Res. Tech., 85: 3895–3907.

Tavabe, K.R., Kuchaksaraei, B.S., and Javanmardi, S. 2020. Effects of ZnO nanoparticles on the Giant freshwater prawn (*Macrobrachium rosenbergii*, de Man, 1879): Reproductive performance, larvae development, CHH concentrations and anti-oxidative enzymes activity. Anim. Reprod. Sci., 221: 106603.

Tauseef, A., Hisamuddin, Gupta, J., Rehman, A., and Uddin, I. 2021a. Differential response of cowpea towards the CuO nanoparticles under *Meloidogyne incognita* stress. S. Afr. J. Bot., 139: 175–182.

Tauseef, A., Hisamuddin, Khalilullah, A., and Uddin, I. 2021b. Role of MgO nanoparticles in the suppression of *Meloidogyne incognita*, infecting cowpea and improvement in plant growth and physiology. Exp. Parasitol., 220: 108045.

Tedesco, P.A., Beauchard, O., Bigorne, R., Blanchet, S., Buisson, L., Conti, L. et al. 2017. A global database on freshwater fish species occurrence in drainage basins. Sci. Data, 4: 170141.

Temiz, Ö., and Kargın, F. 2022. Toxicological impacts on antioxidant responses, stress protein, and genotoxicity parameters of aluminum oxide nanoparticles in the liver of *Oreochromis niloticus.* Biol. Trace Elem. Res., 200: 1339–1346.

Thiagarajan, V., Seenivasan, R., Chandrasekaran, N., and Mukherjee, A. 2021. The toxicological effects of titanium dioxide nanoparticles on marine microalgae. pp. 479–493. *In*: Tsatsakis, A.M. (Ed.). Toxicological Risk Assessment and Multi-Systen Health Impacts from Exposure. Elsevier, Amterdam, the Netherlands.

Thomas, R. 2019. Marine Biology: An Ecological Approach. ED-Tech Press, Waltham Abbey, UK.

Thummala, H., Raju, N.V., Manasa, B., Paritala, V., Srikanth, K., and Nutalapati, V. 2022. Sublethal effects of zinc oxide nanoparticles induced toxicity and oxidative stress in *Pila virens*: A validation of homology modelling and docking. Mater. Sci. Eng. B., 283: 115842.

Thwala, M., Klaine, S.J., and Musee, N. 2016. Interactions of metal-based engineered nanoparticles with aquatic higher plants: A review of the state of current knowledge. Environ. Toxicol. Chem., 35: 1677–1694.

Torbati, S., Khataee, A., and Saadi, S. 2017. Comparative phytotoxicity of undoped and Er-doped ZnO nanoparticles on *Lemna minor* L.: changes in plant physiological responses. Turk. J. Biol., 41: 575–586.

Torres-Garcia, D., Faria, M., Soares, A.M.V.M., Barata, C., Montes, R., Baeza, M. et al. 2020. Lethal and sub-lethal effects of nanosized titanium dioxide particles on *Hydropsyche exocellata* Dufour, 1841 insect. Aquat. Insects, 41: 85–103.

Trevisan, R., Delapedra, G., Mello, D.F., Arl, M., Schmidt, E.C., Meder, F. et al. 2014. Gills are an initial target of zinc oxide nanoparticles in oysters *Crassostrea gigas*, leading to mitochondrial disruption and oxidative stress. Aquat. Toxicol., 153: 27–38.

Tryfon, P., Kamou, N.N., Ntalli, N., Mourdikoudis, S., Karamanoli, K., Karfaridis, D. et al. 2022. Coated Cu-doped ZnO and Cu nanoparticles as control agents against plant pathogenic fungi and nematodes. NanoImpact, 28: 100430.

Tuncsoy, M. 2021. Impacts of titanium dioxide nanoparticles on serum parameters and enzyme activities of *Clarias gariepinus*. Bull. Environ. Contamin, Toxicol., 106: 629–636.

Ucar, A., Parlak, V., Ozgeris, F.B., Yeltekin, A.C., Arslan, M.E., Alak, G. et al. 2022. Magnetic nanoparticles-induced neurotoxicity and oxidative stress in brain of rainbow trout: Mitigation by ulexite through modulation of antioxidant, anti-inflammatory, and antiapoptotic activities. Sci. Total Environ., 838: 155718.

Van der Schatte, A.O., Jones, L., Le Vay, L, Christie, M., Wilson, J., and Malham, S.K. 2018. A global review of the ecosystem services provided by bivalve aquaculture. Rev Aquac., 12: 3–25.

Vaughn, C.C., and Hoellein, T.J. 2018. Bivalve impacts in freshwater and marine ecosystems. Annu. Rev. Ecol. Evol. Syst., 49: 183–208.

Wang, Z.Y., Yin, L.Y., Zhao, J., and Xing, B.S. 2016. Trophic transfer and accumulation of TiO_2 nanoparticles from clamworm (*Perinereis aibuhitensis*) to juvenile turbot (*Scophthalmus maximus*) along a marine benthic food chain. Water Res., 95: 250–259.

Wang, F., Guan, W., Xu, L., Ding, Z.Y., Ma, H.L., Ma, A.Z. et al. 2019. Effects of nanoparticles on algae: Adsorption, distribution, ecotoxicity and fate. Appl. Sci., 9: 1534.

Wang, W.W., Yang, Y.Z., Yang, L.H., Luan, T.G., and Lin, L. 2021. Effects of undissociated SiO_2 and TiO_2 nano-particles on molting of *Daphnia pulex*: Comparing with dissociated ZnO nano particles. Ecotoxicol. Environ. Saf., 222: 112491.

Wang, X.G., and Wang, W.X. 2022. Cu-based nanoparticle toxicity to zebrafish cells regulated by cellular discharges. Environ. Pollut., 292: 118296.

Wang, Y., Dimkpa, C., Deng, C.Y., Elmer, W.H., Gardea-Torresdey, J., and White, J.C. 2022a. Impact of engineered nanomaterials on rice (*Oryza sativa* L.): A critical review of current knowledge. Environ. Pollut., 297: 118738.

Wang, X.P., Liu, X.J., Yang, X., Wang, L.Q., Yang, J., Yan, X.L. et al. 2022b. *In vivo* phytotoxic effect of yttrium-oxide nanoparticles on the growth, uptake and translocation of tomato seedlings (*Lycopersicon esculentum*). Ecotoxicol. Environ. Saf., 242: 113939.

Wani, R.A., Ganai, B.A., Shah, M.A., and Uqab, B. 2017. Heavy metal uptake potential of aquatic plants through phytoremediation technique – A review. J. Bioremediat. Biodegrad., 8: 404.

Wu, D., Zhang, J., Du, W. Yin, Y., and Guo, H.Y. 2022. Toxicity mechanism of cerium oxide nanoparticles on cyanobacteria *Microcystis aeruginosa* and their ecological risks. Environ. Sci. Pollut. Res., 29: 34010–34018.

Xia, J., Zhao, H.Z., and Lu, G.H. 2013. Effects of selected metal oxide nanoparticles on multiple biomarkers in *Carassius auratus*. Biomed. Environ. Sci., 26: 742–749.

Xia, B., Chen, B., Sun, X., Qu, K., Ma, F., and Du, M. 2015. Interaction of TiO2 nanoparticles with the marine microalga *Nitzschia closterium*: Growth inhibition, oxidative stress and internalization. Sci. Total Environ., 508: 525–533.

Xia, B., Zhu, L., Han, Q., Sun, X.M., Chen, B.J., and Qu, K.M. 2017. Effects of TiO$_2$ nanoparticles at predicted environmental relevant concentration on the marine scallop *Chlamys farreri*: An integrated biomarker approach. Environ. Toxicol. Pharmacol., 50: 128–135.

Xia, B., Sui, Q., Sun, X.M., Han, Q., Chen, B.J., Zhu, L. et al. 2018. Ocean acidification increases the toxic effects of TiO$_2$ nanoparticles on the marine microalga *Chlorella vulgaris*. J. Hazard. Mater., 346: 1–9.

Xiong, T.T., Zhang, T., Xian, Y.H., Kang, Z.Z., Zhang, S.S., Dumat, C. et al. 2021. Foliar uptake, biotransformation, and impact of CuO nanoparticles in *Lactuca sativa* L. var. ramosa Hort. Environ. Geochem. Health, 43: 423–439.

Xu, C., Peng, C., Sun, L.J., Zhang, S., Huang, H.M., Chen, Y.X. et al. 2015. Distinctive effects of TiO$_2$ and CuO nanoparticles on soil microbes and their community structures in flooded paddy soil. Soil Biol. Biochem., 86: 24–33.

Xu, J.B., Luo, X.S., Wang, Y.L., and Feng, Y.Z. 2018. Evaluation of zinc oxide nanoparticles on lettuce (*Lactuca sativa* L.) growth and soil bacterial community. Environ. Sci. Pollut. Res., 25: 6026–6035.

Xu, K., Li, Z., Juneau, P., Xiao, F.S., Lian, Y.L., Zhang, W. et al. 2021. Toxic and protective mechanisms of cyanobacterium *Synechocystis* sp. in response to titanium dioxide nanoparticles. Environ. Pollut., 274: 116508.

Xu, Z.X., Long, X., Jia, Y., Zhao, D.M., and Pan, X.J. 2022. Occurrence, transport, and toxicity of nanomaterials in soil ecosystems: A review. Environ. Chem. Lett., 20: 3943–3969.

Yang, Z.Z., Chen, J., Dou, R.Z., Gao, X., Mao, C.B., and Wang, L. 2015. Assessment of the phytotoxicity of metal oxide nanoparticles on two crop plants, maize (*Zea mays* L.) and rice (*Oryza sativa* L.). Int. J. Environ. Res. Public Health, 12: 15100–15109.

Yang, Y.R., Zhang, C., Huang, X.J., Gui, X.W., Luo, Y.F., and Li, Z.L. 2020. Exogenous Fe^{2+} alleviated the toxicity of CuO nanoparticles on *Pseudomonas tolaasii* Y-11 under different nitrogen sources. PEERJ., 8: e10351.

Yang, Y.Y., Fan, X.L., Zhang, J.K., Qiao, S.Y., Wang, X., Zhang, X.Y. et al. 2022. A critical review on the interaction of iron-based nanoparticles with blue-green algae and their metabolites: From mechanisms to applications. Algal Res., 64: 102670.

Yao, Y.S., Zhang, T., and Tang, M. 2022. A critical review of advances in reproductive toxicity of common nanomaterials to *Caenorhabditis elegans* and influencing factors. Environ. Pollut., 306: 119270.

Yin, Y., Hu, Z.X., Du, W.C., Ai, F.X., Ji, R., Gardea-Torresdey, J.L. et al. 2017. Elevated CO$_2$ levels increase the toxicity of ZnO nanoparticles to goldfish (*Carassius auratus*) in a water-sediment ecosystem. J. Hazard. Mater., 327: 64–70.

Yin, J.A., Huang, G.H., An, C.J., and Feng, R.F. 2022. Nanocellulose enhances the dispersion and toxicity of ZnO NPs to green algae *Eremosphaera viridis*. Environ. Sci. Nano., 9: 393–405.

You, T.T., Liu, D.D., Chen, J., Yang, Z.Z., Dou, R.Z., Gao, X. et al. 2018. Effects of metal oxide nanoparticles on soil enzyme activities and bacterial communities in two different soil types. J. Soils Sediments, 18: 211–221.

Yuan, T., Gao, L., Zhan, W., and Dini, D. 2022. Effect of particle size and surface charge on nanoparticles diffusion in the brain white matter. Pharm. Res., 39: 767–781.

Yue, L., Zhao, J., Yu, X.Y., Lv, K.M., Wang, Z.Y., and Xing, B.S. 2018. Interaction of CuO nanoparticles with duckweed (*Lemna minor* L.): Uptake, distribution and ROS production sites. Environ. Pollut. 243, Part A, 543–552.

Zein, R., Sharrouf, W., and Selting, K. 2020. Physical properties of nanoparticles that result in improved cancer targeting. J. Oncol., 2020, 5194780.

Zhai, Y.J., Hunting, E.R., Wouterse, M., Peijnenburg, W.J.G.M., and Vijver, M.G. 2017. Importance of exposure dynamics of metal-based nano-ZnO, -Cu and -Pb governing the metabolic potential of soil bacterial communities. Ecotox. Environ. Saf., 145: 349–358.

Zhai, Y.J., Hunting, E.R., Liu, G., Baas, E., Peijnenburg, W.J.G.M., and Vijver, M.G. 2019. Compositional alterations in soil bacterial communities exposed to TiO_2 nanoparticles are not reflected in functional impacts. Environ. Res., 178: 108713.

Zhai, Y.J., Chen, L.H., Liu, G., Song, L., Arenas-Lago, D., Kong, L.C. et al. 2021. Compositional and functional responses of bacterial community to titanium dioxide nanoparticles varied with soil heterogeneity and exposure duration. Sci Total Environ., 773: 144895.

Zhang, D.Q., Hua, T., Xiao, F., Chen, C.P., Gersberg, R.M., Liu, Y. et al. 2014. Uptake and accumulation of CuO nanoparticles and CdS/ZnS quantum dot nanoparticles by *Schoenoplectus tabernaemontani* in hydroponic mesocosms. Ecol. Eng., 70: 114–123.

Zhang, W.C., Xiao, B.D., and Fang, T. 2018a. Chemical transformation of silver nanoparticles in aquatic environments: Mechanism, morphology and toxicity. Chemosphere, 191: 324–334.

Zhang, Q,Y., Guo, R., Ai, S.W., Yang, Y., Ding, J., and Zhang, Y.M. 2018b. Long-term heavy metal pollution varied female reproduction investment in free-living anura, *Bufo raddei*. Ecotoxicol. Environ. Saf., 159: 136–142.

Zhao, J.A., Cao, X.S., Liu, X.Z., Wang, Z.Y., Zhang, C.C., White, J.C. et al. 2016. Interactions of CuO nanoparticles with the algae *Chlorella pyrenoidosa*: Adhesion, uptake, and toxicity. Nanotoxicology, 10: 1297–305.

Zhao, J., Ren, W.T., Dai, Y.H., Liu, L.J., Wang, Z.Y., Yu, X.Y. et al. 2017. Uptake, distribution, and transformation of CuO NPs in a floating plant *Eichhornia crassipes* and related stomatal responses. Environ. Sci. Technol., 51: 7686–7695.

Zhao, S.Y., Su, X.X., Wang, Y.Y., Yang, X.Y., Bi, M., He, Q. et al. 2020. Copper oxide nanoparticles inhibited denitrifying enzymes and electron transport system activities to influence soil denitrification and N_2O emission. Chemosphere, 245: 125394.

Zhao, X.P., Liu, Y.B., Jiao, C.L., Dai, W.Q., Song, Z.D., Li, T. et al. 2021. Effects of surface modification on toxicity of CeO_2 nanoparticles to lettuce. NanImpact, 24: 100364.

Zhao, Y.M., Yang, Q.X., Liu, D., Liu, T.Q., and Xing, L.Y. 2022a. Neurotoxicity of nanoparticles: Insight from studies in zebrafish. Ecotoxicol. Environ. Saf., 242: 113896.

Zhao, Y.L., Chen, J.Y., Wang, R., Pu, X.X., and Wang, D.Y. 2022b. A review of transgenerational and multigenerational toxicology in the *in vivo* model animal *Caenorhabditis elegans*. pp. 73–101. *In*: Sharma, N., and Sahi, S. (Eds.). Nanomaterial Biointeractions at the Cellular, Organismal and System Levels. Springer Nature, Cham, Switzerland.

Zhou, S., Qian, W., Ning, Z.G., and Zhu, X.S. 2021. Enhanced bioaccumulation and toxicity of arsenic in marine mussel *Perna viridis* in the presence of CuO/Fe_3O_4 nanoparticles. Nanomaterials, 11: 2769.

Zhu, X., Tan, L., Zhao, T., Huang, W.Q., Guo, X., Wang, J.Y. et al. 2022a. Alone and combined toxicity of ZnO nanoparticles and graphene quantum dots on microalgae *Gymnodinium*. Environ. Sci. Pollut. Res., 29: 47310–47322.

Zhu, L., Booth, A.M., Feng, S.L., Shang, C.C., Xiao, H., Tang, X.X. et al. 2022b. UV-B radiation enhances the toxicity of TiO_2 nanoparticles to the marine microalga *Chlorella pyrenoidosa* by disrupting the protection function of extracellular polymeric substances. Environ. Sci. Nano., 9: 1591–1604.

Zuo, S.P., Yang, H., Jiang, X.F., and Ma, Y.Q. 2021. Magnetic Fe_3O_4 nanoparticles enhance cyanobactericidal effect of allelopathic *p*-hydroxybenzoic acid on *Microcystis aeruginosa* by enhancing hydroxyl radical production. Sci. Total. Environ., 770: 145201.

6

Hydrogel Materials in Sandy Soil

Mujeebat Bashiru and *Noureen Siraj**

Introduction

Man is entirely reliant on food produced by agriculture. There is increased demand for food due to the exponential human population growth which has caused continuous fear and tension in Man. Lack of crucial resources such as water necessary to increase crop yield has also been listed as an aggravated problem that prevented agriculture from producing enough food to feed the globe (Huang and Hartemink 2020, Spiertz and Ewert 2009). In addition, agriculture is also severely impacted by drought and the swift depletion of groundwater reserves, leading to special management practices that require only a selected group of crops to be grown in arid and semi-arid places with insufficient or unpredictable rainfall (Grum et al. 2016).

One of the commonly known types of soil, along with loamy and clay, is 'sandy soil'. It is composed of an average sand percentage of more than 50% and an average clay content less than 20% (Huang and Hartemink 2020). Sand-filled soil makes up about one third of the Earth's surface. Only about 4% of sandy soil was used for cultivation, with 35% being barren and not recommended for plant growth; approximately 59% was categorized as "land cover categories" (covered with grassland, shrubland, savanna, and woodland) (Huang and Hartemink 2020). Using India as a workable case study, reports indicate that 277.49 million tons of food grains were produced in the 2017–2018 fiscal year. By 2025, farmers would have to produce more than 300 million tons to satisfy the rising demand for food, according to predictions (Kumar et al. 2020). Due to the continued rise in global population and ongoing reduction in cropland and water supplies, boosting crop productivity with sandy soil for cultivation has become crucial. The 2030 Water Resources Group predicts that water demand would increase by 50% by the year 2030, despite agriculture continuing to be one of the largest consumers of water, accounting for over 70% of total consumption and utilization worldwide. As a result of the need for increased calorie output to feed 9.6 billion people by 2050, agriculture will account for approximately half of this growth in demand (Kumar et al. 2020, Lal 2015). Thus, the need to employ and optimize water conservation methods should be addressed to meet the demand. It is evident that crop viability increases with effective water and

Department of Chemistry, University of Arkansas at Little Rock, Little Rock, AR 72204, USA. Email: mobashiru@ualr.edu
* Corresponding author: nxsiraj@ualr.edu

fertilizer application (Du et al. 2020). Rainwater is the most well-known source of water for plant growth (Thombare et al. 2018). While conventional irrigation systems have been useful, droughts reduce their productivity. Sewage water has also been employed to conserve and maximize the amount of water and nutrients available particularly in areas with low rainfall (Shaheen et al. 2017). Despite the fact that sewage is a rich source of nutrients that can successfully promote crop growth, it may also pose environmental risks (Shaheen et al. 2017). These sewage effluent-irrigated soils may have long-term detrimental impacts on people and ecosystems due to high concentrations of potentially hazardous elements (PTE). With recurrent irrigation of sewage effluent build up, some elements including Cd, Cr, Cu, Ni, Pb, and Zn tend to accumulate over time until their levels exceed the current environmental concern. In particular, if these soils are arable and utilized for cultivation of vegetables, citrus, and cereals, the elevated PTE levels in crops may present significant health risks to individuals through the food chain (Shaheen et al. 2017). Based on the above reasons, it is of utmost importance to consider safe and effective techniques in providing solution to the problem of water or nutrients application for maximizing plant growth. Hydrogels have been explored for agricultural application due to their unique water retention capabilities (Thombare et al. 2018). They can also serve as soil conditioner meant to introduce some form of nutrient/drug to optimize plant growth. Figure 1 shows hydrogel application to plant. In addition to agriculture, hydrogels have potential use for a variety of purposes due to their excellent properties which includes hydrophilicity, high swelling capacity, and biocompatibility. Examples of these applications include production of hygiene products, additives for drilling fluid, sealing material and drug delivery (Singh and Singhal 2012).

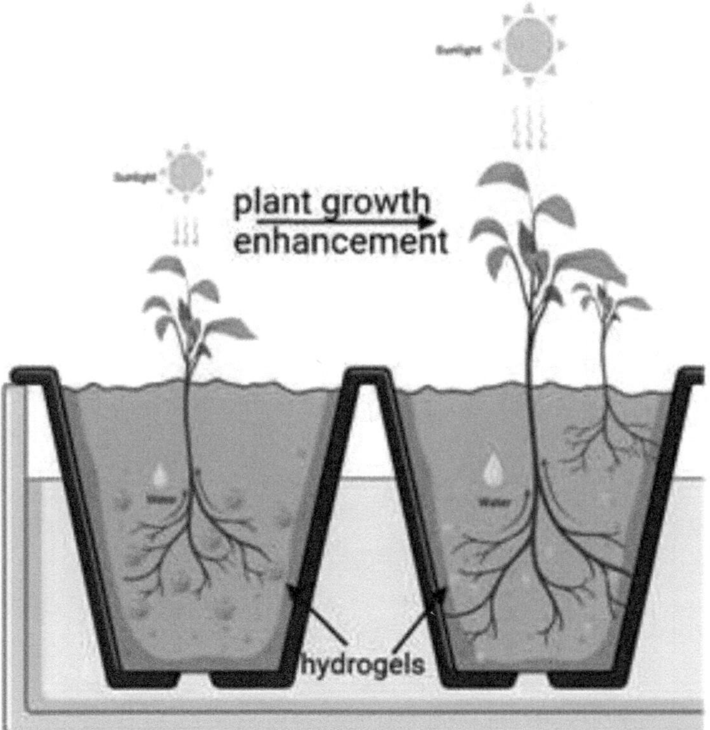

Figure 1. Pictorial representation of hydrogel application to plant. Created from Biorender

Brief overview of sandy soil characteristics and limitations

For a clear understanding of how to manage sandy soils, it is crucial to examine the impact of soil texture, soil properties and soil processes. Sandy soil in comparison to other soil types behaves differently from other soil types. Physical properties of sandy soil include soil bulk density, total porosity, microporosity and aggregates, all of which can vary depending on the size and arrangement of the grains, the type of clay, natural processes (such as biological activity) or human activities (such as tillage) (Huang and Hartemink 2020). Sandy soil is known to be highly coarse with small surface area and does not retain water (soil water repellency) or nutrients. Additional sandy soil properties and limitations are stated in Figure 2. Due to the nature of the sandy soil, animal production facilities frequently apply solid and liquid animal wastes like manure and sewage as a waste management technique, which can also enhance the physical and fertility of the land (Kang et al. 2011). In addition, inorganic or organic fertilizer and irrigation are also utilized to improve crop productivity with the use of sandy soil. As a result, the soil located at the surface remains loaded with lots of manure, which raises the risk of nutrient loss to ground, surface water and environment (Du et al. 2020), subsequently resulting in increasing and accumulating phosphorus levels in aquatic habitat, affecting freshwater algae development (Kang et al. 2011, Soinne et al. 2014, Awad et al. 2012). Soil amendment processes are thus required to eliminate the toxic products from the application of nutrients to the soil. It has been found that biochar has the ability to help with soil improvement, greenhouse gas reduction, removal of animal waste from soil and phosphorus loss to aquatic field through a sorption process (Soinne et al. 2014, Schulze et al. 2016, Wang et al. 2015). To avoid the needless processes for the removal of surplus nutrients in the soil, alternative and more effective method of fertilizer application must be utilized.

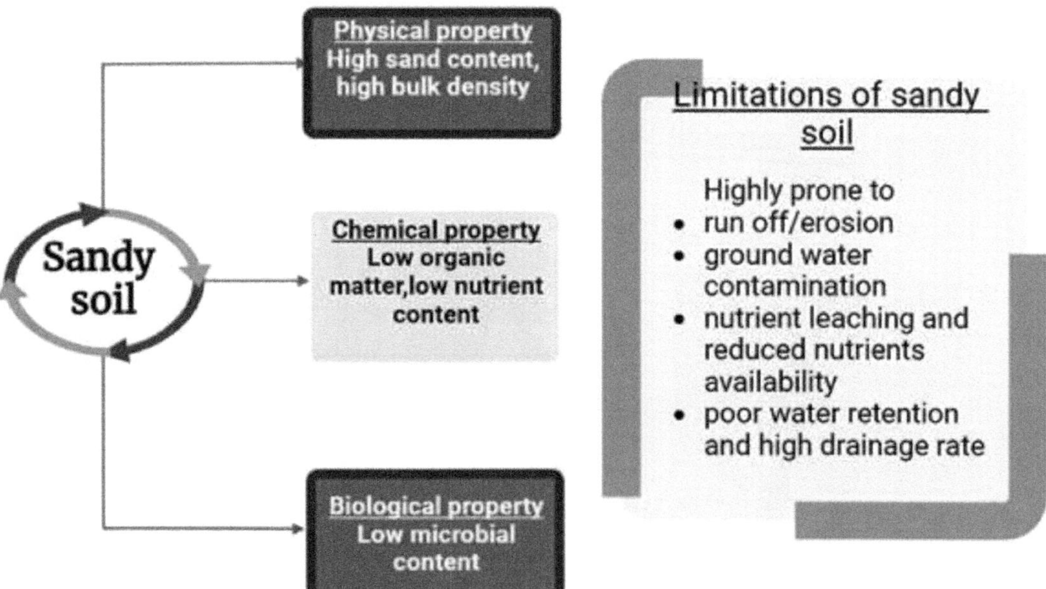

Figure 2. Chart showing properties and limitations of sandy soil. Created with BioRender.com

Hydrogel material and its potential applications in soil improvement

Hydrogels are soft, porous materials consisting of three-dimensional networks formed by cross-linked hydrophilic polymers. They are composed of 90% water with unique property to swell and maintain their shape. Hydrogels are composed of mainly hydrophilic and hydrophobic groups assisting with the water absorption, swelling, and releasing profile (Jansen-Van Vuuren et al. 2021, Michalik and Wandzik 2020). They have extremely high swelling ratio due to their ability to retain water up to a thousand times their dry weight (Khushbu and Kumar 2019). Utilizing hydrogel is a more practical, handy, and eco-friendly way to increase sandy soil production without leaving any hazardous residue in the soil once it degrades. Hydrogels have been referred to by many names such as soil conditioners, water reservoirs, yield enhancers and superabsorbent materials due to their ability to take in, hold onto, and then release water as needed by plant roots (Kumar et al. 2020, Narjary et al. 2012, de Vasconcelos et al. 2020, Michalik and Wandzik 2020). Hydrogels are mostly incorporated into sandy soil during cultivation, applied as soil amendment, introduced in dosage, timing and frequency considerations.

The main purpose of hydrogels is to increase irrigation effectiveness. They represent smart delivery materials that can serve as water and nutrient retentive agent (Michalik and Wandzik 2020). Particularly, drug or water can easily be controlled and discharged at the point of need, thus having a positive effect for conservation of resources. Plant diseases can also be prevented by treating the soil with hydrogel containing pesticide (Kumar et al. 2020). Scientists have been developing various water saving materials to boost soil penetration rates and decrease water runoff, required to maximize output in sandy soil in order to sustain the existing level of food and to fulfill the demands of the future (Kumar et al. 2020).

Another advantage of hydrogel is that it can be utilized as either gravitational or capillary water (Kumar et al. 2020). Water in the soil that percolates down till it reaches the water table while being pulled downward by gravity is referred to as gravitational water. Contrarily, capillary water is the volume of water that is retained in minuscule interstitial spaces as thin films that cover the soil particles. Hydrogels are also suitable materials for reinstating/conserving the low microbial biological properties and physico-chemical features in sandy soil (Awad et al. 2012). It is also reasonable to load nutrients in the soil for easy accessibility to plant roots when applying manure using hydrogel. This also assists in surmounting obstacles posed by the challenging characteristics of tropical soils as well as avoiding erosion responsible for frequent removal of the nutrients added to the soil. Interestingly, studies have shown that using hydrogels can cut irrigation time by 20–40% and reduce cost of water and labor, which is absolutely beneficial from economic and environmental perspective (de Vasconcelos et al. 2020).

Generally, the role of hydrogel material in sandy soil improvement can be categorized into three main categories:

(a) Water retention and moisture management—this includes the water absorption and release by hydrogel material, effect of hydrogel material impact on plants' ability to get water and decrease water-stress environment.

(b) Nutrient retention and availability—this includes the nutrient absorption and release properties of hydrogel material, impact on nutrient leaching and increased nutrient availability, role in promoting healthy plant growth and improved crop yield.

(c) Soil structure and root development—this includes the influence of hydrogel material on soil structure and porosity, effects on root growth, root penetration, and anchorage, enhancement of soil aeration and reduction of compaction.

Classification of hydrogels

Hydrogels are majorly classified based on sources, synthesis, or generation. For sources, they can be derived naturally or bio-based from natural polymers such as carbohydrates, cellulose, polysaccharides (alginate, chitosan and dextran) or prepared through chemical synthesis scheme such as polymerization of synthetic monomers (Bashir et al. 2020). Synthetic based hydrogels include poly (ethylene glycol) diacrylate, poly (acrylic amide) and poly (vinyl alcohol) (Adjuik et al. 2022, Agnihotri and Singhal 2017). According to hydrogel generation, the first generation is majorly reliant on water-soluble synthetic polymers, and cellulose based natural macromolecules. They are majorly synthesized from natural substances like polysaccharides and polypeptides. Synthetic materials for first generation hydrogels include acrylic acid (AA) and acrylamide known for high biodegradability, biocompatibility, non-toxicity, and renewability (Hua and Wang 2009, Rizwan et al. 2021). Synthesis of the second generation of hydrogels include Michael addition, polymerization including redox-based radical polymerization, photo-polymerization, gamma radiation polymerization and atom-transfer radical polymerization (Rizwan et al. 2021). In comparison to hydrogels made from natural materials, second generation hydrogels are mechanically stronger and durable. Lastly, the third generation of hydrogels, introduced in 2010, are simply referred to as smart hydrogels. Mostly, they are synthesized *in situ* using chemically cross-linked hydrogels by enzymes, radical polymerization, double network hydrogels, and combination of natural and synthetic polymers and composite hydrogels (Rizwan et al. 2021).

Types of hydrogel materials used for soil applications

Superabsorbent polymers (SAPs) have attracted attention recently as a cutting-edge tool to increase water effectiveness in agricultural operations. These synthetic materials were created to enhance the physical characteristics of soil, including its water-holding capacity (WHC), permeability, and infiltration rates, which aided in the development of plants, particularly in arid regions with structurally devoid soils. Adding soil amendments to minimize water stress or to enhance the physical characteristics of the soil has a significantly positive impact on agriculture. Mainly, the commonly employed practice utilized in refilling the soils with nutrients exposes the soil and environment to potentially toxic elements for long term effect.

Acrylic acid (AA) and some natural based hydrogels are among the hydrogel materials that are frequently investigated for soil applications due to their affordability, biodegradability, ease of handling requiring less technical know-how, their porous structure, their hydrophilicity, and morphology. Alginate-based hydrogels have also been extensively explored due to their ability to gelatinize (Shen et al. 2020). Although polyacrylamide and biopolymeric based hydrogels are useful for soil conservation technologies and prevention from soil erosion (Awad et al. 2012, Loo et al. 2021), the release of free acrylamide into aquatic environment poses potential risks to human health (Chen and Chen 2019). Cellulose-based hydrogels are very biodegradable and may be readily converted freely to carbon dioxide and water as end products. Non-toxic, renewable, environmentally responsive hydrogels developed using green synthesis/processes are mostly preferred. Although synthetic polymers possess excellent physical, chemical and mechanical durability but they are negatively constrained by their lack of biodegradability leading to increased waste products (Saruchi et al. 2019).

Hydrogel that has been taken into consideration for use in agricultural industry must possess specific criteria including being economical, having a high capacity for water absorption and retention, having a high biodegradability, providing sustained delivery or a constant rate of delivery, flexibility with pH that can be controlled and tolerance to high temperatures which aids to increase the rate of drug release.

Hydrogel material with potential relevance to soil improvement

The production of hydrogels for agricultural usage has utilized a wide variety of material types, including polymers, polyacrylamide, cellulose, chitosan, alginate, and many more (Agnihotri and Singhal 2017, Rizwan et al. 2021, Shen et al. 2020, Loo et al. 2021). Nowadays, carbohydrates and cellulose are primarily used in synthesizing superabsorbent hydrogels (SAH) and these are very common and easily available in nature. In general, SAH has capacity to influence soil permeability, structure, texture, density, rates of water evaporation and penetration through the soils (Agnihotri and Singhal 2017). The SAHs lessens the frequency of irrigation, stop erosion and water runoff, improve air aeration and promote nutrient availability (Agnihotri and Singhal 2017).

Synthetic hydrogels require the need of crosslinking agents because it prevents the hydrogel from dissolving in water. Bhaskar and co-worker (Narjary et al. 2012) developed "Pusa hydrogel" from polyacrylate via free-radical polymerization, grafting and crosslinking of acrylamide with addition of about 50% cellulose derivative. The impact of Pusa hydrogel samples on various types of soil was examined by incubating the soil sample with various concentrations of gels. Results revealed a significant improvement in water retention ability of sandy soil for a longer period of time as opposed to other types of soil such as black soil (Narjary et al. 2012).

Major characterization techniques for 3D hydrogel materials are Fourier transform infrared spectroscopy (FTIR), which is useful in identifying the presence of different functional groups, and Xray diffractometry (XRD), which is useful in determining a material's crystallographic structure. Rheological instrument is used for determining the elasticity or viscosity of the hydrogels, scanning electron microscopy (SEM) is used for quantitative element analysis, surface morphology examination for determining the mean diameter of the hydrogel, and thermal gravimetric analysis (TGA) for examination of thermal stability of samples (Jabrail and Mahmood 2020, Ahmad et al. 2023, Heise et al. 2019).

The effect of different variables necessary for the alteration of the swelling ability of the polymeric materials was also investigated (Sadovnikova et al. 2014). Hydrogel materials were synthesized using high ionizing energy "radiation-grafted method" such as gamma ray for the cross linking of acrylamide and acrylic acid (Sadovnikova et al. 2014). To provide information regarding specific thermodynamic performance in different soil (water retention capacity) types, various polymer (material) rates and fractions of produced hydrogel materials were examined. Variations in the composition of the hydrogel yielded an enhancement of water retention capacity of sandy soil, total capillary and field capacities (Sadovnikova et al. 2014).

Initially, research on the fabrication of SAH composite based on rice husk ash (RHA) and starch were conducted by Azevedo et al. (de Azevedo et al. 2017). These composite were potentially used for the adsorption of dye from water and it has remarkable mechanical and swelling qualities as well as its high liquid uptake capacity, low cost, biodegradability, easy handling, porous and hydrophilic nature. The remarkable properties of this material inspired Sadovnikova's team to explore its potential application as a soil conditioner. It is well recognized that soil conditioners can help to maintain or improve soil structure, release nutrients, and enhance water retention property of the soil (de Vasconcelos et al. 2020). This hydrogel was produced using free radical polymerization of sodium acrylate monomer and graft of starch under aqueous media. RHA has also been used in the preparation of slow-release fertilizer. This low-cost biodegradable hydrogel composite synthesized from starch and rice husk ash was very promising as soil conditioner. It was tested on *Cucumis melo L.* plant, which is known to be grown in sandy soil with drip irrigation that consequently results in increased cost of production (de Vasconcelos et al. 2020).

Soil hydraulic characteristics are crucial for the flow and distribution of water in soils. The texture, structure, and types of soil amendments added to the soil can all affect how quickly plants can take water from the soil. This led to the examination of synthetic and bio-based hydrogel for modification of the soil. Both bio-based and synthetic hydrogels upon application to soil lowered saturated hydraulic conductivity, infiltration, and soil evaporation. In comparison to unaltered soil, hydrogels made from synthetic and biological materials boosted soil water retention while lowering the soil water pressure head (Adjuik et al. 2022).

SAH with crosslinked hydrophilic polymers can absorb and retain water, and have been reported useful for irrigation purposes and fertilizer applications of soil to improve plant growth (Zhang et al. 2005). The high cost of some polymeric superabsorbent made from poly (sodium acrylate) which are unsuitable for soil and water containing salt, led to the utilization of inexpensive, salt resistant materials such as inorganic clays (kaolin, montmorillonite, attapulgite, mica, bentonite and sercite) incorporated into polymeric superabsorbent (Zhang et al. 2005, Shahid et al. 2012). Attapulgite (APT) is a less salt-sensitive magnesium aluminum silicate absorbent material with presence of -OH functional groups (Zhang et al. 2005). Previous synthesis with poly (acrylic acid acrylamide) back bone has been greatly enhanced with the intervention of introducing sodium humate and APT (Zhang et al. 2005). Commonly used crosslinking agents and initiator for the synthesis of these SAH hydrogels include N,N-methylene-bis-acrylamide (MBA) and ammonium persulfate (Wang and Wang 2010, Agnihotri and Singhal 2017, Hua and Wang 2009).

Due to the low swelling ratio of certain naturally occurring SAH derived from agricultural waste (hazelnut, orange peels, maize cob), research shows that the introduction of sodium alginate has been useful to improve the swelling ratio due to its gelatinization capability (Agnihotri and Singhal 2017). Effect of sodium humate (SH) on poly (acrylic acid/ sodium alginate) SAH hydrogels was investigated. It was observed that alteration of SH concentration changed the hydrogels properties. For example, a higher swelling ratio was observed with lower SH concentration hydrogels. Additional SH concentration reduced the hydrogel's ability to absorb water once equilibrium swelling ratio had been established (Agnihotri and Singhal 2017). SH is made up of multifunctional aliphatic components and aromatic constituents. SH are produced from organic matter deposition from peat, lignites, sapropels and non-living organic matter of soil and water (Shahid et al. 2012). SH has a positive influence on plant growth in terms of regulation, root development, photosynthesis, improve soil cluster structures and promote nutrient absorption (Zhang et al. 2005, Hua and Wang 2009, Shahid et al. 2012). Humic compounds/humate has received extensive attention for use in agriculture as soil conditioner and fertilizer. Humate is generally available in the form of soluble salts- potassium humates, sodium humate. Potassium-humate are advantageous due to their aromatic nature and the presence of carboxylic and phenolic group responsible for important chemical reactions, biological activity, increased pH buffering, improves soil structure and enhance soil water transportation (Shahid et al. 2012). Numerous SAH have been further stimulated or enhanced using fertilizers proven to promote crop quality and production (Zn and Fe) (Shahid et al. 2012). When potassium humate superabsorbent hydrogel nanocomposite (PHNC) was applied to soil, the hydrophilic polymer network increased PHNC's hydrophilicity and affected soil's water retention ability. Higher hydrogel application rates, a drop in soil pH, and increased electrical conductivity were all found to have a positive impact on water retention (Shahid et al. 2012). Other properties of soil such as bulk density was altered, porosity greatly increased, hydraulic conductivity decreased, and seed germination and seedling growth (measured by the length and weight of the shoot, among other things) were enhanced as compared to the control (Shahid et al. 2012). The wilting effect has also been reduced due to the slowed rate of soil water loss, which increased water retention in comparison to control. This is related to a rise in storage pores in sandy loam soil, which can be used as a marker for better soil structure. Acrylic acid/ sodium alginate/sodium humate has also been previously explored for elimination of metal ion and dyes from waste water (Singh and Singhal 2012).

Saruchi et al. (Saruchi et al. 2019) explored the biodegradation of hydrogel composite by imbibing the hydrogel sample into the soil and investigated the degradation over time. These composites served as a delivery vehicle for fertilizer application at desired concentration, with controlled release of the nutrient leading to increased efficacy and reduced environmental pollution. Gum tragacanth (GT) is considerably used for hydrogel synthesis due to its long shelf life and its stability in heat as well as at wide range of pH. Generally, grafting reactions are conducted using chemical initiators but, in this work, lipase—an enzyme—was utilized for the initiation of graft co-polymerization (Saruchi et al. 2019).

Despite being the most popular hydrogels, acrylamide-based hydrogels lack hydrolytic stability. This can be fixed by replacing acrylamides with alkyl or hydroxyl alkyl groups. The SAP frequently uses natural gums like tara gum, and acacia gum which are mainly manufactured using various techniques like gamma radiation and graft polymerization (Jabrail and Mahmood 2020, Lal 2015, Liu et al. 2022). Physical crosslinked hydrogels are also useful in agriculture but they generally possess low thermal stability and mechanical strength (Jabrail and Mahmood 2020) as a result of their electrostatic contacts and ideal three-dimensional network, which make them more suited SAPs and useful in agriculture.

Cellulose based hydrogels derived from rice-straw have been recently investigated by various groups (Ibrahim et al. 2015, Abou-Baker et al. 2020, Solieman et al. 2023). In the recent follow-up study conducted by Solieman et al. (Solieman et al. 2023), rice straw was subjected to different step-by-step processes such as chemical pulping, bleaching and polymerization. The straw-based hydrogel was synthesized using a heterogeneous polymerization method that included cellulosic fibers and acrylic acid. Rice straw-based hydrogels in comparison to acrylamide hydrogels, both were tested on sandy soil cultivated with beans and maize. Interestingly, the cost of producing hydrogels based on rice straw was discovered to be generally low as compared to acrylamide hydrogels. In addition, the hydro-physical properties, physiochemical properties, and biological properties of the soil used for testing were also significantly enhanced.

Another recent study reported the use of wastepaper to synthesize a cellulose-based hydrogel with the use of epichlorohydrin as a crosslinker (Ahmad et al. 2023). These inexpensive hydrogels having a porous network were further tested on different soils to deliver both moisture and fertilizer onto the soil. It was observed that the synthesized hydrogels facilitated the controlled release of about 72% fertilizer in sandy soil for a period of 42 days as opposed to the free fertilizer which was about 99% released in sandy soil for just a period of 7 days. In other words, the fertilizer content loaded into the hydrogel was released for a longer period in the soil of interest than when the fertilizer was directly introduced in the sandy soil.

Wheat straw-based hydrogels have also been investigated for the synthesis of hydrogel materials that are suitable for cultivation of plants due to their water retention properties in soil (Heise et al. 2019). These hydrogels were prepared using cross linking method involving lignocellulosic matrix and citric acid serving as the cross linker. Examination of these hydrogels in sandy soil substrates revealed the enhancement of the water retention capacity of the soil (Heise et al. 2019).

Galactomannan hydrogel with a high water retaining capability was recently synthesized from a natural polyssacharide known as *Gleditsia microphylla* gum (Liu et al. 2022). Galactomannan hydrogel was synthesized by introducing "borax solution", a crosslinker to the *Gleditsia microphylla* gum solution upon stirring at room temperature. Investigation of the water retention ability of this hydrogel when mixed with different proportions of the sandy soil significantly improved the water holding capacity of the soil tested. This also increased the water retention holding time of the sandy soil.

Hydrogels have tremendous potential in sandy soil for agriculture application. Table 1 shows a summary of research that has been published regarding different hydrogel materials utilized for agricultural applications in sandy soil. It will be very beneficial for future research in this area.

Conclusion

Hydrogel materials with potential for application in agriculture have been investigated from a variety of sources, from natural to synthetic. They have been seen to alter the soil performance in a variety of ways that could raise crop output. It's interesting to note that the impact of hydrogels as soil conditioners will offer remedies to the current issue with agriculture's inability to meet human food demand due to scarcity of natural resources such as water. From the promotion of soil structure, and

Table 1. Hydrogel materials explored for agricultural applications in sandy soil.

Hydrogel material	Hydrogel synthesis	Soil type	Plant used for testing	Applications	References
Poly acrylic acid/acrylamide/ sodium humate	Cross linking free radical copolymerization	Not specified	Not specified	Increased water absorbency of soil. Also studied for other applications including wastewater metal ion and dye removal.	(Singh and Singhal 2012)acrylamide (AM
Polyacrylic acid/ sodium alginate/sodium humate	Free radical co-polymerization	Sandy	Seeds of corn and white gourd	Swelling ratio, diffusion coefficient, water retention capability	(Agnihotri and Singhal 2017)
Acrylic acid/sodium alginate/ sodium humate	Graft co-polymerization	Not specified	Not specified	Water absorbency, swelling behavior, hydrogel in solution	(Hua and Wang 2009)
Cellulose based hydrogel	Direct addition of K_2SO_4 to aqueous MC and HPMC	Not specified	Chinese gabbage	Gelation temperatures, equilibrium swelling ratio, pore sizes, mechanical properties, water holding and retention capacity, fertilizer release	(Chen and Chen 2019)use of polyacrylamide-based hydrogel in agriculture pose potential human health hazards. Hydrogel systems were prepared from temperature-responsive methylcellulose (MC
Tragacanth gum polysaccharide/ acrylamide and methacrylic acid	Grafting and crosslinking method	Sandy-loam	Not specified	Improved soil fertility, biodegradation of hydrogels, water retention capacity,	(Saruchi et al. 2019)
Poly (acrylic acid co-acrylamide) $AlZnFe_2O_4$/ potassium humate SAH	Crosslinking acrylic acid and acrylamide co-polymer	Sandy-loamy	*Triticum aestivum* L.	Water retention of soil, pH of soil, electrical conductivity	(Shahid et al. 2012)
Acrylic acid/acrylamide/ sodium humate/attapulgite	Aqueous solution polymerization	Not specified	*Elymusdahuricus Turcz*	Water absorbency	(Zhang et al. 2005)
Sodium alginate/sodium acylate/poly vinyl pyrrolidone	Free radial solution polymerization	Not specified	Not specified	Not specified	(Wang and Wang 2010)
Gum Arabic/acrylic acid/ acrylamide/N-vinyl pyrrolidone	Graft polymerization	Sandy soil	Not specified	Water absorbency, water retention, thermal stability, mechanical properties	(Jabrail and Mahmood 2020)
Guar gum-polyacrylic acid polyaniline	Two-step aqueous graft polymerization	Sandy, sandy-loamy, sandy clay	Not specified	Soil moisture retention, biodegradability of hydrogel	(Kaith et al. 2015)
Rice straw-based hydrogel	Heterogeneous polymerization	Sandy soil, calcareous soil	Beans and maize	Hydro-physical, physiochemical, and biological properties	(Solieman et al. 2023)
Cellulose fibers	Crosslinking method	Sandy soil, clayey soil, topsoil	Not specified	Water retention capability of soil	(Ahmad et al. 2023)
Galactomannan hydrogel	Borax-crosslinking method	Sandy soil	Not specified	Water retention capability, self-repair capacity	(Liu et al. 2022)

root development, hydrogel offers very promising potential for moisture management, availability, and retention of water and nutrients. It's crucial, however, that researchers examine structural constraints and develop risk-free substitutes for the toxic chemicals that are currently in use for crosslinking techniques.

References

Abou-Baker, N.H., Ouis, M., Abd-Eladl, M., and Ibrahim, M.M. 2020. Transformation of lignocellulosic biomass to cellulose-based hydrogel and Agriglass to improve beans yield. Waste and Biomass Valorization, 11(7): 3537–51.

Adjuik, T.A., Nokes, S.E., Montross, M.D., and Wendroth, O. 2022. The impacts of bio-based and synthetic hydrogels on soil hydraulic properties: a review. Polymers, 14(21): 1–23.

Agnihotri, S., and Singhal, R. 2017. Effect of sodium humate on the swelling characteristics and agricultural application of superabsorbent hydrogels of poly (acrylic acid/sodium alginate/sodium humate). J. Polym. Mater, 34(4): 663–80.

Ahmad, D.F.B.A., Wasli, M.E., Tan, C.S.Y., Musa, Z., and Chin, S.F. 2023. Eco-friendly cellulose-based hydrogels derived from wastepapers as a controlled-release fertilizer. Chemical and Biological Technologies in Agriculture, 10(1): 1–10.

Awad, Y.M., Blagodatskaya, E., Ok, Y.S., and Kuzyakov, Y. 2012. Effects of polyacrylamide, biopolymer, and biochar on decomposition of soil organic matter and plant residues as determined by 14C and Enzyme Activities. European Journal of Soil Biology, 48: 1–10.

Azevedo, A.C.N.de., Vaz, M.G., Gomes, R.F., Pereira, A.G.B., Fajardo, A.R., and Rodrigues F.H.A. 2017. Starch/rice husk ash based superabsorbent composite: high methylene blue removal efficiency. Iranian Polymer Journal (English Edition), 26(2): 93–105.

Bashir, S., Hina, M., Iqbal, J., Rajpar, A.H., Mujtaba, M.A., Alghamdi, N.A. et al. 2020. Fundamental concepts of hydrogels: synthesis, properties, and their applications. Polymers, 12(11): 1–60.

Chen, Y.C., and Chen, Y.H. 2019. Thermo and PH-responsive methylcellulose and hydroxypropyl hydrogels containing K2SO4 for water retention and a controlled-release water-soluble fertilizer. Science of the Total Environment, 655: 958–67.

Du, Y., Cui, B., Zhang, Q., Wang, Z., Sun, J., and Niu, W. 2020. Effects of manure fertilizer on crop yield and soil properties in China: A Meta-Analysis. Catena, 193 (April).

Grum, B., Hessel, R., Kessler, A., Woldearegay, K., Yazew, E., Ritsema, C. et al. 2016. A decision support approach for the selection and implementation of water harvesting techniques in arid and semi-arid regions. Agricultural Water Management, 173: 35–47.

Heise, K., Kirsten, M., Schneider, Y., Jaros, D., Keller, H., Rohm, H. et al. 2019. From agricultural byproducts to value-added materials: wheat straw-based hydrogels as soil conditioners? ACS Sustainable Chemistry and Engineering, 7(9): 8604–12.

Hua, S., and Wang, A. 2009. Synthesis, characterization and swelling behaviors of sodium Alginate-g-Poly(acrylic acid)/sodium humate superabsorbent. Carbohydrate Polymers, 75(1): 79–84.

Huang, J., and Hartemink, A.E. 2020. Soil and environmental issues in sandy soils. Earth-Science Reviews, 208 (September 2019): 103295.

Ibrahim, M.M., Abd-Eladl, M., and Abou-Baker, N.H. 2015. Lignocellulosic biomass for the preparation of cellulose-based hydrogel and its use for optimizing water resources in agriculture. Journal of Applied Polymer Science, 132(42).

Jabrail, F.H., and Mahmood, M.E. 2020. Hydrophilic monomers graft gum arabic hydrogels prepared in different cross-linking nature and study of their effects on irrigation of agriculture hydrophilic monomers graft gum arabic hydrogels prepared in different cross-linking nature and study of their effects on irrigation of agriculture. Research Journal of Chemical Science, 6(1): 1-19.

Jansen-Van Vuuren, R.D., Vilela, G.D., Ramezani, M., Gilbert, P.H., Watson, D., Mullins, N. et al. 2021. CO_2-responsive superabsorbent hydrogels capable of > 90% dewatering when immersed in water. ACS Applied Polymer Materials, 3(4): 2153–65.

Kaith, B.S., Sharma, R., and Kalia, S. 2015. Guar gum based biodegradable, antibacterial and electrically conductive hydrogels. International Journal of Biological Macromolecules, 75: 266–75.

Kang, J., Amoozegar, A., Hesterberg, D., and Osmond, D.L. 2011. Phosphorus leaching in a sandy soil as affected by organic and inorganic fertilizer sources. Geoderma, 161(3-4): 194–201.

Khushbu, S.G.W., and Kumar, A. 2019. Synthesis and assessment of carboxymethyl tamarind kernel gum based novel superabsorbent hydrogels for agricultural applications. Polymer, 182 (July): 121823.

Kumar, R., Patel, S.V., Yadav, S., Singh, V., Kumar, M., and Kumar, M. 2020. Hydrogel and its effect on soil moisture status and plant growth: a review. ~ 1746 ~ Journal of Pharmacognosy and Phytochemistry, 9(3): 1746–53.

Lal, R. 2015. Restoring soil quality to mitigate soil degradation. Sustainability (Switzerland), 7(5): 5875–95.

Liu, C., Tang, M., Zhang, F., Lei, F., Li, P., Wang, K. et al. 2022. Facile access to gleditsia microphylla galactomannan hydrogel with rapid self-repair capacity and multicyclic water-retaining performance of sandy soil. Polymers, 14(24).

Loo, S.L., Vásquez, L., Athanassiou, A., and Fragouli, D. 2021. Polymeric hydrogels—a promising platform in enhancing water security for a sustainable future. Advanced Materials Interfaces, 8(24).

Michalik, R., and Wandzik, I. 2020. A mini-review on chitosan-based hydrogels with potential for sustainable agricultural applications. Polymers, 12(10): 1–16.

Narjary, B., Aggarwal, P., Singh, A., Chakraborty, D., and Singh, R. 2012. Water availability in different soils in relation to hydrogel application. Geoderma, 187188: 94–101.

Rizwan, M., Gilani, S.R., Durani, A.I., and Naseem, S. 2021. Materials diversity of hydrogel: synthesis, polymerization process and soil conditioning properties in agricultural field. Journal of Advanced Research, 33: 15–40.

Sadovnikova, N.B., Smagin, A.V., and Sidorova, M.A. 2014. Thermodynamic assessment of the effect of strongly swelling polymer hydrogels on the water retention capacity of model porous media. Eurasian Soil Science, 47(4): 287–96.

Saruchi, V.K., Mittal, H., and Alhassan, S.M. 2019. Biodegradable hydrogels of tragacanth gum polysaccharide to improve water retention capacity of soil and environment-friendly controlled release of agrochemicals. International Journal of Biological Macromolecules, 132: 1252–61.

Schulze, M., Mumme, J., Funke, A., and Kern, J. 2016. Effects of selected process conditions on the stability of hydrochar in low-carbon sandy soil. Geoderma, 267: 137–45.

Shaheen, S.M., Shams, M.S., Khalifa, M.R., El-Dali, M.A., and Rinklebe, J. 2017. Various soil amendments and environmental wastes affect the (im)mobilization and phytoavailability of potentially toxic elements in a sewage effluent irrigated sandy soil. Ecotoxicology and Environmental Safety, 142(April): 375–87.

Shahid, S.A, Qidwai, A.A., Anwar, F., Ullah, I., and Rashid, U. 2012. Effects of a novel poly (AA-Co-AAm)/AlZnFe 2O 4/potassium humate superabsorbent hydrogel nanocomposite on water retention of sandy loam soil and wheat seedling growth. Molecules, 17(11): 12587–602.

Shen, Y., Wang, H., Li, W., Liu, Z., Liu, Y., Wei, H. et al. 2020. Synthesis and characterization of double-network hydrogels based on sodium alginate and halloysite for slow release fertilizers. International Journal of Biological Macromolecules, 164: 557–65.

Singh, T., and Singhal, R. 2012. Poly(Acrylic Acid/Acrylamide/Sodium Humate) superabsorbent hydrogels for metal ion/dye adsorption: effect of sodium humate concentration. Journal of Applied Polymer Science, 125(2): 1267–83.

Soinne, H., Hovi, J., Tammeorg, P., and Turtola, E. 2014. Effect of biochar on phosphorus sorption and clay soil aggregate stability. Geoderma, 219-220: 162–67.

Solieman, N.Y., Afifi, M.M.I., Abu-ElMagd, E., Abou Baker, N., and Ibrahim, M.M. 2023. Hydro-physical, biological and economic study on simply, an environment- friendly and valuable rice straw-based hydrogel production. Industrial Crops and Products, 201(May): 116850.

Spiertz, J.H.J., and Ewert, F. 2009. Crop production and resource use to meet the growing demand for food, feed and fuel: opportunities and constraints. NJAS - Wageningen Journal of Life Sciences, 56(4): 281–300.

Thombare, N., Mishra, S., Siddiqui, M.Z, Jha, U., Singh, D., and Mahajan, G.R. 2018. Design and development of guar gum based novel, superabsorbent and moisture retaining hydrogels for agricultural applications. Carbohydrate Polymers, 185(October 2017): 169–78.

Vasconcelos, M.C. de., Gomes, R.F., Sousa, A.A.L, Moreira, F.J.C., Rodrigues, F.H.A., Fajardo, A.R. et al. 2020. Superabsorbent hydrogel composite based on starch/rice husk ash as a soil conditioner in melon (*Cucumis Melo* L.) Seedling Culture. Journal of Polymers and the Environment, 28(1): 131–40.

Wang, W., and Wang, A. 2010. Synthesis and swelling properties of ph-sensitive semi-IPN superabsorbent hydrogels based on sodium Alginate-g-Poly(Sodium Acrylate) and Polyvinylpyrrolidone. Carbohydrate Polymers, 80(4): 1028–36.

Wang, Y., Lin, Y., Chiu, P.C., Imhoff, P.T., and Guo, M. 2015. Phosphorus release behaviors of poultry litter biochar as a soil amendment. Science of the Total Environment, 512-513: 454–63.

Zhang, J.P., Li, A., and Wang A.Q. 2005. Study on superabsorbent composite. v. synthesis, swelling behaviors and application of poly(Acrylic Acid-Co-Acrylamide)/sodium humate/attapulgite superabsorbent composite. Polymers for Advanced Technologies, 16(11-12): 813–20.

7

Nano-agrochemicals' Transformation in the Soil-plant System

Iqra Naseer,[1] Sumera Javad,[1,] Ajit Singh,[2] Nimrah Azam,[1] Khajista Jabeen[1] and Mohammad Faizan[3]*

Introduction

Recent agricultural advances have stimulated the agricultural pollution worldwide. The extensive use of chemical pesticides and fertilizers contaminate the soil and has hazardous effects on soil structure, texture, soil biota and ecosystem. The use of these agrochemicals negatively affects the human health as well. The use of nanotechnology in agriculture is currently being explored in plant hormone delivery, seed germination, water management, transfer of target genes, nanobarcoding, nanosensors, and controlled release of agrichemicals. The nano-agrochemicals are also known as engineered nanomaterials (ENMs) or nanostructured materials (NSMs). Nanoparticles because of their unique properties have proven to be effective as biopesticide, and bio-fungicides. Nano-agrochemicals, because of their unique properties, have proven to be effective fungicides in the management of environmental problems caused by the use of synthetic agrochemicals. The use of nanotechnology in agriculture is currently being explored in plant hormone delivery, seed germination, water management, transfer of target genes, nanobarcoding, nanosensors, and controlled release of agrichemicals. The applications of nanomaterials increase the productivity of soil. Nanotechnology facilitates and supports the efficient utilization of nutrients for promoting sustainable agriculture. Nanotechnology has provided new inventive era to the agriculture area by providing applied strategies and practices. The use of nanotechnology in terms of nano-agrochemicals causes substantial impact on crop growth, food security and sustainable agricultural ecosystem. Nanoparticles have a great potential to be used directly as nano-agrochemicals. They can also be used to enhance the efficacy of already used agrochemicals. Applications of these

[1] Department of Botany, Lahore College for Women University, Lahore Pakistan.
[2] School of Biosciences, University of Nottingham Malaysia, Semenyih 43500, Selangor, Malaysia.
[3] Department of Botany, School of Sciences, Maulana Azad National Urdu University, Hyderabad, Telangana, India.
* Corresponding author: zif_4@yahoo.com

nanomaterials in soil increase the crop yield and lessen the use of synthetic fertilizers/pesticides. Nanoparticles are able to enter the soil directly by absorption of gaseous compounds, atmospheric precipitants and via sedimentation (dust and aerosols). However, soil physiological processes can be altered by long term accumulation of nanoparticles as nano-agrochemicals which can affect the cell structure of plants, microbial colonies in rhizosphere and their interaction with plant roots. This chapter summarizes such interactions and transformation of nanoparticles after their application.

Nano-agrochemicals

For developing countries, agriculture has always played a primary role to sustain their economy and to increase their exports. Agricultural products and waste materials have been very useful raw materials for the survival of the fast-growing global population of 7.5 billion approximately (Dethier and Effenberger 2011). While discussing the agricultural sustainability, it is clear that a sustained agriculture system of any country means that it is concerned with both global and national issues. A sustainable system cannot be attained by ignoring any one of these at micro or macro level. At national level, one has to see the financial resources for farmers while adopting any advanced technology from global level (Fuglie et al. 2017). Similar is the case with nanotechnology where researchers from all over the world are suggesting that nanotechnology and related technological developments have the great capacity to increase the agricultural outputs using the same available resources. It can help countries to develop a real sustainable agricultural systems in rapidly changing climate systems. Nanotechnology and related products can help farmers to increase the crop yield while maintaining environmental sustainability, economic stability, and ecological balance (Acharya and Pal 2020).

It is challenging for farmers to generate enough money while using conventional methods of farming and conventional fertilizers, pesticides, etc. They are forced to use synthesized fertilizers and chemicals (herbicides, pesticides, fungicides, etc.) that enhance crop yield but also have negative impacts on the environment. It causes loss of biodiversity, degradation of the quality of soil, pollution, and bio-magnification of hazardous materials. This problem can be resolved by nanotechnology because nano-agrochemicals can be used in lesser quantities where they have their own specific way to reach the target (Qazi and Dar 2020).

Nanoagrochemical is the term used now for the agrochemicals that are either based on nanoparticles or being prepared by using nanotechnology and these include nano-nutrients, nano-fertilizers, nano-herbicides, nano-fungicides and nano-pesticides that are more specific in their targeted operations. These agriculture related nano-chemicals are known for their economic viability along with the environment friendly nature (Qazi and Dar 2020). Use of nanomaterials in a nano-agrochemical delivery system has significant potential for improving the efficiency of agricultural inputs, minimizing environmental pollution, and saving labor costs which play a significant role in the improvement of food security and maintaining the system of agricultural sustainability (Pulizzi 2019) as explained in Figure 1.

Effective role of nano-agrochemicals in plants

Nanotechnology has provided advancement of opportunities in different fields of sciences related to plants such as agriculture. The interaction of the NMs with plants leads to an effect on the physiology and morphology of plant organs (Hassanisaadi et al. 2022). Nanotechnology has the potential for the precise delivery of agrochemicals for improving plant growth, disease resistance and nutrient use (Pramanik et al. 2020).

- **Nanofertilizer**

Nanoparticles encapsulated materials that transport nutrients slowly to crops or NPs directly used as nano-nutrients are called nano-fertilizers (DeRosa et al. 2010). Nano-fertilizers are smart fertilizers

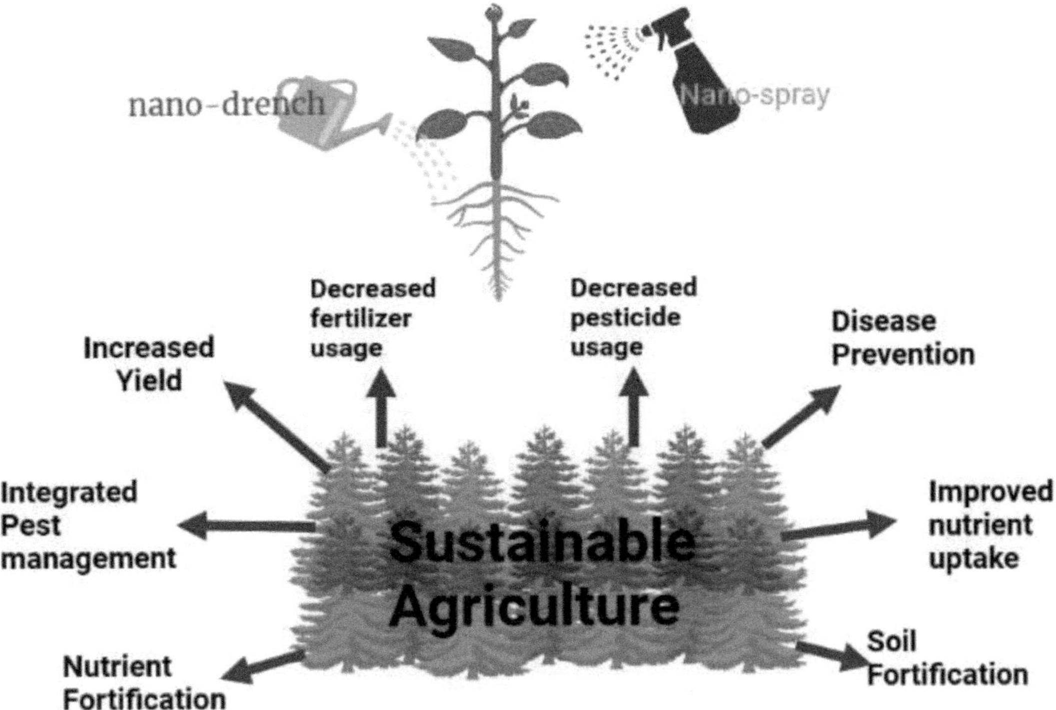

Figure 1. Nano-agrochemicals for sustainable agriculture.

because they enhance soil fertility, quality of agricultural products, and productivity and act at a faster pace due to their small size (Meena et al. 2017). The properties of nano-fertilizers that make them different from conventional fertilizers are their size, large surface area, and penetration capacity. Nano-fertilizers are synthesized to release nutrients according to crops requirement, at a slow but steady pace (Liu et al. 2006).

Nano-fertilizers can be applied as foliar spray or soil application. But studies are still being done to find out which method would be more effective for nutrient utilization for different crops in different environmental conditions and soil types. It also depends upon the type of nanomaterial or nutrient and its pathway in a plant. Each way of application has its own advantages, e.g., foliar application of nano-fertilizers helps to penetrate them in plants more easily and directly to targeted sites. Furthermore, it does not interact with soil environment, decreasing the chances of soil pollution.

Nano-fertilizers have opened up new possibilities to improve input and usage efficiency, to reduce cost, and to reduce environmental deterioration in some possible ways. In the coming years, the application of nano-fertilizers in the agricultural systems should be prioritized to sustain soil health, enhance crop production, and sustain the quality of environment through promoting the use of nanoparticles in fertilizers and nano-sensors (Meghana et al. 2021).

- **Nanopesticide**

The common practice to control pests and weeds in agriculture is the use of pesticides. Pesticides have a long history of use for a common farmer in any part of the world. But the chemicals being used as pesticides have their own side effects. With the passage of time, research has proved more negative effects of these conventional pesticides as compared to their positive impacts on crops and environment. In recent era, research in nanotechnology has helped researchers for producing some effective pesticides. One of the methods include the encapsulation of some present pesticides that prevents the harmful spread of these chemicals in the environment. Such nano-encapsulating methods help in the controlled release of the pesticides at the targeted sites as per crop needs (Alfadul et al. 2017).

Nano plant-protection products or nano-pesticides are an emerging technique related to pesticides, and offer a lot of benefits including increased efficacy, durability, and reduction in the number of active ingredients that need to be used (Kookana et al. 2014). Nano-insecticides, nano-fungicides, nano-weedicides, and nano-bactericides are all nano-pesticides that are used for their easier application, availability, and low cost (Baker et al. 2019).

Nano-insecticides (mainly nano-silicates, alumina, and inorganic nanoparticles) lead to the development of sustainable techniques for the pest management (Stadler et al. 2018). Nano-weedicides and nano-herbicides are specifically produced to target weeds and herbs. These weeds and herbs are causing losses to main crops by competing with them for nutrients. But again herbicides are the chemicals that may not be very suitable for environment and other beneficial plants in the environment. They may also persist in the food chains. Here, nano-encapsulation of the herbicides presents a solution again, where their application can be smarter and more target oriented. This way of application of herbicides may also decrease their phytotoxic effects in crops (Jalil and Ansari 2020).

- **Nano-sensor**

Nano-sensors is another application and product of nanotechnology in the field of sustainable agriculture. Researchers and agriculturists all around the world are showing serious interests in nano-sensors due to their multi-dimensional applications in smart agriculture systems, and other food-related industrial sector. Nano-sensors can significantly improve crop yield by managing the application of water and other chemicals according to the need of the soil and crop grown. They are smart enough to work efficiently, utilizing the minimum resources. They also detect the presence of pesticides, moisture content, chemical fertilizers, soil nutrient, and pollutants such as large amounts of fertilizers, pesticide residues, plant pathogens, and heavy metals around (Shawon et al. 2020). Nano-sensors have also been used as a part of soil analysis as well in quantifying the carbon content, organic matter and phosphates of the soil (Li et al. 2020). Nano-sensors have several advantages over conventional sensors. Nano-sensors have shown their higher efficiencies as they contain nanoparticles that have a larger surface-to-volume ratio. These nano-sensors can show even more sensitivity, selectivity, and higher stability with a very compact and smart size. They have modern approach of real time detection of nutrient requirement of crops or any type of infection making them even smarter devices (Kaushik 2019). These sensors are important in smart agriculture because they can identify infectious plant diseases even before symptoms occur (Dar et al. 2020).

Nano-sensor approach has made the way for smart agriculture based on sustainable development, environmental safety, and enhancement in productivity of crops, reduction in input costs, prevention of overuse of water, and smart management of fertilizers. Despite these unique advantages, the majority of nano-sensors have been developed on a laboratory scale and more work is needed to design the sensing system for on-field agricultural practices (Singh et al. 2021).

- **Nano-products for remediation of contaminated soils and water systems**

Poisonous inorganic and organic contaminants and improper agricultural waste management have put groundwater and soil at great risk. For the removal of biphenyls, dyes, volatile organic compounds, heavy metals, and polyaromatic hydrocarbons from wastewater, nanotechnology provides cost-effective methods (Ali and Ahmad 2020).

All over the world, contaminated soils contain heavy metals, organic contaminants, and antibiotic residues, etc., that can be degraded using nanomaterials (Jain et al. 2020). Therefore, there is a hope that nanotechnology can help to decrease the unwanted dumping off waste and agricultural residues on agricultural lands. It can help to reclaim the soils by nano-remediation. Research is needed in this field as well to fill the gaps of technological use (Sangeetha et al. 2017).

Nano-remediation strategies have the advantages of reducing the time of clean-up procedures, and elimination of difficult practices, required for the disposal and treatment of polluted soil. It also reduces the cost of cleaning polluted soil and decreases the concentration of pollutants to near-zero low levels (Pramanik et al. 2020).

Advantages of nano-agrochemicals

- Nano-agrochemicals have greater effect as compared to conventional agrochemicals. They are known to have environment friendly nature and also economically viable.
- According to available data in the form of research articles, it has been found that nano-agrochemicals are more efficient in increasing the quality and yield of crops along with the improvement of soil fertility. They are also non-toxic and economically feasible as they lower the input costs and increase the profit overall.
- Nano-devices and materials have a significant role in agriculture, as nano-biosensors can detect the status of nutrient and moisture content in soil instantly, and thus agriculturists find its application in site-specific water and nutrient management (Qureshi et al. 2018).
- The ultimate goal of nano-agrochemicals is to enhance the crop production and to minimize agricultural inputs by using nanoscale products through continuous monitoring systems (Pramanik et al. 2020).
- Nano-agrochemicals play an important role in sustainable development in agriculture (Sekhon 2014).

Transformation of nano-agrochemicals

Transformation of nanomaterials is when properties of nanomaterials are changed by the physical, chemical, and biological processes, subsequently influencing their behavior, including transport in soil, uptake and translocation in the plant, and their toxicity to organisms. These transformations can be beneficial, toxic, or even fatal for the environment around them.

Types of transformation

There are three types of transformation of nano-agrochemicals, namely physical, chemical, and biotransformation.

(a) Physical transformation—It is more common and known to happen to almost all types of nanomaterials. Aggregation of nanomaterials in soil takes place in both forms, which include homo-aggregation and hetero-aggregation (Hotze et al. 2010). When an electrical double layer of particle surface compresses in high ionic strength, homo-aggregation of particles of the same nature takes place. Particle size becomes larger and clusters are formed due to the continuation of this procedure. But hetero-aggregation is the aggregation of nanomaterials and larger soil particles; as a result, nano-materials become part of it. They tend to accumulate and form a structure known as "corona" with certain biological and environmental macromolecules, i.e., polysaccharides, proteins, humic acid, and nucleic acids that could cause surface changes in them (Lundqvist et al. 2011).

(b) Chemical transformation—occurs when there are changes in the structure and composition of the nanomaterials, known as chemical transformation. Chemical transformation of chemicals is caused by various chemical reactions where more prominent are redox reactions, chelation, phosphorylation, or sulfidation. The most common example of a chemical transformation is the dissolution of metal based NMs.

(c) Biotransformation—is a kind of transformation involving biological processes. It includes both previously mentioned methods of transformation, which are physical and chemical transformation. Nanoparticles might form clusters near the root surface, in the rhizosphere, or in biological solution (cell sap). Intracellular chemical transformations (redox reaction, chelation, or degradation) occur in the vacuole, cell membrane, cell wall, and cytoplasm while extracellular transformation may take place through ROS (reactive oxygen species) or with the help of enzyme-assisted reactions. An example is the oxidation of graphene in plants by reacting with the hydroxyl group while releasing carbon dioxide (Huang et al. 2018).

Location and transformation of nano-agrochemicals in a plant soil system

Nanoparticles, when they enter into the plant system or are applied to the soil, can be transformed and their transformation may be controlled by different factors. Since nanoparticles are entered into plants mostly through the soil, therefore, the nature of soil is very crucial in this regard. All the soil types have their own chemical and physical characteristics. These characteristics of the soil can regulate and drive the nano-materials' behavior in the terrestrial ecosystem. Generally, the soil is equally made up of particles (organic matter and minerals) and voids (including pore water and air) (Ondrasek et al. 2019).

Nanomaterials are transformed in soil due to the following factors:

- Soil humidity level
- Ionic strength
- Quality of organic matter content
- Quantity of organic matter content
- Redox potential
- pH of soil solution and,
- Soil texture (mineral composition),

Additionally, the soil is teamed up with other invertebrates (nematodes) that are quite large in size with some small size population like protozoa, bacteria, algae, and fungi. Physical and chemical characteristics of soil at the micro-environment stage like the rhizosphere might be changed considerably due to the occurrence of this biota; as a consequence, spatial heterogeneity of soil takes place (Frey 2015). Soil has the largest population in the form of bacteria, which are important in various physiological and geochemical procedures like the nitrogen and carbon cycle (Ren et al. 2018). Their respiration (Salas et al. 2010) and excretion of organic acids (Guo et al. 2017) might activate various chemical transformations of nanomaterials. One of the important places to transform nanomaterials is nano-leaf and nano-root interface before they enter into plants. Properties of nanomaterials are altered at nano-plant interfaces because plants excrete metabolites that may react with them, resulting in changes in physico-chemical properties of nano-materials (Zhang et al. 2012a). **Rhizosphere** is a vital and narrow area around the root surface which has an excessive amount of plant exudates. This region of roots environment also contains colonies of bacteria and fungi as well (Philippot et al. 2013). This area is of considerable importance because physicochemical properties and microbial community of this zone are quite different from other zones of soil. Soil areas farther from roots have lesser and different population of microbes because plant roots secrete certain chemicals due to which more microbial colonies are attracted there chemotactically. Physicochemical properties of rhizosphere are also quite different from plant itself. These diverse characters of rhizosphere are very important in transforming the fate of nanomaterials.

Although little research has been done in this area, there is a possibility of the transformation of nanomaterials inside plant tissues or fluids. The transformation of NMs at each of these sites will be covered in this part, along with the main factors influencing these processes in the soil-plant system.

Nano-agrochemicals' transformation due to soil composition

The interface of plant surface and soil particles is also a very important region for nanomaterial transformation. When these NMs are applied to the soil, they come in contact with plant surface and due to the presence of a variety of organic and inorganic compounds such as natural organic matter, minerals, and ions in the soil, applied nanomaterials may be transformed (Baalousha 2017, Duan et al. 2017). Many factors are involved in the transformation of nanomaterials in the soil. Some of them are given below:

Soil texture

Soil texture is defined as the relative content of minerals/particles present in it, including clay, silt, sand and gravel. These particles of soil make soil texture as well as other soil environments. Whenever soil texture is changed, the aggregation behavior of nanomaterials is also changed. Therefore, there is a direct relationship of soil texture and transformation of nanomaterials (Jahn et al. 2006). This can be explained with various examples. Silty and clayey soils are known for the presence of negative sites that can be used by NMs for binding (Zhang et al. 2020). Hence, this causes stronger hetero aggregation of nanomaterials in these soils as compared to sandy soils because these negative sites form strong bonds with the positive charged nanomaterials. Redox potential and humidity of soil are also affected by the soil texture, which has an indirect effect on the amount of water and air in the soil (Adamu and Aliyu 2012) (Figure 2). Clay-rich soils have a lesser pore space with lesser air, whereas its water holding capacity is higher than sandy soils. It results in a reducing environment that favors the reduction of nanomaterials. Sandy soils have more oxygen, which may facilitate the oxidative transformation of nanomaterials such as silver (Ag) and copper (Cu) (Li 2017).

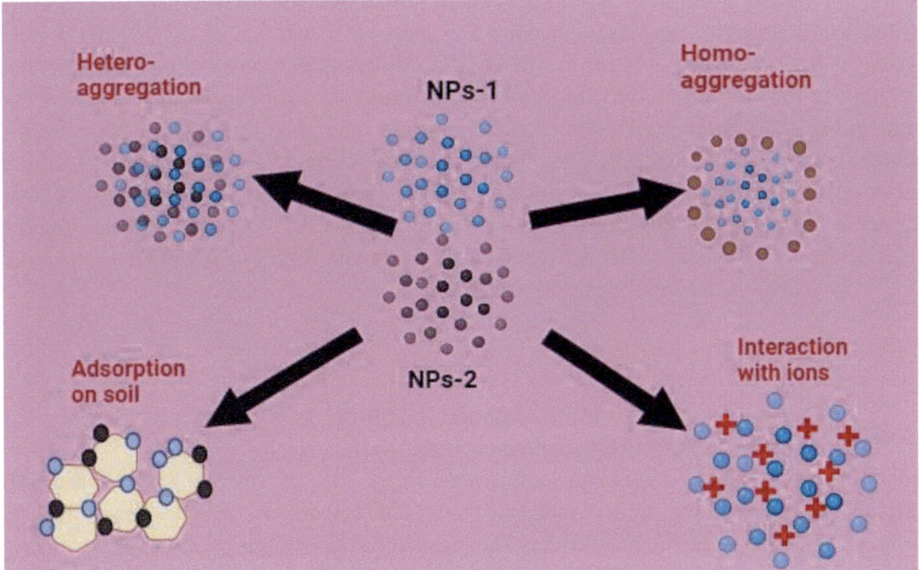

Figure 2. Effects of soil texture.

Natural organic matter (NOM)

Agricultural soil is composed of almost 1 to 5% of natural organic matter (NOMs) (Brady and Weil 2008). These are natural macromolecules (which include polysaccharides, proteins, humic acid, etc.), produced when dead bodies or remains of plants and animals are broken down by microbes. These chemicals may be of varying nature depending upon the stage of decomposition. This organic matter forms a layer or corona-like coating on the particle surface, which in consequence changes the action and fate of the nanomaterials. NOMs in soil are present in large amounts as compared to nanomaterials; therefore, characteristics of nanomaterials are considerably affected by them in the soil environment. It should also be considered while studying the transformation of nanomaterials that the soil texture and composition of the NOMs individually affect the NM's transformation. But the interaction of these two factors, i.e., soil texture and composition of NMs in soil is also a critical factor, affecting transformation and characters of NMs differently (Lin et al. 2017). Natural organic matter in soil has two types, i.e., hydrophilic and hydrophobic. About 60 to 80% of natural organic matter is made up of humic acid, which has a hydrophilic fraction (Zhang et al. 2020). When hydrophilic and hydrophobic fractions of NOMs react with NMs surfaces, they affect their surface in a very diverse way, depending upon the nature of NOMs present with humus. These NOMs have the higher molecular weight due to which they might completely or partially cover the surface of the nanomaterials. This covering might cause electrostatic stabilization of nanomaterials, being dependent on characteristics of the NOMs (Degenkolb et al. 2019). Organic materials also enhance the aggregation of nanomaterials because these organic molecules have very higher affinities for each other (Luo et al. 2018).

Ionic strength

One more factor which may affect the stability of nanomaterials in soils is the ionic strength of the soil. The stability of colloidal particles is dependent upon electrostatic repulsion and van der Waals attraction according to the classic Derjaguin, Landau, Verwey, and Overbeek (DLVO) theory (Ohshima 2012). In normal situation, when the electrostatic double layer (EDL) of soil is sufficiently charged, NMs are stable. An electrostatic double layer is directly correlated to certain parameters such as pH, ion valences, and ionic strength present in that environment. When EDL is compressed by the high ionic strength, the nanomaterials are aggregated (Zhang et al. 2012b). Furthermore, the efficacy of high-valent ions is more intense than those of low-valent ions. For instance, when ionic strength is the same, calcium ions (Ca^{2+}) are found to form more aggregates of the citrate functionalized negatively charged ions of Ag nanoparticles than Na^+ (Tang and Cheng 2018). Citrate has carboxyl groups that binds with Ca^{2+}, as a result, reduction of Ag NMs occurs following their aggregation. But, the zeta potential was not changed by the Ca^{2+} or Na^+ and un-charged PVP-coated Ag NMs were aggregated irrespective of the pH. However, it is suggested that the stability of all nanomaterials is not affected by the presence of ions because some of the NMs resist aggregation due to their own physicochemical characteristics (Badawy et al. 2010).

pH

Another significant factor in NMs' transformation is soil pH as it affects aggregation. Isoelectric point of NMs is known as the pH value when there is no net charge on their surface and they can easily form aggregates. The values of isoelectric points are different for different types of nanoparticles ranging from acidic to basic pH. Hence, pH has a very varied impact on the stability of NMs. pH of the soil significantly affects the chemical transformation of nanoparticles. Soil pH is a very important factor for metallic nanomaterials. In acidic soil, their dissolution is increased significantly. Chemical speciation of nanomaterials is also affected by the pH of soil. Effect of pH on NM transformation can be further explained with example of zinc oxide nanoparticle's

(ZnO NMs) transformation. When pH of soil solution is low, i.e., < 6, applied ZnO NMs will release bivalent zinc ions (Zn^{2+}). But when pH is higher, transformation of ZnO NMs is altered due to formation of a $Zn(OH)_2$ layer on the surface of nanoparticles. When pH of soil is even higher, i.e., > 12, then these ZnO NMs can be transformed to zincate ions, $Zn(OH)n\ (n^{-2})^-$ (Tang et al. 2002).

Redox reactions

Redox reactions take place frequently and are very common in soil environment and are crucial to many biogeochemical processes. These processes are crucial for the transformation of NMs because some of the NMs are very sensitive to the changes in redox potential (E_h). E_h determines the extent of these redox reactions. The value of E_h is determined by the chemical environment of soil. Some of the widespread oxidants in soil are CO_2, oxygen, iron, sulfate, nitrate, and nitrite, etc. (Bartlett and James 1993). Iron ions (Fe^{2+}), for instance, can reduce graphene oxide (Pearson 1963). NMs are more likely to be oxidized in oxygen-rich, aerated soil. As a consequence, this oxide layer formed on the surface of NMs become protective and restrain the further transformation of the particle core. On the other hand, some nanoparticles like Ag NPs may form Ag_2O and release Ag^+ ions. Nanomaterials are reduced in anoxic conditions where the environment is reduced like organic-rich soil, paddy soils, and wetlands, etc. Li et al. (2017) utilized paddy soils under different conditions to compare the Ag NP transformation. After 2 days of incubation, 90% of silver remained as silver nanoparticles under aerobic conditions, while in an anaerobic environment, 95% of it was converted to Ag_2S as a result of oxidation. Nonetheless, the soil microenvironment is greatly diverse. Even paddy soil might have aerobic areas which are oxygen-rich and affect the transformation of nanomaterials. Here the microbial composition of soil again plays its role. In all of the previously mentioned circumstances, the reality of soil complexity must be taken into account (Qiu et al. 2021).

Inorganic ligands on NM surface

Metal-based nanomaterials are mostly chemically transformed by the inorganic ligands which involve phosphate, chloride, and sulfide, etc. According to the Pearson acid-base idea, sulfide is very common in the environment and found everywhere; therefore, metal NMs face sulfidation in the environment very frequently (such as ZnO, PbO, Ag, and Cu) (Chen 2018, Reinsch et al. 2012). Sulfide formation of nanoparticles also decrease their solubility. There are different effects of inorganic ligands on various NPs. For example, a low sulfidation level has a minute impact on the release of zinc ions (Zn^{2+}) from zinc oxide nanoparticles (ZnONPs), while greater than 50% of sulfidation (higher level) will significantly decrease the solubility of zinc oxide nanoparticles (Chen et al. 2021a). Due to the higher affinity of zinc ions with phosphate ($Ksp = 2.0 \times 10^{-25}$) than to sulfide ($Ksp = 2.0 \times 10^{-25}$), the ZnS possibly transforms to $Zn_3(PO_4)_2$ further. Similarly, in silver nanoparticles, dominant species is dependent upon the soil pH. Under alkaline or neutral conditions, sulfidation of silver occurs whereas acidic conditions favor AgCl formation (Reinsch et al. 2012, Yang et al. 2021). Again same problem is faced due to higher stability and lesser solubility of silver sulfide for even a longer time period. Nanomaterials' synthesis and re-synthesis can occur due to changes in the chemical formation of the soil, e.g., cerium ions (Ce^{3+}) may be released from cerium oxide (CeO_2) nanoparticles due to reduction caused by bacteria in the soil. This may form $CePO_4$ due to weathering of apatite and also the representative of phosphate mineral (rhabdophane) in the soil. When biogenic organic acids like ascorbic acid, oxalic acid, and citric acid in soil come in contact with $CePO_4$, its re-dissolution takes place (Schwabe et al. 2015). Trivalent cerium ions are released and converted into tetravalent cerium ions Ce^{4+} by the process of oxidation. In this stage, their conversion to CeO_2 (nanoscale) happens again over the surface of cerium phosphate. The transformation of metal ions to metallic nanoparticles such as silver and gold directly in plants or by plant extracts are some examples (Roy et al. 2017, Torresdey et al. 2002).

Microorganisms

(1) Microorganisms are the main source of biomass because these are present everywhere in the soil. Therefore, nanoparticles are transformed in the soil environment due to the biological activity of bacteria. All of the aforementioned physical and chemical transformation processes may be mediated by bacteria or may be indirectly related to the microbial intervention.

(2) They can cause the reduction, oxidation, and hydrolysis of nanomaterials.

(3) The main sites of these reactions are the cell membrane, and cell cytoplasm. These reactions can also happen exterior to the main cells. Many biological processes like decomposition, and mineral (carbon, nitrogen, sulfur, and phosphorus) cycle are initiated by the enzymes released by microorganisms into the soil. This would facilitate the release of inorganic ions that are used by plants as nutrients for their growth (Zhou and Staver 2019, Tian et al. 2020). That is the natural function of these released enzymes including hydrolases, phosphatases, reductases, oxidases and lyases. These enzymes can also cause the transformation of applied NMs by the process of oxidation, reduction, sulfidation, etc. (Kandeler and Dick 2006, Josko et al. 2019).

(4) For example, CeO_2 NMs are one example as these can be reduced at the surface of a bacteria named *Bacillus subtilis* which would release cerium ions (Ce^{3+}) and a further carboxyl group on the bacterial cell wall surface would chelate them (Xie et al. 2019). Enzymes or radicals produced by bacteria have the potential to oxidize carbon nanotubes (CNTs), leading to the nanomaterial's oxidative destruction (Allen et al. 2009). During the respiration process, some facultative anaerobic bacteria (*Shewanella*) can reduce the nanomaterials. In doing so, these bacteria donate electrons to the material like graphene oxide and they extract these electrons from their cytochromes (Salas et al. 2010).

(5) As previously mentioned, soil environment is heterogeneous and diverse, so the procedure of transformation of nanomaterials in the soil is extremely complicated and may take place in a sequence or at the same time. Therefore, it can be concluded here that which type of NMs will stay in the soil depends upon a number of factors. These factors include the life of NMs in soil, environmental factors of soil (including water content, microbial population, pH, and oxygen level, etc.) and the type and amount of ligands at surface.

- **Nano-agrochemicals' transformation in plants**

Transformations of nanomaterials are not just restricted to the soil-nanomaterials interface, but once they are absorbed by plants, transformation occurs at plant-NM interface.

Nano-agrochemicals' transformation at plant interfaces

One of the most important sites for the transformation of nanomaterials is the **rhizosphere**. Many reactions between plants and other organisms (invertebrates, microorganisms) occur in rhizosphere (1–3 mm area from roots' surface) (Kuzyakov and Razavi 2019). According to studies, NMs undergo more transformation in the rhizosphere than in the soil away from the roots (Zhang et al. 2012a, Rico et al. 2018). The chemistry of the rhizosphere also affects the changes and alterations of NMs in rhizosphere. Nanomaterials take various paths of absorption and movements when inside the plants, which also affect their mode of transformation (Figure 3).

    Plant root exudates are mainly of two types: high molecular weight and low molecular weight. The former category includes mucilage and cellulose type macro molecules while lower molecular weight includes sugars, organic acids, phenols, amino acids, and proteins, etc. These are secondary metabolites and can be used by the associated microbes. Root cells are protected by the mucilage which is a polysaccharide and helps the roots in penetrating deep into the soil. The surface properties of nanomaterials are changed significantly because they may form a coating on NMs. In the

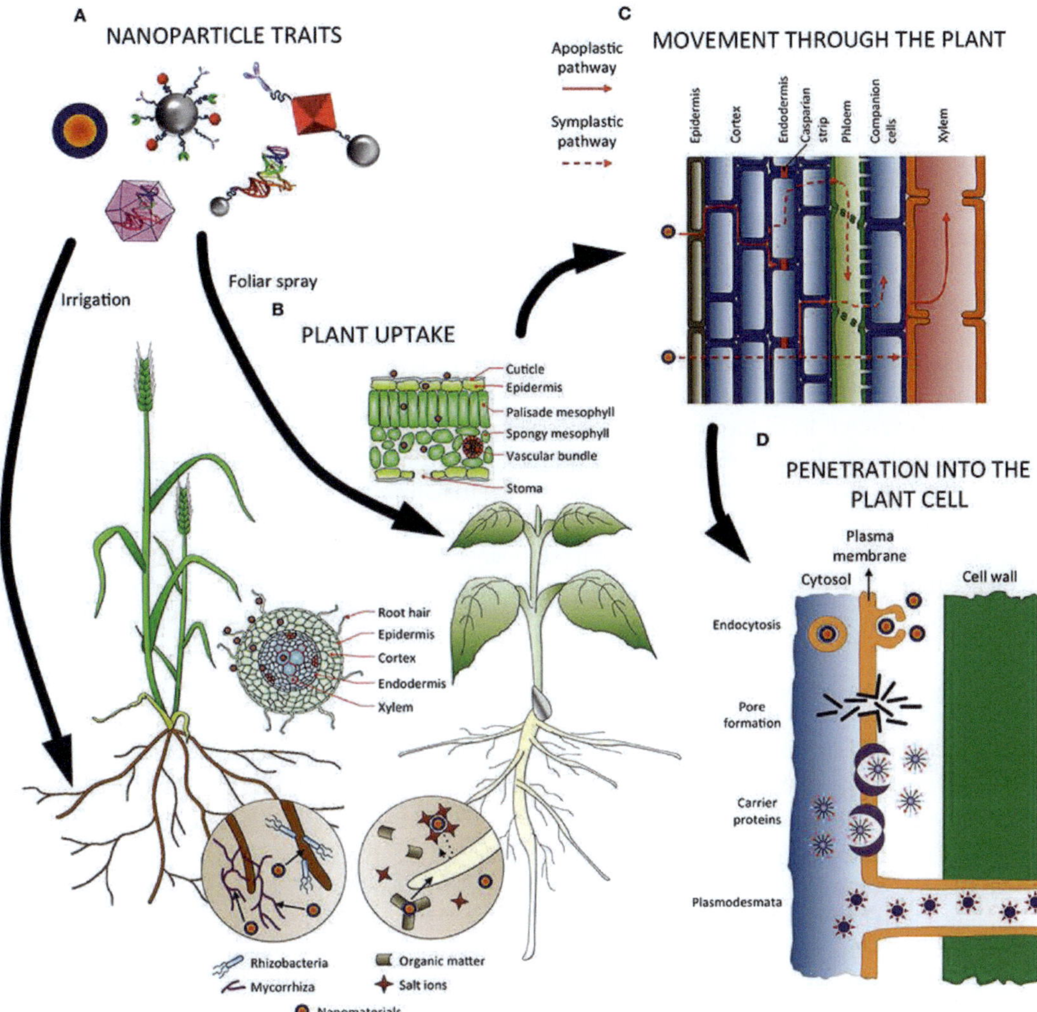

Figure 3. Various interactions of plant cells and nanoparticles of various shapes absorbed by them. Nanoparticles take various pathways when inside plants depending upon their size and charge on the surface (Perez-de-Luque 2017).

rhizosphere, the chemical transformation of nanomaterials takes place more significantly by small organic molecules. When organic acids are released near the rhizosphere, these may cause changes in the pH. Due to these released acids, pH of the soil in rhizosphere is lowered in comparison to the surroundings. This lowered pH may cause an increase in the solubility of the NMs (Huang et al. 2017). Moreover, the redox reaction also takes place in the rhizosphere. Phenols and ascorbic acid (plant-based reducing agents), for example, when released can cause reduction of cerium oxide NMs to form cerium ions. These cerium ions then can form cerium carboxylates or phosphates on or near root surface (Zhang et al. 2017). For some types of NMs, these oxidation reduction reactions are not simple like copper nanomaterials (Cu NMs). These CuNPs can produce divalent or univalent copper cations or copper atoms with these plant based reducing agents being released from roots. The chemical structure of the root exudates depends upon plant species like mugineic acids, which are particularly released by the gramineous species, and helps in the dissolution of NMs (McManus et al. 2018).

Leaf interface

At the leaf interface, many transformation processes, which include dissolution, aggregation, and redox reactions, may occur. Therefore, type of plant species is a very important and crucial factor while considering the fate of applied NMs and their transformation.

Phyllosphere

The term "phyllosphere" refers to the portion of a plant that is above the earth and is inhabited by microorganisms (Vorholt 2012).

Transformation of nanoparticles in air is also common as they can interact with aerosol, dust, and even with the microbes present on plant leaf surface (phyllosphere) (Kim et al. 2010). Other factors may also contribute to the aggregation of nanomaterials such as irrigation and rainwater or due to water excreted from the leaf surface. Furthermore, epiphytes (yeast, bacteria, fungi), which are present in the phyllosphere, are another major cause of the chemical transformation of NMs at leaf surfaces. Although the complete chemistry of the transformation of NMs in the phyllosphere is not known, there are some theories that explain the transformations. For example, microbes present at the surface of the leaves secrete some extracellular polymeric substances (EPS). These EPS can interact with the NMs and can form a corona at their surface. Thus, they can change the chemistry of NMs' surface.

Nano-agrochemicals' transformation inside plants

There are not many studies that have distinguished the processes of transformation occurring inside or outside the plants. There is a need to study and understand the fate of NMs and their alteration inside the plant tissues.

- The transformation on the outside of plants comes about at the coordination of NMs and roots or in soil; on the other hand, transformation on the inside takes place in sap, inside the cells of plants, etc. The main location of transformation of CeO_2 NMs was shown by Ma et al. (2017) in plants of cucumber by employing the methods of exposure of petiole and split root. The latter method involved the manual two-part separation of the root and their placement into two different vessels, where one of them contained the suspensions of CeO_2 NMs. The plant was exposed for three hours for the CeO_2 NMs to be taken up by it. They studied the Ce's chemical species after severing CeO_2 NMs' exposed roots and continuing for seven days for the remainder of the seedlings to grow. CeO_2 NMs' transformation didn't occur after they were exposed for three hours and seven days and showed that there was no transformation through short-spanned exposure of the root and there cannot occur an abrupt transformation in particles of CeO_2 that were taken up by the plant. On the other hand, Wang et al. (2012) reported that by the use of the same method, CuO transformation occurred in the plants of maize compared to CeO_2 NMs. Whereas during translocation through xylem and phloem tissues, NMs may be transformed like CuO NMs to Cu_2O, in the plants. Ce-NPs may behave differently and it is not a surprise that the fate of the CuO and CeO_2 NMs may differ as type of transformation is dependent upon type of NMs.

- Celma et al. (2016) reported that the NMs' transformation may also involve the components of plant sap, i.e., enzymes and reducing sugars. The sap of the xylem, which helps in movement of nutrients acquired by soil and water via roots and into the shoots, contains inorganic ions and water primarily as well as a smaller quantity of biomolecules, organic acids, and hormones. Aggregation is one of the possibilities for the transformation of NMs in the xylem, mostly because other inorganic ions are present in it.

- Celma et al. (2016) also reported that the phloem sap has the greater constitution of hormones, organic acids, proteins, amino acids, sugars, and inorganic ions. Hence, there may be an occurrence of transformation more readily in both physical and chemical methods. The NMs may be dissolved by the organic acids, while fructose and glucose like reducing sugars can produce the transformation of NMs through ions (Wang et al. 2012). Due to presence of sugars and other metabolites in phloem, there is more probability of NMs' transformation in phloem as compared to xylem, although researchers do not have direct proof of how and whether NMs are transformed by sap.

- The rate of flow of sap can also influence the transformation that most importantly controls the time of residence. As stated by Su et al. (2019), the rate of flow of sap in the xylem (0.47–4.8 mms^{-1}) is greater than that of the phloem's (0.07–0.58 mms^{-1}). Not enough time is allowed for the interface of the components of sap with NMs as when the roots take up the NMs into the xylem, some particles get distributed into tissues by quickly traveling in the upward direction. In the case of phloem, it is the other way around. Ma et al. (2020) also stated that there cannot be any transformation of CeO_2 NMs in the root of rhizosphere through a shortly spanned method of exposure, leaving the desirable suggestion of a reaction-time that is critical for the particle transformation of CeO_2, as they are innately stable.

- The method of NMs' green synthesis has reported that the leaves of the plant can act as a factory that synthesizes many types of NMs, i.e., CuO NMs, ZnO, Au, and Ag NMs (Torresdey et al. 2002, Qu et al. 2011, Sharma et al. 2015, Ahmed et al. 2016). For the synthesis of the NMs in metallic form, there is a widely used method in which the phytochemicals like fructose, glucose, and flavonoids, of reactive nature, are extracted from the leaf of the plants to be used as the precursor of NMs with metal salt (Singh et al. 2018).

- It will be a reasonable conclusion that those substances generated by plants can reduce the NMs. For instance, while undergoing the same phenomena in the vascular fluids of the plant, reduction of CuO to univalent copper or elemental copper can happen. On the other hand, many physiological processes in a plant can generate free radicals, which drive the NMs' oxidation in the leaf. Radicals of hydroxyl (OH•) were reported to transform graphene to CO_2 in the leaves of rice (Huang et al. 2018, Hu et al. 2021).

- Location of NMs inside the organism is of great importance to bring the precise kind of transformation. The subject is studied less, but the NMs' presence and their species that have been transformed in the spaces between cells (Zhang et al. 2012a), chloroplasts (Shabnam et al. 2016), and vacuoles and cytoplasm (Ma et al. 2015) were documented; especially, these transformations are driven by the processes that are hardly explored yet.

Effect of nano-agrochemicals' transformation on mycorrhizal sphere

There is a need for studying the effects of NMs on plants and their rhizosphere as well because not only nano-agrochemicals are affected and transformed by plants, soil, and microbes, but these factors are influenced by the nano-agrochemicals as well.

The symbioses of rhizobia and mycorrhiza are the most significant factors of the terrestrial environments, and play a significant role in the nutrients' recycling. These symbiotic relations are also responsible to mineralize the organic matter and ultimately strive to preserve the resilience and functionality of ecosystems (Finlay 2008, van der Heijden et al. 2008). In short, a symbiotic and collaborative interface called "mycorrhiza" has come through the evolution of both groups of ecto-mycorrhiza and endo-mycorrhiza, and their association is found with more than ninety percent of higher plants (Sharma et al. 2014).

Various studies have been carried out about the interaction of nanotechnology and mycorrhiza, especially the arbuscular one, but on the other hand, there are only a few reports are there about relationships of NMs and other types of mycorhizae. The symbiosis of arbuscular one is constituted by approximately eighty percent of species of vascular plants that comprise bryophytes, pteridophytes, gymnosperms, and angiosperms, with Mucoromycota's substratum, and Glomeromycotina's fungi (Spatafora et al. 2016). There are various advantages of this symbiosis; like it enhances the growth of the plant, resists the biotic and abiotic hindrances, i.e., drought, the toxicity of metal, and salinity, along with access to the nutrients of soil including phosphorus, potassium, and nitrogen (Oldroyd et al. 2011).

- The nanoparticles of ZnO, TiO_2, and Ag, after being added to soil, produce different effects on the development of plants and this depends on the use of dose, kind, and size of NPs during the experimentation (Ahmed et al. 2016). Moreover, there are many fungi and bacteria against which the NPs, stated previously, can have antimicrobial use. They can destroy and change the structure of fungal hyphae due to their nature and size. The influence of nano-formulations and nano-materials on essential symbioses of plants and microbes, i.e., rhizobia and mycorrhiza, could not be evaluated easily because of the critical results generated by the use of NPs in the promotion of the growth of a plant. The connection of various fungi of mycorrhiza and NPs was found to be producing both negative and positive effects; that both are important for sustainable and efficient agriculture. There are types of NPs that help in the colonization of fungi and also those that hinder it. There are various applications—nano-carriers, nano-fertilizers, nano-pesticides, and stimulators of plants—that influence growth based on nanoparticles like TiO_2 and SiO_2, and nanotubes of carbon, having control over the discharge of agrochemicals. The advantages of these chemicals include the enhancement of the development of crops and their defensiveness to stressors; on the other hand, they can also lessen the quantity of conventional use of agrochemicals. It may have a positive effect on reducing the contamination in the environment, wastage, and crops (Tian et al. 2019). But there is a strict need to study the mutual relationship of mycorrhizae with each NP used. In fact, there should be protocols for finalizing the dose of NPs for any crop, and it should also be tested for its effects on rhizobia and mycorrhizae.

- The growth of the plant and its interaction with the soil fungus is influenced by various nanoparticles of metal. It can primarily be illustrated by the fungal part of mycorrhiza, especially arbuscular. As Bowles et al. (2016) stated, plants that are inoculated with AMF, show more fungal resistance which becomes the reason for healthy roots and a considerable improvement in the uptake of nutrients, i.e., phosphate and of those nutrients that are present in lower concentration and are not mobile (Begum et al. 2019). Alternatively, it is found that the growth of the fungi in soil, which are helpful in the growth of plants, gets affected by various NPs of metallic nature.

Up till now, various formulations based on nanoparticles that are metallic, i.e., nano-fertilizers, nano-pesticides, and nano-herbicides, have been utilized and researched for their encouraging function in the development of the plant. A few studies are also available on the effects of NPs on mycorrhizae. Cao et al. (2017) reported that the pace of the development of AMF and its ecological role has been reduced (p 0.05) when it is exposed to AgNPs. It also decreases the availability, activity, and nutrition of phosphorus and alkaline phosphatase in soil. Li et al. (2015) studied the colonization of AMF *(Funneli formismosseae)* in the plants of maize and demonstrated the influence of the combined ($ZnSO_4$ and ZnONPs) as well as the individual exposure to it. The results were remarkable in showing that colonization of AMF in plants of maize declined through the treatment of 500 mg kg^{-1} of $ZnSO_4$ and ZnONPs; nevertheless, the growth was enhanced by the combination of substances $ZnSO_4$ and ZnONPs. The reason behind this may be the increase in the nitrogen concentration and diminishing Zn quantity in the plants of maize.

- The knowledge of the processes involved in the interactions of NP-rhizobial/mycorrhiza is still limited and needs to be explored. Various studies show the qualities of NPs being antifungal including the discharge of the metallic ions, i.e., Ag and Zn^{+2}, root boundedness, enhanced concentration of NPs inside, and the cessation of growth of roots; these are the harmful influences of NPs on the colonization of mycorrhiza. The antifungal effect of the NMs include adhesiveness to the surfaces of the cell and physical harm to the wall of the fungal cell and its membrane, enhancing the permeability of the membrane, the water channel obstruction, and also the death of the cell as an outcome of the deposition and penetration of NPs into the cells. Their edges are sharp and so can cut the structures of fungi and walls of the cell, hinder the germinating spore by constructing a barrier, cell death by entering, and the deposition of NPs in cells. The forces of van der Waals can also hinder the germination of the spore by the creation of the aggregates of NPs. The disturbance in the regulation and cycle of glutathione, release of ion from the nanoparticles of metallic nature, and a few NPs' properties that are photocatalytic like TiO_2, accumulate active oxygen species (ROS) in fungal cells (Tian et al. 2019).

Phytotoxicity of transformed nano-agrochemicals

Phytotoxicity

When a toxic compound is built up in plant tissues to such an extent that it interferes with the normal functions (growth and development) of the plants, it is known as phytotoxic while the process is called phytotoxicity (Bould et al. 1983). The use of nanomaterials in agriculture due to their remarkable properties is growing in recent years. Nonetheless, their toxic behavior and bioavailability make them vulnerable to be used at large scale. Their highly reactive traits, which include larger surface area and their chemical structure, are also linked to their toxic behavior (Nel et al. 2006). Phytotoxicity is also caused by some vital metal nanoparticles by an increased discharge of important ions or by dissolution in plants (Anjum et al. 2015).

Phytotoxicity of transformed nano-agrochemicals

Due to the wider application of nanomaterials or nano-agrochemicals in agriculture as nano-fertilizer, nano-pesticides, etc., there is a great concern among researchers, users, and environmental experts about their phytotoxic effects (Sun et al. 2017, Li et al. 2019). The interaction of nanomaterials with their adjacent environment is unavoidable so it will necessarily react with plants. Nanomaterials may attach to the roots of the plant, thereby causing chemical and physical toxicity in plants. These NMs, once entered into the plants by the cell wall, can be moved through the plasmodesmata. RNAs and various regulatory proteins are transported by plasmodesmata through non-selective and selective routes for shorter pathways (Xingmao et al. 2010).

The phytotoxic behavior of nanoparticles is dependent upon size, shape, structure, and even the structure of coating material. It may be caused by the toxic nature of the compounds which are being utilized for the synthesis of nano-materials. Moreover, the type and nature of plant, and its environment also affect the phytotoxic behavior of the nanomaterials. In any part of a plant life cycle, NMs may cause detrimental and harmful effects on plant growth (Figure 4). They might move from the root to the upper portion of the plant body with the vascular system while some may remain attached to the roots. However, it depends upon the anatomy of the plant and characteristics of nanoparticles such as size, shape, and chemical structure.

Phytotoxicity of nanoparticles may be caused by two factors:

- Discharge of toxic ions which might cause chemical toxicity depending upon chemical structure.
- The size, shape, and surface area of NMs cause stimuli or stress. The absorption of NMs significantly impacts the reaction in cell culture (Brunner et al. 2006).

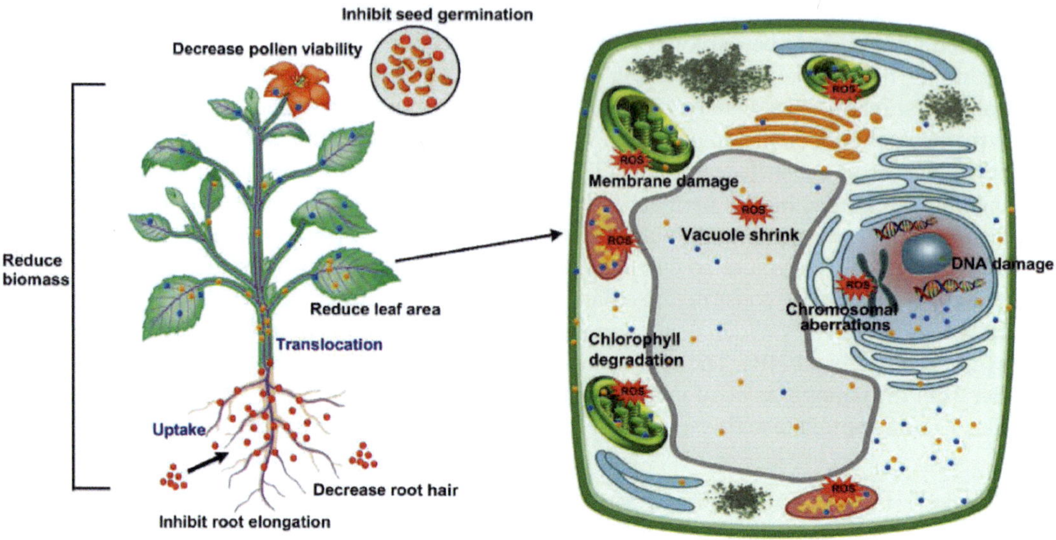

Figure 4. Possible phytotoxic effects of NPs on plants (Yan and Chen 2019).

Plant cell

NMs have to pass through the cell wall and cell membrane of the epidermis of the root system to enter the xylem and be transported to leaves via the stem to interact with plants. The cell wall is made up of polysaccharide fiber which is a porous network. Almost all solutes and water molecules are transported through it to the roots. Aggregated NMs of small sizes may get entry through this pathway to the cell membrane while aggregates of larger size particles might not pass through it to the plant cells (Ma et al. 2010).

Plant seedlings

It should be noted that plant development should not be stopped by chemical phytotoxicity because at the germination stage, plants require water to kickstart metabolism in the mature seeds as they are comparatively dry. Rather, phytotoxicity may be caused as a consequence of plant cell transport routes and nanoparticles (physical reactions), which would stop the apoplastic pathway by blocking cell wall pores and intercellular spaces. The symplastic pathways may also be affected by the blocking of nano-sized plasmodesmata, which depends upon the dose and size of nanoparticles and plant species (Clement et al. 2013). This kind of toxicity is caused by the activation of the antioxidant defense system and reactive oxygen species. It contains hydrogen peroxide, singlet oxygen, superoxide radical, and hydroxyl radical (Gill and Tuteju 2010).

Transformed nanoparticles are also involved in the downregulation of photosynthetic enzymes. NPs upregulate the defense mechanism but a balance is required in plant responses. This elevated defense system also causes an increase in the biomass of the plant by increasing carbohydrate synthesis. Amino acids are metal ion chelating agents and their synthesis is upregulated in plants treated with metal nanomaterials (Vecerova et al. 2016). But there are also research reports that silver nanoparticles reduce the biosynthesis of glutamine and asparagine in plants. They are also considered responsible for the reduced glycine and serine ratio (Hu et al. 2014). Graphene nanoparticles have shown phytotoxic effects on alfalfa plants with the conclusion that whenever NPs are in abundance, they pose toxic effects to plants (Hong et al. 2014, Chen et al. 2021b). Sometimes, NPs are also responsible for the enhancement of the cytotoxic effects of other contaminants present in soil including heavy metals, like arsenic (Cao et al. 2021).

Nanoparticles released in the water bodies are absorbed by plants and these nanomaterials of the natural water cause inhibition in the synthesis of essential amino acids resulting in a decrease or inhibition of cell division in plants (Hu et al. 2018). Cell membrane fluidity is also decreased in presence of some nanomaterials that ultimately destabilize and disrupt the membrane structures (Vecerova et al. 2016). Higher doses of nanomaterials also disrupt the phenylpropanoid pathway, thus disturbing the secondary metabolite production as well. Therefore, metabolite markers can also be used to check the toxic effects of nanomaterials in plants.

Generally, reactive oxygen species in plants are produced as signaling molecules as a byproduct of aerobic metabolism. However, high level of these molecules may disturb the normal functions of the plant and cause oxidative stress which would damage the cell membrane, DNA, proteins, and lipids. These also cause electrolyte leakage and cell death (Sharma et al. 2012). Hence, it is clearly understood that environmental modification is very important in this regard. There is a great need for understanding the post-application of modification of nanoparticles in soil, air, and plants. Whether the interaction of plants and NMs will be positive, negative or impartial, it all depends upon the size and types of NPs, type and growth stage of the plant, soil type and condition, and method and season of application of NPs (Li et al. 2019, 2020, Chen et al. 2021a).

Conclusion

In the present era, new technologies are being rapidly adopted to enhance food production from available resources for a rapidly growing population. Nanomaterials have played a vital role in this regard. There are multiple advantages of using NPs as they are active with higher absorbing potential onto the plant cell surface. They are known to increase plant growth and yield, both in monocots and dicots. Meanwhile, their phytotoxic effects can't be neglected, particularly when the transformation of nano-agrochemicals is considered after their application to plants and soils. This transformation can be physical, chemical, or biological; furthermore, these transformations can be either beneficial or hazardous to plants. These transformations of nano-agrochemicals greatly depend upon soil texture, pH, ionic strength, redox potential, humidity, organic matter, and biota present in the rhizosphere. Nano-agrochemicals are also transformed under the effect of plant root exudates in soil. They are also changed inside the plant after absorption. The fate of these NPs and their interaction with plants greatly depend upon their mode of transformation inside and outside of plants and this is also dose-dependent. Therefore, there is a dire need to study the effects of these transformations relative to the individual NPs for individual crops.

References

Acharya, A., and Pal, P.K. 2020. Agriculture nanotechnology: Translating research outcome to field applications by influencing environmental sustainability. NanoImpact, 19: 100232.

Adamu, G., and Aliyu, A. 2012. Determination of the influence of texture and organic matter on soil water holding capacity in and around Tomas irrigation scheme, Dambatta local government Kano state. Res. J. Environ. Earth Sci., 4: 1038–1044.

Ahmed, S., Ahmad, M., Swami, B.L., and Ikram, S. 2016. A review on plants extract mediated synthesis of silver nanoparticles for antimicrobial applications: A green expertise. J. Adv. Res., 7: 17–28.

Alfadul, S.M., Altahir, O.S., and Khan, M. 2017. Application of nanotechnology in the field of food production. Academia J. Sci. Res., 5(7): 143–154.

Ali, Z., and Ahmad, R. 2020. Nanotechnology for water treatment. Environ. Nanotech., 3: 143–163.

Allen, B.L., Kotchey, G.P., Chen, Y., Yanamala, N.V.K., Klein-Seetharaman, J. et al. 2009. Mechanistic investigation of horseradish peroxidase-catalysed degradation of single-walled carbon nanotube. J. Am. Chem. Soc., 131: 17194–17205.

Anjum, N.A., Adam, V., Kizek, R., Duarte, A.C., Pereira, E., Iqbal, M. et al. 2015. Nanoscale copper in the soil-plant system-toxicity and underlying potential mechanisms. Environ Res., 138: 306–325.

Baalousha, M. 2017. Effect of nanomaterial and media physicochemical properties on nanomaterial aggregation kinetics. NanoImpact, 6: 55–68.

Badawy, A.M.E., Luxton, T.P., Silva, R.G., Scheckel, K.G., Suidan, M.T., and Tolaymat, T.M. 2010. Impact of environmental conditions (pH, ionic strength, and electrolyte type) on the surface charge and aggregation of silver nanoparticles suspensions. Environ. Sci. Technol., 44: 1260–1266.

Baker, S., Satish, S., Prasad, N., and Chouhan, R.S. 2019. Nano-agromaterials: Influence on plant growth and crop protection. Ind. Appl. Nanomat., pp. 341–363.

Bartlett, R.J., and James, B.R. 1993. Redox chemistry of soils. Adv. Agron., 50: 151–208.

Begum, N., Qin, C., Ahanger, M.A., Raza, S., Khan, M.I., Ashraf, M. et al. 2019. Role of arbuscular mycorrhizal fungi in plant growth regulation: implications in abiotic stress tolerance. Front. Plant Sci., 10: 1068.

Bould, C., Hewitt, E.J., and Needham, P. 1983. Diagnosis of mineral disorders in plants. Volume 1. Principles. pp 174.

Bowles, T.M., Barrios-Masias, F.H., Carlisle, E.A., Cavagnaro, T.R., and Jackson, L.E. 2016. Effects of arbuscular mycorrhizae on tomato yield, nutrient uptake, water relations, and soil carbon dynamics under deficit irrigation in field conditions. Sci. Total Environ., 566: 1223–1234.

Brady, N.C., and Weil, R.R. 2008. The nature and properties of Soils, Ed. 14. Prentice Hall, Upper Saddle River, N J.

Brunner. T.J., Wick, P., Manser, P., Spohn P., Grass, R.N., Limbach, I.K. et al. 2006. *In vitro* cytotoxicity of oxide nanoparticles: comparison to asbestos, silica and the effect of particle solubility. Environ. Sci. Technol., 40: 4374–4381.

Cao, J., Feng, Y., He, S., and Lin, X. 2017. Silver nanoparticles deteriorate the mutual interaction between maize (*Zea mays* L.) and arbuscular mycorrhizal fungi: a soil microcosm study. Appl. Soil Ecol., 119: 307–316.

Cao, X., Ma, C., Chen, F., Luo, X., Musante, C., White, J.C. et al. 2021. New insight into the mechanism of graphene oxide-enhanced phytotoxicity of arsenic species. J. Hazard. Mat., 410: 124959.

Celma, J., Ceballos-Laita, L., Grusak, M.A., Abadía, J., and López-Millán, A.F. 2016. Plant fluid proteomics: delving into the xylem sap, phloem sap and apoplastic fluid proteomes. Biochim. Biophys. Acta, Proteins Proteomics, 1864: 991–1002.

Chen, C., Unrine, J.M., Hu, Y., Guo, L., Tsyusko, O.V., Fan, Z. et al. 2021a. Responses of soil bacteria and fungal communities to pristine and sulfidized zinc oxide nanoparticles relative to Zn ions. Journal of Hazardous Materials, 405: 124258.

Chen, H. 2018. Metal based nanoparticles in agricultural system: behavior, transport, and interaction with plants. Chemical Speciation & Bioavailability, 30(1): 123–134.

Chen, Z., Niu, J., Guo, Z., Sui, X., Xu, N., Kareem, H.A. et al. 2021b. Integrating transcriptome and physiological analyses to elucidate the essential biological mechanisms of graphene phytotoxicity of alfalfa (*Medicago sativa* L.). Ecotox. Environ. Safety, 220: 112348.

Clement, L., Hurel, C., and Marmier, N. 2013. Toxicity of TiO nanoparticles to cladocerans algae, rotifers and plants-Effects of size and crystalline structure, Chemosphere, 90(3): 1083–10903.

Dar, F.A., Qazi, G., and Pirzadah, T.B. 2020. Nano-biosensors: NextGen diagnostic tools in agriculture. Nanobiotech. Agr., pp. 129–144.

Degenkolb, L., Kaupenjohann, M., and Klitzke, S. 2019. The variable fate of Ag and TiO_2 nanoparticles in natural soil solutions-sorption of organic matter and nanoparticle stability. Water, Air, Soil Pollut., 230: 62.

DeRosa, M.C., Monreal, C., Schnitzer, M., Walsh, R., and Sultan, Y. 2010. Nanotechnology in fertilizers. Nat. Nanotech., 5(2): 91.

Dethier, J., and Effenberger, A. 2011. Agriculture and development: A brief review of the literature. World Bank Policy Research Working Paper, 5553.

Duan, L., Hao, R., Xu, Z., He, Adeleye, A.S., and Li, Y. 2017. Removal of graphene oxide nanomaterials from aqueous media via coagulation: effects of water chemistry and natural organic matter. Chemosphere, 168: 1051–1057.

Finlay, R.D. 2008. Ecological aspects of mycorrhizal symbiosis: with special emphasis on the functional diversity of interactions involving the extraradical mycelium. J. Exp. Bot., 59: 1115–1126.

Frey, S.D. 2015. Soil distribution of soil organisms. Soil Microbiology, Ecology and Biochemistry, Elsevier, Amsterdam., pp. 283–299.

Fuglie, K., Clancy, M., Heisey, P., and MacDonald, J. 2017. Research, productivity, and output growth in US agriculture. J. Agr. Appl. Econ., 49(4): 514-554.

Gill, S.S., and Tuteja, N. 2010. Reactive oxygen species and antioxidant machinery in abiotic stress tolerance in crop plants. Plant Physiol. Biochem., 48: 909–930.

Guo, Z., Xie, C., Zhang, P., Zhang, J., Wang, G., He, X. et al. 2017. Toxicity and transformation of graphene oxide and reduced graphene oxide in bacteria biofilm. Sci. Total Environ., 580: 1300–1308.

Hassanisaadi, M., Barani, M., Rahdar, A., Heidary, M., Thysiadou, A., and Kyzas, G.Z. 2022. Role of agrochemical-based nanomaterials in plants: Biotic and abiotic stress with germination improvement of seeds. Plant Growth Reg., pp. 1–44.

Hong, J., Peralta, J.R., Rico, C., Sahi, S., Viveros, M.N., Bartonjo, J. et al. 2014. Evidence of translocation and physiological impacts of foliar applied CeO_2 nanoparticles on cucumber (*Cucumis sativus*) plants. Environ. Sci. Technol., 48: 4376–4385.

Hotze, E.M., Phenrat, T., and Lowry, G.V. 2010. Nanoparticle aggregation: challenges to understanding transport and reactivity in the environment. J. Environ. Qual., 39: 1909–1924.

Hu, H., Ou, J.Z., Xu, X., Lin, Y., Zhang, Y., Zhao, H. et al. 2021. Graphene-assisted construction of electrocatalysts for carbon dioxide reduction. Chemical Engineering Journal, 425: 130587.

Hu, X., Lu, K., Mu, L., Kang, J., and Zhou, Q. 2014. Interactions between graphene oxide and plant cells: regulation of cell morphology, uptake, organelle damage, oxidative effects and metabolic disorders. Carbon, 80: 665–676.

Hu, X., Ren, C., Kang, W., Mu, L., Liu, X., Li, X. et al. 2018. Characterization and toxicity of nanoscale fragments in wastewater treatment plant effluent. Sci. Total Environ., 626: 1332–1341.

Huang, C., Xia, T., Niu, J., Yang, Y., Lin, S., Wang, X. et al. 2018. Transformation of ^{14}C-labeled graphene to $^{14}CO_2$ in the shoots of rice plant. Angew. Chemie, 130: 9759–63.

Huang, Y., Zhao, L., and Keller, A.A. 2017. Interactions, transformations and bioavailability of nanocopper exposed to root exudates. Environ. Sci. Technol., 51: 9774–9783.

Jahn, R., Blume, H., Asio, V., Spaargaren, O., and Schad, P. 2006. Guidelines for Soil Description, FAO, Rome.

Jain, M., Mudhoo, A., Ramasamy, D.L., Najafi, M., Usman, M., Zhu, R. et al. 2020. Adsorption, degradation, and mineralization of emerging pollutants (pharmaceuticals and agrochemicals) by nanostructures: a comprehensive review. Environ. Sci. Poll. Res., 27(28): 34862–34905.

Jalil, S.U., and Ansari, M.I. 2020. Role of nanomaterials in weed control and plant diseases management. Nanomat. Agr. Forestry Appl., pp. 421–434.

Josko, I., Oleszczuk, P., Dobrzyńska, J., Futa, B., Joniec, J., and Dobrowolski, R. 2019. Long-term effect of ZnO and CuO nanoparticles on soil microbial community in different types of soil. Geoderma, 352: 204–212.

Kandeler, E., and Dick, R.P. 2006. In Biodiversity in Agricultural Production Systems (Eds: G. Benckiser, S. Schnell), Taylor and Francis, London. p. 263.

Kaushik, A. 2019. Advances in nanosensors for biological and environmental analysis: Book Review. Akash Deep, Sandeep Kumar (Eds.). ISBN: 978-0-12-817456-2.

Kim, H.S., Ahn, J.Y., Hwang, K.Y., Kim, I.K., and Hwang, I. 2010. Atmospherically stable nanoscale zero-valent iron particles formed under controlled air contact: Characteristics and reactivity. Environ. Sci. Technol., 44: 1760–1766.

Kookana, R.S., Boxall, A.B.A., Reeves, P.T., Ashauer, R., Beulke, S., Chaudhry, Q. et al. 2014. Nanopesticides: guiding principles for regulatory evaluation of environmental risks. J. Agr. Food Chem., 62(19): 4227-4240.

Kuzyakov, Y., and Razavi, B.S. 2019. Rhizosphere size and shape: temporal dynamics and spatial stationarity. Soil Biol. Biochem., 135: 343–360.

Lewis, R.W., Bertsch, P.M., and McNear, D.H. 2019. Nanotoxicity of engineered nanomaterials (ENMs) to environmentally relevant beneficial soil bacteria–a critical review. Nanotoxicol., 13(3): 392–428.

Li, M., Wang, P., Dang, F., and Zhou, D.M. 2017. The transformation and fate of silver nanoparticles in paddy soil: effects of soil organic matter and redox conditions. Environ. Sci.: Nano., 4: 919–928.

Li, X., Peng, T., Mu, L., and Hu, X. 2019. Phytotoxicity induced by engineered nanomaterials as explored by metablomics: Perspectives and challenges. Ecotox. Environ. Safety, 184: 109602.

Li, X., Xie, Y., Jiang, F., Wang, B., Hu, Q., Tang, Y. et al. 2020. Enhanced phosphate removal from aqueous solution using resourceable nano-CaO_2/BC composite: Behaviors and mechanisms. Sci. Total Environ., 709: 136123.

Lin, D., Story, S.D., Walker, S.L., Huang, Q., Liang, W., and Cai, P. 2017. Role of pH and ionic strength in the aggregation of TiO_2 nanoparticles in the presence of extracellular polymeric substances from *Bacillus subtilis*. Environ. Pollut., 228: 35–42.

Liu, M., Liang, R., Zhan, F., Liu, Z., and Niu, A. 2006. Synthesis of a slow-release and superabsorbent nitrogen fertilizer and its properties. Polymers for Adv. Tech., 17(6): 430–438.

Lundqvist, M., Stigler, J., Cedervall, T., Berggard, T., Flanagan, M.B., Lynch, I. et al. 2011. The evolution of the protein corona around nanoparticles: a test study. ACS Nano, 5: 7503–7509.

Luo, M., Huang, Y., Zhu, M., Tang, Y., Ren, T., Ren, J. et al. 2018. Properties of different natural organic matter influence the adsorption and aggregation behavior of TiO_2 nanoparticles. J. Saudi Chem. Soc., 22: 146–154.

Ma, X., Geiser-Lee, J., Deng, Y., and Kolmakov, A. 2010. Interactions between engineered nanoparticles (ENPs) and plants: phytotoxicity, uptake and accumulation. Sci. Total Environ., 408(16): 3053–3061.

Ma, Y., He, X., Zhang, P., Zhang, Z., Ding, Y., Zhang, J. et al. 2017. Xylem and phloem based transport of CeO_2 nanoparticles in hydroponic cucumber plants. Environ. Sci. Technol., 51: 5215–5221.

Ma, Y., Xie, C., He, X., Zhang, B., Yang, J., Sun, M. et al. 2020. Effects of ceria nanoparticles and CeCl3 on plant growth, biological and physiological parameters, and nutritional value of soil grown common bean (*Phaseolus vulgaris*). Small, 16(21): 1907435.

Ma, Y., Zhang, P., Zhang, Z., He, X., Zhang, J., Ding, Y. et al. 2015. Where does the transformation of precipitated ceria nanoparticles in hydroponic plants take place? Environ. Sci. Technol., 49: 10667–10674.

Marchioni, M., Battocchio, C., Joly, Y., Gateau, C., Nappini, S., Pis, I. et al. 2020. Thiolate-capped silver nanoparticles: discerning direct grafting from sulfidation at the metal–ligand interface by interrogating the sulfur atom. The Journal of Physical Chemistry C, 124(24): 13467–13478.

McManus, P., Hortin, J., Anderson, A.J., Jacobson, A.R., Britt, D.W., Stewart, J. et al. 2018. Rhizosphere interactions between copper oxide nanoparticles and wheat root exudates in a sand matrix: Influences on copper bioavailability and uptake. Environ. Toxicol. Chem., 37: 2619–2632.

Meena, D.S., Gautam, C., Patidar, O.P., Meena, H.M., Prakasha, G., and Vishwajith, M. 2017. Nano-fertilizers is a new way to increase nutrients use efficiency in crop production. Int. J. Agr. Sci., 9(7): 75–91.

Meghana, K.T., Wahiduzzaman, M.D., and Vamsi, G. 2021. Nanofertilizers in Agriculture. Acta Sci. Agr., 5(3): 35–46.

Nel, A., Xis, T., Midler, L., and Li, N. 2006. Toxic potential of materials at the nanolevel. Sci., 311: 622–627.

Ohshima, H. 2012. Electrical Phenomena at Interfaces and Biointerfaces: Fundamentals and Applications in Nano-, Bio-, and Environmental Sciences, John Wiley & Sons, Inc., New York, 27.

Oldroyd, G.E.D., Murray, J.D., Poole, P.S., and Downie, J.A. 2011. The rules of engagement in the legume-rhizobial symbiosis. Ann. Rev. Genetics, 45: 119–144.

Ondrasek, G., Begić, H.B., Zovko, M., Filipović, L., Meriño-Gergichevich, C., Savić, R., and Rengel, Z. 2019. Biogeochemistry of soil organic matter in agroecosystems & environmental implications. Science of the Total Environment, 658: 1559–1573.

Pearson, R.G., 1963. Hard and soft acids and bases. J. Am. Chem. Soc., 85: 3533–3539.

Perez-de-Luque, A. 2017. Interaction of Nanomaterials with Plants: What do we need for real applications in agriculture? Front. Environ. Sci., 5, DOI=10.3389/fenvs.2017.00012.

Philippot, L., Raaijmakers, J.M., Lemanceau, P., and Van Der Putten, W.H. 2013. Going back to the roots: the microbial ecology of the rhizosphere. Nat. Rev. Microbiol., 11: 789–799.

Pramanik, P., Krishnan, P., Maity, A., Mridha, N., Mukherjee, A., and Rai, V. 2020. Application of nanotechnology in agriculture. Environ. Nanotech., 4: 317–348.

Pulizzi, F. 2019. Nano in the future of crops. Nat. Nanotech., 14(6): 507.

Qazi, G., and Dar, F.A. 2020. Nano-agrochemicals: economic potential and future trends. Nanobiotech. Agr., pp. 185–193.

Qiu, L., Zhang, Q., Zhu, H., Reich, P.B., Banerjee, S., van der Heijden, M.G. et al. 2021. Erosion reduces soil microbial diversity, network complexity and multifunctionality. The ISME Journal, 15(8): 2474–2489.

Qu, J., Luo, C., and Hou, J. 2011. Synthesis of ZnO nanoparticles from Zn-hyperaccumulator (*Sedum alfredii* Hance) plants. Micro Nano Lett., 6: 174.

Qureshi, A., Singh, D.K., and Dwivedi, S. 2018. Nano-fertilizers: a novel way for enhancing nutrient use efficiency and crop productivity. Int. J. Curr. Microbiol. Appl. Sci., 7(2): 3325–3335.

Reinsch, B.C., Levard, C., Li, Z., Ma, R., Wise, A., Gregory, K.B. et al. 2012. Sulfidation of silver nanoparticles decreases Escherichia coli growth inhibition. Environmental Science & Technology, 46(13): 6992–7000.

Ren, M., Zhang, Z., Wang, X., Zhou, Z., Chen, D., Zeng, H. et al. 2018. Diversity and contributions to nitrogen cycling and carbon fixation of soil salinity shaped microbial communities in Tarim Basin. Front. Microbiol., 9: 431.

Rico, C.M., Johnson, M.G., and Marcus, M.A. 2018. Cerium oxide nanoparticles transformation at the root–soil interface of barley (*Hordeum vulgare* L.). Environ. Sci.: Nano, 5: 1807.

Roy, P., Das, B., Mohanty, A., and Mohapatra, S. 2017. Green synthesis of silver nanoparticles using *Azadirachta indica* leaf extract and its antimicrobial study. Appl. Nanosci., 7: 843–850.

Salas, E.C., Sun, Z., Luttge, A., and Tour, J.M. 2010. Reduction of graphene oxide via bacterial respiration. ACS Nano, 4: 4852–4856.

Sangeetha, J., Thangadurai, D., Hospet, R., Purushotham, P., Manowade, K.R., Mujeeb, M.A. et al. 2017. Production of bionanomaterials from agricultural wastes. Nanotech., pp. 33–58.

Schwabe, F., Tanner, S., Schulin, R., Rotzetter, A., Stark, W., von Quadt, A. et al. 2015. Dissolved cerium contributes to uptake of Ce in the presence of differently sized CeO_2-nanoparticles by three crop plants. Metallomics, 7(3): 466–477.

Sekhon, B.S. 2014. Nanotechnology in agri-food production: an overview. Nanotech. Sci. Appl., 31–53.

Sekine, R., Brunetti, G., Donner, E., Khaksar, K., Vasilev, A., Jämting, K.G. et al. 2015. Speciation and lability of Ag-, AgCl-, and Ag_2S-nanoparticles in soil determined by X-ray absorption spectroscopy and diffusive gradients in thin films. Environ. Sci. Technol., 49: 897–905.

Shabnam, N., Sharmila, P., Kim, H., and Pardha-Saradhi, P. 2016. Light mediated generation of silver nanoparticles by spinach thylakoids/chloroplasts. PLoS One, 11: e0167937.

Sharma, J.K., Akhtar, M.S., Ameen, S., Srivastava, P., and Singh, G. 2015. Green synthesis of CuO nanoparticles with leaf extract of *Calotropis gigantea* and its dye-sensitized solar cells applications. J. Alloys Compd., 632: 321-325.

Sharma, P., Jha. A.B., Dubey, R.S., and Pessarakli, M. 2012. Reactive oxygen species. Oxidative damage, and antioxidative defense mechanism in plants under stressful conditions. J. Bot., p. 26.

Sharma, Y.P., Watpade, S., and Thakur, J.S. 2014. Role of mycorrhizae: a component of integrated disease management strategies. J. Mycol. Plant Pathol., 44(1): 12–20.

Shawon, Z.B.Z., Hoque, M.E., and Chowdhury, S.R. 2020. Nanosensors and nanobiosensors: Agricultural and food technology aspects. Nanofabrication for Smart Nanosensor Applications, pp. 135–161.

Singh, H., Sharma, A., Bhardwaj, S.K., Arya, S.K., Bhardwaj, N., and Khatr, M. 2021. Recent advances in the applications of nano-agrochemicals for sustainable agricultural development. Environ. Sci.: Impact and Process, 23: 213.

Singh, J., Dutta, T., Kim, K.H., Rawat, M., Samddar, P., and Kumar, P. 2018. Green synthesis of metals and their oxide nanoparticles: applications for environmental remediation. J. Nanobiotechnol., 16: 1–24.

Spatafora, J.W., Chang, Y., Benny, G.L., Lazarus, K., Smith, M.E., and Berbee, M.L. 2016. A phylum-level phylogenetic classification of zygomycete fungi based on genome-scale data. Mycologia, 108: 1028–1046.

Stadler, T., Buteler, M., Valdez, S.R., and Gitto, J.G. 2018. Particulate nano-insecticides: a new concept in insect pest management. Insecticides: Agriculture and Toxicology, 10.5772/intechopen.72448.

Su, Y., Ashworth, V., Kim, C., Adeleye, A.S., Rolshausen, P., Roper, C. et al. 2019. Delivery, uptake, fate, and transport of engineered nanoparticles in plants: a critical review and data analysis. Environ. Sci.: Nano, 6: 2311–2331.

Sun, T.Y., Mitrano, D.M., Bornhoft, N.A., Scheringer, M., Hungerbuhler, K., and Nowack, B. 2017. Envisioning nano release dynamics in a changing world: using dynamic probabilistic modeling to assess future environmental emissions of engineered nanomaterials. Environ. Sci. Technol., 51: 2854–2863.

Tang, F., Uchikoshi, T., and Sakka, Y. 2002. Electrophoretic deposition behavior of aqueous nanosized zinc oxide suspensions. J. Am. Ceram. Soc., 85: 2161–2165.

Tang, Z., and Cheng, T. 2018. Stability and aggregation of nanoscale titanium dioxide particle (nTiO2): effect of cation valence, humic acid, and clay colloids. Chemosphere, 192: 51–58.

Tian, H., Kah, M., and Kariman, K. 2019. Are nanoparticles a threat to mycorrhizal and rhizobial symbioses? A critical review. Front. Microbiol., 10: 1660.

Tian, P., Razavi, B.S., Zhang, X., Wang, Q., and Blagodatskaya, E. 2020. Microbial growth and enzyme kinetics in rhizosphere hotspots are modulated by soil organics and nutrient availability. Soil Biology and Biochemistry, 141: 107662.

Torresdey, J., Parsons, E., Gomez, J., Peralta-Videa, H., Troiani, P., Santiago, M. et al. 2002. Formation and growth of Au nanoparticles inside live alfalfa plants. Nano Lett., 2: 397–401.

van der Heijden, M.G.A., Bardgett, R.D., and van Straalen, N.M. 2008. The unseen majority: soil microbes as drivers of plant diversity and productivity in terrestrial ecosystems. Eco. Lett., 11: 296–310.

Vecerova, K., Vecera, Z., Docekal, B., Oravec, M., Pomeiano, A., Triska, J. et al. 2016. Changes of primary and secondary metabolites in barley plants exposed to CdO nanoparticles. Environ. Pollut., 218: 207–218.

Visser, S., and Parkinson, D. 1992. Soil biological criteria as indicators of soil quality: soil microorganisms. Am. J. Altern. Agric., 7: 33-37.

Vorholt, J.A. 2012. Microbial life in the phyllosphere. Nat. Rev. Microbiol., 10: 828–840.

Wang, Z., Xie, X., Zhao, J., Liu, X., Feng, W., White, J.C. et al. 2012. Xylem-and phloem-based transport of CuO nanoparticles in maize (*Zea mays* L.). Environ. Sci. Technol., 46: 4434–4441.

Xie, C., Zhang, J., Ma, Y., Ding, Y., Zhang, P., Zheng, L. et al. 2019. *Bacillus subtilis* causes dissolution of ceria nanoparticles at the nano–bio interface. Environ. Sci.: Nano, 6: 216–223.

Xingman, M., Geiser-lee, J., Deng, Y., and Kolmakov, A. 2010. Interactions between engineered nanoparticles ENPs and plants: Phytotoxicity, uptake and accumulation. Sci. Total Environ., 408: 3053–3061.

Yan, A., and Chen, Z. 2019. Impacts of silver nanoparticles on plants: A focus on the phytotoxicity and underlying mechanism. Int. J. Mol. Sci., 20(5): 1003.

Yang, Y., Zheng, S., Li, R., Chen, X., Wang, K., Sun, B. et al. 2021. New insights into the facilitated dissolution and sulfidation of silver nanoparticles under simulated sunlight irradiation in aquatic environments by extracellular polymeric substances. Environmental Science: Nano, 8(3): 748–757.

Zhang, P., Guo, Z., Zhang, Z., Fu, H., White, J.C., and Lynch, I. 2020. Nanomaterial transformation in the soil–plant system: implications for food safety and application in agriculture. Small, 16(21): 2000705.

Zhang, P., He, X., Ma, Y., Lu, K. Zhao, Y., and Zhang, Z. 2012b. Distribution and bioavailability of ceria nanoparticles in an aquatic ecosystem model. Chemosphere, 89: 530–535.

Zhang, P., Ma, Y., Liu, S., Wang, Zhang, G.J., He, X., Zhang, J. et al. 2017. Phytotoxicity, uptake and transformation of nano-CeO$_2$ in sand cultured romaine lettuce. Environ. Pollut., 220: 1400–1408.

Zhang, P., Ma, Y., Zhang, Z., He, X., Zhang, J., Guo, Z. et al. 2012a. Biotransformation of ceria nanoparticles in cucumber plants. ACS Nano, 6: 9943–9950.

Zhou, Y., and Staver, A.C. 2019. Enhanced activity of soil nutrient-releasing enzymes after plant invasion: a meta-analysis. Ecology, 100(11): e02830.

8

Effect of Nano-based Agrochemicals on Soil Microbiome

Tayyaba Samreen,[1,] Muhammad Ahmad,[1] Sehar Rasool,[1]
Muhammad Zulqernain Nazir,[1] Samia Arshad,[1]
Faisal Nadeem[2] and Sehrish Kanwal[3]*

Introduction

Enhancing sustainable agricultural productivity presents formidable challenges in light of the burgeoning global population. Although the green revolution of the past has significantly increased global food production, a substantial upscaling effort is imperative to accommodate an additional estimated 3 billion individuals. To illustrate, a staggering rise in cereal production of approximately 1 billion tons per year must be attained by 2050 (De Vries et al. 2020). This escalation of agriculture, however, raises significant environmental concerns due to the indiscriminate practice of agrochemicals, which pose hazards to miscellaneous ecosystems and their functions. Pesticides are widely used for agricultural purposes worldwide, but it affects human beings and the environment. The expansion of new technologies is needed to overcome the drastic effect of pesticides by using them in low amounts. Thus, nanotechnology provides solutions to all these problems (Sharma et al. 2019). Nanotechnology provides solutions to enhance soil quality and pest control. Recently, it has been used in different agricultural fields to provide essential nutrients, herbicides and pesticides etc. (Scott 2007, Barani et al. 2014). Nano-based agrochemicals are used to increase agricultural production. The nanotechnology used for these purposes is known as the "agri-tech revolution". Biomolecules can be applied as sustainable agrochemicals through the use of ecologically friendly nanoparticles as carriers which increase their biological efficiency while reducing ecosystem degradation.

[1] Institute of Soil and Environmental Sciences, University of Agriculture, Faisalabad-38040, Pakistan.

[2] Department of Soil Science, University of Punjab, Lahore, Pakistan.

[3] DDSDP (Data-Driven Smart Decision Platform), PMAS University of Arid Agriculture Rawalpindi, Rawalpindi 46000, Pakistan.

* Corresponding author: tybasamreen@gmail.com; tayyaba.samreen@uaf.edu.pk

The nanocarrier biomolecule hybrids are perfect for topical treatments to improve output and quality (Li et al. 2021). The injudicious submission of agrochemicals results in release of a substantial portion (1–25%) of pesticides into the environment, rather than reaching their intended targets. Consequently, it becomes paramount to devise delivery methods that are both targeted and responsive to stimuli. This approach aims to increase the efficiency of crop production while mitigating ecological risks. Likewise, the excessive use of chemical fertilizers adversely influences plant nutrient utilization, leading to a reduction in different nutrients uptake efficiency (Sachdev et al. 2023). Nano-agrochemicals are gaining popularity due to being biodegradable and eco-friendly. They create nanofertilizers, pesticides, herbicides, and fungicides. Carrier vehicles made from nanomaterials control agrochemical release due to low cost, toxicity, high production rates, and easy water uptake. Nano-based agrochemicals, using carriers like carbon tubes and metal-based nanoparticles, have been shown to improve crop yield, disease resistance, soil fertility, productivity, and water availability. According to An et al. (2022), these nanomaterials have positive impacts on plant health and vitality. Their distinct physicochemical characteristics allow to easily pass biological barriers/membranes and go through new, more efficient pathways to their desired locations (Xu et al. 2006). They moreover improve the erection and performance of agrochemicals, leading to improved solubility, defense against hydrolysis and photodecomposition, and precise control over release (Zong et al. 2022).

While comprehensive assessments have documented the encouraging, neutral, and adverse possessions of nanomaterials on plants, their application hinges on factors such as type, size, application rate and method (Sarkar et al. 2022). Conversely, conventional agrochemicals exhibit undesirable consequences, including long-term environmental stability, alterations in soil properties, the evolution of insect's resistance and potential acquaintance to humans through the food chain. Consequently, the sustainability of conventional agrochemicals in agricultural growth is increasingly questionable (Zhang and Goss 2022).

If nano agrochemicals are employed without taking appropriate care, there are dangers such as water pollution, the transfer of chemical residues into food items, disturbance of plant functions, and changes in the balance of microbial communities (Prasad et al. 2017, Rajput et al. 2022). Microorganisms are crucial to the wholesome operation and structural integrity of plant-soil systems in both agricultural and terrestrial ecosystems (Wang et al. 2023). Beneficial soil-dwelling microorganisms like bacteria and fungi are significantly responsible for the health of plants and soil (Jacoby et al. 2017). They are the primary source of nutrition for plants and play a significant role in the carbon, nitrogen, phosphorus, and sulfur cycles in soil (Sokol et al. 2022). Furthermore, these bacteria are essential for the decomposition of organic matter, the remediation of contaminated soil, and the evaluation of the soil's health (Frac et al. 2018, Kulikova and Perminova 2021). Perhaps most important is the potential of plant-microbe symbiosis to enhance plant growth and defend plants against illness (Berg 2009). This chapter explores the impact of nano-based agrochemicals on soil microbiome, reviewing existing research and studies. It provides an overview of various types of nano-based agrochemicals and their potential effects on soil microorganisms. The chapter emphasizes the importance of maintaining a balanced soil microbiome for sustainable agriculture and identifies knowledge gaps.

Nano-agrochemicals

Nano-enabled products are becoming more and more common in agricultural applications due to their distinctive physicochemical properties, which include an extensive surface area to volume ratio, improved reactivity and durability, adjustable physical/chemical properties, and the ability to manipulate molecules (Khan et al. 2019). These unique nano-specific assets can be harnessed to ripen effective nano-agrochemicals that contribute to sustainable agriculture. Nano-agrochemicals, often referred to as MNOs (Metal Nanoparticles or Metal-Based Nanoparticles), typically exhibit sizes ranging from 1 to 100 nanometers, which endow them with remarkable surface area, chemical traits, and quantum attributes that facilitate diverse interactions within living tissues (Khan et al. 2019).

Nevertheless, comprehending the precise outcomes of these interactions remains challenging, contingent upon factors such as the method of MNO delivery and the developmental stage at which they engage with the host organism. When compared to traditional agrochemicals, these MNOs are recognized for their powerful antibacterial capabilities, which result from adhesion, dissolution, cytotoxicity, and the production of reactive oxygen species (ROS), finally leading to cell death (Albalghiti et al. 2021). Despite their abuse and ineffective absorption by plants, traditional agrochemicals have several specific benefits, such as simplicity of standardization, cost-effectiveness (as opposed to certain expensive MNOs), quick pest control, and improved development within short timeframes (Su et al. 2022). Nano-pesticides are categorized into two main groups according to recent meta-analysis findings (Wang et al. 2022). Type I nano pesticides consist of MNOs directly employed as active ingredients (AIs). Silver-based nanomaterials are frequently employed as nano bactericides, nano fungicides, and nano insecticides, while Ti and Cu-based variants serve as nano bactericides and nano fungicides, respectively (Chen et al. 2019). These nano Cu and Ag materials can effectively combat endogenous plant pathogens, as suggested by several *in vitro* experiments (Wang et al. 2022). An et al. (2020) give example of Ag's capacity to protect citrus plants from the *Candidatus liberibacter* infection, which causes citrus Huanglongbing. Despite the fact that nano pesticides generally cost 50% less than traditional pesticides, cost-effectiveness is still a key factor for these newly released MNOs for crop improvement (Su et al. 2022).

Materials including polymers and zein nanoparticles, which serve as nano transporters for active chemicals, are included in type II nano-pesticides. Type II nano-pesticides are usually biocompatible, sensitive to stimuli, and cost-effective. Common nanocarriers for AIs include mesoporous silica nanoparticles and montmorillonite, which have high encapsulation capacities, as well as nano-capsules, nano-spheres, and nano-micelles made of biopolymers like cellulose, and polylactides (He et al. 2019). Additionally, advanced nanocarriers encompass nanocomposites, nanotubes, and 2D nanomaterials, all of which offer improved loading capacity (Haris et al. 2023). According to Wang et al. (2022, nanoscale insecticides are generally 31.5% more effective than their conventional or non-nanoscale equivalents.

Introduction of soil microbial community

Soil has a vast variety of soil microbes like bacteria, archaea, fungi, algae, plants, and insects (as shown in Figure 1). Soil provides essential nutrients to these organisms that support their life above and below ground (Dominati et al. 2010). Archaea and bacteria are the single cell, minutest independently living beings. The diameters of these are ranging from 0.5–1 μm. Archaea are also known as "extremophiles" because they live in harsh environments. The morphology of bacteria and archaea is the same (Woese et al. 1990). They are eukaryote and heterotrophic, and feed on dead matter (Aislabie et al. 2013). The nutrient cycling and organic matter production are greatly influence by the soil microbes in order to support the growth of plants. Soils regulate a variety of atmospheric elements, affecting air quality as well as regional and global temperatures. Carbon is retained by the soils as stable organic material, which helps to offset CO_2 emissions and bacteria that produce nitrous oxide (N_2O) and methane (CH_4). Microbes have a noteworthy role in determining the soil ability to store carbon because they mineralize soil carbon and nutrients. Nitrous oxide (N_2O) and methane (CH_4) release from soil is regulated by denitrifying bacteria, bacteria that produce and consume methane, and denitrifying fungus (Dominati et al. 2010).

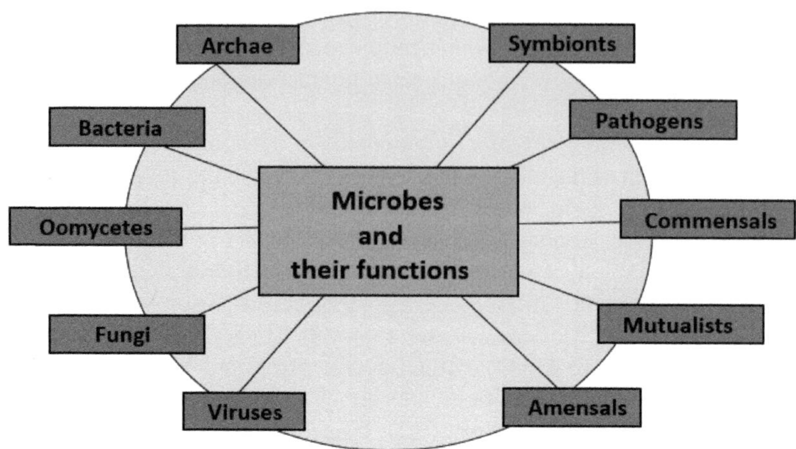

Figure 1. Microbes and their functions.

Nano-agrochemicals' effects on soil microbial diversity

Scientific developments that increase agricultural output always come with dangers and advantages for the health of people and plants (Majumdar and Keller 2021). Changes in the variety of soil microbes throw off the delicate balance of the soil ecosystem, making it less able to control infections (De Varies et al. 2020). Modified nanoparticles have potential to resume normal soil microbial activity, thus bolstering resistance against pathogenic assaults (Wei et al. 2021). There are concerns about how using nano-agrochemical extensively can affect the variety of microorganisms in soil and host plants, boost crop quality and yield during disease outbreaks, or replenish microbial populations that have been reduced. It is challenging to evaluate the impact of nano-agrochemicals on soil microbiomes across a variety of soil environments due to the complexity of soil structures and varying moisture levels (Schimel and Schaeffer 2012). Therefore, an in-depth, long-term study is urgently required to understand the changes in the soil ecology caused by MNOs, including both the advantages and disadvantages of these changes.

The size, type, concentration, length of exposure, and particular soil type all play a role in how resilient soil microbial populations are to MNOs. In the presence of imported MNOs, the microbiome's ability to operate effectively, whether as a single species or a hub species, is crucial. When exposed to different MNOs, the responses of soil microbes in investigated agricultural soils differ. For instance, silver nanoparticles (Ag nanoparticles) at a concentration of 550 mg/kg did not exhibit any toxicity to native soil microbial communities, including significant species of nitrogen-fixing bacteria like bradyrhizobium and rhizobium; rather, they increased the richness of the community (Shah et al. 2014). In different research, proteobacteria that support plant development and defense, showed enhanced population dynamics after the submission of silver nanoparticles, even after a 30-day period. Understanding the critical traits of biosynthesized nanoparticles for improving soil health is essential to comprehending the structure and function of the soil microbiome, including potential functional groups (Mishra et al. 2017). The direct application of minute amounts (0.5–5 g/kg) of single-walled carbon nanotubes boosted the relative abundance of proteobacteria and bacteroidetes, which are typically the dominant phyla in soil environments (Wu ct al. 2019). Without influencing plant growth, sulfate-modified polystyrene nano-spheres drastically changed the bacterial populations in lettuce plants (Kibbey and Strevett 2019). CuO nanoparticles (100 mg/L), on the other hand, exhibited higher efficiency in Triticum aestivum plants by altering soil microbes (Hosseinpour et al. 2022). When exposed to 6.68 mg/L $Cu(OH)_2$ for a year, using deionized water as a control treatments, a field research found very little deleterious impacts on soil microbial diversity (Carley et al. 2020).

With multiple strains commercialized for broad usage in agriculture, bacillus bacteria constitute a well-researched category of biocontrol agents against different plant diseases (Ahmed et al. 2020). These results raise the possibility that adding ZnO nanoparticles to commercialized bacillus strains may improve their ability to control invasive plant diseases. However, it is known that ZnO nanoparticles produce reactive oxygen species (ROS) that may be harmful to bacteria. A growing body of research demonstrates that certain plant metabolites can reduce the harm done to both targeted and non-targeted microbes by ROS produced by introduced MNOs (Wang et al. 2016). Readings have shown positive aids of mixing ZnO nanoparticles with bacteria that promote plant development, suggesting that beneficial microorganisms may have developed methods to regulate ROS levels (Abhilash et al. 2016). Our understanding of the destiny of imported nanoparticles and their capacity to draw in a diversity of microbial populations in the soil ecosystem that can fend against pathogen invasions through plant defense mechanisms and the synthesis of secondary metabolites, however, is still restricted.

It is critical to establish the proper MNO concentrations before considering the widespread use of nano-agrochemicals in soil ecosystems. According to Wang et al. (2022), these concentrations are influenced by the specific MNO, crop, and soil characteristics such as texture, pH, moisture content, electrical conductivity, soil organic carbon, and critical nutrients and minerals. Because these MNOs may have negative consequences on the soil ecology, it is important to determine the specific places where they can penetrate as well as how they affect non-target bacteria (Prado et al. 2001). In two separate studies, the direct application of 50 mg/kg metallic silver to pastureland soil and suburban garden soil with no MNOs added to the soil as controls led to a significant reduction in the amount of acidobacteria (McGee et al. 2017). Molybdenum nanoparticles at complex concentrations have deleterious effects.

Effect of soil microbes on soil ecosystem

Soil microbial effects on mineral metabolism

The most significant and vital component of soil is its microbial population. They are very crucial for the alteration and storage of several. The root system of plants has the ability to absorb microbes, successfully giving a range of nutrients to the plant. They have two effects. The first one is that they have some nutrients like C, P, and N, which are good sources of C, P, and N that can regulate and store the nutrients and the second one is that microorganisms can enhance inorganic element flow by transforming and promoting the system's metabolic process (Swift and Anderson 1994). Microbes are a significant part of nutrient cycling and decay of C and other nutrients.

Microbial effect on plant growth

The microbes play an important role in the provision of nitrogen to plants by converting the atmospheric nitrogen into the plant's available ionic form and ensuring the plant's nutrition. The hyphae of some fungi and plant root zone have symbiotic associations. This kind of symbiotic interaction is designed to shield plants from dangerous environmental conditions and speed up their growth and development. By destroying various organic and inorganic contaminants, microorganisms also shield plants from their damaging effects and promote healthy growth (Wu and Lin 2003).

Several types of organic acids such as humic acid by microbes in turn again aid in enhancing plant development. The plant growth promoting rhizo-bacteria (PGPRs) increase the evolution by producing growth hormones, enhancing the specific enzymatic activities, fixation of atmospheric nitrogen, antibiotics production, and reducing the pathogen attack through the production of chelating agents, etc. (as shown in Figure 2). Some bio fertilizers along with the provision of phosphorus also

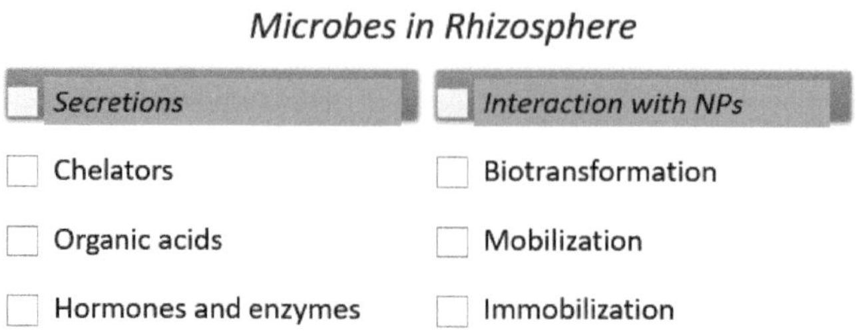

Microbes in Rhizosphere

Secretions	Interaction with NPs
Chelators	Biotransformation
Organic acids	Mobilization
Hormones and enzymes	Immobilization

Figure 2. Interaction of microbes in Rhizosphere.

improved the plant development by improving their root systems, and as a result, trace elements such as iron and zinc were more readily available (Gyaneshwar et al. 2002).

Microbial effect on the soil structure

The soil has various physio-chemical properties that vary from soil to soil. Microbes have the ability to acclimatize to the plant micro-environment and make symbiotic relationships with its roots. In the formation of good soil structure, microbial community plays a chief role. The mycelia that produced actinomycetes help to bind the soil particles together thus improving its flocculation. That is why more actinomycetes are found in productive and fertile soils in contrast to non-fertile soils (Waldrop et al. 2000). In the different types of soils, microbial community plays an important role. When the polysaccharides (secreted by microbes), microbes themselves, and soil particles combine, they form the flocculated soil structure that must improve the soil's physical condition. Production of soil humus results in the microbial decomposition. This would help the soil to retain a sufficient volume of water and soil provision to retain a good soil structure (Xu et al. 2006).

Application of nano-based agrochemicals

Technology related to nanotechnology has considerably improved in order to reduce environmental pollution. Due to their biodegradable effects, cheap cost, toxicity, increased production, quick and simple water absorption, and production procedures, several types of nano-produced materials are utilized as carrier vehicles to regulate the release of various agrochemicals. The chief aim of this system is to gradually control the release of these agrochemicals for a longer time to increase agricultural crop production and protection (Aouada and Moura 2015). Nano-agrochemicals are a combination of nano-technology and agromaterials. That in turn is utilized to synthesize nano fertilizers, pesticides, herbicides, fungicides, etc. The use of nano-agrochemicals has attained popularity these days as compared to conventional agrochemicals due to their biodegradable and environment-friendly nature (Qazi and Dar 2020).

In modern agro-technology, nano-based agrochemicals such as non-fertilizers, nano pesticides, and other substances have created a revolution (as described in Figure 3). The benefits obtained from these chemicals are higher crop yield, disease resistance, increased fertility status of soil, productivity of crops, and availability of water. Different types of carriers and sites as active ingredients are used for the nano-based agrochemicals, i.e., nano-based carbon tubes and their oxides, different metal-based nanoparticles (Fe, Zn, Cu, Ag, TiO_2), and different nanoparticle composites (Mishra and Khare 2021). Due to the ecofriendly nature of the agrochemicals, it helps the farmers to synthesize more nano-based agrochemicals to get more yield by minimizing the input of fertilizers, thus improving the crop yield both qualitatively and quantitatively (Qazi and Dar 2020). But their toxicity, residual effect, the leftover amount in foodstuff, and their toxic effects on plants should also be managed carefully. In order to manage the agroecosystem, their exposure, interaction with different components of the environment, their characterization, and their carryover effect in food and feedstuff should also be managed (Mishra and Khare 2021).

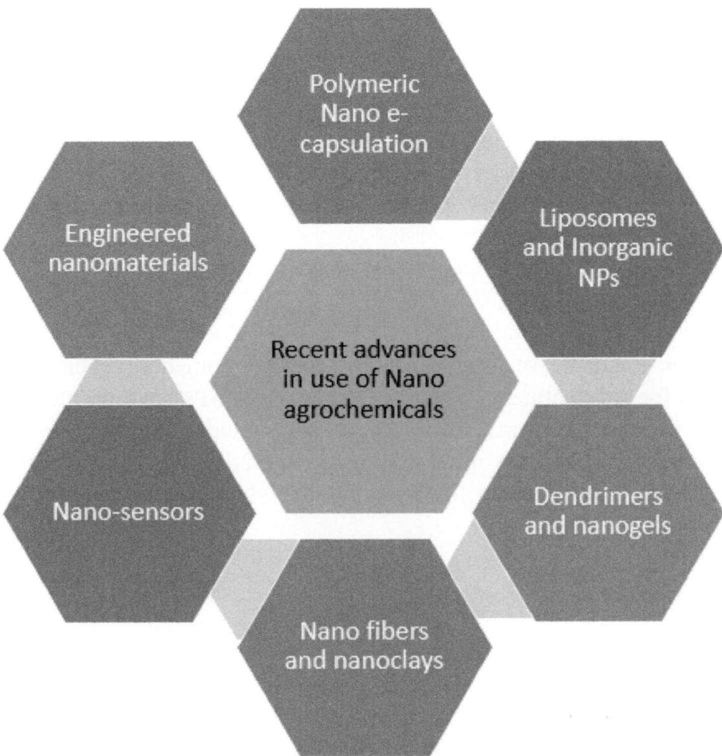

Figure 3. Advances in use of nano agrochemicals.

Beneficial effect of nano agrochemicals against harmful soil microbes

In agriculture, agrochemicals are essential because they stop production losses brought on by insects, pests, and plants. The increased demand for food by the expanding population is met by current agricultural practices, which use a lot of agrochemicals like pesticides, herbicides, and insecticides to maintain optimal production. To protect crops from weeds, diseases, and insects, active ingredients (AIs) are used in nano pesticides, which demonstrate effective scientific progress by offering a number of advantages, notably enhanced longevity, less need for AIs, and improved usefulness (Kibbey and Strevett 2019). So, to lessen the losses and improve insect targeting, controlled distribution is required (Simonin and Richaume 2015). In agriculture, nano-pesticides with nano-carriers are opening up new distribution channels for hydrophobic pesticides along with the protection of crops. The consequence of engineered-based nanomaterials (ENMs) on the soil microbes depends on the type of nanomaterials used (Simonin and Richaume 2015). TiO$_2$ NMs can change the community of prokaryotes except fungi (Moll et al. 2017).

Carbon-based nanoparticles

The nanomaterials of carbon are extensively used but different nanomaterials have different effects on soil microbes. The fullerene enhances the contents of bio-carbon and the multi-walled carbon nanotubes (MWCNTs) at their higher concentration, which enhance growth of more tolerant microbes and also increase the atrazine-degrading microbes at its lower amount (Schlich and Hund 2015). The single-walled carbon nanotubes (SWCNTs) at their lower concentration enhance the metabolic activity and growth of tolerant soil microbes. The effect of graphene increases the biomass of microbes, its proliferation and enzymatic activity of microbes when used in lower amount (Chen et al. 2019).

Magnesium-based nanoparticles

MgO-NPs plays an effective part against plant pathogens as antimicrobial agents. By using an extract of Pisonia alba, simple ecological green-produced MgO-NPs showed its effect as a fungicide alongside *F. solani* (Sharmila et al. 2019). The fungicidal and antioxidant effects of nanoparticles have been used in various practices of agriculture. It was discovered that the green MgO nano-flowers produced from rosemary extract were effective at preventing rice from infection caused by bacteria (Abdallah et al. 2019).

Zinc-based nanoparticles

Nano-fungicides that are made of zinc oxide are an environmentally benign alternative to all synthetic fungicides for the treatment of plant diseases. ZnO-NPs have fungicidal efficacy against pathogenic fungi that grow after harvesting. These may include *Botrytis cinerea* and *Penicillium expansum* (Saratale et al. 2018). Grey mold disease (both seed as well as soil borne and foliar plant fungi) has also been successfully controlled by ZnO-NPs when used in a low amount (Malandrakis et al. 2019).

Copper-based nanoparticles

Cu-based nanoparticles have demonstrated antifungal and antibacterial effects. Commercial pesticides have previously used $Cu(OH)_2$ based compounds like Kocide 3000 (Giannousi et al. 2013, Patra and Baek 2017). The environment is endangered by the widespread use of this conventional agrochemical. Recent studies have demonstrated the Cu-NPs produced from biological sources with antibacterial capabilities against gram positive and gram negative harmful microbes. For example, Giannousi et al. (2013) created Cu-based NPs and examined the antifungal activity of these particles for the disease of tomato late blight. It was discovered to be successful in preventing disease while posing no risk to the plants. A fascinating discovery from the literature is that a pesticide's ionic and nano forms might have different effects, possibly connected to the release of ions. Some researchers have already noted similarities and differences between ionic and nanoforms in terms of antibacterial activity and their effects on microbial communities exposed to Cu-NPs.

Silver-based nanoparticles

Silver-NPs are widely used as antibacterial agents and boost plant immunity, which makes them a potential nano pesticide. Ag-NPs are known to be one of the most potent antimicrobial agents against different pathogens. Ag-NPs showed their activity against *Fusarium culmorum, Megnaporthe grisea, Sphaerotheca pannasa, Rhizoctonia solani, Botrytis cinerea, Trichoderma* sp. and *Scalerotinia sclerotiorum* (Saratale et al. 2018). Without altering the morphological characteristics of the leaf, a different study found that Ag-NPs were efficient at preventing the spread of disease caused by fungus in strawberry and tomato leaves that were already diseased (Khandelwal et al. 2019). All the data point to the possibility that silver (Ag) nanoparticles (NPs) might have a major impact on the makeup of microbial societies and modify activity of extracellular enzymes intricate in elemental cycle processes. Apart from a study by Simonin and Roddy (2018), it is important to note that majority of studies in this area have concentrated on evaluating the environmental effects of NPs rather than specifically evaluating the effects of Ag and Cu based nano pesticides on non-goal soil microbiota (You et al. 2018).

Silicon-based nanoparticles

SiO$_2$-NPs, synthesized with the help of biological waste materials including sugar beet bagasse, bamboo, and rice husk, are crucial for increasing plant resistance to diseases and pests (Snehal et al. 2018). To protect the wheat seedling from Ultraviolet-B abiotic stress and ultimately crops from diverse microbial diseases, SiO$_2$-NPs improve the antioxidant defense mechanism (Maruyama et al. 2016).

Titanium-based nanoparticles

TiO$_2$-NPs are broadly utilized in agriculture to prevent plant diseases and promote growth. They are frequently employed as antibacterial agents against various plant bacteria whether applied as foliar spray or in the form of an amendment in soil (Raliya et al. 2015). Following are the examples of different nano-agrochemicals shown in Table 1.

Table 1. Different nano agrochemicals and their effects.

Nano Agrochemicals	Effects	Microbial species	References
Ag-NPs	Antimicrobial	*B. cinerea, B. sorokinniana, C. gloeosporioides, S. pannasa, R. solani*	(Salatale et al. 2018)
Bio-synthesized Ag-NPs	Antibacterial	*E.coli and V. cholera*	(Khandelwal et al.2019)
Ag-NPs	Antifungal	*S. sclerotinia, A. alternate, C. lunata*	(Khandelwal et al.2019)
Cu-NPs	Antifungal	*Tomato late blight* by *P. infestans*	(Giannousi et al. 2013)
ZnO-NPs	Antifungal	*P. expansum,*	(Saratale et al. 2018)
ZnO-NPs	Antifungal	*Grey mold* by *B. cinerea*	(Malandrakis et al. 2019)
Green synthesized MgO-NPs	Antifungal	*F. solani*	(Sharmila et al. 2019)
MgO-NPs	Antibacterial	*Rice bacterial infection* by *B. cereus*	(Abdallah et al. 2019)
MgO-NPs	Antibacterial	*Bacterial disease* by *R. solanacearum*	(Imada et al. 2016)

Impact of Nano-agrochemicals on Microbial Enzymatic activities

The soil and pesticides' interaction can lead to modified biochemical processes influenced by microorganisms. Soil harbors a plethora of enzymes, present in various forms such as free enzymes, immobilized enzymes, and both extracellular and intracellular enzymes. These soil enzymes aid as valuable gauge of soil quality and are pivotal in numerous roles, comprising the cycling of nutrient elements and the decay of organic matter (Karaca et al. 2011). Consequently, monitoring changes in soil enzyme activities provides valuable biological insights into how soil responds to stressors like pesticides. Soil enzymes are necessary for the functioning of the carbon cycle, which includes glycosyl hydrolases, oxidases, and peroxidases, the nitrogen cycle, which includes proteases, peptidases, urease, and chitinase, the phosphorus cycle, which is influenced by phosphatases, and the sulfur cycle, which is influenced by arylsulfatase Therefore, all the nutrient cycles are affected by the application of nano-agrochemicals due to their impact on microbial enzyme activities.

Impact of nano-based agrochemicals on soil microbiome

The soil is a three-dimensional body and is not static but dynamic. It is heterogeneous, having a mixture of different organic and inorganic materials but well organized in composition.

The rhizosphere of soil contains plenty of root exudates and carbohydrates on which microbes feed. Thus, the soil is considered the habitat for microbiomes (Pascoli et al. 2019). So, the microbiomes serve many exosystemic roles in agroecosystems through the recycling of many macro and micronutrients in the soil such as nitrogen, potassium, phosphorus, carbon, sulfur, zinc, and iron, etc. It is known that nano-based agrochemicals are active substances that increase the properties of agro-based chemicals such as herbicides, fungicides, and fertilizers (Sirelkhatim et al. 2015).

The nano-based agrochemicals that are the innovation of nanotechnology boost the efficiency of nano-based agrochemicals, but these have an impact on the non-targeted soil microbiomes (Santaella and Plancot 2020). Large quantity of nano-based materials is engineered and produced commercially. Large amount of these materials are disposed of into the soil. Their behavior and their persistence in soil depend on the physiochemical properties of soil and the chemical nature of nanomaterials. Several minerals such as zinc (Zn) and silver (Ag) nanoparticles have antimicrobial properties (Sun et al. 2014). These nanoparticles have the power to alter how the soil microbiome functions. These include modifications to the soil's capacity to fix nitrogen, mineralization, and other traits that support plant development. Understanding how these agro-based nanomaterials affect the soil microbiota and its related chemical and physiological processes is critical. The appropriate use of these nano-based agrochemicals is necessary to maintain the agroecosystem and improve agriculture production as well (Khan 2020).

Toxic effects of nanomaterials on microbes

In continuing studies, the poisonousness of NPs appears to be kinetics-driven and is seemingly associated with dissolution or transformation events within the soil, resulting in temporary adjustments and adaptations within the microbial community. Additionally, Chhipa (2017) have highlighted the potential of biogenic nanoparticles, which have minimal impacts on cells but demonstrate sturdy activity against certain plant pathogenic fungi. Denitrification, a critical step in the cycling of nitrogen in soil, is one of the microbial processes that can be impacted by NPs. Soil structure provides many niches suitable for bacteria with varying oxygen demands (aerobic and anaerobic). The unavailability of oxygen, the presence of nitrate and carbon as electron donors, pH, and hydric conditions are all factors that favor denitrification (Shah et al. 2017, Reidy et al. 2013). Denitrification is preferred in soils with inferior redox potential values, which are influenced by soil texture. Sandy soils tend to have higher redox values, which are less conducive to denitrification, whereas clayey soils provide the necessary conditions for this biological transformation (Zhai et al. 2021).

Toxicity mechanisms of nano-agrochemicalnano-agrochemicals/MNOs

The arena of nano toxicology refers to the learning of the toxicity of nanomaterials towards living organisms, including animals, plants, and humans. This terminology has since evolved, and it is now denoted as "nano safety" (Santaella and Plancot 2020). Concerning the toxicity of Modified Nanoparticles (MNOs), it can be categorized into two distinct approaches: experimental nano toxicology, encompassing both *in vitro* and *in vivo* studies, and computational studies conducted through *in silico* approaches (Fraceto et al. 2016). Direct submission of MNOs as nano-agrochemicalnano-agrochemicals to soil may have a variety of unanticipated effects on microbial populations. The detrimental effects of MNOs on soil bacteria necessitate a thorough study to improve their performance and avoid unintended results. Because MNOs are persistent, they can build up in the environment and eventually interfere with the physiological and metabolic processes of living things (Tripathi et al. 2023).

Nano chemicals can exert noticeable toxic impacts on non-targeted organisms, and the extent of these effects depends on their properties, concentration, application method, exposure duration, and environmental factors. AgO and CuO-based commercial nanomaterials have demonstrated negative impacts on soil microbiota at the physiological, and genetic levels. However, when utilized at quantities relevant to agriculture over prolonged trials, these benefits often tend to be fleeting (Fraceto et al. 2016). We can give a comprehensive summary of the circumstances leading to toxicity, despite the fact that the particular processes behind microbial toxicity for the majority of MNOs are not yet well understood. Additional hallmarks of toxicity include membrane rupture, oxidation, lipid peroxidation, and disruption of energy-related activities (Abbas et al. 2020). Oxidative stress is caused by the interaction of MNOs with bacteria. According to Moreno-Garrido et al. (2015), reactive oxygen species produced by MNOs interfere with DNA replication, ATP generation, and mitochondrial activity.

MNOs can alter soil fungus and cause toxicity, just as bacteria do; however, the exact pathways are yet unknown. Since chitin, glycoproteins, and glycans make up the majority of fungal cell walls, they differ greatly from bacterial cell walls (Van Der Weerden and Anderson 2013). It has been hypothesized that MNOs may aim chitin synthase and prevent production of the components of cell walls. Inhibiting the creation of proteins, microtubules, and nucleic acids is another proposed method (Kulikova and Purminova 2021). Nanoparticles may penetrate and distort fungal hyphae due to their size and shape, which alters their morphology. In sustainable agriculture, soil yeast is a key player. The accumulation of nano diamonds in the bulbs of lignin-degrading fungus was recently shown to be caused by cell wall breakdown and cytoplasm loss (Ahmed et al. 2023). Algae are an important but less common category of soil microflora that are essential to preserving the health of the soil and interacting with other helpful bacteria (Crouzet et al. 2019). MNOs accumulate in the periplasmic region after crossing the cell membrane to harm algal cells. Once within the cytoplasm, they interact with organelles and cause ROS to be produced, which injures the organelles' structure and functionality (Melegari et al. 2013, Wang et al. 2016).

Nano-pesticides have a sequential method of action, according to Wang and colleagues (2022). Nanoparticles first cling to the cell wall and membrane, damaging them through lipid peroxidation and changing the permeability and structure of the membrane, which causes cell components to seep-out. The cell's organelles and biomolecules are then damaged by metal ions or nanoparticles that enter the cell. Protein denaturation, ribosomal instability, DNA damage, and mitochondrial dysfunction are a few examples of this harm. In parallel, the production of reactive species like OH and $O_2{}^{\bullet}$-, as well as associated chemicals like H_2O_2 and HOCl, causes cells to experience oxidative stress and cytotoxicity. Finally, genotoxicity affects microbial signal transduction pathways, which results in cell death.

Conclusion

The integration of nano-based agrochemicals into modern agriculture holds immense promise for addressing the food security and sustainability challenges. These innovations offer precision and efficiency in nutrient delivery and pest management. Our exploration of the research reveals that nano-based agrochemicals can induce shifts in soil microbial populations, which can be both valuable and detrimental. While some formulations enhance microbial diversity and soil functions, others may disrupt these delicate ecosystems. To harness the full potential of nanotechnology in agriculture while safeguarding soil health, multidisciplinary research efforts, responsible application, and stringent regulations are vital. Striking the right balance between agricultural innovation and environmental protection will be key to ensuring the long-term sustainability of our farming practices and the health of our ecosystems.

Future research priorities

The future of agriculture holds great promise with the potential impact of nano-based agrochemicals on soil microbiomes. As technology continues to advance, these innovative solutions offer the prospect of enhancing crop yields while minimizing environmental impacts. Nano-agrochemicals can precisely target pests and diseases, reducing the need for excessive chemical applications. Moreover, they have the potential to influence the soil microbiomes by promoting beneficial microorganisms that improve nutrient uptake and soil health. However, careful research and monitoring are necessary to ensure the protection and long-lasting effects of these materials on soil ecosystems. Harnessing the potential of nano-based agrochemicals may revolutionize sustainable agriculture in the years to come.

References

Abbas, Q., Yousaf, B., Ali, M.U., Munir, M.A.M., El-Naggar, A., Rinklebe, J. et al. 2020. Transformation pathways and fate of engineered nanoparticles (ENPs) in distinct interactive environmental compartments: A review. Environ. Int., 138: 105646.

Abdallah, Y., Ogunyemi, S.O., Abdelazez, A., Zhang, M., Hong, X., and Ibrahim, E. 2019. The green synthesis of MgO nano-flowers using *Rosmarinus officinalis* L. (Rosemary) and the antibacterial activities against *Xanthomonas oryzae* pv. *oryzae*. Biomed. Res. Int., 17: 1–8.

Abhilash, P.C., Dubey, R.K., Tripathi, V., Gupta, V.K., and Singh, H.B. 2016. Plant growth-promoting microorganisms for environmental sustainability. Trends Biotechnol., 34(11): 847–850.

Ahmed, A., Munir, S., He, P., Li, Y., He, P., Yixin, W. et al. 2020. Biocontrol arsenals of bacterial endophyte: An imminent triumph against clubroot disease. Microbiol. Res., 241: 126565.

Ahmed, T., Noman, M., Gardea-Torresdey, J.L., White, J.C., and Li, B. 2023. Dynamic interplay between nano-enabled agrochemicals and the plant-associated microbiome. Trends Plant Sci.

Aislabie, J., Deslippe, J.R., and Dymond, J. 2013. Soil microbes and their contribution to soil services. Ecosystem services in New Zealand–conditions and trends. Manaaki Whenua Press, Lincoln, New Zealand., 1(12): 143–161.

Albalghiti, E., Stabryla, L.M., Gilbertson, L.M., and Zimmerman, J.B. 2021. Towards resolution of antibacterial mechanisms in metal and metal oxide nanomaterials: a meta-analysis of the influence of study design on mechanistic conclusions. Environ. Int., 8(1): 37–66.

An, C., Sun, C., Li, N., Huang, B., Jiang, J., Shen, Y., and Wang, Y. 2022. Nanomaterials and nanotechnology for the delivery of agrochemicals: strategies towards sustainable agriculture. J. Nanobiotechnology, 20(1): 1–19.

Aouada, F.A., and Moura, M.R.D. 2015. Nanotechnology applied in agriculture: controlled release of agrochemicals. In Nanotechnologies in Food and Agriculture, pp. 103–118.

Berg, G. 2009. Plant–microbe interactions promoting plant growth and health: perspectives for controlled use of microorganisms in agriculture. Microbiol Biot., 84: 11–18.

Bharani, R.A., Namasivayam, S.K.R., and Shankar, S.S. 2014. Biocompatible chitosan nanoparticles incorporated pesticidal protein Beauvericin (Csnp-Bv) preparation for the improved pesticidal activity against major groundnut defoliator Spodoptera litura (Fab.) (Lepidoptera; Noctuidae). Int. J. ChemTech Res., 6(12): 5007–5012.

Carley, L.N., Panchagavi, R., Song, X., Davenport, S., Bergemann, C.M., McCumber, A.W. et al. 2020. Long-term effects of copper nanopesticides on soil and sediment community diversity in two outdoor mesocosm experiments. Environ. Int., 54(14): 8878–8889.

Chen, M., Sun, Y., Liang, J., Zeng, G., Li, Z., Tang, L. et al. 2019. Understanding the influence of carbon nanomaterials on microbial communities. Environ. Int., 126: 690–698.

Chhipa, H. 2017. Nanofertilizers and nanopesticides for agriculture. Environ. Chem. Lett., 15: 15–22.

Crouzet, O., Consentino, L., Pétraud, J.P., Marrauld, C., Aguer, J.P., Bureau, S. et al. 2019. Soil photosynthetic microbial communities mediate aggregate stability: influence of cropping systems and herbicide use in an agricultural soil. Front. Microbiol., 10: 1319.

De, Vries, F.T., Griffiths, R.I., Knight, C.G., Nicolitch, O., and Williams, A. 2020. Harnessing rhizosphere microbiomes for drought-resilient crop production. Science, 368(6488): 270–4.

Dominati, E., Patterson, M., and Mackay, A. 2010. A framework for classifying and quantifying the natural capital and ecosystem services of soils. Ecol. Econ., 69(9): 1858–1868.

Frąc, M., Hannula, S.E., Bełka, M., and Jędryczka, M. 2018. Fungal biodiversity and their role in soil health. Front. Microbiol., 9: 707.

Fraceto, L.F., Grillo, R., de Medeiros, G.A., Scognamiglio, V., Rea, G., and Bartolucci, C. 2016. Nanotechnology in Agriculture: Which innovation potential does it have. Front. Environ. Sci., 4.

Giannousi, K., Avramidis, I., and Dendrinou-Samara, C. 2013. Synthesis, characterization and evaluation of copper based nanoparticles as agrochemicals against Phytophthora infestans. RSC Adv., 3(44): 21743–21752.

Gyaneshwar, P., Kumar, G.N., Parekh, L.J., and Poole, P.S. 2002. Role of soil microorganisms in improving P nutrition of plants. Plant soil., 245(1): 83–93.

Haris, M., Hussain, T., Mohamed, H.I., Khan, A., Ansari, M.S., Tauseef, A. et al. 2023. Nanotechnology–A new frontier of nano-farming in agricultural and food production and its development. Sci. Total Environ., 857: 159639.

He, X., Deng, H., and Hwang, H. 2019. The current application of nanotechnology in food and agriculture. J. Food Drug Anal., 27: 1–21.

Hosseinpour, A., Ilhan, E., Özkan, G., Öztürk, H.İ., Haliloglu, K., and Cinisli, K.T. 2022. Plant growth-promoting bacteria (PGPBs) and copper (II) oxide (CuO) nanoparticle ameliorates DNA damage and DNA Methylation in wheat (*Triticum aestivum* L.) exposed to NaCl stress. J. Plant Biochem. Biotechnol., 31(4): 751–764.

Jacoby, R., Peukert, M., Succurro, A., Koprivova, A., and Kopriva, S. 2017. The role of soil microorganisms in plant mineral nutrition—current knowledge and future directions. Front. Plant Sci., 8: 1617.

Karaca, A., Cetin, S.C., Turgay, O.C., and Kizilkaya, R. 2011. Soil enzymes as indication of soil quality. pp. 119–148. *In*: Shukla G., and Varma, A. (Eds.). Soil Enzymology. Springer, Berlin, Heidelberg.

Khan, I., Saeed, K., and Khan, I. 2019. Nanoparticles: Properties, applications and toxicities. Arab. J. Chem., 12(7): 908–931.

Khan, S.T. 2020. Interaction of engineered nanomaterials with soil microbiome and plants: their impact on plant and soil health. Sustain. Agric. Res., 41: 181–199.

Khandelwal, A., Joshi, R., Mukherjee, P., Singh, S.D., and Shrivastava, M. 2019. Use of biobased nanoparticles in agriculture. pp. 89–100. *In*: Nanotechnology for Agriculture: J. Sustain. Agric.

Kibbey, T.C., and Strevett, K.A. 2019. The effect of nanoparticles on soil and rhizosphere bacteria and plant growth in lettuce seedlings. Chemosphere, 221: 703–707.

Kulikova, N.A., and Perminova, I.V. 2021. Interactions between humic substances and microorganisms and their implications for nature-like bioremediation technologies. Molecules, 26(9): 2706.

Li, P., Huang, Y., Fu, C., Jiang, S.X., Peng, W., Jia, Y. et al. 2021. Eco-friendly biomolecule-nanomaterial hybrids as next-generation agrochemicals for topical delivery. EcoMat., 3(5): e12132.

Majumdar, S., and Keller, A.A. 2021. Omics to address the opportunities and challenges of nanotechnology in agriculture. Crit. Rev. Environ., 51(22): 2595–2636.

Malandrakis, A.A., Kavroulakis, N., and Chrysikopoulos, C.V. 2019. Use of copper, silver and zinc nanoparticles against foliar and soil-borne plant pathogens. Sci. Total Environ., 670: 292–299.

Maruyama, C.R., Guilger, M., Pascoli, M., Bileshy-José, N., Abhilash, P.C., Fraceto, L.F. et al. 2016. Nanoparticles based on chitosan as carriers for the combined herbicides imazapic and imazapyr. Sci. Rep., 6: 1–15.

McGee, C.F., Storey, S., Clipson, N., and Doyle, E. 2017. Soil microbial community responses to contamination with silver, aluminium oxide and silicon dioxide nanoparticles. Ecotoxicol. Environ. Saf., 26(3): 449–458.

Melegari, S.P., Perreault, F., Costa, R.H.R., Popovic, R., and Matias, W.G. 2013. Evaluation of toxicity and oxidative stress induced by copper oxide nanoparticles in the green alga *Chlamydomonas reinhardtii*. Aquat. Toxicol., 142: 431–440.

Mishra, D., and Khare, P. 2021. Emerging nano-agrochemicals for sustainable agriculture: benefits, challenges and risk mitigation. Sustain. Agric. Res., 50: 235–257.

Mishra, S., Keswani, C., Abhilash, P.C., Fraceto, L.F., and Singh, H.B. 2017. Integrated approach of agri-nanotechnology: challenges and future trends. Front. Plant Sci., 8: 471.

Moll, J., Klingenfuss, F., Widmer, F., Gogos, A., Bucheli, T.D., Hartmann, M. et al. 2017. Effects of titanium dioxide nanoparticles on soil microbial communities and wheat biomass. Soil Biol. Biochem., 111: 85–93.

Moreno-Garrido, I., Pérez, S., and Blasco, J. 2015. Toxicity of silver and gold nanoparticles on marine microalgae. Mar. Environ. Res., 111: 60–73.

Pascoli, M., Jacques, M.T., Agarrayua, D.A., Avila, D.S., Lima, R., and Fraceto, L.F. 2019. Neem oil based nanopesticide as an environmentally-friendly formulation for applications in sustainable agriculture: An ecotoxicological perspective. Sci. Total Environ., 677: 57–67.

Patra, J.K., and Baek, K.H. 2017. Antibacterial activity and synergistic antibacterial potential of biosynthesized silver nanoparticles against foodborne pathogenic bacteria along with its anticandidal and antioxidant effects. Front. Microbiol., 8: 167.

Prado A.G., and Airoldi, C. 2001. The effect of the herbicide diuron on soil microbial activity. Pest Manag. Sci., 57: 640–644.

Prasad, R., Bhattacharyya, A., and Nguyen, Q.D. 2017. Nanotechnology in sustainable agriculture: recent developments, challenges, and perspectives. Front. Microbiol., 8: 1014.

Qazi, G., and Dar, F.A. 2020. Nano-agrochemicals: economic potential and future trends in Nanobiotechnology in Agriculture, pp. 185–193.

Rajput, V.D., Faizan, M., Upadhyay, S.K., Kumari, A., Ranjan, A., Sushkova, S. et al. 2022. Influence of nanoparticles on the plant *Rhizosphere microbiome*. pp. 83–102. In: The Role of Nanoparticles in Plant Nutrition under Soil Pollution: Nanoscience in Nutrient Use Efficiency. Cham: Springer International Publishing.

Raliya, R., Biswas, P., and Tarafdar, J.C. 2015. TiO$_2$ nanoparticle biosynthesis and its physiological effect on mung bean (*Vigna radiata* L.). Biotechnol. Reports, 5(1): 22–26.

Reidy, B., Haase, A., Luch, A., Dawson, K.A., and Lynch, I. 2013. Mechanisms of silver nanoparticle release, transformation and toxicity: a critical review of current knowledge and recommendations for future studies and applications. Materials, 6(6): 2295–2350.

Sachdev, D., Jha, P.K., Rani, R., Verma, G., Kaur, N., and Sahu, O. 2023. Structural and optical investigation of highly fluorescent tartaric acid derived from the tamarind pulp. Mater. Chem. Phys., 296: 127294.

Santaella, C., and Plancot, B. 2020. Interactions of nanoenabled agrochemicals with soil microbiome. In Nanopesticides. 137–16.

Saratale, R.G., Saratale, G.D., Shin, H.S., Jacob, J.M., Pugazhendhi, A., and Bhaisare, M. 2018. New insights on the green synthesis of metallic nanoparticles using plant and waste biomaterials: current knowledge, their agricultural and environmental applications. Environ. Sci. Pollut. Res., 25(11): 10164–10183.

Sarkar, M.R., Rashid, M.H.O., Rahman, A., Kafi, M.A., Hosen, M.I., Rahman, M.S. et al. 2022. Recent advances in nanomaterials based sustainable agriculture: An overview. Environ. Nanotechnol. Monit. Manag., 18: 100687.

Schimel, J.P., and Schaeffer, S.M. 2012. Microbial control over carbon cycling in soil. Front. Microbiol., 3: 348.

Schlich, K., and Hund-Rinke, K. 2015. Influence of soil properties on the effect of silver nanomaterials on microbial activity in five soils. Environ. Pollut., 196: 321–330.

Scott, N.R. 2007. Nanotechnology opportunities in agriculture and food systems. In Biological & Environmental Engineering, Cornell University NSF Nanoscale Science & Engineering Grantees Conference December, 5: 2007.

Shah, F.M., Razaq, M., Ali, A., Han, P., and Chen, J. 2017. Comparative role of neem seed extract, moringa leaf extract and imidacloprid in the management of wheat aphids in relation to yield losses in Pakistan. PloS One. 12: e0184639.

Shah, V., Collins, D., Walker, V.K., and Shah, S. 2014. The impact of engineered cobalt, iron, nickel and silver nanoparticles on soil bacterial diversity under field conditions. Environ. Res. Lett., 9(2): 024001.

Sharma, A., Sood, K., Kaur, J., and Khatri, M. 2019. Agrochemical loaded biocompatible chitosan nanoparticles for insect pest management. Biocatal. Agric. Biotechnol., 1: 101079.

Sharmila, G., Muthukumaran, C., Sangeetha, E., Saraswathi, H., Soundarya, S., and Kumar, N.M. 2019. Green fabrication, characterization of Pisonia alba leaf extract derived MgO nanoparticles and its biological applications. Available from Nano-Structures and Nano-Objects, 20: 100–380.

Simonin, K.A., and Roddy, A.B. 2018. Genome downsizing, physiological novelty, and the global dominance of flowering plants. PLoS Biology, 16(1): 2003706.

Simonin, M., and Richaume, A. 2015. Impact of engineered nanoparticles on the activity, abundance, and diversity of soil microbial communities: a review. Environ. Sci. Poll. Res., 22(18): 13710–13723.

Sirelkhatim, A., Mahmud, S., Seeni, A., Kaus, N.H.M., Ann, L.C., Bakhori, S.K.M. et al. 2015. Review on zinc oxide nanoparticles: antibacterial activity and toxicity mechanism. Nanomicro Lett., 7: 219–242.

Snehal, S., Sc, M.A., Lohani, P., Correspondence, S., Snehal, M., and Sc, A. 2018. Silica nanoparticles: Its green synthesis and importance in agriculture. J. Pharmacogn. Phytochem., 7(5): 3383–3393.

Sokol, N.W., Slessarev, E., Marschmann, G.L., Nicolas, A., Blazewicz, S.J., Brodie, E.L. et al. 2022. Life and death in the soil microbiome: how ecological processes influence biogeochemistry. Nat. Rev. Microbiol., 20(7): 415–430.

Su, L., Feng, H., Mo, X., Sun, J., Qiu, P., Liu, Y. et al. 2022. Potassium phosphite enhanced the suppressive capacity of the soil microbiome against the tomato pathogen *Ralstonia solanacearum*. Biol. Fertil. Soils, 58(5): 553–563.

Sun, C., Shu, K., Wang, W., Ye, Z., Liu, T., Gao, Y. et al. 2014. Encapsulation and controlled release of hydrophilic pesticide in shell cross-linked nanocapsules containing aqueous core. Int. J. Pharm., 463(1): 108–114.

Swift, M.J., and Anderson, J.M. 1994. Biodiversity and ecosystem function in agricultural systems. In Biodiversity and ecosystem function. Springer, Berlin, Heidelberg, pp. 15–41.

Tripathi, S., Mahra, S., Tiwari, K., Rana, S., Tripathi, D.K., Sharma, S. et al. 2023. Recent Advances and Perspectives of Nanomaterials in Agricultural Management and Associated Environmental Risk: A Review. J. Nanomater., 13(10): 1604.

Van der Weerden, N.L., and Anderson, M.A. 2013. Plant defensins: common fold, multiple functions. Fungal Biol. Rev. 26(4): 121–131.

Waldrop, M.P., Balser, T.C., and Firestone, M.K. 2000. Linking microbial community composition to function in a tropical soil. Soil Biol. Biochem., 32(13): 1837–1846.

Wang, C.Y., Qin, J.C., and Yang, Y.W. 2023. Multifunctional Metal–Organic Framework (MOF)-based nanoplatforms for crop protection and growth promotion. J. Agric. Food Chem., 71(15): 5953–5972.

Wang, D., Saleh, N.B., Byro, A., Zepp, R., Sahle-Demessie, E., Luxton, T.P. et al. 2022. Nano-enabled pesticides for sustainable agriculture and global food security. Nat. Nanotechnol., 17(4): 347–60.

Wang, P., Lombi, E., Zhao, F.J., and Kopittke, P.M. 2016. Nanotechnology: a new opportunity in plant sciences. Trends Plant Sci., 21(8): 699–712.

Wei, X., Cao, P., Wang, G., Liu, Y., Song, J., and Han, J. 2021. CuO, ZnO, and γ-Fe2O3 nanoparticles modified the underground biomass and rhizosphere microbial community of *Salvia miltiorrhiza* (Bge.) after 165-day exposure. Ecotoxicol. Environ. Saf., 217: 112232.

Woese, C.R., Kandler, O., and Wheelis, M.L. 1990. Towards a natural system of organisms: proposal for the domains Archaea, Bacteria, and Eucarya. Proc. Natl. Acad. Sci., 87(12): 4576–4579.

Wolińska, A., Stępniewska, Z., and Pytlak, A. 2015. The effect of environmental factors on total soil DNA content and dehydrogenase activity.

Wu, F., You, Y., Zhang, X., Zhang, H., Chen, W., Yang, Y. et al. 2019. Effects of various carbon nanotubes on soil bacterial community composition and structure. Environ. Sci. Technol., 53(10): 5707–5716.

Wu, J.F., and Lin, X.G. 2003. Effects of soil microbes on plant growth. Soil., 35(1): 18–21.

Xu, L., Li, Q., and Jiang, C. 2006. Diversity of soil actinomycetes in Yunnan, China. Appl. Environ. Microbiol., 62(1): 244–248.

You, T., Liu, D., Chen, J., Yang, Z., Dou, R., Gao, X. et al. 2018. Effects of metal oxide nanoparticles on soil enzyme activities and bacterial communities in two different soil types. J. Soils Sediments, 18: 211–221.

Zhai, Y., Chen, L., Liu, G., Song, L., Arenas-Lago, D., Kong, L. et al. 2021. Compositional and functional responses of bacterial community to titanium dioxide nanoparticles varied with soil heterogeneity and exposure duration. Sci. Total Environ., 773: 144895.

Zhang, Y., and Goss, G.G. 2022. Nanotechnology in agriculture: Comparison of the toxicity between conventional and nano-based agrochemicals on non-target aquatic species. J. Hazard. Mater., 439: 129559.

Zong, Q., Wang, Q., Liu, C., Tao, D., Wang, J., Zhang, J. et al. 2022. Potassium ammonium vanadate with rich oxygen vacancies for fast and highly stable Zn-ion storage. ACS nano., 16(3): 4588–4598.

9

Nanomaterial Pollution in Agricultural Soils

Farah Noshin Chowdhury[1] and *Md. Mostafizur Rahman*[1,2,*]

Introduction

The usage of Nanomaterials (NM) has grown significantly in recent years, with the estimated output of synthetic nanomaterials totalling over 58,000 tons between 2012 and 2020 (Concha-Guerrero et al. 2014). The number of NM products is rising every day and is expected to double several times during the next years (Ameen et al. 2021, Boyes 2018). The scientific field of nanotechnology is concerned with the alteration of substances at the nanoscale. In general, it refers to the creation, handling, and characterization of building blocks and materials with a number of parameters at the level of the nanoscale, or less than 100 nm (Ameen et al. 2021). NMs usually have an average diameter lesser than 100 nm and possess size-dependent physical and chemical characteristics that set them apart compared to their mass or sub-micron/micron-sized equivalents. Quantum confinement and increased surface energy are NPs' fundamental characteristics that significantly set them apart in terms of how they behave and fare in various settings (Ameen et al. 2021). The use of NM in industry and the development of these products have significantly risen during the last twenty years (Nafisi and Maibach 2017, Parisi et al. 2015). Numerous industrial, residential, agricultural, and biological applications are suited for NMs. Due to their special characteristics, which include a substantial surface area relative to volume proportion, greater reactivity, surface potential, adjustable physicochemical characteristics, molecular alteration, etc., in contrast to their counterparts in salt as well as bulk counterparts, NM are capable of and preferred for larger applications (Ameen et al. 2021). NMs are being produced, used, and abused excessively, and this has led to a fast introduction of their discharge into numerous natural regions (Ameen et al. 2021, González-Gálvez et al. 2017). This rising use of NMs might lead to accidental NM accumulation in the environment via a variety of pathways, with unidentified impacts on soil, water, and organisms (Eduok and Coulon 2017).

Through increased crop output, monetary stability, and environmental sustainability, nanotechnology is additionally benefiting the agricultural industry. The manufacturing and

[1] Laboratory of Environmental Health and Ecotoxicology, Department of Environmental Sciences, Jahangirnagar University, Dhaka 1342, Bangladesh.

[2] Department of Environmental Sciences, Jahangirnagar University, Dhaka 1342, Bangladesh.

* Corresponding author: rahmanmm@juniv.edu

preservation of crops might be revolutionized by nanotechnology-based farming products like nano-fertilizers or nano-pesticides. Additionally, nanotechnology plays a role in a number of food industry processes, including processing food, wrapping, and storage. Using sensors for disease and pollutant identification, nano-sensors in the food sector offer security for the manufacturing, processing, and shipping of food items (Ameen et al. 2021). Moreover, the world population is expected to reach 9.7 billion around 2050, necessitating a sixty percent rise in crop production over the level at present. As a result, a 2.4% yearly increase is required to attain food security over the long term (Sun et al. 2021). NMs may enhance agrochemical utilization effectiveness by providing recovery of resources as well as target administration and gradual release approaches, enhance both plant and soil health and operates via microbiome improvement, enhance the cultivation process through the integration of nanosensors into crops, and reduce damage through rendering crops less susceptible and effective. These methods can boost yields while using fewer supplies (compost, herbicide) and have a lower environmental effect (Giraldo et al. 2019, Kah et al. 2018). Nanopesticides outperform traditional pesticides due to the benefits of high adsorption, decreased volatilization, enhanced tissue penetration, controlled dispersion, and so on. Yet, research have revealed nanopesticides' possible toxic effects in non-target organisms as well as their environmental concern (Kannan et al. 2023).

The structural integrity of agricultural soil is maintained in large part by helpful soil microorganisms including bacteria and fungus. These may involve support by rhizobacteria, detoxification or minimization and degradation of contaminants, the biogeochemical cycling as well as carbon cycling. These are necessary to maintain the robustness and functioning of agricultural ecosystems (Ameen et al. 2021). In view of the critical role that beneficial soil microbial communities play in soil fertility and development of plants, metal nanoparticles are produced, applied, and released into the soil system. However, this proved to have an effect on a variety of soil microbes, particularly rhizobacteria. They also have an adverse effect on single-celled, mycorrhizal fungi that degrade cellulose and lignin. The molecular changes in bacteria and fungi were brought due to NMs (Ameen et al. 2021). The long-term viability of this ecosystem and the wide range of invaluable microbial communities have been gravely challenged by the unrestricted build-up of NMs containing metals in soil ecosystems, notably in agricultural systems. NMs continuously deposited in soils, and their poor biological degradation and a longer permanence, have a negative effect on the population of good bacteria and leads to nanotoxicity (Ameen et al. 2021).

During the course of the past years, research has examined the infiltration of NMs into soil ecosystems from different sources as well as their effects on soil microbes, plants and nematodes (Abbas et al. 2020, Ameen et al. 2021, Lead et al. 2018). The effects of NMs on the organisms in the soil depend on the characteristics and complexities of the soil, including its ability to function as a buffer, the amount of natural organic matter present, the way that NM aggregate and immobilize, how they deposit, and how they create an environmental corona. NM in soils have been found to have negative impacts on agricultural crops (Pittol et al. 2017), soil microbiology (Yanga et al. 2017). Due to these factors, it is crucial to evaluate how NM affect soil microorganism development and physiology, especially those that are essential for the health of plants and soil (Ameen et al. 2021).

Engineered NMs offer enormous promise for enhancing pesticide usage effectiveness, crop yield, and soil health; nevertheless, the behavior and destination of these NMs, as well as the possibility for deleterious cumulative impacts on agroecosystems, are poorly unexplored. There is a specific paucity of knowledge regarding the modification of NMs across both land and plant divisions (P. Zhang et al. 2020). In this study a holistic approach was taken to understand and document all aspects NMs in agricultural soil - from their sources and types, to the transformation and transport of the NMs in the soil and ultimately the fate and manifested effects of the NMs presence in agricultural soil.

Source and use of nanomaterial in agricultural soil

On the basis of their source, NMs in soil may exist both naturally and artificially (Ameen et al. 2021, Bakshi et al. 2015). Considering them as the major subgroups of NMs present in agricultural soil, they are explained in the following discussion.

Naturally occurring nanomaterials in agricultural soil

NMs are plentiful in soil because of the abundant inorganic and organic material in soil with a nanoscale diameter. Due to high dispersion along with reduced dimensions and the slower sedimentation in the Earth's gravitation field, NMs are copious. Soils and sediments contain a variety of NMs: (1) secondary mineral silicates (which includes imogolite and allophone, which are plentiful in volcano soils), (2) nanoparticulate goethite, hematite, along with ferrihydrite, also, (3) soil humic compounds (humic acid, fulvic acid, and humin). Continuous physicochemical degradation and restructuring of geogenic elements, as well as significant biological activity that alters biological material and minerals, facilitates the continuous production of NMs in soil (Bakshi et al. 2015). Natural NMs in plant-soil-water systems can come from pedogenic or geological origins, as well as plant and microbiological sources. Mineral weathering products, as well as biogenic items resulting from microbial activity, are examples of geologic sources (Bakshi et al. 2015). Many mineral NMs in nature are the result of mechanical grinding near cracks following earthquakes. These mineral NMs combine with organic materials in the soil to produce aggregates ranging in size from 1–1000 nm. Weathering processes in soils can result in the production of layers on the outermost layer of pyrite composed exclusively of or combined oxides and hydroxide compounds of Fe along with additional elements. Eruptions of volcanoes, photochemical processes, air storms laden with dust, wildfires, plants, and animals are examples of generic routes (Bakshi et al. 2015). A further origin of NMs in environments is the microbial population. Microorganisms produce biogenic NMs both directly (by metabolic activities associated to motility) and indirectly (the precipitation of nano-crystalline Fe- and Mn oxides throughout microbially-mediated redox reactions). Mineralization and NM formation can also be aided by microbial shell interfaces and additional metabolites that function as organic templates. Pollen and virus fragments are two more organic sources of NMs (Bakshi et al. 2015).

Additional studies show that natural NMs (crystalline through amorphous solid substances) may be discovered in all environments. Many studies have revealed that a considerable percentage of ambient NMs persist in solid form at nanoscale sizes for lengthy periods of time. However, some of the aforementioned natural NMs can have negative effects, and the danger from manufactured NMs stems courtesy of their novelty, which includes possibly having an entirely distinct structure and activity than natural NMs. Given the significantly larger concentrations of natural colloids, natural NMs may be anticipated to govern the behavior and destiny of synthetic NMs (Ju-Nam and Lead 2016). Natural NMs are engaged in a variety of soil functions, including aggregate creation, retaining nutrients, microbial activity, purifying water, and pollution reduction, and so have an impact on the condition of soil and human well-being. They are often monodisperse owing to natural size determination, sparse in toxicity, rich with water and hydroxyl groups, owing to the availability of hydrous setting, aggregated into complex forms, and void of particular ligands linked to their surfaces (Bakshi et al. 2015).

Engineered NMs in agricultural soil

There are hundreds of intentionally produced engineered NM products for specific purposes in a variety of industries and used in agricultural soil, which has consequently become a major priority for technological development and research (Ameen et al. 2021, Tolaymat et al. 2017). Designed NMs are particles created purposefully for their beneficial qualities. They can be made up of a single element, such as C or Si, or a combination of components (Bakshi et al. 2015). Engineered NMs are often divided into three groups depending on their chemical-based compositions: (1) Carbon NMs, which include a single- or multiwalled nanotubes made of carbon (CNTs), fullerenes (C60, C70, etc.), (2) Polymer NMs, and (3) Metal NMs, which includes metal along with metal oxide. Other than these, engineered NMs of varied topographical structures include nano-polymers, dendrimers, polymeric nanotubes, nanowires, nanorods, nanofibers, nanosheets and nanocellulose, as well as nanostructured polymer films, nanoplates, quantum dots, nano clays, and nanocomposites. Second generation, third generation, and fourth generation NMs are novel kinds of NMs that are currently being developed (Ameen et al. 2021, Javed et al. 2019). Some of the popular NMs which are frequently used are: titanium (Ti), silver (Ag), zinc (Zn), aluminium (Al), copper (Cu), gold (Au), indium (In), bismuth (Bi), cobalt (Co), iron (Fe), nickel (Ni), silica (Si), tin (Sn), and molybdenum (Mo). The most frequently generated metal-oxide among these is titanium dioxide (TiO_2), followed by aluminium oxide (Al_2O_3), silicon dioxide (SiO_2), cuprous oxide (Cu_2O), cerium oxide (CeO_2), magnesium oxide (MgO_2), nickel oxide (NiO_2), indium oxide (Un_2O_3), lanthunum oxide (La_2O_3) and zirconium oxide (ZrO_2). The metallic NMs may be created and altered using a variety of chemical, physical, and biological techniques to have a broad spectrum of acceptable forms and sizes (Ameen et al. 2021, Pantidos 2014). NMs enter the soil by agricultural spread, waterways, surface precipitation, and industrial emissions, influencing plant development and the soils' microenvironment (Shi et al. 2021). Engineered NMs are directly exposed to soils through immediate contact routes when they are utilized as fertilizer, insecticides, for rehabilitation of polluted soils, or by unintentional discharge (Cornelis et al. 2014).

Figure 1 here shows a simplified view of nanomaterial in agricultural soil.

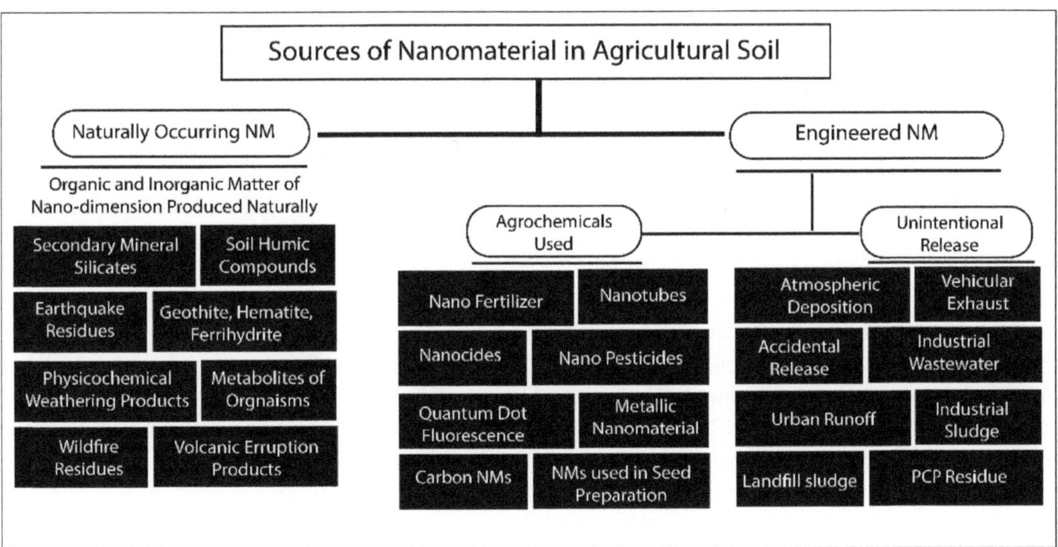

Figure 1. Sources of Nanomaterial in Agricultural Soil.

Intentional use of NMs in agricultural soil

The primary source of NMs in agriculture is the utilization of a variety of NM agricultural chemicals, including the cultivation of crops (nanofertilizers, nanopesticides, development boosters, protection for plants, priming of seeds, biological sensing), livestock husbandry, and water and soil remediation (Dwivedi et al. 2016, Mukherjee et al. 2016, Rastogi et al. 2017, Spanos et al. 2021). The soil system turns into a significant sink for explored and unadulterated NMs once these substances enter the ecosystem as dry powders as well as aerosolized sprays and are linked with litter, biosolids and effluents (Ameen et al. 2021, Geert Cornelis Kerstin Hund-Rinke and Nickel 2014). The majority of NMs employed in crop protection are anti-microbial in nature, with silver nanoparticles being the most common. They might have a negative impact on the soil ecology. NMs used for fertilization in products that are commercially available might be primarily made of calcium, potassium, zinc, magnesium, silver, and iron (Perea Vélez et al. 2021). The worldwide nanopesticide industry is expected to expand 14.6% each year from 2020 through 2027 (Bratovcic et al. 2021). The benefits of employing NMs as agrochemicals include higher yields of crops with smaller quantities of materials utilized, reduced chemical spread, increased availability of nutrients for plants, reduced loss of nutrients, and, as a result, less contamination of water and soil (Perea Vélez et al. 2021, Prasad et al. 2017).

Because many NMs include important micronutrients, they may be able to substitute traditional sources of nutrients as slow-releasing fertilizer. Zinc oxide nanoparticles (ZnONPs) act as an effective, slow-releasing Zn fertilizer to alleviate prevalent Zn shortage in agricultural soils (Dimkpa et al. 2017). Copper oxide nanoparticles-CuONPs are also being studied as a slow-releasing supply of Cu for lasting agricultural demands (Servin et al. 2015). NMs have the potential to be innovative transporters of important macronutrients. For example, silicon oxide nanoparticles (SiO_2NPs) were developed as a transporter of key macronutrients (N, P, and K) for plants as well as a slow-releasing Si supplement to reduce salt stress (Mushtaq et al. 2018). Some NMs have been investigated as more effective insecticides and disease prevention agents due to their antibacterial characteristics (Wang et al. 2019). Silver nanoparticles (AgNPs), were efficient against a number of plant-associated diseases including *Biploaris sorokinniana* and *Botrytis cinerea* (Chhipa 2017). Similarly, Cu-based NMs demonstrated antibacterial and antifungal activity against a variety of plant-related diseases (Sun et al. 2021). Cu-based nanoparticles (CuONPs and Cu_2ONPs) proved stronger against *Phytophthora infestans* in greenhouse research with diseased tomato plants than many marketed Cu-containing insecticides. Furthermore, Cu-based nanoparticles had no deleterious impact on plant development (Giannousi et al. 2013). ZnONPs, or titanium oxide nanoparticles, exhibit antibacterial characteristics as well (Chhipa 2017). TiO_2NPs, SiO_2NPs, and metal-based nanocomposites are examples of these materials. NM applications unavoidably enhance their discharge into the environment, allowing them to infiltrate wastewater streams. Most of these (> 90%) remain in biosolids after treatment of wastewater, with 60% being discarded in agricultural fields in the United States. The estimated amount of ZnONPs in biosolids may go as high as 80 mg/kg (Keller and Lazareva 2014). Whenever these biosolids are added to soils used for farming, the amount of ZnONPs can approach 50 mg/kg, which is significantly greater than the anticipated average concentration in soil (Sun et al. 2021).

Copper-based NM are employed in agricultural chemicals for effective pesticide and fertilizer distribution as well as regulated discharge. Their antifungal efficacy was demonstrated by penetration of chitosan in Cu NPs solution, which inhibited the plant pathogenic fungus *Sclerotium rolfsii* alongside *Rhizoctonia solani* (Adisa et al. 2019, Rubina et al. 2017, Zhao et al. 2016). The antifungal and growth promotion capacities of tomato plants against *Alternaria solani* as well as *Fusarium oxysporum* fungus were investigated. Both fungi were found to have much decreased disease severity. Cu-chitosan NPs at 0.10 and 0.12% concentrations were as effective as commercial fungicides in suppressing the corresponding fungal infections. Another study found that chitosan-copper (CS-Cu) nanocomposite created without the use of any chemical agents has outstanding antibacterial characteristics over a total of 5 of Gram-positive and -negative bacteria, as well as one fungus (Arjunan et al. 2017). The increased use of CuNPs in agricultural chemicals and agricultural goods may result in their unexpected spread in soil systems. CuO NPs prevented soil denitrification, elevated NO3 build-up, and lowered N_2O release rates by 10–24% in 60 hours. Cu NMs were observed to affect the constitution of microbial population following 160 days of administration, with the bacterial populations of *Flavobacteriales* and *Sphingomonadales* being more vulnerable to their existence. CuO NPs had a greater impact on the composition of the bacteria population in two distinct kinds of soil. When compared to untouched control soil, the bacterial community profile for CuO NP treated soil revealed less than 60% similarity (Bakshi and Kumar 2021).

TiO_2 NMs' photocatalytic ability aids in the breakdown of insecticides and the suppression of plant diseases. Furthermore, residual pesticides in the earth may be identified using TiO_2 NM detectors (Thiagarajan and Ramasubbu 2021). Zinc is a necessary micronutrient, but too much can produce chlorosis and limit plant development. Zinc NMs comprise metallic particles having a large specific surface area and the chemical formula ZnO. ZnO nanostructures, nanoparticles, nanowires, nanorods, nanotubes, nanobelts, and other sophisticated shapes are created using a variety of processes (Lund 2018). Iron is a necessary micronutrient for plant development; however, it is inaccessible to plants in particular soils, such as calcareous soil. As a result, the use of iron NMs has lately grown. The zero-valent Fe NMs have been successfully utilized for the remediation of a wide range of environmental toxins, which enhances the likelihood of their retention in the soil system (Lund 2018, Trujillo-Reyes et al. 2014). The incorporation of NMs into soils has the potential to influence rhizospheric microorganisms or agriculturally essential microbes and increase their functioning, increasing the supply of nutrients for vegetation and boosting the root system and crop development in general, thereby providing a fresh opportunity for soil health enhancement (Rajput et al. 2023). At mild to moderate administration rates, water dispersible carbon nanomaterials promote lettuce (Latuca sativa) production and soil biochemical status (Nepal et al. 2023). In a study, graphene, a novel efficient nanomaterial, was employed in plant-soil operations to increase plant nutrient absorption, minimize chemical fertilizer contamination by alleviating insufficient soil nutrient conditions, and enhance soil absorption of nutrients. With rising use, graphene increased surface total nitrogen (TN), the total amount of phosphorus (TP), and total potassium (TK) components, which increased surface fresh weight, dry weight, plant height, and stalk thickness. By increasing the fertility of the soil and optimizing the soil environment, applying graphene to the ground may boost maize plant biomass (Wang et al. 2023). NM are utilized as an adjunct approach to bioremediation for *in situ* treatment of heavy metals in polluted soils. Iron, nickel, palladium demonstrated favourable outcomes concerning sites polluted with a number of hazardous materials, including chromium and arsenic, and even dehalogenation of persistent organic compounds suggested the notion of nanobioremediation (Pérez-Hernández et al. 2020). NMs may speed up the response rate of remediation compared to standard *in-situ* remediation approaches, reducing remediation time (Delfani et al., 2014).

Inadvertent release of NMs in agricultural soils

Other than through direct agricultural application, NMs enter the agroecosystem via inadvertent release, air fallout, water supply, and sewage-sludge usage, all of which has a negative influence on crop yields. If NMs are introduced into the soil ecosystem via various channels without suitable treatment, it is reasonable to predict both favourable and negative effects on soil property and productivity of crops (Thiagarajan and Ramasubbu 2021). The concentration of NM has increased due to unintentional manufactured NM releases, ongoing misuse, and other factors in various contexts (Ameen et al. 2021, Nowack et al. 2015). 63–91% of NMs produced globally wind up in landfills, opening the door for their release into the water, soil, and environment (Ameen et al. 2021, Rizwan et al. 2017). Additionally, NMs are used in pesticides as additives to improve the solubility of vital components or shield them from accelerated deterioration. In addition, they are also unintentionally transported into the soil system from other environmental niches (Ameen et al. 2021).

Waste water purification facility sludge usage is an essential inadvertent ENM exposure route for soils used for farming. Since, on the basis of mass, over ninety percent of silver (Ag), zinc oxide (ZnO), ceria (CeO_2), titania (TiO_2) NMs, along with fullerenes (nC60) increased in the 'WWTP remains' within the sludge, and comparable separation is anticipated for carbon nanotubes. Because of its substantial organic matter and high accessibility to nitrogen and phosphorus, WWTP sludge is utilized as a land treatment in some countries, unintentionally subjecting soils to ENM (Cornelis et al. 2014).

Moreover, it is improbable that NMs will stay linked to products at the conclusion of their life span. NMs are found in the effluents of wastewater (Azimzada et al. 2017, Jiang et al. 2018), sludge from sewers (Wigger et al. 2015), and landfill discharges (Hennebert et al. 2013). The release of sewage and wastewater sludge, incorporating solid waste in soil used for agriculture causes NM contamination. According to estimations, 270,000 metric tonnes of the most commonly utilized NMs are manufactured each year. In soil, the projected release scenarios range from 8 to 28% (Bundschuh et al. 2018). Agriculture and amendments to soil consume 55% of sewage sludge. Because of them, aged-NMs may enter the environment mostly through the usage of wastewater (Eduok and Coulon 2017). This happens as a result of the improper handling of biosolids following the process of wastewater treatment and its utilization in agricultural areas to increase soil fertility (Ameen et al. 2021).

Around 94% of the NM released into soil and landfills in the United States came from the usage of personal care goods, with TiO_2-NPs accounting for 0.87 to 1.0×10^3 metric tonnes per year along with ZnO-NPs contributing 1.8–2.1×10^3 metric tonnes per year. Amongst these, it was predicted that 24–36% of NMs from personal care goods (PCPs) were released into soil systems. Around 81–82% of the overall discharge is made up of ZnO and TiO_2 NMs, which are used in sunscreen as ultraviolet-blocking agents. Particularly, TiO_2-NPs are regularly released into the local ecosystem due to their huge use in the nanotechnology sectors and routine usage in everyday life, including in beauty products, medicines, additives to food, and pigments (Ameen et al. 2021, Frazier et al. 2014). Next in line are face lotion (7.5%) and cosmetics (5.7%) (Ameen et al. 2021, Keller et al. 2013). During waste disposal procedures, Ag-NPs were released into wastewater discharges in amounts between 0.1 and 1.3 mg L^{-1} (Ameen et al. 2021, Kaegi et al. 2013, Kiser et al. 2012, Wang et al. 2012). These environmental release quantities may have negative toxicological effects. Additionally, the establishment of systematized methodologies for the investigation of NMs in intricate environments is constrained by the paucity of understanding concerning the physicochemical characteristics of NMs. Unavoidably, an enormous generation of NMs will cause them to build up in nature (Ameen et al. 2021, Ottofuelling et al. 2011). Ag NMs are frequently found in a variety of goods, owing to their antibacterial qualities. The soils are the principal recipient media of silver pollution due to the surface utilization of sewage sludge with farming or remedial reasons (Courtois et al. 2019).

Transport and transformation of NMs in agricultural soil

The interaction of NMs, plants, and microbes (NPM) is critical for soil productivity. This nexus has a number of internal connections or linkages that influence the bioavailability of substances, agrochemicals, or contaminants in farmed plants. The NPM connection is also influenced by a number of factors relating to soil productivity and restoration (El-Ramady et al. 2022). The physical and chemical characteristics of both the NMs as well as the surrounding matrix influence their mobility (Chen 2018). NMs adhesion to microorganisms appeared charge-dependent, exhibiting positively charged MNPs possessing better attachment effectiveness and toxicity to microbes (Sun et al. 2021). The movement of NMs in porous medium is regulated by the rate of filtration of the media's pores. In theory, particle mobility decreases as particle size increases (Chen 2018). The outer layer chemistry of NMs influences their ability to move in porous media. NMs with hydrophilic surfaces may be disseminated relatively readily, whereas NMs with hydrophobic surfaces tend to clump and separate from the solution phase (Chen 2018). Ionic strength is an important component that determines both the durability and movement of TiO_2 NMs in soil. Increased ionic strength enhances homo and hetero-aggregation by compressing the electrostatic dual layers surrounding NMs, which adds to their decreased stability. In this regard, it has been observed that a very low ionic strength promotes the adsorption of hydrophobic Natural Organic Matter (NOM), which lowers the surface potential of TiO_2 NM, reduces aggregation, and stabilizes TiO_2 NMs in natural soil mixtures

(Thiagarajan and Ramasubbu 2021). In contrast, when ionic strength increases, multivalent cations span the hydrophilic NOM, which leads to the aggregating and destabilization of TiO_2 NMs in naturally occurring soil solutions. TiO_2 NMs' use in agroecosystems might jeopardize food crop nutritional value and yield. These harmful impacts include changes in crop anatomy and physiology morphology, and expression of genes (Thiagarajan and Ramasubbu 2021). TiO_2 NM have little movement in soils used for agriculture with varied textures and nutrients concentration. CuO nanoparticles have been discovered to be transportable in all soils. This is most likely due to their lower apparent size and limited homo-aggregation capability in all soil fluids (Simonin et al. 2021).

NMs endure many transformations after discharge, including homo- or hetero-agglomeration, dissolution, sedimentation, adhesion, oxidation, reduction, sulphidation, photochemical and physiologically driven reactions. The ultimate destination and availability of these to various species are determined by these transformation mechanisms (Amde et al. 2017, Bhatt and Tripathi 2011). Physical, chemical, biological transformations of NM may take place in soil, at the plant's interface and within the plant. They might develop new traits unique from their initial profile as a result of these extremely dynamic mechanisms; their conduct, destiny, and biological impacts may also alter dramatically, including NM transformation mechanisms and sites in the soil-plant system, as well as the impacts of conversion on analyte absorption, transfer, and toxicity (Zhang et al. 2020). The physicochemical conditions within the soil drives and regulates the evolution of NMs in the agricultural soil. Soil is a very complicated combination of fifty percent solids and fifty percent voids, which are about equal amounts of pore air and water. The alteration of NMs in soil matrices may be influenced by soil texture, pH, ionic capacity, redox potential, moisture, and organic matter concentration. The existence of a community of organisms can result in significant variation in the spatial distribution of soil; the physical and chemical characteristics of soil in different microbial environments, like the rhizosphere, can vary drastically. The existence of diverse organic as well as inorganic components in soil is a significant driver in triggering physical alteration of ENMs. Soil texture variations include gravel, silt, sand and clay, among others also influence this process (Zhang et al. 2020). The modification of NMs is a continuous phenomenon. They can disintegrate at their root surfaces and be carried through plant tissues, where reducing conditions can cause small particles to regenerate in intracellular gaps or within plant cells. Some modifications may occur fast at the plant's root surface, leading to instantaneous ion absorption and consequent acute plant toxicity. Such quick transition may result in the loss of desired NM features for agricultural applications. A corona can develop on the surface of an NM and temporarily affect material translocation, but it will eventually dissolve inside the plant. Knowing the following behaviour and eventual fate of NMs will need knowing the kinetics of each stage of these transformations (Zhang et al. 2020).

Silver, copper, Silver and zinc oxide, for example, are metal-based NMs that may disintegrate swiftly. On the outer layer of these NMs, a corona may form, which either speeds or delays disintegration. The metal ions produced by the plant might be absorbed directly or in complexes involving other environmental components. Other NMs, including silicon dioxide and titanium dioxide, retain their chemical makeup in nature, but physical alterations such as aggregation, agglomeration, as well as corona formation may take place, altering the surface's charge and chemical reactions and influencing future behaviour and bio availability. The function of the originally planned analyte may be affected as a result of these transformations within the soil or across the nanoparticle-plant interface (Zhang et al. 2020). Research on Ag, ZnO, and CuO NPs show that as the size of particles reduces, solubility increases. While size is the most important physicochemical parameter influencing solubility, this tendency may be reversed, particularly by coating the surface (Chen 2018).

Types of transformation of NMs in the environment

NMs undergo physical, chemical and biological transformation in agricultural soil.

Physical transformation

All forms of NMs undergo physical change. NMs generally form aggregate in soil, resulting in homoaggregation as well as heteroaggregation. Aggregation can cause NPs to grow in size, and aggregate strain is a frequent method for capturing NPs in porous medium. Particle dimensions, flow rate, and the size of grains can all be critical factors in NM filtering (Chen 2018). Homoaggregation refers to the aggregation of similar particles caused by the contraction of the electrical two-layer structure on the molecule's surfaces under high ionic strength circumstances. Homoaggregation rates diminish if an electrostatic boundary prevents collisions between particles to the point that the NM suspension is seen as stable in kinetic terms and is not evidently aggregating during an objectively defined time limit. Aluminium oxide NM of various sizes were found to have comparable reduced movement in a soil gradient (Cornelis et al. 2014). On account of actual NP concentrations, hetero-agglomeration involving natural colloids dominates homo-agglomeration in the ambient matrix (Abbas et al. 2020).

Heteroaggregation takes place in soil among NMs and bigger soil particles, thereby incorporating the NMs into the soil particle. NMs may also bind and develop a developing coating (referred to as a "corona") containing environmental or organic macromolecules such as humic compounds, nucleic acids, proteins, and polysaccharides, which can significantly alter the material's surface characteristics (Zhang et al. 2020). The rates of heteroaggregation seem to differ based on the soil colloid that surrounds the NM heteroaggregates, although the rates appear to be greater throughout a wide pH range when contrasted with homoaggregation rates. In the environment, heteroaggregates formed by organic colloids are significantly more common than homoaggregates. Heteroaggregation is likely to result in decreased rather than enhanced NM movement (Cornelis et al. 2014). Silty and clay-based soils have more negative spots for analyte sorption and interaction that can firmly attach positive-charged NMs and generate more heteroaggregation of NMs. The soils with a substantial content of clay have a greater capacity to hold water as well as reduced pore volume for air, resulting in an environment that reduces that promotes NM reduction. Sandy soils have more oxygen, which may enhance oxidative conversions of NMs (P. Zhang et al. 2020).

Chemical transformation

In agricultural soil, NMs can go through chemical changes (Chen 2018). Chemical conversion entails changes in the chemical makeup or structure of the NMs. The Dissolution is a typical chemical transition, and it can often be followed by additional chemical changes such Reduction, Oxidation, Sulfidation, Phosphorylation and Chelation Process. Oxidation may be used to liberate Ag ions coming from Ag NMs, and nanoscale CeO_2 is capable of being Reduced to generate Ce3+ ions (Stabryla et al. 2018). Other methods of transformation can take place, such as Deposition, Phosphorylation of Ce and Zn, Chlorination of Ag, and Sulfidation of Fe, Cu, Zn, and Ag. Organic ligand can attach to the metal ions that have been liberated (Zhang et al. 2020). Metal NMs are prone to ion release and dissolution. This is an evolving procedure that is heavily influenced by NM reactivity as well as toxicity to the environment. The inherent characteristics of particles, along with the matrices of adjacent media, influence both dissolution and solubility percentages. Solubility of metal-based NMs having a similar chemical makeup might change depending on shape (Chen 2018).

Metal-based NM breakdown can be greatly accelerated in acidic soil. The chemical differentiation of NMs can also be determined by soil pH. Oxidation of NMs is more favoured in an oxygenated soil with an elevated oxygen concentration, resulting in the production of an oxide layer on the surface of some NMs that can function as a protective covering for the particle's core. For other NMs, such as Ag, the creation of the Ag_2O layer causes the disintegration and release of Ag. The environment is largely decreasing in anoxic circumstances such as paddy soil, marshes, and organic-rich soils, which causes the decrease of NMs. Under aerobic circumstances, 90% of the

Ag NM in rice field soil persisted as Ag NMs, but under anaerobic circumstances, 95% was oxidized to Ag_2S (PZhang et al. 2020).

However, the soil microclimate can be quite varied. Even rice field soil may have very specific oxygen-rich patches that influence NM transformation. Such soil complexity must be acknowledged as a fact in each of the instances indicated above. Inorganic ligands including sulphide, chloride, as well as phosphate play important roles in the chemical conversion of metal-based NMs. NM production or reformation may also occur based on the chemical makeup of the soil. For example, microorganisms in soil might diminish CeO_2 NMs and perhaps create $CePO_4$. $CePO_4$ is a phosphate mineral (rhabdophane) that occurs in soil through the process of weathering of apatite (Zhang et al. 2020). Certain metals (Ag, Cu, and nZVI- Nano Zero Valent Iron) including metal oxide constituted NPs may go through oxidation-reduction (Mitrano et al. 2016).

Biological transformation

Biological transformation is unavoidable since the environment is teeming with organisms such as plants, fungus, protozoa, and the aforementioned chemical transformation processes, but it is controlled by biological mechanisms. NMs can clump together in the rhizosphere, upon root surfaces, or in physiological fluids. Dissolution, Deterioration, Redox reactions, or Chelation can occur in the cell's wall/membrane, cytoplasm, the vacuole or outside the cell as a result of enzyme-assisted chemical reactions or interactions with oxygen species that are reactive. Carbon-based NM may undergo redox processes. Moreover, bacterial respiratory activities can diminish graphene oxide in soil (Zhang et al. 2020). Bacteria are the most numerous organisms found in soil. Bacteria play an important role in many geochemical and metabolic procedures. Their metabolic processes, respiration, and organic acid discharge can cause NMs to undergo a variety of chemical changes. Nanoplant interaction points, like nanoroot, are also key sites for NM transformation before they enter plants. Plants can secrete metabolites at their roots and leaf surface, which can react with NMs and alter their physicochemical characteristics. A huge colony of bacteria and other microbes populate the root surface, where waste products abound, establishing a limited but essential zone known as the "rhizosphere" (Zhang et al. 2020). Natural organic materials (NOMs) can promote NM conversion by producing a NOM covering or "corona" on the outermost layer of particles, altering the surface of NMs and influencing the resulting behaviour and fate. The relative levels of NOMs vs NMs in the soil's environment are quite considerable, and NOM alteration of NM characteristics might be significant and dynamic. However, the consequences of this change on the characteristics of NMs are heavily influenced by the composition of the NOMs as well as other soil variables (Zhang et al. 2020).

NOMs with a higher molecular weight can create consistent and dense coverings that completely or nearly completely cover the exterior of NMs. This coating may promote electrostatic stability of NMs or flocculation and aggregation owing to linking effects among NOM molecules, based on the characteristics of the NOMs (Zhang et al. 2020). Carbon-based ENMs such as CNTs and fullerene can biodegrade the outer coating or inner component of ENPs. In thirty-two weeks, the white-rot basidiomycete fungus may enzymatically breakdown hydroxylated fullerol via the oxidation reaction and mineralize it to CO_2 (Schreiner et al. 2009). Additionally, the adhesion of Dissolved Organic Matter (DOM) on the NM exterior dramatically affects its physicochemical properties and, as a result, its environmental behaviour. Humic and fulvic acids, transitional products of decomposition or discharges from soil organisms, including smaller acids or biopolymers, can interact with NMs. The adhesion of DOM to chemical NM surfaces happens primarily via ligand transfer with the DOM's epidermal hydroxyl groups. Soil DOM frequently improves the measured ecotoxicity (and hence bioavailability) by enhancing the portability of NM, but also by being biocompatible, particularly if the organic covering is biocompatible (Cornelis et al. 2014). The following Figure 2 shows the pathway of NMs in agricultural soil.

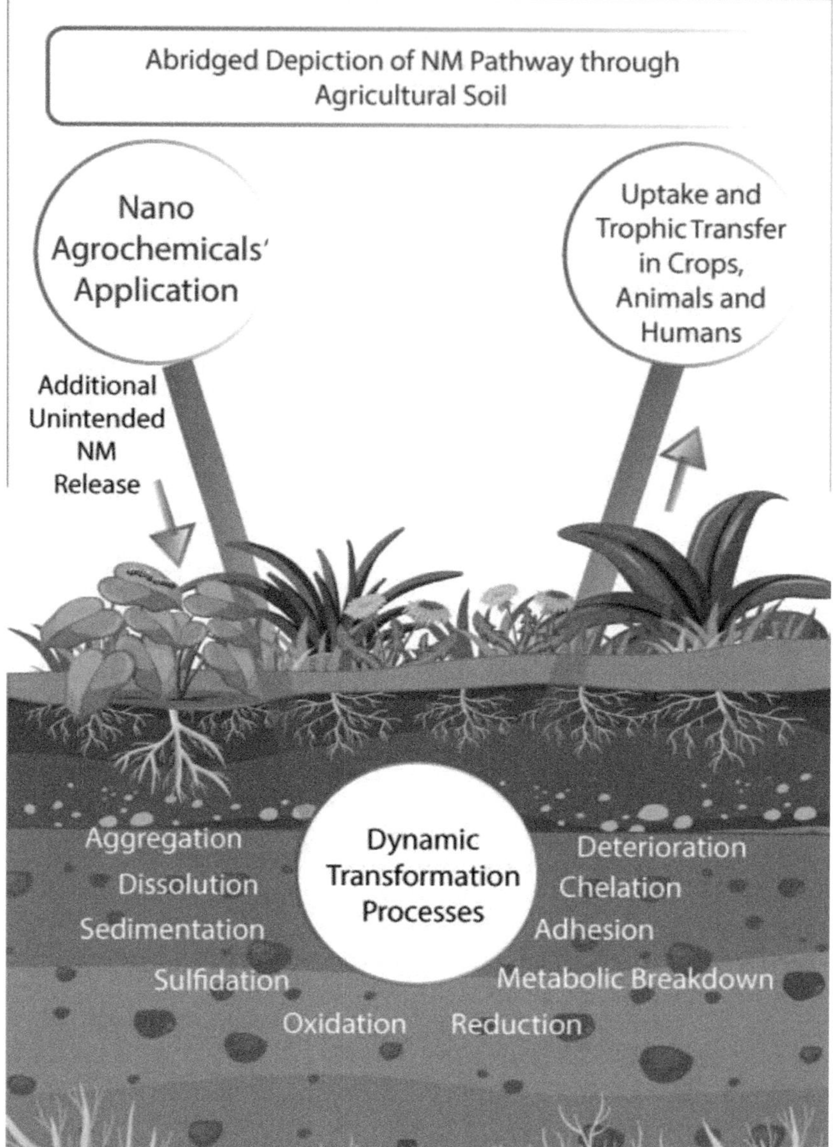

Figure 2. Abridged Depiction of Pathway of NMs in Soil.

Fate and effects of the NMs

NMs' fate in soils is determined via their reactivity, transportation, and availability (Bakshi et al. 2015). They have the potential to interact with live beings, influencing crucial life processes (Delfani et al. 2014). The shape, dimension, agglomeration duration and magnitude, composition of elements, surface chemistry, area, charge, structure of crystals, solubility and dispersibility, and capping component represent a few of the basic characteristics that characterize nanoparticles (Ameen et al. 2021, Das and Chatterjee 2019, Ingale and Chaudhari 2013). The toxicity profiling or harmful consequence of a nanoparticle can be predicted after evaluating its physicochemical features (Ameen et al. 2021, Santos et al. 2015, Sezer et al. 2019).

Physicochemical aspects

There have been limited investigations on the effects of MNPs on soils' physical properties. A study found porosity and hydraulic conductivity of 2 cropping soils were not affected by as much as five percent by CuONPs or magnetite NMs (Fe_3O_4NPs) (Ben-Moshe et al. 2013). Another investigation found that injecting 6000 mg/kg of CeO_2 NMs to a soil during six recurrent watering cycles had no effect on soil porosity, but it did change water distribution in various pore sizes and plant accessible water. Nevertheless, the NM concentrations employed in these researches are orders of magnitude larger than the predicted NM concentrations in agricultural soil. NM injection at present and anticipated levels is not expected to impact soil physical qualities (Sun et al. 2021). Additionally, it is found that waste water treatment plants' sludge is dug through the agricultural soil as fertilizer. Soil bacterial breakdown and biological degradation of the organic content in the sludge leads to the re-emergence of NM aggregate within the sludge matrices into soil pore fluid. It can take place identically to the way portable colloids dissolve as aggregates or separate particles into soil pore fluid by hydrodynamic shear. Additionally through rainfall, sharp decrease in ion strength, redox shift in which iron oxides which cement colloids collectively are reduced and dispersed, or adherence of ions that include phosphate or dispersed organic macromolecules such NMs can be embedded in the soil pore fluid. Phosphate, amphiphilic proteins, as well as humic acids are common in WWTP sludge and may enhance the interparticle repellent forces between NM after adsorption (Cornelis et al. 2014).

Because of the substantial mobility of the nanoparticle in the Earth's soil, the extensive use of zinc in agricultural land has had several negative consequences on the ecosystem, posing a larger danger of pollution through leakage of the metal into water bodies (Bao et al. 2022, Gao et al. 2018). Furthermore, the presence of several organic and inorganic substances in aquatic and terrestrial ecosystems, including surfactants, colorants, and microplastics, might have a major impact on zinc oxide transformation from agricultural operations (Ricardo et al. 2021). NM conversions can result in the creation of more dangerous chemicals or enhance their continued existence in the environment (Ali et al. 2022). In a manure-soil-plant framework, the incorporation of iron oxide NM to the surrounding soil enhanced the quantity of iron in the microbial biomass. Additionally, it caused a reduction in nitrogen availability to plants (spinach), resulting in a decline in nitrogen cycling (Kamran et al. 2020), although some studies demonstrate that metals in soil have little influence on soil microbiota in the near term (Bao et al. 2022).

NMs' operation is largely influenced by the soil's pH and the capacity for cation exchange (CEC). Some NM can change soil pH; however, the effects vary depending on the soil. TiO_2, CeO_2, or $Cu(OH)_2$ NMs raised the pH level of a loam soil through 0.4 unit while lowering the pH of a soil made of sandy loam by 0.1–0.4 unit. NMs can change soil pH via interacting with plant roots when plants are present. The pH of the rhizosphere was reduced by CeO_2 NMs (Rossi et al. 2018). The effect of NM-induced plant drainage of roots on bulk soil pH change is unclear (Sun et al. 2021). Soil CEC is significantly connected with its ability to store plant nutrients and environmental contaminants, making it an important indication of soil chemical quality. There has been little research on the effects of NMs on soils CEC. In an experiment, Fe_3O_4 NMs boosted rhizosphere CEC (De Souza et al. 2019). NMs have the ability to change soil qualities via modifying SOM (Soil Organic Matter) properties. Adding 1–5% CuO or Fe_3O_4 NMs reduced the amounts of fluorescent organic molecules in two soils while leaving the total SOM unchanged. Humic compounds' carboxyl (-COOH) or phenolic hydroxyl (-OH) groups with functional properties generate powerful bonds with TiO_2 NMs, altering their chemical characteristics (Sun et al. 2021).

Biological aspects

The biological health of the soil metrics is more vulnerable to NM effect than soil physical and chemical characteristics are. In general, the impacts of NMs and their converted constituents on soil health differ depending on soil, NM, duration of exposure and dose, and soil type (Sun et al. 2021). Unique microbial taxa may respond differently to NMs. The existence of nanoparticles of metal on soil microbiota is a concern with multiple irreversible consequences (Feng et al. 2013). NMs can have an impact both promptly and in an indirect manner: direct particle-microbe contact and higher toxicology or decreased nutrient absorption (Bundschuh et al. 2018). Soil health is closely connected to the presence of microorganisms, which can be employed as reliable indicators of soil quality (Avidano et al. 2005). Other research has found that NMs can reduce the variety and amount of organisms in the soil, as well as impact the synthesis of important enzymes (Tang et al. 2022). These changes in soil microbiota can have a negative influence on biogeochemical cycles of numerous nutrients (Blinova et al. 2021, de Oliveira et al. 2022, Dutta et al. 2022, Ricardo et al. 2021). Non-target species in the ground on the other hand, may be harmed by Ag-NMs. After being exposed to Ag-NMs, the number of *Acidobacteria*, *Actinobacteria*, *Cyanobacteria*, and *Nitrospirae* reduced considerably as the level of Ag-NPs increased (Wang et al. 2017).

Metal nanoparticles continuously deposited in soils, which have a poor biological degradation and a longer permanence, have a negative effect on the population of good bacteria and fungi. Fe NMs and TiO_2 NMs used in water purification and environmental remediation both restrict and encourage the development of target species (Ameen et al. 2021, Lecoanet and Wiesner 2004, Mueller and Nowack 2010, Yavuz et al. 2006). However, at the same concentrations, non-target microorganisms and other living entities are also poisoned by Fe and TiO_2 NMs. nZVI, on the other hand, had solely negative impacts on soil microbes (Ameen et al. 2021, Cullen et al. 2011). Other NMs, including ZnO, CuO, Ag, FeO, and TiO_2, exhibit varied short- and long-term harmful effects on soil microorganisms and unadulterated microbial cultures. Size, surface charges, capping agent, the existence of divalent anions/cations, as well as the make-up and charges within the bacterial cell wall are other elements that affect how NMs or bacteria interface (Acharya et al. 2018, Ameen et al. 2021, Sondi and Salopek-Sondi 2004). The surface characteristics, bioavailability, and diversification of NMs to plants can be changed by microbial compounds that include extracellular polymeric substances (EPSs), enzymes, organic acids, hormones in addition to the plants' root discharges in the rhizosphere (Ameen et al. 2021).

NMs' effect on bacteria

Numerous pure culture investigations have revealed that bacterial interactions with NMs can cause cytotoxicity in multiple parts of the cell, including membrane disarray, denaturation of thiol-containing proteins in the membrane, damage to DNA, and oxidation exceeding ROS generation (Concha-Guerrero et al. 2014). CuO nanoparticles have been investigated on distinct strains of bacteria (Bacilli, Flavobacteria, as well as Gammaproteobacteria) from soil from agriculture. 11 of the 56 strains tested positive for high sensitivity. CuO NMs are extremely poisonous to soil microbes (Concha-Guerrero et al. 2014). Copper oxide nanoparticles' adverse impacts on carbon and nitrogen cycling microbiological processes in varying farming soils and in plants were investigated. CuO-NPs significantly reduced each of the bacterial processes tested (denitrification, nitrification, alongside soil respiration) at 100 mg/kg dry soil, while the effects were restricted at lower doses (0.1 and 1 mg/kg) (Simonin et al. 2018).

The relationships between NMs and rhizobacteria that promote plant growth (PGPR) are essential, given the significance of PGPR to plant viability (Ameen et al. 2021, Mesa-Marín et al. 2018). Comparable to other xenobiotics, the detrimental impact of NMs on beneficial soil bacteria is just now becoming apparent and continues to be not completely understood (Ameen et al. 2021, Duhan et al. 2017, Wagner et al. 2016). Inhibition of bioactive molecules and cellular toxicity towards useful soil bacteria is usually determined by the nanoparticle's form, dimension, chemistry, quantity, and length of exposure. The cells of bacteria may eventually become entirely or even partially immune to some nanoparticles, but not all of them. In one instance, 4 crop growth-promoting species of rhizobacteria—*Pseudomonas mosselii, Bacillus thuringiensis, Sinorhizobium meliloti,* and *Azotobacter chroococcum*—adapted to as much as 3000 µg/ml doses of CuO, TiO_2, and Al_2O_3 nanoparticles, but had been intolerant of < 1500 µg/ml doses of Ag and ZnO (Ahmed et al. 2020, Ameen et al. 2021). The toxicity of Ag and ZnO nanoparticles might be caused by a couple of factors: diminished bacterial cell respiratory activity and damage from oxidation and cellular breakdown (Ameen et al. 2021, Choi et al. 2008, Zhang et al. 2018).

When juxtaposed to an unprocessed control, the nanoparticles produced from Ag and ZnO increased roughness of the surface in advantageous bacteria, but decreased bacterial adhesion to a solid terrain, Extracellular Polymeric Substance (EPS), and colonization by bacteria (Ahmed et al. 2020, Ameen et al. 2021). In a comparable manner, the harmful effects of AgNPs on a pair of useful rhizobacteria, *Azotobacter vinelandii* and *Bacillus subtilis*, has been described (Gambino et al. 2015). Aside from destroying cell shape and the amount of viable cells, NMs also prevent the formation of certain bioactive chemicals that are essential for plant development and the fertility of the soil. As an example, ZnO and Ag nanoparticles inhibited the synthesis of indole-3-acetic acid through 3 strains of PGPR in a way that depends on the concentration, which was totally eliminated at a level of 1000 µg/ml (Ameen et al. 2021, Seneviratne et al. 2016).

TiO_2 and ZnO NMs generated changes in the two primary soil communities of bacteria, *Rhizobiales, Bradyrhizobiaceae,* and *Bradyrhizobium* (associated with fixation of N) and *Sphingomonadaceae* along with *Streptomycetaceae* (associated with biological pollutants and biological polymer decomposition). In reaction to TiO_2 and ZnONPs, there was a noticeable decline in *Rhizobiales, Bradyrhizobiaceae,* and *Bradyrhizobium* and an increase in *Sphingomonadaceae* and *Streptomycetaceae* microbiological taxa. Later, it was discovered that ZnO NM-mediated toxicity had a negative impact on ammonification, respiration, and dehydrogenase processes among soil microorganisms (Ameen et al. 2021).

The mutually beneficial association of *Vicia faba* grown in soil and *R. leguminosarum, Glomus aggregatum,* or a mix of the two cultures has been observed to be impacted by Ag NMs. Around 800 mg kg^{-1}, Ag-NPs slowed down the activity of the microbes in the soil and, as a result, the symbiosis. Furthermore, nitrogenase activity, nodulation, mycorrhizal colonization, and glomalin content were all markedly reduced by the Ag-NPs. Ag-NPs also caused the breakdown of bacteroids and autophagy, which degraded cytoplasmic components inside cells (Abd-Alla et al. 2016, Ameen et al. 2021). In a study of Ag NMs to 3 bacteria, *B. amyloliquefaciens* GB03, *S. meliloti* 2011, and *P. putida* UW4, it suggests that *P. putida* and *B. amyloliquefaciens* both showed excessive or medium oxygen consumption, correspondingly (Lewis et al. 2017). Furthermore, 94 days following the use of Ag NMs, 95.7% of the silver was still in the soil, which led to significant changes in the composition of the microbial community (Cao et al. 2018). In an evaluation of the effect of Ag NMs on the development of a pair of strains of *P. putida* on agarose beads, it significantly reduced the production of biofilms and dramatically decreased the potential for survival of the cells in produced biofilms (Hartmann et al. 2013). Research on the proteome of *B. subtilis* biofilms infused with Ag NM showed that the expression of several proteins linked to adverse reactions, redox sensing, and sensing of quorum was elevated. The generation of EPS and inorganic phosphate solubilization was likewise increased in reaction to the NMs stress (Ameen et al. 2021, Gambino et al. 2015).

ZnO NMs have been shown to have substantial negative impacts on plants that are edible in addition to changing the development of soil microbes. For instance, in comparison to the control group, acetate stabilized ZnO NMs induced diverse global protein expression patterns within the bacteria *Cupriavidus necator* JMP134 that lives in soil (Ameen et al. 2021, Neal et al. 2011). In one investigation, it was shown that 306 mM ZnO NMs totally stopped *Pseudomonas* species from growing, whereas 7.7 mM ZnO-NPs of a size < 100 nm decreased colony-forming units of *P. chlororaphis* O6 by 12.5% (Ameen et al. 2021, Dimkpa et al. 2011, Soni et al. 2017).

Moreover, ZnO NMs (50 nm) nearly totally stopped *P. aeruginosa* PAO1 and *P. putida* KT2440 from building biofilms. Furthermore, when ZnO NMs were exposed to humic acid, the elimination of EPSs enhanced the anti-biofilm phenomena by reducing the metabolism by over 75% (Ameen et al. 2021, Lee et al. 2014, Ouyang et al. 2017). The Garden pea with the related bacteria *Rhizobium leguminosarum* were used to test the impact of ZnO NMs on Rhizobium-legume symbiosis. ZnO NM exposure altered the morphology of rhizobial cells and caused damage to the surface to the bacteria, disrupted the procedure of root nodulation, slowed the start of nitrogen fixation, and induced early senescence of nodules (Ameen et al. 2021, Fan et al. 2014). The symbiotic bacteria *S. meliloti* and N_2 fixing were both negatively impacted by 10 nm ZnO NMs. The ZnO NM on bacterial cell exteriors and their subsequent internalization into periplasmic regions resulted in modifications to the polysaccharide structure and proteins of cell walls. It was determined that crop output may be endangered by ZnO NMs existence in the surroundings and their interactions with beneficial microorganisms in the soil (Ameen et al. 2021, Bandyopadhyay et al. 2015).

Two perspectives exist regarding the impact of TiO_2 NMs upon soils bacteria: toxicity evaluation following photo-activation (vide infra), alongside the absence of photo-activation. The majority of research refutes the idea of photo-activation. Even at greater quantities, TiO_2 NMs are not very harmful, according to certain research. At a dosage concentration of 418 mM, the TiO_2 NM (50 nm) lacking photo-activation were not harmful to bacteria (Ameen et al. 2021, Jiang et al. 2009). TiO_2 NMs were shown to have an effect on PGPR metabolism in soils, and the growth of *T. aestivum* increased upon inoculation with mutant strains of *Paenibacillus polymyxa, Alcaligenes faecalis*, and *Bacillus thuringiensis* (Ameen et al. 2021). The seedlings of wheat co-inoculated with *P. polymyxa, B. thuringiensis*, or *A. faecalis*, TiO_2 NMs can promote the development of PGPR (Ameen et al. 2021, Timmusk et al. 2018). Damage to the morphology of the symbiotic bacteria *R. leguminosarum* bv. viciae was additionally caused by TiO_2 NMs. TiO_2 NMs interfered with the symbiotic connection between *R. leguminosarum* and *P. sativum*. There was an obstruction to nodule development and eventual nitrogen fixation. The host (*P. sativum*) experienced a systemic reaction, and the nodule's polysaccharide composition changed. TiO_2-NPs also have an effect on plant development. TiO_2 NMs were found to dramatically decrease by 54% the nitrogen fixation by *Rhizobium trifolii* (Ameen et al. 2021, Fan et al. 2014, Moll et al. 2016).

NMs effect on beneficial fungi

The potential of soils to support biological production and the health of plants is defined by total soil health rather than soil quality. Scientists have recognized the function of fungal variety in promoting the condition of the soil, crop yield, and the farming sector altogether (Ameen et al. 2021, Nielsen et al. 2015). NMs can enter fungal hyphae and alter and harm native anatomy due to their dimensions and nature. The interaction of different NMs with mycorrhizal fungus was discovered to impact their growth and demonstrated both positive and negative effects. Some NMs aid in the colonizing of fungus, whereas others inhibit colonization (Ameen et al. 2021). *Arbuscular mycorrhizal* fungi (AMF) are a typical impact of different metals NMs on the growth and plant development-endorsing functions of this soil fungus. Trees or crops immunized with AMF show improved susceptibility to the fungal root-rot infection (Ameen et al. 2021, George et al. 2016) as well as considerable increases in nutritional intake such as phosphate (P), along with additional nutrients that are generally immovable and accessible in small concentrations in the soil (Ameen et al. 2021, Begum et al. 2019).

Wide range of metallic NMs were shown to have a significant impact on the development of soil fungi that are helpful to plant health. Several metallic NMs-based products, for example nano-herbicides, nano-pesticides, nano-fertilizers, as well as a means for target-specific distribution in crops, have been employed and studied to date. The relative dose-related biological impacts of bulk as well as nano types Ag and FeO nanoparticles on AMF mycorrhizal clover (*Trifolium repens*) have been evaluated (Ameen et al. 2021, Feng et al. 2013). The results showed that AMF colonization had a considerably favourable influence on plant growth-promoting activities. A substantial reduction (p 0.05) in AMF development and ecological function was detected after Ag NMs contact, as measured by root mycorrhizal colony percentage, soil alkaline phosphatase action, accessible phosphorus, and P intake in plants (Ameen et al. 2021, Cao et al. 2017). A similar study found that AMF (*Funneliformis mosseae*) colonization in maize plants was affected by both single and combined (ZnO NMs and $ZnSO_4$) treatment. AMF colonization and maize growth were shown to be decreased when exposed to ZnO NMs and $ZnSO_4$ individually; however, a combined (ZnO NMs-$ZnSO_4$) treatment had a beneficial effect on growth. Reduced Zn concentrations and enhanced nitrogen absorption in maize plants might be the cause (Ameen et al. 2021).

Yeast treatments have a variety of significant functions in maintaining ecologically sound agriculture (Ameen et al. 2021, Yurkov 2018). In this regard, the yeast *Saccharomyces* cerevisiae is a suitable unicellular eukaryotic specimen for toxicological research on NMs (Ameen et al. 2021, Kasemets et al. 2009). Furthermore, a variety of metallic NMs, including TiO_2, Fe_2O_3, HfO_2, SiO_2, Al_2O_3, and CeO_2, showed little suppression of O_2 and damage to membranes in *S. cerevisiae*. Mn_2O_3 NMs, on the other hand, were shown to limit O_2 absorption by 50% in *S. cerevisiae* while causing 30% damage to cell membranes (Ameen et al. 2021, García-Saucedo et al. 2011, Otero-González et al. 2013). A recent research found minor toxicological consequences in white-rot fungi, which were exacerbated by direct build-up of nanodiamonds (NDs) inside. Several filamentous fungus and lignin-degrading fungi responsible for white rot, including *Phanerochaete chrysosporium*, *P. sordid*, and *Trametes hirsute*, were shown to interact with NMs, altering their metabolic activities. The contact of ZnONPs to a local fungus community greatly lowered the rate of litter degradation (Ameen et al. 2021, Andries et al. 2016, Du et al. 2020, Ma et al. 2020). Furthermore, a recent research suggested that ZnO and CuO NM-treated soil altered dehydrogenase activity in many fungi as well as oligo and copiotrophic bacteria (Jośko et al. 2019).

Several pieces of data suggest that a variety of mycorrhizal fungus species may respond differently to varying NPs. *Glomus caledonium*, a fungus, managed to survive ZnO NMs toxicity better than *G. versiforme*, indicating a detrimental influence on root colonization. Furthermore, a soil microcosm investigation of two distinct reference soils, Lufa 2.2 with Lufa 2.4 following Al_2O_3-NPs and Ag-NPs, found that the microbes of Lufa 2.4 were more susceptible, as evidenced by a degree shift in microbial structure. Ag-NPs had a minor impact on the phylogenetic makeup of microbial populations of both soils. As a result, the subclass *Cytophaga-Flavobacterium* decreased in Lufa 2.4 soil and -Proteobacteria decreased in Lufa 2.2 soil (Ameen et al. 2021).

Table 1 represents a part of agricultural soil microorganisms that are effected by NMs.

Impact on soil enzymes' activity

NMs have a significant influence on the activity of soil enzymes. The activity of b-glucosaminidase in soils was lowered after one hour of being exposed to Ag NMs, but rose after a week of treatment (EIVAZI et al. 2018). Furthermore, the chemical makeup of NMs plays a significant role in controlling soil enzymatic activity. ZnO NMs inhibited urease activity greater than TiO_2 or SiO_2 NMs. Enzymes of various types react differently to MNP exposure. ZnO NMs boosted soil b-glucosidase activities but had no impact on phosphatase. Citrate-coated gold nanoparticles (Au NMs) stimulated enzymatic activity more than polyvinylpyrrolidone-coated Au NMs. Varying coating materials most likely have varying surface charges, which influences their relationship with minerals from clay, SOM, and local microbes (Asadishad et al. 2018, Sun et al. 2021). Smaller AuNPs stimulated soil enzymes more strongly. Yet, at greater concentration, the size impact vanished (Asadishad et al. 2017). CuO NMs' impacts on enzymatic activity in C and N cycling in 5 soils were investigated, and the influence of CuO NMs on the enzymes were unable to reveal an evident soil texture-dependent trend. The major explanation is attributed to differences in the converted products of CuO NMs and their varying availability owing to various soil conditions (Simonin et al. 2018). NiO NMs have a

Table 1. A Short Overview of Soil Bacteria and Fungi Affected by Nanomaterials.

SI No.	Nanomaterial	Organism	References
1.	Zn	**Soil Bacteria** *Firmicutes, Proterobacteria, Actinobacteria, Verrucomicrobia, Acidobacteria, Ascomycota, Chytridiomycota*	(Fajardo et al. 2019) (Chen et al. 2014) (da Silva Júnior et al. 2022)
2.	ZnO	**Soil Bacteria** **Soil Fungi** *Mycena citricolor, Collectotrichum, Glomus versiforme, Funneliformis caledonium, Glomus caledonium, Funneliformis mosseae*	(Ge et al. 2011) (Wang et al. 2018) (Arciniegas-Grijalba et al. 2019) (Wang et al. 2016) (Wang et al. 2018)
3.	TiO$_2$	**Soil Bacteria** **Soil Fungi** AMF Rice Consortium	(Ge et al. 2011) (Priyanka et al. 2017)
4.	Ag	**Soil Bacteria** *Bradyrhizobium canariense,* *Aztobacter Vinelandii,* *Pseudomonas stutzeri,* *Nitrosomonas europea, Nitrospira multiformis,* *Nitrosococcus oceani* **Soil Fungi** *Aspergillus niger* *Trifolium repens* *Glomus aggregatum,* AMF-Tomato	(Kumar et al. 2011) (Zhang et al. 2018) (Feng et al. 2013) (Abd-Alla et al. 2016) (Noori et al. 2017) (Yang et al. 2013) (Beddow et al. 2014)
5.	Pb	**Soil Bacteria** *Firmicutes, Proterobacteria, Actinobacteria, Verrucomicrobia, Acidobacteria, Ascomycota, Chytridiomycota*	(Fajardo et al. 2019) (Chen et al. 2014)
6.	Cd	**Soil Bacteria** *Firmicutes, Proterobacteria, Actinobacteria, Verrucomicrobia, Acidobacteria, Ascomycota, Chytridiomycota*	(Fajardo et al. 2019) (Chen et al. 2014) (Wang et al. 2018)
7.	Cu	**Soil Bacteria** *Acidobacteria, Ascomycota, Chytridiomycota*	(Chen et al. 2014)
9.	Zerovalent Ag	**Soil Bacteria** *Nitrosomonas europea*	(Yuan et al. 2013)
10.	nZVI	**Soil Bacteria** *Paracoccus* sp.	(Jiang et al. 2015)
11.	Fe$_2$O$_3$	**Soil Fungi** *Trifolium repens*	(Feng et al. 2013)
12.	CuO	**Soil Bacteria** Nitrifying Bacteria, Bacteria in CN cycles	(Simonin et al. 2018) (VandeVoort and Arai 2018)
13.	Nanodiamonds	**Soil Fungi** *Phenerochaete chrysosporium*	(Ma et al. 2020)
14.	MoO$_3$	**Soil Fungi** *Aspergillus flavus,* *A. niger*	(Chaves-Lopez et al. 2018)
15.	FeO	**Soil Fungi** *Trifolium repens* *Glomus caledonium*	(Feng et al. 2013)

deleterious impact on the soil microbial population and soil enzymes such as b-glucosidase (Avila-Arias et al. 2019).

Impacts on earthworms and nematodes

Earthworms along with nematodes play a crucial role in soil health. There have been few studies on the effects of NMs on nematodes. Studies over the root knot pathogenic nematode, Meloidogyne incognita, demonstrate that tomato roots soaked in Ag NMs had substantially reduced infection, demonstrating the NMs powerful nematocidal action on *M. incognita*. A recent study employing agricultural soil found that CeO_2 NMs increased nematode counts in the soybean rhizosphere, but at higher concentrations reduced nematode development (Sun et al. 2021). Earthworms (*Eisenia fetida*) were ten times more susceptible to Ag ions than Ag NMs. The study also discovered that soil type had a greater impact on Ag NMs' toxic effects to earthworms compared to nanoparticle size. It was discovered that Ag NMs encapsulated with polyvinylpyrrolidone (hydrophilic) and oleic acid (amphiphilic) had equal earthworm toxicity. A study demonstrated that biochar amendment significantly reduced the harmful effects of CeO_2 NMs to earthworms (Sun et al. 2021). Nickel oxide NPs cause physical, ultrastructural, as well as oxidative impairment in the earthworm Eisenia fetida. Lesser quantities of NiO-NPs had no effect on mature worm survival, procreation, or development rate. Greater exposure, on the other hand, inhibited reproduction by inducing oxidative stress, which resulted in DNA damage, anomalies in the epithelial layer, microvilli, and mitochondria, as well as underlying diseases of the dermis and muscle tissue, as well as negative effects on the intestinal barrier (Adeel et al. 2019). Zero mortality was found in nematodes as well as earthworms subjected to Ag NMs (*Eisenia fetida, Enchytraeus albidus, Enchytraeus crypticus*, and *Lumbricus rubellus*). For *E. fetida*, fatality was found at extremely high concentrations along with tiny NMs (5 nm). The harmful effect of Ag NMs is most likely influenced by the size of the nanoparticles. Ionic silver proved two times as hazardous to *Eisenia andrei* as Ag NMs. Less Ag NMs is found to bioaccumulate in earthworms compared to $AgNO_3$ (Courtois et al. 2019).

Other effects of NM in soil

ZnO NMs hindered pokeweed root development (*Phytolacca Americana* L.), damaged soil acid-base equilibrium, and influenced mineral cycling by altering soil enzyme function and microbial community composition, according to a research. Acidobacteria Subgroup_10 may be utilized as an indicator for ZnONPs or Zn content in soil (Shi et al. 2021). Copper nanoparticles (NCu) might interact with other contaminants. The combined existence of the Herbicide Atrazine (ATZ) along with NCu caused significant changes in bacterial richness. In the instance of Cu NM, the ATZ's half-life became approximately 2-fold longer (Parada et al. 2019). In an outside greenhouse, the impact of TiO_2 and ZnO nanoparticles upon soil microbes and the enantioselective alteration of racemic-metalaxyl in soil used for agriculture containing or not containing *Lolium perenne* were studied. Microbial biomass carbon fell by 66%, and the makeup of the bacterial population altered dramatically. Meanwhile, ZnO reduced *L. perenne* chlorophyll accumulation by 34%. *L. perenne* increased absorbed TiO_2 and ZnO, and decreased their inhibitory influence (X. Zhang et al. 2020). A research found that adding Mn oxide nanoparticles (Mn_2O_3 NMs) to wheat shoots lowered N, P, and K levels (Dimkpa et al. 2018). ZnO NMs sprinkled on mung beans increased plant P contents by 11% (Sun et al. 2021). After C or N base modification, ZnO NMs increased the discharge of N_2O from soils. The combination of nitrogen and carbon increased total release of N_2O by 6.28–8.35 instances weighed against the standard therapy lacking ZnO NMs (Feng et al. 2021). The incorporation of MCs (modified clays), zeolite and bentonite, waste sludge, and NMs reduced the present portion of cadmium (Cd), copper (Cu), nickel (Ni), and zinc (Zn), and the researchers proposed bentonite and MgO as new, intriguing, and achievable adsorbents with distinctive properties in alleviating metal toxicity (Feizi et al. 2018, Pérez-Hernández et al. 2020). ZnO NPs boosted development, the process of photosynthesis as well as production of grain, but Cd and Zn levels dropped and rose in wheat (Triticum aestivum) shoots, roots, and grains, correspondingly. In comparison against the control, soil treatment with NPs reduced Cd contents in grain by 16–78% (Hussain et al. 2018, Pérez-Hernández et al. 2020). CeO_2-NPs caused modifications in *Brassica napus* L. (canola) cv. 'Dwarf Essex' development and anatomy, which enhanced the plant reaction to salt stress yet did not totally eliminate canola salt stress (Rossi et al. 2016). TiO_2 nanoparticles have the ability to disrupt the primary step of nitrogen cycling, nitrification, raising worries about potential cascade impacts on other vital ecosystem processes. Furthermore, TiO_2 NMs in conjunction with other minerals might have a negative impact on beneficial microbes in the soil. There were harmful impacts on wheat (*Triticum aestivum*), rice (*Oryza sativa*), and maize (*Zea mays*) (Thiagarajan and Ramasubbu 2021). Soils modified with Ca NM proved to be more crystalline, productive, and had lower salinity levels. M. oleifera showed a rise in heavy metal immobilization percentage, as well as improvements in physiological characteristics (sprouting, vigour indices, comparative hydration contents, root and shoot size, and photosynthetic performance) (Azeez et al. 2021). Recent investigations of CuO NM toxicity on numerous agricultural plants, including lettuce, alfalfa, wheat, mung bean, kidney bean, maize, cilantro, rice, spinach, onion, mustard, barley, cotton, chickpea, radish, pistia, and oak, demonstrate unfavourable effects on seed germination and development of plants (Lund 2018).

Long term effects of NM in agricultural soil on plant translocation and human consumption

Plant absorption of NMs, including particle build-up in plant tissues, particularly in edible sections (such as lettuce leaves, tomatoes fruit), is a major issue with regard to NM danger in agricultural environments. Trophic transmission and biological magnification of NMs throughout ecosystems of agriculture have been recorded, and human contact through dietary intake is possible. The build-up of NMs in plants may potentially have an impact on the nutritional value of plant-derived food. NM leaf absorption, root intake, and NM transfer to the shoots structure via the cells of the xylem or phloem occurs and thus, needs to be studied (Zhang et al. 2020).

NMs serve as carriers for other pollutants. Contaminants found in sediments, earth, or dispersed as solids within water, such as potentially poisonous elements (PTEs), elements that are radioactive, polychlorinated chemicals, and pesticides, sorb on the surface of NMs (Abbas et al. 2020). The phytotoxicity of ENMs ranges from mild to high. However, the majority of current investigations are conducted in basic or model settings (e.g., hydroponic, sand), with brief exposures to rather large and frequently ecologically unrealistic dosages. The prolonged persistence of NMs in soil is conceivable through the continual delivery of biosolids to agricultural land, as well as the prospective recurring use of nanotechnology in farming and remediation of soil situations. As a result, rather than the pure materials, the eventual ecological consequences of NMs will be primarily dictated by the changed products. Long-term effects of NMs and their changed products on the soil and plant microbiota, plant development and nutrition, trophic transmission and probable transgenerational effects, nutrient cycle, water purification, and food production must be assessed (Zhang et al. 2020). Plants grown on polluted media may represent a serious hazard to human wellness through the food chain. For instance, most major NMs, metallic oxide NMs ZnO and CuO, are extremely hazardous to a wide spectrum of species (Rajput et al. 2020).

Future studies on NMs in agricultural soil

NMs have been shown to have a detrimental influence on the development, metabolic processes, and soil advantageous characteristics of bacteria and fungus in agricultural soils. Future study should also focus on the bioavailability of NMs along with the metal ions they liberated in soils, as well as the interactions they have with the relevant microorganisms. It is also required to produce comprehensive picture of the hazardous or non-toxic impact of actual soil quantities of different NMs. Given the physical and operational toxic effects of NMs to favourable soil microbes, methods for proper elimination of metal NMs within soil agricultural ecosystems are supported, and secure by design methods for minimizing NM interaction with beneficial microbes in the soil must be developed. Agro-nanotechnology solutions ought to be tailored to specific applications (Ameen et al. 2021). Moreover, finding if the test environment evaluation is representative of a real-world scenario is necessary. To find effects of traditional and modern farming techniques on the soil's microbiota and agrarian productivity and if the soil has undergone substantial changes. To find the real effects of changes in soil microbiota on biogeochemical cycles and if there is an aberration of the cycle over time. Additionally, finding the most effective ways for preserving soil bacteria that have no negative influence on agricultural productivity (da Silva Júnior et al. 2022).

Water retention is widely recognized as an important function of soil; yet, no research has been conducted to determine if NMs alter soil water storage. The impact of NPs on soil-borne illnesses are essential missing knowledge. Because of their antibacterial capabilities, several NMs may provide unexpected advantages in disease management from soil-borne pathogenic microbes. On the other hand, it was found that NMs selectively concentrated antibiotic resistance genes within soil, which scientists find concerning. The study emphasizes the need of determining the long-term impact of NMs on bacteria with antibiotic resistance in soil (Sun et al. 2021). Cutting-edge analytical approaches must be studied and incorporated further. Because NMs' change is complicated, advanced analytical tools are necessary to explore these occurrences. Synchrotron radiation-based techniques paired with imaging techniques while micro X-ray fluorescence, isotope labelling methods combined with real time methods of detection such as autoradiography, Nano Secondary Ion Mass Spectrometry, and ICPMS are some such tools. Additionally Metabolomics, the field of proteomics exposomics, and secretomics are new potent methods for identifying the elements of exudates from roots, EPS, and plant/microbial component responses to ENMs are also included among these tools (Zhang et al. 2020).

Moreover, plant growth-promoting bacteria, often known as "plant probiotics (PPs)", have garnered broad attention. When applied to seeds, soil, or plant surfaces, PPs act as bio-elicitors, promoting plant growth and colonization of soil or plant tissues. Given their mutually advantageous qualities, PPs and NMs can be utilized in combination to optimize advantages. Nevertheless, the application of combos of NMs and PPs, or even their collaborative use, is in early stages but has shown better crop-modulating implications in regards to crop productivity enhancement, reduction of stresses from the environment (drought, salinity, etc.), soil fertility restoration, and bioeconomy strengthening. The combination can be encased within an appropriate carrier, which assists in regulated and targeted distribution while also extending the shelf life (Upadhayay et al. 2023).

Concluding remarks

Nanomaterial-based strategies and formulation provide a wide range of incredible services including nano-fertilizers, nano-pesticides, nano-carriers for individualized and regulated dispersion of agricultural chemicals, and nano-sensors configured to identify biotic/abiotic pressures in crops, which have transformed present systems of agriculture. The nano-agro-formulations are designed to promote crop development and protection while minimizing waste and contamination. The growth stimulants that utilize NMs have a chance to be more effective than their traditional analogs (Ameen et al. 2021, Fraceto et al. 2016). However, agricultural practices utilizing NMs is heavily reliant on the suitability of the links among nanotechnology and land use (de Oliveira et al. 2014, Salamanca-Buentello et al. 2005). NMs get released into agricultural soil, both purposefully and inadvertently (such as biosolid administration to soil as manure), and can have either good or negative consequences on the ecosystems of agriculture. Moreover, their dynamic transformation makes it hard to pinpoint exact ecological consequence of a specific NM. Owing to both the "obtained" and "real" identities, these dynamic changes challenge our understanding of NM destiny and influence (Abbas et al. 2020). Finally, NPs put into the ecosystem in this manner aggregate in soil over time and impact soil properties for local populations. As a result, it is frequently suggested that this integrated method, i.e., agri-nanotechnology, has various drawbacks in each area, involving bioavailability, transport and NM toxicity. Within the current setting, agro-researchers are attempting to address gaps in existing understanding of agri-nanotechnologies by resolving debates concerning the interface of NMs with critical mechanisms of the agro-ecosystem which include soil, plant and soil biota (Ameen et al. 2021, Simonin and Richaume 2015). More studies need to be done with a holistic approach of life cycle assessment of NM released in agricultural fields utilizing cutting edge modern techniques and realistic test media with materially possible levels of NM.

References

Abbas, Q., Yousaf, B., Amina, Ali, M.U., Munir, M.A.M., El-Naggar, A., Rinklebe, J. et al. 2020. Transformation pathways and fate of engineered nanoparticles (ENPs) in distinct interactive environmental compartments: A review. Environment International, 138: 105646. https://doi.org/https://doi.org/10.1016/j.envint.2020.105646.

Abd-Alla, M.H., Nafady, N.A., and Khalaf, D.M. 2016. Assessment of silver nanoparticles contamination on faba bean-Rhizobium leguminosarum bv. viciae-Glomus aggregatum symbiosis: Implications for induction of autophagy process in root nodule. Agriculture, Ecosystems & Environment, 218: 163–177. https://doi.org/https://doi.org/10.1016/j.agee.2015.11.022.

Acharya, D., Singha, K.M., Pandey, P., Mohanta, B., Rajkumari, J., and Singha, L.P. 2018. Shape dependent physical mutilation and lethal effects of silver nanoparticles on bacteria. Scientific Reports, 8(1): 201. https://doi.org/10.1038/s41598-017-18590-6.

Adeel, M., Ma, C., Ullah, S., Rizwan, M., Hao, Y., Chen, C. et al. 2019. Exposure to nickel oxide nanoparticles insinuates physiological, ultrastructural and oxidative damage: A life cycle study on *Eisenia fetida*. Environmental Pollution, 254. https://doi.org/10.1016/j.envpol.2019.113032.

Adisa, I.O., Pullagurala, V.L.R., Peralta-Videa, J.R., Dimkpa, C.O., Elmer, W.H., Gardea-Torresdey, J.L. et al. 2019. Recent advances in nano-enabled fertilizers and pesticides: a critical review of mechanisms of action. Environ. Sci.: Nano, 6(7): 2002–2030. https://doi.org/10.1039/C9EN00265K.

Ahmed, B., Ameen, F., Rizvi, A., Ali, K., Sonbol, H., Zaidi, A. et al. 2020. Destruction of cell topography, morphology, membrane, inhibition of respiration, biofilm formation, and bioactive molecule production by nanoparticles of Ag, ZnO, CuO, TiO₂, and Al2O₃ toward Beneficial Soil Bacteria. ACS Omega, 5(14): 7861–7876. https://doi.org/10.1021/acsomega.9b04084.

Ali, M., Song, X., Ding, D., Wang, Q., Zhang, Z., and Tang, Z. 2022. Bioremediation of PAHs and heavy metals co-contaminated soils: Challenges and enhancement strategies. Environmental Pollution, 295: 118686. https://doi.org/https://doi.org/10.1016/j.envpol.2021.118686.

Amde, M., Liu, J., Tan, Z.-Q., and Bekana, D. 2017. Transformation and bioavailability of metal oxide nanoparticles in aquatic and terrestrial environments. A review. Environmental Pollution, 230: 250–267. https://doi.org/https://doi.org/10.1016/j.envpol.2017.06.064.

Ameen, F., Alsamhary, K., Alabdullatif, J.A., and ALNadhari, S. 2021. A review on metal-based nanoparticles and their toxicity to beneficial soil bacteria and fungi. Ecotoxicology and Environmental Safety, 213: 112027. https://doi.org/10.1016/j.ecoenv.2021.112027.

Andries, M., Pricop, D., Oprica, L., Creanga, D.-E., and Iacomi, F. 2016. The effect of visible light on gold nanoparticles and some bioeffects on environmental fungi. International Journal of Pharmaceutics, 505(1): 255–261. https://doi.org/https://doi.org/10.1016/j.ijpharm.2016.04.004.

Arciniegas-Grijalba, P.A., Patiño-Portela, M.C., Mosquera-Sánchez, L.P., Guerra Sierra, B.E., Muñoz-Florez, J.E., Erazo-Castillo, L.A. et al. 2019. ZnO-based nanofungicides: Synthesis, characterization and their effect on the coffee fungi *Mycena citricolor* and *Colletotrichum* sp. Materials Science and Engineering: C, 98: 808–825. https://doi.org/https://doi.org/10.1016/j.msec.2019.01.031.

Arjunan, N., Singaravelu, C.M., Kulanthaivel, J., and Kandasamy, J. 2017. A potential photocatalytic, antimicrobial and anticancer activity of chitosan-copper nanocomposite. International Journal of Biological Macromolecules, 104: 1774–1782. https://doi.org/https://doi.org/10.1016/j.ijbiomac.2017.03.006.

Asadishad, B., Chahal, S., Akbari, A., Cianciarelli, V., Azodi, M., Ghoshal, S. et al. 2018. Amendment of agricultural soil with metal nanoparticles: effects on soil enzyme activity and microbial community composition. Environmental Science \& Technology, 52(4): 1908–1918. https://doi.org/10.1021/acs.est.7b05389.

Asadishad, B., Chahal, S., Cianciarelli, V., Zhou, K., and Tufenkji, N. 2017. Effect of gold nanoparticles on extracellular nutrient-cycling enzyme activity and bacterial community in soil slurries: role of nanoparticle size and surface coating. Environ. Sci.: Nano, 4(4): 907–918. https://doi.org/10.1039/C6EN00567E.

Avidano, L., Gamalero, E., Cossa, G.P., and Carraro, E. 2005. Characterization of soil health in an Italian polluted site by using microorganisms as bioindicators. Applied Soil Ecology, 30(1): 21–33. https://doi.org/https://doi.org/10.1016/j.apsoil.2005.01.003.

Avila-Arias, H., Nies, L.F., Gray, M.B., and Turco, R.F. 2019. Impacts of molybdenum-, nickel-, and lithium-oxide nanomaterials on soil activity and microbial community structure. Science of The Total Environment, 652: 202–211. https://doi.org/https://doi.org/10.1016/j.scitotenv.2018.10.189.

Azeez, L., Lateef, A., Adetoro, R.O., and Adeleke, A.E. 2021. Responses of *Moringa oleifera* to alteration in soil properties induced by calcium nanoparticles (CaNPs) on mineral absorption, physiological indices and photosynthetic indicators. Beni-Suef University Journal of Basic and Applied Sciences, 10(1). https://doi.org/10.1186/s43088-021-00128-5.

Azimzada, A., Tufenkji, N., and Wilkinson, K.J. 2017. Transformations of silver nanoparticles in wastewater effluents: links to Ag bioavailability. Environ. Sci.: Nano., 4(6): 1339–1349. https://doi.org/10.1039/C7EN00093F.

Bakshi, M., and Kumar, A. 2021. Copper-based nanoparticles in the soil-plant environment: Assessing their applications, interactions, fate and toxicity. Chemosphere, 281(May): 130940. https://doi.org/10.1016/j.chemosphere.2021.130940.

Bakshi, S., He, Z.L., and Harris, W.G. 2015. Natural nanoparticles: Implications for environment and human health. Critical Reviews in Environmental Science and Technology, 45(8): 861–904. https://doi.org/10.1080/10643389.2014.921975.

Bandyopadhyay, S., Plascencia-Villa, G., Mukherjee, A., Rico, C.M., José-Yacamán, M., Peralta-Videa, J.R. et al. 2015. Comparative phytotoxicity of ZnO NPs, bulk ZnO, and ionic zinc onto the alfalfa plants symbiotically associated with *Sinorhizobium meliloti* in soil. Science of The Total Environment, 515–516: 60–69. https://doi.org/https://doi.org/10.1016/j.scitotenv.2015.02.014.

Bao, S., Xiang, D., Xue, L., Xian, B., Tang, W., and Fang, T. 2022. Pristine and sulfidized ZnO nanoparticles alter microbial community structure and nitrogen cycling in freshwater lakes. Environmental Pollution, 294: 118661. https://doi.org/https://doi.org/10.1016/j.envpol.2021.118661.

Beddow, J., Stolpe, B., Cole, P., Lead, J.R., Sapp, M., Lyons, B.P. et al. 2014. Effects of engineered silver nanoparticles on the growth and activity of ecologically important microbes. Environmental Microbiology Reports, 6(5): 448–458. https://doi.org/https://doi.org/10.1111/1758-2229.12147.

Begum, N., Qin, C., Ahanger, M.A., Raza, S., Khan, M.I., Ashraf, M. et al. 2019. Role of *Arbuscular mycorrhizal* fungi in plant growth regulation: implications in abiotic stress tolerance. Frontiers in Plant Science, 10(September): 1–15. https://doi.org/10.3389/fpls.2019.01068.

Ben-Moshe, T., Frenk, S., Dror, I., Minz, D., and Berkowitz, B. 2013. Effects of metal oxide nanoparticles on soil properties. Chemosphere, 90(2): 640–646. https://doi.org/https://doi.org/10.1016/j.chemosphere.2012.09.018.

Bhatt, I., and Tripathi, B.N. 2011. Interaction of engineered nanoparticles with various components of the environment and possible strategies for their risk assessment. Chemosphere, 82(3): 308–317. https://doi.org/https://doi.org/10.1016/j.chemosphere.2010.10.011.

Blinova, I., Lukjanova, A., Reinik, J., and Kahru, A. 2021. Concentration of lanthanides in the Estonian environment: a screening study. Journal of Hazardous Materials Advances, 4: 100034. https://doi.org/https://doi.org/10.1016/j.hazadv.2021.100034.

Boyes, W.K. 2018. 9.05 - Safety assessment of nanotechnology products. pp. 34–43. *In:* McQueen, C.A. (Ed.). Comprehensive Toxicology (Third Edition) (Third Edit). Elsevier. https://doi.org/https://doi.org/10.1016/B978-0-12-801238-3.99179-7.

Bratovcic, A., Hikal, W.M., Said-Al Ahl, H.A.H., Tkachenko, K.G., Baeshen, R.S., Sabra, A.S. et al. 2021. Nanopesticides and nanofertilizers and agricultural development: scopes, advances and applications. Open Journal of Ecology, 11(04): 301–316. https://doi.org/10.4236/oje.2021.114022.

Bundschuh, M., Filser, J., Lüderwald, S., McKee, M.S., Metreveli, G., Schaumann, G.E. et al. 2018. Nanoparticles in the environment: where do we come from, where do we go to? Environmental Sciences Europe, 30(1): 6. https://doi.org/10.1186/s12302-018-0132-6.

Cao, C., Huang, J., Yan, C., Liu, J., Hu, Q., and Guan, W. 2018. Shifts of system performance and microbial community structure in a constructed wetland after exposing silver nanoparticles. Chemosphere, 199: 661–669. https://doi.org/https://doi.org/10.1016/j.chemosphere.2018.02.031.

Cao, J., Feng, Y., He, S., and Lin, X. 2017. Silver nanoparticles deteriorate the mutual interaction between maize (*Zea mays* L.) and arbuscular mycorrhizal fungi: a soil microcosm study. Applied Soil Ecology, 119: 307–316. https://doi.org/https://doi.org/10.1016/j.apsoil.2017.04.011.

Chaves-Lopez, C., Nguyen, H.N., Oliveira, R.C., Nadres, E.T., Paparella, A., and Rodrigues, D.F. 2018. A morphological{,} enzymatic and metabolic approach to elucidate apoptotic-like cell death in fungi exposed to h- and α-molybdenum trioxide nanoparticles. Nanoscale, 10(44): 20702–20716. https://doi.org/10.1039/C8NR06470A.

Chen, H. 2018. Metal based nanoparticles in agricultural system: Behavior, transport, and interaction with plants. Chemical Speciation and Bioavailability, 30(1): 123–134. https://doi.org/10.1080/09542299.2018.1520050.

Chen, J., He, F., Zhang, X., Sun, X., Zheng, J., and Zheng, J. 2014. Heavy metal pollution decreases microbial abundance, diversity and activity within particle-size fractions of a paddy soil. FEMS Microbiology Ecology, 87(1): 164–181. https://doi.org/10.1111/1574-6941.12212.

Chhipa, H. 2017. Nanofertilizers and nanopesticides for agriculture. Environmental Chemistry Letters, 15(1): 15–22. https://doi.org/10.1007/s10311-016-0600-4.

Choi, O., Deng, K.K., Kim, N.-J., Ross, L., Surampalli, R.Y., and Hu, Z. 2008. The inhibitory effects of silver nanoparticles, silver ions, and silver chloride colloids on microbial growth. Water Research, 42(12): 3066–3074. https://doi.org/https://doi.org/10.1016/j.watres.2008.02.021.

Concha-Guerrero, S.I., Brito, E.M.S., Piñón-Castillo, H.A., Tarango-Rivero, S.H., Caretta, C.A. et al. 2014. Effect of CuO nanoparticles over isolated bacterial strains from agricultural soil. Journal of Nanomaterials, 2014. https://doi.org/10.1155/2014/148743.

Cornelis, G., Hund-Rinke, K., Kuhlbusch, T., Van Den Brink, N., and Nickel, C. 2014. Fate and bioavailability of engineered nanoparticles in soils: a review, Interactions within natural soils have often been neglected when assessing fate and bioavailability of engineered nanomaterials (ENM) in soils. This review combines patch wise ENM resea. Critical Reviews in Environmental Science and Technology, 44(24): 2720–2764. https://doi.org/10.1080/10643389.2013.829767.

Courtois, P., Rorat, A., Lemiere, S., Guyoneaud, R., Attard, E., Levard, C. et al. 2019. Ecotoxicology of silver nanoparticles and their derivatives introduced in soil with or without sewage sludge: A review of effects on microorganisms, plants and animals. Environmental Pollution, 253: 578–598. https://doi.org/10.1016/j.envpol.2019.07.053.

Cullen, L.G., Tilston, E.L., Mitchell, G.R., Collins, C.D., and Shaw, L.J. 2011. Assessing the impact of nano- and micro-scale zerovalent iron particles on soil microbial activities: Particle reactivity interferes with assay conditions and interpretation of genuine microbial effects. Chemosphere, 82(11): 1675–1682. https://doi.org/https://doi.org/10.1016/j.chemosphere.2010.11.009.

da Silva Júnior, A.H., Mulinari, J., de Oliveira, P.V., de Oliveira, C.R.S., and Reichert Júnior, F.W. 2022. Impacts of metallic nanoparticles application on the agricultural soils microbiota. Journal of Hazardous Materials Advances, 7(April): 100103. https://doi.org/10.1016/j.hazadv.2022.100103.

Das, M., and Chatterjee, S. 2019. Chapter 11 - Green synthesis of metal/metal oxide nanoparticles toward biomedical applications: Boon or bane. pp. 265–301. *In*: Shukla, A.K., and Iravani, S. (Eds.). Green Synthesis, Characterization and Applications of Nanoparticles. Elsevier. https://doi.org/https://doi.org/10.1016/B978-0-08-102579-6.00011-3.

de Oliveira, C.R.S., da Silva Júnior, A.H., Immich, A.P.S., and Fiates, J. 2022. Use of advanced materials in smart textile manufacturing. Materials Letters, 316: 132047. https://doi.org/https://doi.org/10.1016/j.matlet.2022.132047.

de Oliveira, J.L., Campos, E.V.R., Bakshi, M., Abhilash, P.C., and Fraceto, L.F. 2014. Application of nanotechnology for the encapsulation of botanical insecticides for sustainable agriculture: Prospects and promises. Biotechnology Advances, 32(8): 1550–1561. https://doi.org/https://doi.org/10.1016/j.biotechadv.2014.10.010.

De Souza, A., Govea-Alcaide, E., Masunaga, S.H., Fajardo-Rosabal, L., Effenberger, F., Rossi, L.M. et al. 2019. Impact of Fe_3O_4 nanoparticle on nutrient accumulation in common bean plants grown in soil. SN Applied Sciences, 1(4): 308. https://doi.org/10.1007/s42452-019-0321-y.

Delfani, M., Firouzabadi, M.B., Farrokhi, N., and Makarian, H. 2014. Some physiological responses of black-eyed pea to iron and magnesium nanofertilizers. Communications in Soil Science and Plant Analysis, 45(4): 530–540. https://doi.org/10.1080/00103624.2013.863911.

Dimkpa, C.O., Calder, A., Britt, D.W., McLean, J.E., and Anderson, A.J. 2011. Responses of a soil bacterium, *Pseudomonas chlororaphis* O6 to commercial metal oxide nanoparticles compared with responses to metal ions. Environmental Pollution, 159(7): 1749–1756. https://doi.org/https://doi.org/10.1016/j.envpol.2011.04.020.

Dimkpa, C.O., Singh, U., Adisa, I.O., Bindraban, P.S., Elmer, W.H., Gardea-Torresdey, J.L. et al. 2018. Effects of Manganese nanoparticle exposure on nutrient acquisition in wheat (*Triticum aestivum* L.). Agronomy, 8(9). https://doi.org/10.3390/agronomy8090158.

Dimkpa, C.O., White, J.C., Elmer, W.II., and Gardea-Torresdey, J. 2017. Nanoparticle and Ionic Zn promote nutrient loading of sorghum grain under low NPK fertilization. Journal of Agricultural and Food Chemistry, 65(39): 8552–8559. https://doi.org/10.1021/acs.jafc.7b02961.

Du, J., Zhang, Y., Yin, Y., Zhang, J., Ma, H., Li, K. et al. 2020. Do environmental concentrations of zinc oxide nanoparticle pose ecotoxicological risk to aquatic fungi associated with leaf litter decomposition? Water Research, 178: 115840. https://doi.org/https://doi.org/10.1016/j.watres.2020.115840.

Duhan, J.S., Kumar, R., Kumar, N., Kaur, P., Nehra, K., and Duhan, S. 2017. Nanotechnology: The new perspective in precision agriculture. Biotechnology Reports, 15: 11–23. https://doi.org/https://doi.org/10.1016/j.btre.2017.03.002.

Dutta, D., Goel, S., and Kumar, S. 2022. Health risk assessment for exposure to heavy metals in soils in and around E-waste dumping site. Journal of Environmental Chemical Engineering, 10(2): 107269. https://doi.org/https://doi.org/10.1016/j.jece.2022.107269.

Dwivedi, S., Saquib, Q., Al-Khedhairy, A.A., and Musarrat, J. 2016. Understanding the role of nanomaterials in agriculture. pp. 271–288. *In*: Singh, D.P., Singh, H.B., and Prabha, R. (Eds.). Microbial Inoculants in Sustainable Agricultural Productivity: Vol. 2: Functional Applications. Springer India. https://doi.org/10.1007/978-81-322-2644-4_17.

Eduok, S., and Coulon, F. 2017. Engineered nanoparticles in the environments: interactions with microbial systems and microbial activity. pp. 63–107. *In*: Cravo-Laureau, C., Cagnon, C., Lauga, B., and Duran, R. (Eds.). Microbial Ecotoxicology. Springer International Publishing. https://doi.org/10.1007/978-3-319-61795-4_5.

Eivazi, F., Afrasiabi, Z., and Jose, E. 2018. Effects of silver nanoparticles on the activities of soil enzymes involved in carbon and nutrient cycling. Pedosphere, 28(2): 209–214. https://doi.org/https://doi.org/10.1016/S1002-0160(18)60019-0.

El-Ramady, H., Brevik, E.C., Fawzy, Z.F., Elsakhawy, T., Omara, A.E.-D., Amer, M. et al. 2022. Nano-restoration for sustaining soil fertility: a pictorial and diagrammatic review article. Plants, 11(18). https://doi.org/10.3390/plants11182392.

Fajardo, C., Costa, G., Nande, M., Botías, P., García-Cantalejo, J., and Martín, M. 2019. Pb, Cd, and Zn soil contamination: Monitoring functional and structural impacts on the microbiome. Applied Soil Ecology, 135: 56–64. https://doi.org/https://doi.org/10.1016/j.apsoil.2018.10.022.

Fan, R., Huang, Y.C., Grusak, M.A., Huang, C.P., and Sherrier, D.J. 2014. Effects of nano-TiO$_2$ on the agronomically-relevant Rhizobium–legume symbiosis. Science of The Total Environment, 466-467: 503–512. https://doi.org/https://doi.org/10.1016/j.scitotenv.2013.07.032.

Fayiga, A. 2017. Nanoparticles in biosolids: effect on soil health and crop growth. Peertechz Journal of Environmental Science and Toxicology, 2: 059–067. https://doi.org/10.17352/pjest.000013.

Feizi, M., Jalali, M., and Renella, G. 2018. Nanoparticles and modified clays influenced distribution of heavy metals fractions in a light-textured soil amended with sewage sludges. Journal of Hazardous Materials, 343: 208–219. https://doi.org/https://doi.org/10.1016/j.jhazmat.2017.09.027.

Feng, Y., Cui, X., He, S., Dong, G., Chen, M., Wang, J. et al. 2013. The Role of Metal Nanoparticles in Influencing Arbuscular Mycorrhizal Fungi Effects on Plant Growth. Environmental Science & Technology, 47(16): 9496–9504. https://doi.org/10.1021/es402109n.

Feng, Z., Yu, Y., Yao, H., and Ge, C. 2021. Effect of Zinc oxide nanoparticles on nitrous oxide emissions in agricultural soil. Agriculture, 11(8). https://doi.org/10.3390/agriculture11080730.

Fraceto, L.F., Grillo, R., de Medeiros, G.A., Scognamiglio, V., Rea, G., and Bartolucci, C. 2016. Nanotechnology in agriculture: Which innovation potential does it have? Frontiers in Environmental Science, 4(MAR): 1–5. https://doi.org/10.3389/fenvs.2016.00020.

Frazier, T.P., Burklew, C.E., and Zhang, B. 2014. Titanium dioxide nanoparticles affect the growth and microRNA expression of tobacco (*Nicotiana tabacum*). Functional & Integrative Genomics, 14(1): 75–83. https://doi.org/10.1007/s10142-013-0341-4.

Gambino, M., Marzano, V., Villa, F., Vitali, A., Vannini, C., Landini, P. et al. 2015. Effects of sublethal doses of silver nanoparticles on Bacillus subtilis planktonic and sessile cells. Journal of Applied Microbiology, 118(5): 1103–1115. https://doi.org/10.1111/jam.12779.

Gao, X., Avellan, A., Laughton, S., Vaidya, R., Rodrigues, S.M., Casman, E.A. et al. 2018. CuO nanoparticle dissolution and toxicity to wheat (*Triticum aestivum*) in rhizosphere soil. Environmental Science & Technology, 52(5): 2888–2897. https://doi.org/10.1021/acs.est.7b05816.

García-Saucedo, C., Field, J.A., Otero-Gonzalez, L., and Sierra-Álvarez, R. 2011. Low toxicity of HfO_2, SiO_2, Al_2O_3 and CeO_2 nanoparticles to the yeast, *Saccharomyces cerevisiae*. Journal of Hazardous Materials, 192(3): 1572–1579. https://doi.org/https://doi.org/10.1016/j.jhazmat.2011.06.081.

Ge, Y., Schimel, J.P., and Holden, P.A. 2011. Evidence for negative effects of TiO_2 and ZnO nanoparticles on soil bacterial communities. Environmental Science & Technology, 45(4): 1659–1664. https://doi.org/10.1021/es103040t.

Geert Cornelis Kerstin Hund-Rinke, T.K.N. van den B., and Nickel, C. 2014. Fate and Bioavailability of Engineered Nanoparticles in Soils: A Review. Critical Reviews in Environmental Science and Technology, 44(24): 2720–2764. https://doi.org/10.1080/10643389.2013.829767.

George, T.S., Dou, D., and Wang, X. 2016. Plant–microbe interactions: manipulating signals to enhance agricultural sustainability and environmental security. Plant Growth Regulation, 80(1): 1–3. https://doi.org/10.1007/s10725-016-0174-y.

Giannousi, K., Avramidis, I., and Dendrinou-Samara, C. 2013. Synthesis{,} characterization and evaluation of copper based nanoparticles as agrochemicals against *Phytophthora infestans*. RSC Adv., 3(44): 21743–21752. https://doi.org/10.1039/C3RA42118J.

Giraldo, J.P., Wu, H., Newkirk, G.M., and Kruss, S. 2019. Nanobiotechnology approaches for engineering smart plant sensors. Nature Nanotechnology, 14(6): 541–553. https://doi.org/10.1038/s41565-019-0470-6.

González-Gálvez, D., Janer, G., Vilar, G., Vílchez, A., and Vázquez-Campos, S. 2017. The life cycle of engineered nanoparticles. pp. 41–69. *In*: Tran, L., M.A. Bañares, and R. Rallo (Eds.). Modelling the Toxicity of Nanoparticles. Springer International Publishing. https://doi.org/10.1007/978-3-319-47754-1_3.

Hartmann, T., Mühling, M., Wolf, A., Mariana, F., Maskow, T., Mertens, F. et al. 2013. A chip-calorimetric approach to the analysis of Ag nanoparticle caused inhibition and inactivation of beads-grown bacterial biofilms. Journal of Microbiological Methods, 95(2): 129–137. https://doi.org/https://doi.org/10.1016/j.mimet.2013.08.003.

Hennebert, P., Avellan, A., Yan, J., and Aguerre-Chariol, O. 2013. Experimental evidence of colloids and nanoparticles presence from 25 waste leachates. Waste Management, 33(9): 1870–1881. https://doi.org/https://doi.org/10.1016/j.wasman.2013.04.014.

Hussain, A., Ali, S., Rizwan, M., Zia ur Rehman, M., Javed, M.R., Imran, M. et al. 2018. Zinc oxide nanoparticles alter the wheat physiological response and reduce the cadmium uptake by plants. Environmental Pollution, 242: 1518–1526. https://doi.org/10.1016/j.envpol.2018.08.036.

Ingale, A.G., and Chaudhari, A.N. 2013. Biogenic synthesis of nanoparticles and potential applications: An eco-friendly approach. Journal of Nanomedicine and Nanotechnology, 4(2): 7. https://doi.org/10.4172/2157-. Journal of Nanomedicine and Nanotechnology, 4(2): 7. https://doi.org/10.4172/2157-7439.1000165.

Javed, Z., Dashora, K., Mishra, M., Fasake, V.D., and Srivastva, A. 2019. Effect of accumulation of nanoparticles in soil health- a concern on future. Frontiers in Nanoscience and Nanotechnology, 5(2). https://doi.org/10.15761/fnn.1000181.

Jiang, C., Xu, X., Megharaj, M., Naidu, R., and Chen, Z. 2015. Inhibition or promotion of biodegradation of nitrate by *Paracoccus* sp. in the presence of nanoscale zero-valent iron. Science of The Total Environment, 530-531: 241–246. https://doi.org/https://doi.org/10.1016/j.scitotenv.2015.05.044.

Jiang, M., Qi, Y., Liu, H., and Chen, Y. 2018. The role of nanomaterials and nanotechnologies in wastewater treatment: a bibliometric analysis. Nanoscale Research Letters, 13(1): 233. https://doi.org/10.1186/s11671-018-2649-4.

Jiang, W., Mashayekhi, H., and Xing, B. 2009. Bacterial toxicity comparison between nano- and micro-scaled oxide particles. Environmental Pollution, 157(5): 1619–1625. https://doi.org/https://doi.org/10.1016/j.envpol.2008.12.025.

Jośko, I., Oleszczuk, P., Dobrzyńska, J., Futa, B., Joniec, J., and Dobrowolski, R. 2019. Long-term effect of ZnO and CuO nanoparticles on soil microbial community in different types of soil. Geoderma, 352: 204–212. https://doi.org/https://doi.org/10.1016/j.geoderma.2019.06.010.

Ju-Nam, Y., and Lead, J. 2016. Part II Environmental release, processes, and modeling of engineered nanoparticles. Engineered Nanoparticles and the Environment: Biophysicochemical Processes and Toxicity, 95–117. https://www.researchgate.net/publication/308544177_Properties_Sources_Pathways_and_Fate_of_Nanoparticles_in_the_Environment_Biophysicochemical_Processes_and_Toxicity%0Ahttps://onlinelibrary.wiley.com/doi/book/10.1002/9781119275855.

Kaegi, R., Voegelin, A., Ort, C., Sinnet, B., Thalmann, B., Krismer, J. et al. 2013. Fate and transformation of silver nanoparticles in urban wastewater systems. Water Research, 47(12): 3866–3877. https://doi.org/https://doi.org/10.1016/j.watres.2012.11.060.

Kah, M., Kookana, R.S., Gogos, A., and Bucheli, T.D. 2018. A critical evaluation of nanopesticides and nanofertilizers against their conventional analogues. Nature Nanotechnology, 13(8): 677–684. https://doi.org/10.1038/s41565-018-0131-1.

Kamran, M., Ali, H., Saeed, M.F., Bakhat, H.F., Hassan, Z., Tahir, M. et al. 2020. Unraveling the toxic effects of iron oxide nanoparticles on nitrogen cycling through manure-soil-plant continuum. Ecotoxicology and Environmental Safety, 205: 111099. https://doi.org/https://doi.org/10.1016/j.ecoenv.2020.111099.

Kannan, M., Bojan, N., Swaminathan, J., Zicarelli, G., Hemalatha, D., Zhang, Y. et al. 2023. Nanopesticides in agricultural pest management and their environmental risks: a review. International Journal of Environmental Science and Technology, 20(9): 10507–10532. https://doi.org/10.1007/s13762-023-04795-y.

Kasemets, K., Ivask, A., Dubourguier, H.-C., and Kahru, A. 2009. Toxicity of nanoparticles of ZnO, CuO and TiO_2 to yeast *Saccharomyces cerevisiae*. Toxicology *in Vitro*, 23(6): 1116–1122. https://doi.org/https://doi.org/10.1016/j.tiv.2009.05.015.

Keller, A.A., and Lazareva, A. 2014. Predicted releases of engineered nanomaterials: from global to regional to local. Environmental Science & Technology Letters, 1(1): 65–70. https://doi.org/10.1021/ez400106t.

Keller, A.A., McFerran, S., Lazareva, A., and Suh, S. 2013. Global life cycle releases of engineered nanomaterials. Journal of Nanoparticle Research, 15(6): 1692. https://doi.org/10.1007/s11051-013-1692-4.

Kiser, M.A., Ladner, D.A., Hristovski, K.D., and Westerhoff, P.K. 2012. Nanomaterial transformation and association with fresh and freeze-dried wastewater activated sludge: implications for testing protocol and environmental fate. Environmental Science & Technology, 46(13): 7046–7053. https://doi.org/10.1021/es300339x.

Kumar, N., Shah, V., and Walker, V.K. 2011. Perturbation of an arctic soil microbial community by metal nanoparticles. Journal of Hazardous Materials, 190(1): 816–822. https://doi.org/https://doi.org/10.1016/j.jhazmat.2011.04.005.

Lead, J.R., Batley, G.E., Alvarez, P.J.J., Croteau, M.N., Handy, R.D., McLaughlin, M.J. et al. 2018. Nanomaterials in the environment: Behavior, fate, bioavailability, and effects—An updated review. Environmental Toxicology and Chemistry, 37(8): 2029–2063. https://doi.org/10.1002/etc.4147.

Lecoanet, H.F., and Wiesner, M.R. 2004. Velocity effects on fullerene and oxide nanoparticle deposition in *Porous Media*. Environmental Science & Technology, 38(16): 4377–4382. https://doi.org/10.1021/es035354f.

Lee, J.-H., Kim, Y.-G., Cho, M.H., and Lee, J. 2014. ZnO nanoparticles inhibit *Pseudomonas aeruginosa* biofilm formation and virulence factor production. Microbiological Research, 169(12): 888–896. https://doi.org/https://doi.org/10.1016/j.micres.2014.05.005.

Lewis, R.W., Unrine, J., Bertsch, P.M., and McNear David, H.J. 2017. Silver engineered nanomaterials and ions elicit species-specific O2 consumption responses in plant growth promoting rhizobacteria. Biointerphases, 12(5): 05G604. https://doi.org/10.1116/1.4995605.

Lund, J.E. 2018. Soil contamination: Sources, assessment and remediation. Soil Contamination: Sources, Assessment and Remediation, 1–189.

Ma, Q., Zhang, Q., Yang, S., Yilihamu, A., Shi, M., Ouyang, B. et al. 2020. Toxicity of nanodiamonds to white rot fungi *Phanerochaete chrysosporium* through oxidative stress. Colloids and Surfaces B: Biointerfaces, 187: 110658. https://doi.org/https://doi.org/10.1016/j.colsurfb.2019.110658.

Mesa-Marín, J., Del-Saz, N.F., Rodríguez-Llorente, I.D., Redondo-Gómez, S., Pajuelo, E., Ribas-Carbó, M. et al. 2018. PGPR reduce root respiration and oxidative stress enhancing spartina maritima root growth and heavy metal rhizoaccumulation. Frontiers in Plant Science, 871(October): 1–10. https://doi.org/10.3389/fpls.2018.01500.

Mitrano, D.M., Limpiteeprakan, P., Babel, S., and Nowack, B. 2016. Durability of nano-enhanced textiles through the life cycle: releases from landfilling after washing. Environ. Sci.: Nano, 3(2): 375–387. https://doi.org/10.1039/C6EN00023A.

Moll, J., Okupnik, A., Gogos, A., Knauer, K., Bucheli, T.D., van der Heijden, M.G.A. et al. 2016. Effects of titanium dioxide nanoparticles on red clover and its rhizobial symbiont. PLOS ONE, 11(5): 1–15. https://doi.org/10.1371/journal.pone.0155111.

Mueller, N.C., and Nowack, B. 2010. Nanoparticles for remediation: solving big problems with little particles. Elements, 6(6): 395–400. https://doi.org/10.2113/gselements.6.6.395.

Mukherjee, A., Majumdar, S., Servin, A.D., Pagano, L., Dhankher, O.P., and White, J.C. 2016. Carbon nanomaterials in agriculture: A critical review. Frontiers in Plant Science, 7(FEB2016): 1–16. https://doi.org/10.3389/fpls.2016.00172.

Mushtaq, A., Jamil, N., Rizwan, S., Mandokhel, F., Riaz, M., Hornyak, G.L. et al. 2018. Engineered silica nanoparticles and silica nanoparticles containing controlled release fertilizer for drought and saline areas. IOP Conference Series: Materials Science and Engineering, 414(1). https://doi.org/10.1088/1757-899X/414/1/012029.

Nafisi, S., and Maibach, H.I. 2017. Chapter 22 - nanotechnology in cosmetics. pp. 337–369. *In*: Sakamoto, K., Lochhead, R.Y., Maibach, H.I., and Yamashita, Y. (Eds.). Cosmetic Science and Technology. Elsevier. https://doi.org/https://doi.org/10.1016/B978-0-12-802005-0.00022-7.

Neal, A.L., Kabengi, N., Grider, A., and Bertsch, P.M. 2011. Can the soil bacterium *Cupriavidus necator* sense ZnO nanomaterials and aqueous Zn2+differentially? Nanotoxicology, 6(4): 371–380. https://doi.org/10.3109/17435 390.2011.579633.

Nepal, J., Xin, X., Maltais-Landry, G., Wright, A.L., Stoffella, P.J., Ahmad, W. et al. 2023. Water-dispersible carbon nanomaterials improve lettuce (*Latuca sativa*) growth and enhance soil biochemical quality at low to medium application rates. Plant and Soil, 485(1-2): 569–587. https://doi.org/10.1007/s11104-022-05852-0.

Nielsen, U.N., Wall, D.H., and Six, J. 2015. Soil biodiversity and the environment. Annual Review of Environment and Resources, 40(1): 63–90. https://doi.org/10.1146/annurev-environ-102014-021257.

Noori, A., White, J.C., and Newman, L.A. 2017. Mycorrhizal fungi influence on silver uptake and membrane protein gene expression following silver nanoparticle exposure. Journal of Nanoparticle Research, 19(2): 66. https://doi.org/10.1007/s11051-016-3650-4.

Nowack, B., Baalousha, M., Bornhöft, N., Chaudhry, Q., Cornelis, G., Cotterill, J. et al. 2015. Progress towards the validation of modeled environmental concentrations of engineered nanomaterials by analytical measurements. Environ. Sci.: Nano, 2(5): 421–428. https://doi.org/10.1039/C5EN00100E.

Otero-González, L., García-Saucedo, C., Field, J.A., and Sierra-Álvarez, R. 2013. Toxicity of TiO2, ZrO2, Fe0, Fe2O3, and Mn2O3 nanoparticles to the yeast, *Saccharomyces cerevisiae*. Chemosphere, 93(6): 1201–1206. https://doi.org/https://doi.org/10.1016/j.chemosphere.2013.06.075.

Ottofuelling, S., Von Der Kammer, F., and Hofmann, T. 2011. Commercial titanium dioxide nanoparticles in both natural and synthetic water: comprehensive multidimensional testing and prediction of aggregation behavior. Environmental Science & Technology, 45(23): 10045–10052. https://doi.org/10.1021/es2023225.

Ouyang, K., Yu, X.-Y., Zhu, Y., Gao, C., Huang, Q., and Cai, P. 2017. Effects of humic acid on the interactions between zinc oxide nanoparticles and bacterial biofilms. Environmental Pollution, 231: 1104–1111. https://doi.org/https://doi.org/10.1016/j.envpol.2017.07.003.

Pantidos, N. 2014. Biological synthesis of metallic nanoparticles by bacteria, fungi and plants. Journal of Nanomedicine & Nanotechnology, 05(05). https://doi.org/10.4172/2157-7439.1000233.

Parada, J., Rubilar, O., Sousa, D.Z., Martínez, M., Fernández-Baldo, M.A., and Tortella, G.R. 2019. Short term changes in the abundance of nitrifying microorganisms in a soil-plant system simultaneously exposed to copper nanoparticles and atrazine. Science of the Total Environment, 670: 1068–1074. https://doi.org/10.1016/j.scitotenv.2019.03.221.

Parisi, C., Vigani, M., and Rodríguez-Cerezo, E. 2015. Agricultural nanotechnologies: What are the current possibilities? Nano Today, 10(2): 124–127. https://doi.org/https://doi.org/10.1016/j.nantod.2014.09.009.

Perea Vélez, Y.S., Carrillo-González, R., and González-Chávez, M. del C.A. 2021. Interaction of metal nanoparticles–plants–microorganisms in agriculture and soil remediation. Journal of Nanoparticle Research, 23(9): 206. https://doi.org/10.1007/s11051-021-05269-3.

Pérez-Hernández, H., Fernández-Luqueño, F., Huerta-Lwanga, E., Mendoza-Vega, J., and Álvarez-Solís José, D. 2020. Effect of engineered nanoparticles on soil biota: Do they improve the soil quality and crop production or jeopardize them? Land Degradation and Development, 31(16): 2213–2230. https://doi.org/10.1002/ldr.3595.

Pittol, M., Tomacheski, D., Simões, D.N., Ribeiro, V.F., and Santana, R.M.C. 2017. Macroscopic effects of silver nanoparticles and titanium dioxide on edible plant growth. Environmental Nanotechnology, Monitoring and Management, 8(July): 127–133. https://doi.org/10.1016/j.enmm.2017.07.003.

Prasad, R., Bhattacharyya, A., and Nguyen, Q.D. 2017. Nanotechnology in sustainable agriculture: Recent developments, challenges, and perspectives. Frontiers in Microbiology, 8(JUN): 1–13. https://doi.org/10.3389/fmicb.2017.01014.

Priyanka, K.P., Harikumar, V.S., Balakrishna, K.M., and Varghese, T. 2017. Inhibitory effect of TiO2 NPs on symbiotic arbuscular mycorrhizal fungi in plant roots. IET Nanobiotechnology, 11(1): 66–70. https://doi.org/https://doi.org/10.1049/iet-nbt.2016.0032.

Rajput, V.D., Kumari, A., Upadhyay, S.K., Minkina, T., Mandzhieva, S., Ranjan, A. et al. 2023. Can nanomaterials improve the soil microbiome and crop productivity? Agriculture (Switzerland), 13(2): 1–18. https://doi.org/10.3390/agriculture13020231.

Rajput, V., Minkina, T., Sushkova, S., Behal, A., Maksimov, A., Blicharska, E. et al. 2020. ZnO and CuO nanoparticles: a threat to soil organisms, plants, and human health. Environmental Geochemistry and Health, 42(1): 147–158. https://doi.org/10.1007/s10653-019-00317-3.

Rastogi, A., Zivcak, M., Sytar, O., Kalaji, H.M., He, X., Mbarki, S. et al. 2017. Impact of metal and metal oxide nanoparticles on plant: A critical review. Frontiers in Chemistry, 5(October): 1–16. https://doi.org/10.3389/fchem.2017.00078.

Ricardo, I.A., Alberto, E.A., Silva Júnior, A.H., Macuvele, D.L.P., Padoin, N., Soares, C. et al. 2021. A critical review on microplastics, interaction with organic and inorganic pollutants, impacts and effectiveness of advanced oxidation processes applied for their removal from aqueous matrices. Chemical Engineering Journal, 424: 130282. https://doi.org/https://doi.org/10.1016/j.cej.2021.130282.

Rizwan, M., Ali, S., Qayyum, M.F., Ok, Y.S., Adrees, M., Ibrahim, M. et al. 2017. Effect of metal and metal oxide nanoparticles on growth and physiology of globally important food crops: A critical review. Journal of Hazardous Materials, 322: 2–16. https://doi.org/https://doi.org/10.1016/j.jhazmat.2016.05.061.

Rossi, L., Sharifan, H., Zhang, W., Schwab, A.P., and Ma, X. 2018. Mutual effects and in planta accumulation of co-existing cerium oxide nanoparticles and cadmium in hydroponically grown soybean (*Glycine max* (L.) Merr.). Environ. Sci.: Nano, 5(1): 150–157. https://doi.org/10.1039/C7EN00931C.

Rossi, L., Zhang, W., Lombardini, L., and Ma, X. 2016. The impact of cerium oxide nanoparticles on the salt stress responses of *Brassica napus* L. Environmental Pollution, 219: 28–36. https://doi.org/10.1016/j.envpol.2016.09.060.

Rubina, M.S., Vasil'kov, A.Y., Naumkin, A.V, Shtykova, E.V, Abramchuk, S.S., Alghuthaymi, M.A. et al. 2017. Synthesis and characterization of chitosan–copper nanocomposites and their fungicidal activity against two sclerotia-forming plant pathogenic fungi. Journal of Nanostructure in Chemistry, 7(3): 249–258. https://doi.org/10.1007/s40097-017-0235-4.

Salamanca-Buentello, F., Persad, D.L., Court, E.B., Martin, D.K., Daar, A.S., and Singer, P.A. 2005. Nanotechnology and the Developing World. PLOS Medicine, 2(5), null. https://doi.org/10.1371/journal.pmed.0020097.

Santos, C.S.C., Gabriel, B., Blanchy, M., Menes, O., García, D., Blanco, M. et al. 2015. Industrial applications of nanoparticles—a prospective overview. Materials Today: Proceedings, 2(1): 456–465. https://doi.org/https://doi.org/10.1016/j.matpr.2015.04.056.

Schreiner, K.M., Filley, T.R., Blanchette, R.A., Bowen, B.B., Bolskar, R.D., Hockaday, W.C. et al. 2009. White-rot basidiomycete-mediated decomposition of C60 Fullerol. Environmental Science & Technology, 43(9): 3162–3168. https://doi.org/10.1021/es801873q.

Seneviratne, M., Gunaratne, S., Bandara, T., Weerasundara, L., Rajakaruna, N., Seneviratne, G. et al. 2016. Plant growth promotion by *Bradyrhizobium japonicum* under heavy metal stress. South African Journal of Botany, 105: 19–24. https://doi.org/https://doi.org/10.1016/j.sajb.2016.02.206.

Servin, A., Elmer, W., Mukherjee, A., De la Torre-Roche, R., Hamdi, H., White, J.C. et al. 2015. A review of the use of engineered nanomaterials to suppress plant disease and enhance crop yield. Journal of Nanoparticle Research, 17(2): 92. https://doi.org/10.1007/s11051-015-2907-7.

Sezer, N., Atieh, M.A., and Koç, M. 2019. A comprehensive review on synthesis, stability, thermophysical properties, and characterization of nanofluids. Powder Technology, 344: 404–431. https://doi.org/https://doi.org/10.1016/j.powtec.2018.12.016.

Shi, Y., Xiao, Y., Li, Z., Zhang, X., Liu, T., Li, Y. et al. 2021. Microorganism structure variation in urban soil microenvironment upon ZnO nanoparticles contamination. Chemosphere, 273(xxxx): 128565. https://doi.org/10.1016/j.chemosphere.2020.128565.

Simonin, M., Cantarel, A.A.M., Crouzet, A., Gervaix, J., Martins, J.M.F., and Richaume, A. 2018. Negative effects of copper oxide nanoparticles on carbon and nitrogen cycle microbial activities in contrasting agricultural soils and in presence of plants. Frontiers in Microbiology, 9(December): 1–11. https://doi.org/10.3389/fmicb.2018.03102.

Simonin, M., Martins, J.M.F., Uzu, G., Spadini, L., Navel, A., and Richaume, A. 2021. Low mobility of CuO and TiO$_2$ nanoparticles in agricultural soils of contrasting texture and organic matter content. Science of the Total Environment, 783: 146952. https://doi.org/10.1016/j.scitotenv.2021.146952.

Simonin, M., and Richaume, A. 2015. Impact of engineered nanoparticles on the activity, abundance, and diversity of soil microbial communities: a review. Environmental Science and Pollution Research, 22(18): 13710–13723. https://doi.org/10.1007/s11356-015-4171-x.

Sondi, I., and Salopek-Sondi, B. 2004. Silver nanoparticles as antimicrobial agent: a case study on E. coli as a model for Gram-negative bacteria. Journal of Colloid and Interface Science, 275(1): 177–182. https://doi.org/https://doi.org/10.1016/j.jcis.2004.02.012.

Soni, D., Gandhi, D., Tarale, P., Bafana, A., Pandey, R.A., and Sivanesan, S. 2017. Oxidative stress and genotoxicity of zinc oxide nanoparticles to *Pseudomonas* species, human promyelocytic leukemic (HL-60), and blood cells. Biological Trace Element Research, 178(2): 218–227. https://doi.org/10.1007/s12011-016-0921-y.

Spanos, A., Athanasiou, K., Ioannou, A., Fotopoulos, V., and Krasia-Christoforou, T. 2021. Functionalized magnetic nanomaterials in agricultural applications. Nanomaterials, 11(11). https://doi.org/10.3390/nano11113106.

Stabryla, L.M., Johnston, K.A., Millstone, J.E., and Gilbertson, L.M. 2018. Emerging investigator series: it{'}s not all about the ion: support for particle-specific contributions to silver nanoparticle antimicrobial activity. Environ. Sci.: Nano, 5(9): 2047–2068. https://doi.org/10.1039/C8EN00429C.

Sun, W., Dou, F., Li, C., Ma, X., and Ma, L.Q. 2021. Impacts of metallic nanoparticles and transformed products on soil health. Critical Reviews in Environmental Science and Technology, 51(10): 973–1002. https://doi.org/10.1080/10643389.2020.1740546.

Tang, B., Xu, H., Song, F., Ge, H., and Yue, S. 2022. Effects of heavy metals on microorganisms and enzymes in soils of lead–zinc tailing ponds. Environmental Research, 207: 112174. https://doi.org/https://doi.org/10.1016/j.envres.2021.112174.

Thiagarajan, V., and Ramasubbu, S. 2021. Fate and behaviour of TiO2 nanoparticles in the soil: their impact on staple food crops. Water, Air, and Soil Pollution, 232(7). https://doi.org/10.1007/s11270-021-05219-8.

Timmusk, S., Seisenbaeva, G., and Behers, L. 2018. Titania (TiO2) nanoparticles enhance the performance of growth-promoting rhizobacteria. Scientific Reports, 8(1): 617. https://doi.org/10.1038/s41598-017-18939-x.

Tolaymat, T., El Badawy, A., Genaidy, A., Abdelraheem, W., and Sequeira, R. 2017. Analysis of metallic and metal oxide nanomaterial environmental emissions. Journal of Cleaner Production, 143: 401–412. https://doi.org/https://doi.org/10.1016/j.jclepro.2016.12.094.

Trujillo-Reyes, J., Majumdar, S., Botez, C.E., Peralta-Videa, J.R., and Gardea-Torresdey, J.L. 2014. Exposure studies of core–shell Fe/Fe3O4 and Cu/CuO NPs to lettuce (*Lactuca sativa*) plants: Are they a potential physiological and nutritional hazard? Journal of Hazardous Materials, 267: 255–263. https://doi.org/https://doi.org/10.1016/j.jhazmat.2013.11.067.

Upadhayay, V.K., Chitara, M.K., Mishra, D., Jha, M.N., Jaiswal, A., Kumari, G. et al. 2023. Synergistic impact of nanomaterials and plant probiotics in agriculture: A tale of two-way strategy for long-term sustainability. Frontiers in Microbiology, 14(May): 1–22. https://doi.org/10.3389/fmicb.2023.1133968.

VandeVoort, A.R., and Arai, Y. 2018. Macroscopic observation of soil nitrification kinetics impacted by copper nanoparticles: implications for micronutrient nanofertilizer. Nanomaterials, 8(11). https://doi.org/10.3390/nano8110927.

Wagner, G., Korenkov, V., Judy, J.D., and Bertsch, P.M. 2016. Nanoparticles composed of Zn and ZnO inhibit *Peronospora tabacina* spore germination *in vitro* and *P. tabacina* infectivity on tobacco leaves. Nanomaterials, 6(3). https://doi.org/10.3390/nano6030050.

Wang, F., Adams, C.A., Shi, Z., and Sun, Y. 2018. Combined effects of ZnO NPs and Cd on sweet sorghum as influenced by an arbuscular mycorrhizal fungus. Chemosphere, 209: 421–429. https://doi.org/https://doi.org/10.1016/j.chemosphere.2018.06.099.

Wang, F., Jing, X., Adams, C.A., Shi, Z., and Sun, Y. 2018. Decreased ZnO nanoparticle phytotoxicity to maize by arbuscular mycorrhizal fungus and organic phosphorus. Environmental Science and Pollution Research, 25(24): 23736–23747. https://doi.org/10.1007/s11356-018-2452-x.

Wang, F., Liu, X., Shi, Z., Tong, R., Adams, C.A., and Shi, X. 2016. Arbuscular mycorrhizae alleviate negative effects of zinc oxide nanoparticle and zinc accumulation in maize plants – A soil microcosm experiment. Chemosphere, 147: 88–97. https://doi.org/https://doi.org/10.1016/j.chemosphere.2015.12.076.

Wang, J., Shu, K., Zhang, L., and Si, Y. 2017. Effects of silver nanoparticles on soil microbial communities and bacterial nitrification in suburban vegetable soils. Pedosphere, 27(3): 482–490. https://doi.org/https://doi.org/10.1016/S1002-0160(17)60344-8.

Wang, S., Liu, Y., Wang, X., Xiang, H., Kong, D., Wei, N. et al. 2023. Effects of concentration-dependent graphene on maize seedling development and soil nutrients. Scientific Reports, 13(1): 1–13. https://doi.org/10.1038/s41598-023-29725-3.

Wang, X., Sun, W., and Ma, X. 2019. Differential impacts of copper oxide nanoparticles and Copper(II) ions on the uptake and accumulation of arsenic in rice (*Oryza sativa*). Environmental Pollution, 252: 967–973. https://doi.org/https://doi.org/10.1016/j.envpol.2019.06.052.

Wang, Y., Westerhoff, P., and Hristovski, K.D. 2012. Fate and biological effects of silver, titanium dioxide, and C60 (fullerene) nanomaterials during simulated wastewater treatment processes. Journal of Hazardous Materials, 201-202: 16–22. https://doi.org/https://doi.org/10.1016/j.jhazmat.2011.10.086.

Wigger, H., Hackmann, S., Zimmermann, T., Köser, J., Thöming, J., and von Gleich, A. 2015. Influences of use activities and waste management on environmental releases of engineered nanomaterials. Science of The Total Environment, 535: 160–171. https://doi.org/https://doi.org/10.1016/j.scitotenv.2015.02.042.

Yang, Y., Wang, J., Xiu, Z., and Alvarez, P.J.J. 2013. Impacts of silver nanoparticles on cellular and transcriptional activity of nitrogen-cycling bacteria. Environmental Toxicology and Chemistry, 32(7): 1488–1494. https://doi.org/https://doi.org/10.1002/etc.2230.

Yanga, J., Cao, W., and Rui, Y. 2017. Interactions between nanoparticles and plants: Phytotoxicity and defense mechanisms. Journal of Plant Interactions, 12(1): 158–169. https://doi.org/10.1080/17429145.2017.1310944.

Yavuz, C.T., Mayo, J.T., Yu, W.W., Prakash, A., Falkner, J.C., Yean, S. et al. 2006. Low-field magnetic separation of monodisperse Fe3O4 nanocrystals. Science, 314(5801): 964–967. https://doi.org/10.1126/science.1131475.

Yuan, Z., Li, J., Cui, L., Xu, B., Zhang, H., and Yu, C.-P. 2013. Interaction of silver nanoparticles with pure nitrifying bacteria. Chemosphere, 90(4): 1404–1411. https://doi.org/https://doi.org/10.1016/j.chemosphere.2012.08.032.

Yurkov, A.M. 2018. Yeasts of the soil – obscure but precious. Yeast, 35(5): 369–378. https://doi.org/10.1002/yea.3310.

Zhang, L., Wu, L., Si, Y., and Shu, K. 2018. Size-dependent cytotoxicity of silver nanoparticles to *Azotobacter vinelandii*: Growth inhibition, cell injury, oxidative stress and internalization. PLOS ONE, 13(12): 1–18. https://doi.org/10.1371/journal.pone.0209020.

Zhang, P., Guo, Z., Zhang, Z., Fu, H., White, J.C., and Lynch, I. 2020. Nanomaterial transformation in the soil–plant system: implications for food safety and application in agriculture. Small, 16(21): 1–13. https://doi.org/10.1002/smll.202000705.

Zhang, X., Zhou, Q., and Wu, Z. 2020. Impact of TiO2 and ZnO nanoparticles on soil bacteria and the enantioselective transformation of racemic-metalaxyl in agricultural soil with lolium perenne: A wild greenhouse cultivation. Journal of Agricultural and Food Chemistry, 68(40): 11242–11252. https://doi.org/10.1021/acs.jafc.0c03959.

Zhao, L., Ortiz, C., Adeleye, A.S., Hu, Q., Zhou, H., Huang, Y. et al. 2016. Metabolomics to detect response of lettuce (*Lactuca sativa*) to Cu(OH)2 nanopesticides: oxidative stress response and detoxification mechanisms. Environmental Science & Technology, 50(17): 9697–9707. https://doi.org/10.1021/acs.est.6b02763.

10

Hybrid Magnetic Nanostructures for Wastewater Treatment

Nimra Nadeem,[1,2] Muhammad Zahid,[1,] Usman Zubair,[2] Palwasha Tehseen[3] and Zulfiqar Ahmad Rehan[4]*

Introduction

In the face of escalating environmental concerns and the imperative to address the ever-growing challenges associated with wastewater treatment, the exploration and implementation of advanced technologies become paramount (Nadeem et al. 2021a). This book chapter delves into the promising realm of hybrid magnetic nanostructures, offering a comprehensive exploration of their potential applications in revolutionizing wastewater treatment methodologies.

Wastewater production is a complex issue influenced by various factors, with industrial, agricultural, and domestic activities serving as major contributors. According to the World Bank, industrial processes account for a significant share of wastewater generation, with estimates suggesting that industrial discharges contribute to the pollution of approximately 70% of the world's rivers and lakes. Agriculture is another major contributor, as irrigation practices and the use of fertilizers and pesticides lead to runoff, contaminating water sources. In urban areas, domestic activities, including residential sewage and runoff from paved surfaces, contribute substantially to wastewater production. The United Nations reports that globally, over 80% of wastewater generated by society flows back into the ecosystem without adequate treatment, exacerbating water scarcity and environmental pollution. Rapid urbanization and population growth further amplify these challenges. Tackling wastewater production necessitates a holistic approach that addresses diverse sources and engages in sustainable water management practices across sectors. Traditional wastewater treatment methods, while effective to a certain extent, often fall short of addressing the diverse and persistent contaminants present in modern wastewater streams (Ashiq et al. 2021). The pressing need for innovative and efficient treatment approaches has driven researchers to explore

[1] Department of Chemistry, University of Agriculture Faisalabad, Faisalabad, Pakistan.

[2] Department of Textile Engineering, School of Engineering and Technology, National Textile University, Faisalabad 37610, Pakistan.

[3] Department of Chemistry, University of Agriculture, Sub-campus Toba Tek Singh, Pakistan, 36050.

[4] Department of Chemistry began with the establishment of Sultan Qaboos University, Oman.

* Corresponding author: rmzahid@uuaf.edu.pk; zahid595@gmail.com

the unique properties of nanotechnology, particularly the integration of magnetic nanostructures (Zahid et al. 2019).

Conventional wastewater treatment employs a multi-stage process aimed at effectively removing impurities and contaminants from water before its discharge into the environment. The primary treatment phase involves the physical removal of large solids through processes like screening and sedimentation, where gravitational settling separates suspended particles from the water. Subsequently, secondary treatment focuses on the biological breakdown of organic matter by microorganisms, often utilizing activated sludge treatment, followed by sedimentation to separate the microbial biomass. The tertiary treatment phase further refines the water quality through processes like adsorption, flocculation, and advanced techniques such as reverse osmosis. Adsorption involves the attachment of pollutants to surfaces like activated carbon, while flocculation promotes the aggregation of fine particles into larger, easily removable masses. Reverse osmosis, on the other hand, employs semi-permeable membranes to selectively remove remaining contaminants. The treated water undergoes a final disinfection step, commonly using chlorination or ultraviolet irradiation, targeting any remaining pathogens. Despite the efficacy of conventional methods, the growing challenges posed by micro-pollutants and increased water demand necessitate ongoing exploration of innovative technologies, including the incorporation of magnetic nanostructures, to enhance the overall efficiency and sustainability of wastewater treatment processes. The motivation behind this exploration stems from the limitations inherent in existing wastewater treatment technologies. Conventional methods face challenges in effectively removing emerging contaminants, heavy metals, and other pollutants that persist in wastewater effluents. Magnetic nanostructures, owing to their tunable properties and high surface area, emerge as a promising avenue for overcoming these challenges, offering a platform for enhanced contaminant removal and improved treatment efficiency.

This book chapter aims to provide a comprehensive understanding of hybrid magnetic nanostructures and nanotechnology, particularly the utilization of magnetic nanostructures, which plays a pivotal role in revolutionizing wastewater treatment processes. Magnetic nanostructures exhibit unique properties, such as high surface area, controllable reactivity, and magnetic responsiveness, making them highly effective in the removal of pollutants from water. These nanostructures can be employed as efficient adsorbents, capturing contaminants through physical or chemical interactions. Additionally, their magnetic nature facilitates easy separation from water matrices, enabling a simple and effective recovery process. Magnetic nanostructures are also utilized in catalytic applications, promoting advanced oxidation processes for the degradation of organic pollutants. The integration of these nanostructures into wastewater treatment systems enhances the overall efficiency, selectivity, and sustainability of the processes, contributing to the development of more advanced and environmentally friendly water purification technologies. As the field of nanotechnology continues to advance, the tailored design and synthesis of magnetic nanostructures hold great promise in addressing the evolving challenges of water pollution and ensuring the responsible management of water resources. By combining the unique characteristics of magnetic materials with the versatility of nanotechnology, these hybrid structures present new opportunities for targeted and efficient removal of contaminants. This chapter explores the fundamentals of wastewater treatment, delves into the synthesis and properties of magnetic nanostructures, and examines their application in addressing the complexities of contemporary wastewater challenges.

As we embark on this exploration, the overarching goal is to contribute to the knowledge base that informs sustainable and effective solutions for wastewater treatment, aligning with the global commitment to environmental stewardship and water resource conservation.

Fundamentals of wastewater treatment

Wastewater treatment is a complex process that involves the removal of contaminants and pollutants from water to make it safe for discharge into the environment or for reuse. The fundamentals of wastewater treatment encompass various physical, chemical, and biological processes. Here are the key fundamentals of wastewater treatment:

Screening

Screening is a primary process in wastewater treatment that involves the removal of large objects and debris from the influent wastewater before it undergoes further treatment. The purpose of screening is to protect downstream treatment processes, such as pumps, pipes, and biological treatment units, from damage and clogging caused by large and solid materials. The screening process typically involves the use of screens or bar racks with openings of a specific size. Wastewater flows through these screens and materials that are larger than the openings are physically retained and removed. The removed materials may include items such as sticks, leaves, plastics, papers, rags, and other debris. There are different types of screens used in wastewater treatment, including coarse screens, fine screens, rotating drum screens, and step screens.

Efficient screening is crucial for preventing damage to equipment and ensuring the proper functioning of subsequent treatment processes. After screening, the wastewater undergoes additional treatment steps, such as grit removal, primary sedimentation, biological treatment, and final clarification, to further purify the water before discharge or reuse.

Primary treatment

The primary treatment process is designed to remove settleable and floating solids from raw sewage or influent wastewater. While it is effective in reducing the overall pollutant load, primary treatment is considered a preliminary or basic level of treatment compared to more advanced processes like secondary and tertiary treatment. Here are the main components of the primary treatment process: grit removal (this helps prevent abrasion and damage to pumps and other equipment in the treatment plant).

The primary treatment process primarily focuses on the physical separation of solids from the wastewater (Figure 1). It does not effectively address dissolved pollutants or provide a high degree of biological treatment. As a result, many wastewater treatment plants combine primary treatment with secondary and tertiary treatment processes to achieve more comprehensive removal of contaminants.

Secondary treatment (biological treatment)

Secondary treatment involves the use of microorganisms to break down organic matter present in the wastewater. Common methods include activated sludge processes, trickling filters, and rotating biological contactors. Microorganisms consume organic pollutants, converting them into biomass, carbon dioxide, and water.

Tertiary Treatment

Tertiary treatment is the third and final stage in the conventional wastewater treatment process. This stage is designed to further improve the quality of the treated effluent from secondary treatment processes, making it suitable for discharge into water bodies or for reuse in various applications. Tertiary treatment targets the removal of remaining impurities, including fine particles, dissolved solids, nutrients, and trace contaminants. Key processes and technologies used in tertiary treatment include:

Disinfection

To eliminate or reduce the concentration of harmful microorganisms (bacteria, viruses, and parasites), disinfection is applied. Common disinfection methods include chlorination, ultraviolet (UV) irradiation, and ozonation. Disinfection ensures that the treated water is safe for human and environmental contact.

Reverse Osmosis (RO) and membrane filtration

These membrane-based processes can be used to remove dissolved salts, organic compounds, and other contaminants at the molecular level, providing a high level of purification.

Advanced Oxidation Processes (AOPs)

Advanced Oxidation Processes (AOPs) refer to a set of chemical treatment techniques used in wastewater treatment to remove or degrade organic and inorganic pollutants that may be resistant to conventional treatment methods (Zahid et al. 2020). AOPs are characterized by the generation of highly reactive hydroxyl radicals (•OH), which are powerful oxidizing agents. These radicals can effectively break down a wide range of pollutants into simpler, less harmful substances. AOPs are particularly useful for treating recalcitrant and persistent pollutants. Common AOPs include: (1) **Ozonation** (Ozone (O_3), which is a strong oxidizing agent and reacts with organic and inorganic substances, leading to the formation of hydroxyl radicals). (2) **Photocatalysis**—photocatalytic processes that involve the use of a semiconductor catalyst (such as titanium dioxide, TiO_2) activated by ultraviolet (UV) light (Arfa et al. 2023). The catalyst generates electron-hole pairs that react with water and oxygen to produce hydroxyl radicals. (3) **Fenton and Fenton-like**

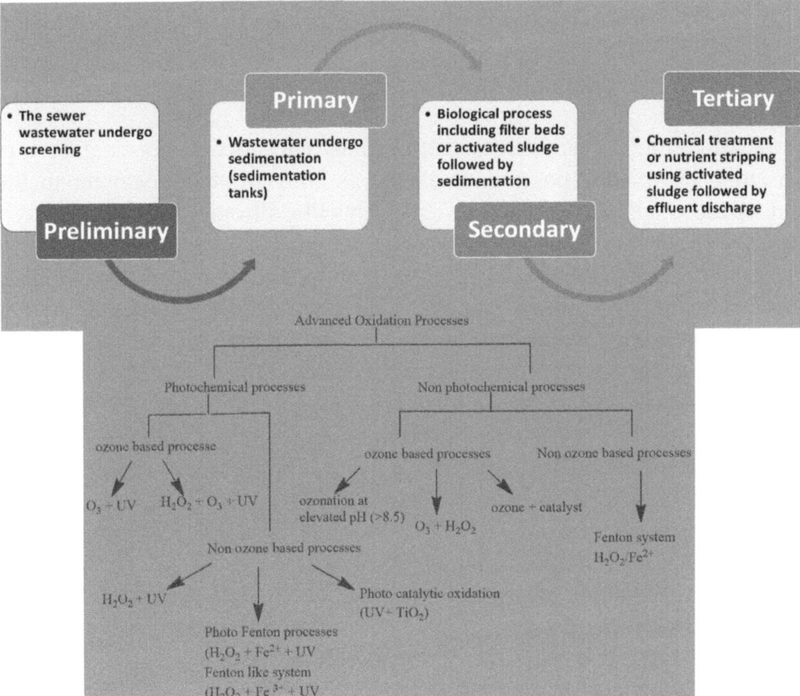

Figure 1. A schematic representation of key steps involved in wastewater treatment including primary, secondary, tertiary, and Advanced Oxidation Processes.

Processes—the Fenton process involves the addition of ferrous ions (Fe^{2+}) and hydrogen peroxide (H_2O_2) to the wastewater (Tabasum et al. 2020). The reaction between Fe^{2+} and H_2O_2 generates hydroxyl radicals. Fenton-like processes use other transition metals and peroxides. (4) **UV/H_2O_2** (Ultraviolet/Hydrogen Peroxide)—ultraviolet light, in the presence of hydrogen peroxide, generates hydroxyl radicals through photolysis. This process is effective in degrading organic compounds. (5) **Sonolysis** (Ultrasound)—sonolysis involves the application of ultrasound waves to water, leading to the formation of cavitation bubbles. The collapse of these bubbles generates shockwaves and high temperatures, creating hydroxyl radicals. (6) **Electrochemical Oxidation/Reduction**—electrochemical AOPs involve the use of electrodes to generate reactive species. Examples include electro-Fenton and anodic oxidation processes. (7) **Peracetic Acid (PAA)**—peracetic acid is a strong oxidizing agent that can be used as an AOP in wastewater treatment for the degradation of various pollutants.

Applications of AOPs in wastewater treatment include the removal of persistent organic pollutants, pharmaceuticals, personal care products, pesticides, and industrial wastewater contaminants. AOPs are particularly useful when conventional treatment methods are insufficient or when stringent effluent quality standards must be met.

The specific combination of tertiary treatment processes employed depends on the quality standards required for the treated water and the intended use of the effluent. Tertiary treatment is particularly important when the treated water is discharged into sensitive environments or when it is intended for applications like irrigation, industrial processes, or potable water supply. The goal is to achieve compliance with regulatory standards and ensure that the discharged or reused water is environmentally safe and poses minimal risks to human health. These fundamentals form the basis of conventional wastewater treatment processes. However, advancements in technology and a growing emphasis on sustainability and resource recovery are driving innovation in wastewater treatment to improve efficiency, reduce energy consumption, and maximize the reuse of treated water and by-products.

Challenges in current wastewater treatment methods

The presence of emerging contaminants, such as pharmaceuticals, personal care products, and endocrine-disrupting chemicals, poses a challenge as conventional treatment methods may not effectively remove these substances. Excessive nutrients like nitrogen and phosphorus in wastewater can lead to nutrient pollution, causing issues such as eutrophication in receiving water bodies. Removing these nutrients efficiently can be challenging. Many wastewater treatment processes require a significant amount of energy. Finding sustainable and energy-efficient treatment methods is crucial to reduce the environmental impact of wastewater treatment. The handling and disposal of sludge generated during wastewater treatment can be challenging. Finding environmentally friendly and economically viable ways to manage sludge is an ongoing concern.

In some regions, centralized wastewater treatment plants might not be practical due to geographical constraints or population distribution. Developing effective decentralized treatment options is essential for these areas. Changes in climate patterns can affect the quantity and quality of wastewater. Extreme weather events, such as floods or droughts, can strain existing treatment facilities and infrastructure. Extracting valuable resources from wastewater, such as energy or reusable materials, is a challenge. Maximizing resource recovery can enhance the sustainability of wastewater treatment processes. In some regions, the cost of implementing advanced wastewater treatment technologies may be prohibitive. Ensuring that cost-effective solutions are accessible to communities, especially in developing regions, remains a challenge.

Ensuring the effective removal of pathogens from wastewater is crucial to prevent the spread of waterborne diseases. Developing robust methods for pathogen removal is an ongoing challenge. Meeting stringent environmental regulations and standards for wastewater discharge is a constant challenge for treatment facilities. Compliance with evolving regulations requires ongoing updates and improvements in treatment processes. Researchers and engineers continually work on addressing these challenges through innovations in wastewater treatment technologies, process optimization, and the development of more sustainable and efficient treatment methods.

Emerging trends in wastewater treatment

Nano-sized adsorbents, such as graphene oxide, carbon nanotubes, and various nanoparticles, are being employed to enhance the adsorption capacity of composite adsorbents. These materials offer high surface areas and specific interactions, improving the removal of pollutants from water. Nanostructured membranes, often incorporating materials like graphene or nanoparticle coatings, are being developed to improve the efficiency of membrane filtration processes (Nadeem et al. 2020). These membranes can enhance selectivity, fouling resistance, and overall performance in separating contaminants from water (Rehan et al. 2023, Zahid et al. 2023, Zubair et al. 2022). Nano-catalysts, including metal and metal oxide nanoparticles, are used to catalyze the degradation of organic pollutants in advanced oxidation processes (Saeed et al. 2021). These materials can promote the generation of reactive species, such as hydroxyl radicals, for more effective degradation of contaminants.

Magnetic nanomaterial, such as iron oxide nanoparticles, are utilized for magnetic separation of contaminants. This allows for easy recovery of the nanoparticles and the contaminants, simplifying the overall treatment process. Nano-sized materials are being explored to enhance bioremediation processes. Nano-bioremediation involves the use of nanoparticles to deliver nutrients, enhance microbial activity, or facilitate the removal of specific contaminants by microorganisms. Semiconductor nanoparticles, like titanium dioxide and zinc oxide, are used in photocatalytic processes for the degradation of organic pollutants under light irradiation (Tabasum et al. 2021). This technology harnesses the photoactivity of nano-materials to break down contaminants.

Researchers are developing composite materials by combining nanosized adsorbents with other materials to create hybrid structures with improved adsorption properties. This approach aims to address the limitations of individual materials (Nadeem et al. 2021b, Rubab et al. 2021). Antimicrobial nanomaterials, such as silver nanoparticles, are explored for their effectiveness in disinfecting water and removing pathogenic microorganisms (Nadeem et al. 2021c). It's important to note that while these trends show promise, the environmental and health impacts of nanomaterials must be carefully evaluated. Ongoing research is essential to understand the long-term implications and optimize the use of these materials in wastewater treatment applications.

Magnetic nanostructures

Magnetic nanostructures refer to materials or structures at the nano-scale that exhibit magnetic properties. These structures are typically composed of magnetic elements, such as ferromagnetic, antiferromagnetic, or ferrimagnetic materials, and have dimensions on the order of nanometers (1 to 100 nanometers). The properties of these magnetic nanostructures can differ significantly from their bulk counterparts due to quantum effects and increased surface-to-volume ratios. Here are some common types of magnetic nanostructures:

Magnetic nanoparticles

Magnetic nanoparticles are typically single-domain magnetic particles with dimensions in the nanometer range. These particles can be composed of various magnetic materials, such as iron oxide (e.g., magnetite, maghemite), cobalt, nickel, and others (Nadeem et al. 2021c). They find applications in biomedical imaging, drug delivery (Dongsar et al. 2023), magnetic storage devices, and wastewater treatment.

Magnetic nanowires and nanotubes

Magnetic nanowires and nanotubes are elongated structures with diameters in the nanometer range. These structures can exhibit unique magnetic properties and are studied for their potential applications in wastewater treatment. When the dimensions of nanowires are reduced to the nanoscale, they may display superparamagnetic behavior, where their magnetic moments fluctuate between alignment and random orientation in the absence of an external magnetic field (Fert et al. 1999). Magnetic nanowires can be composed of various magnetic materials, such as iron, cobalt, nickel, or their alloys. The choice of material affects their magnetic behavior.

Magnetic nano disks and nanodots

Nano disks and nanodots are flat or spherical magnetic structures with dimensions in the nanoscale range (Nemati et al. 2017). These structures are often used in studies of magnetic phenomena at the nanoscale and in applications such as magnetic data storage.

Magnetic nanoclusters

Magnetic nanoclusters are nanostructured materials composed of magnetic nanoparticles that aggregate or cluster together. These nanoclusters typically have dimensions in the nanometer range and exhibit unique magnetic and surface properties. While magnetic nanoclusters may not be as extensively studied as some other nanostructures, they hold potential applications in various fields, including wastewater treatment. Here are some potential applications and considerations for the use of magnetic nanoclusters in wastewater treatment. Magnetic nanoclusters can be functionalized to enhance their adsorption capabilities. Their clustered structure provides a high surface area, making them effective for the adsorption of contaminants such as heavy metals, organic pollutants, and dyes from wastewater.

Magnetic nanocomposites

Magnetic nanocomposites involve embedding magnetic nanoparticles within a matrix of non-magnetic material. These composites can have tailored properties and find applications in areas such as magnetic sensors, contrast agents in medical imaging, and catalysts.

Magnetic nanorods

Magnetic nanorods are elongated structures with magnetic properties, often used in applications like targeted drug delivery and imaging.

Magnetic nanolayers and thin films

Magnetic nanolayers and thin films are thin coatings with magnetic properties. These can be used in magnetic sensors, magnetic recording media, and other applications.

Research in the field of magnetic nanostructures is dynamic and ongoing, with applications spanning a wide range of fields including electronics, medicine, energy, and information storage. The unique properties of magnetic nanostructures make them promising candidates for innovative technologies. Figure 2 represents the configurational illustration of various magnetic nanostructures.

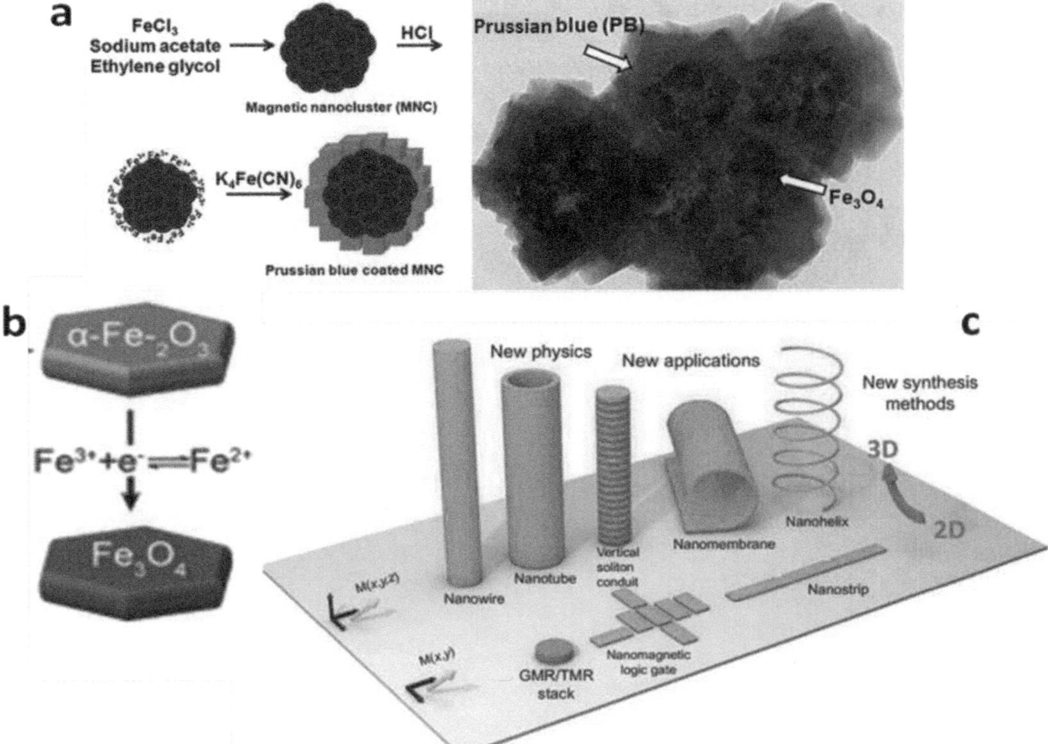

Figure 2. Various configurations of magnetic nanostructures (a) Magnetic Nano clusters (Yang et al. 2016) Copyrights© (2016) Elsevier, (b) Magnetic Nano discs (Gregurec et al. 2020) Copyrights© (2020) American Chemical Society, and (c) 3D curved magnetic nanostructures (Open access).

Synthesis of magnetic nanostructures

The synthesis of magnetic nanostructures involves creating materials with specific magnetic properties and nanosized dimensions. Magnetic nanostructures, such as nanoparticles, nanowires, nanotubes, and nanoclusters, can be synthesized using various methods. Here are some common techniques for the synthesis of magnetic nanostructures:

Chemical precipitation

A precursor solution containing metal salts is mixed with a reducing agent or a base. The reaction results in the precipitation of magnetic nanoparticles (Nahar et al. 2023). Synthesis of magnetite (Fe_3O_4) nanoparticles from iron salts in the presence of a reducing agent like sodium borohydride is an example of chemical precipitation.

Sol-gel method

In the sol-gel method, a sol containing metal precursor is converted into a gel, followed by drying and annealing to form the desired nanostructure (Kammar et al. 2023). For example, the synthesis of iron oxide (Fe_2O_3) nanoparticles using a sol-gel approach.

Hydrothermal/solvothermal synthesis

In hydrothermal/solvothermal synthesis, the reaction of metal precursors in a high-pressure, high-temperature aqueous or organic environment is achieved to facilitate controlled growth. The hydrothermal synthesis of iron oxide nanoparticles using iron salts in an autoclave (Rubab et al. 2021).

Co-precipitation

In the co-precipitation process, the simultaneous precipitation of multiple metal ions to form alloy or core-shell structures takes place under controlled reaction conditions (Shah et al. 2023). Co-precipitation of iron and cobalt salts to synthesize magnetic alloy nanoparticles.

Mechanical milling/ball milling

In this process, the mechanical grinding or milling of bulk materials is achieved to produce nanoparticles (Abada et al. 2023). For example, the ball milling of iron powder to produce iron nanoparticles.

Template-assisted synthesis

In template-assisted synthesis, the magnetic materials are deposited within or on the surface of a template structure, and then the template is removed. For example, the synthesis of magnetic nanotubes or nanowires using templates like alumina or porous membranes.

Microemulsion method

The versatile microemulsion method utilizes surfactant-stabilized microemulsions to control the size and shape of nanoparticles (Muhiuddin et al. 2023). For example, the synthesis of magnetic nanoparticles by adjusting the composition of the microemulsion.

Green synthesis

Green synthesis included the use of natural sources such as plant extracts or microorganisms to synthesize magnetic nanoparticles (Esther Nimshi et al. 2023). For example, the green synthesis of iron oxide nanoparticles using plant extracts.

The choice of synthesis method influences the size, shape, and properties of the magnetic nanostructures. Surface modification with ligands or coatings can enhance stability and functionalization for specific applications.

The selection of a specific synthesis method depends on the desired properties of the magnetic nanostructure and the intended applications in fields such as catalysis, biomedical applications, sensors, and wastewater treatment.

Hybridization in nanostructures

Hybrid nanostructures refer to materials or structures that combine two or more different types of nanomaterials, often with distinct properties, to create a composite material with enhanced or synergistic functionalities. Hybrid nanostructures can be designed by combining nanoparticles, nanotubes, nanowires, or other nanoscale components made of different materials. The integration of diverse materials allows for tailoring the properties of the resulting hybrid structure to meet specific requirements for various applications. Hybrid nanostructures find applications in a wide range of fields, including electronics, sensors, catalysis, medicine, and energy storage. There are various types of hybrid nanostructures based on the combination of different nanomaterials.

Core-shell nanostructures

Core-shell nanostructures consist of a core material surrounded by a shell of another material. The core and shell materials can have different properties, and this design is often used to enhance stability, functionality, or reactivity (Khan et al. 2023). Example: gold nanoparticles as the core with a silica shell.

Nanostructured composites

Nanostructured composites involve combining different types of nanoparticles or nanomaterials to form a composite material with improved properties. Example: carbon nanotube-polymer composites, where the carbon nanotubes enhance the mechanical and electrical properties of the polymer (de Tuesta et al. 2023).

Nanocomposites with inorganic and organic components

Hybrid nanostructures can be created by combining inorganic and organic nanomaterials to exploit the unique properties of both components. Example: Metal-organic frameworks (MOFs) incorporating metal nanoparticles for enhanced catalytic properties (Huang et al. 2024).

Janus nanostructures

Janus nanostructures are hybrid structures with two distinct faces or regions made of different materials. Example: Janus nanoparticles with one side composed of gold and the other side composed of silica (Milian et al. 2023).

Heterostructures

Heterostructures involve the integration of different materials in a layered or structured form, often with well-defined interfaces between the materials. Example: semiconductor heterostructures in which layers of different semiconducting materials are stacked for multipurpose applications (Ashiq et al. 2021).

Bimetallic nanoparticles

Bimetallic nanoparticles involve combining two different metal nanoparticles to create a composite material with unique electronic and catalytic properties. Example: platinum-gold bimetallic nanoparticles for enhanced catalytic activity (Li 2023).

Hybrid nanowires

Hybrid nanowires involve combining nanowires made of different materials to create structures with unique electrical, mechanical, or optical properties. Example: silicon nanowires combined with metal oxide nanowires for improved performance in sensors or energy storage devices (Li et al. 2023).

Graphene-based hybrids

This category includes hybrid structures of graphene and other nanomaterials that exploit the exceptional properties of graphene. Example: graphene oxide combined with metal nanoparticles for various applications, including catalysis and sensing (Arabkhani et al. 2023).

The design and synthesis of hybrid nanostructures offer a versatile platform for achieving tailored properties and functionalities, opening up new possibilities for innovation in nanotechnology and materials science.

Synthesis strategies of hybrid magnetic nanostructures

Hybrid magnetic nanostructures combine magnetic components with other materials, resulting in structures with diverse functionalities. The synthesis of hybrid magnetic nanostructures often involves the integration of magnetic nanoparticles with organic, inorganic, or other functional components. Here are some common methods for the synthesis of hybrid magnetic nanostructures:

Co-precipitation with organic molecules

Magnetic nanoparticles are co-precipitated with organic molecules or polymers to form hybrid structures.

Example: co-precipitation of magnetic nanoparticles with a polymer or surfactant during the synthesis process (Munir et al. 2023).

Encapsulation in polymer nanospheres

Magnetic nanoparticles are encapsulated within polymeric nanospheres or microspheres.

Example: synthesis of polymer-coated magnetic nanoparticles using emulsion polymerization (Hoang et al. 2023).

Layer-by-layer assembly

Alternating layers of magnetic nanoparticles and other materials are deposited on a substrate using methods like electrostatic self-assembly.

Example: layer-by-layer assembly of magnetic nanoparticles with polymers, biomolecules, or other functional materials (Ahmadi et al. 2023).

Functionalization of magnetic nanoparticles

Magnetic nanoparticles are functionalized with ligands, polymers, or biomolecules to impart specific properties.

Example: surface modification of magnetic nanoparticles with biocompatible polymers (Munir et al. 2023).

In situ growth

Additional components are grown *in situ* on the surface of magnetic nanoparticles.

Example: *In situ* growth of metal or metal oxide shells onto magnetic nanoparticles (Nithya et al. 2023).

Sol-gel process with magnetic nanoparticles

Magnetic nanoparticles are incorporated into a sol-gel matrix during the gelation process.

Example: incorporation of magnetic nanoparticles into silica gel matrices for enhanced stability and functionality.

Hybrid core-shell structures

Magnetic nanoparticles serve as the core, surrounded by a shell of another material.

Example: Core-shell structures with a magnetic core and a shell of polymer, silica, or another material.

Chemical vapor deposition (CVD) or physical vapor deposition (PVD)

Magnetic nanoparticles are deposited onto a substrate using CVD or PVD, and other materials are subsequently deposited on the magnetic layer.

Example: CVD deposition of magnetic nanoparticles followed by the deposition of a metal or semiconductor layer (Oyegoke 2023).

The choice of a particular synthesis method for hybrid magnetic nanostructures depends on the specific requirements of the intended application and the desired properties of the resulting material. A schematic illustration of some synthesis strategies of iron oxide nanostructures is presented in Figure 3.

Analytical methods for nanostructure characterization

Characterizing magnetic nanostructures is crucial to understanding their properties and optimizing their performance for various applications. Several fundamental characterization techniques are employed for this purpose. Magnetic measurements, such as vibrating sample magnetometry (VSM) and superconducting quantum interference device (SQUID) magnetometry, provide information about the material's magnetic behavior, including magnetic moment, coercivity, and saturation magnetization. Transmission electron microscopy (TEM) and scanning electron microscopy (SEM) offer insights into the morphological features, size distribution, and overall structure of the nanostructures. X-ray diffraction (XRD) is utilized to determine the crystalline structure and phase composition, aiding in identifying different magnetic phases. Fourier-transform infrared spectroscopy (FTIR) helps in identifying functional groups on the surface of the nanostructures, which is crucial for understanding their chemical nature and potential interactions in various environments. Mössbauer spectroscopy provides detailed information about the oxidation state and coordination environment of magnetic atoms within the nanostructures. Additionally, dynamic light scattering (DLS) can be employed to assess the size distribution in solution, essential for applications where colloidal stability is crucial. These characterization techniques collectively contribute to a comprehensive understanding of the magnetic nanostructures, enabling researchers to tailor their properties for specific applications in fields such as medicine, catalysis, and wastewater treatment.

Wastewater contaminants and removal mechanisms

Engineered magnetic nanoparticles play a pivotal role in water treatment through diverse mechanisms when employed either as direct purification agents or as carriers for other treatment phases. As direct agents, magnetic nanoparticles with functionalized surfaces exhibit impressive adsorption capabilities, effectively binding to a wide range of pollutants such as heavy metals,

Co-precipitation

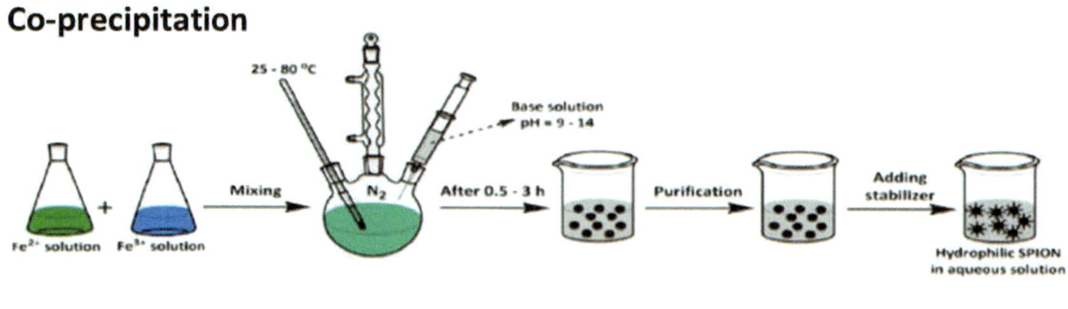

Thermal decomposition

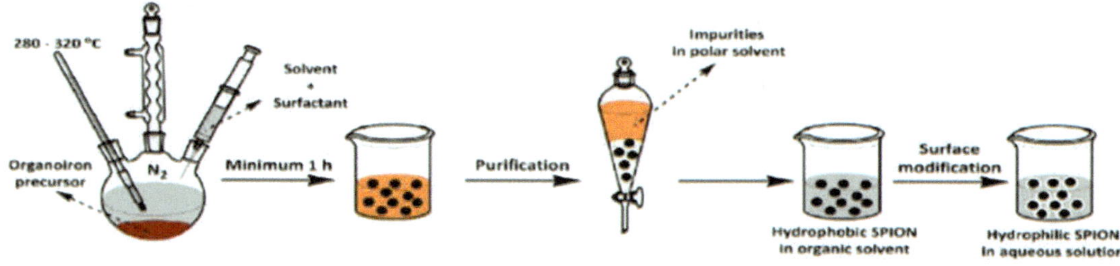

Microemulsion

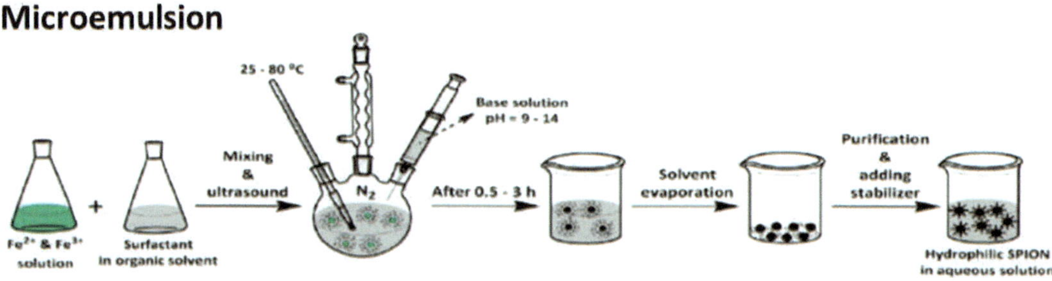

Sol-gel

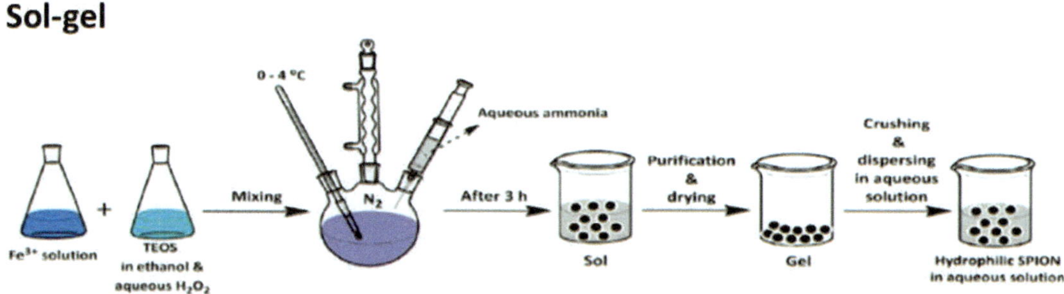

Figure 3. Illustrative representation of frequently employed chemical methodologies for synthesizing iron oxide nanoparticles (Dadfar et al. 2019) Copyrights© (2019) Elsevier

organic contaminants, and dyes (Figure 4a). Additionally, their magnetic properties facilitate easy separation from water matrices, ensuring efficient pollutant removal. Moreover, these nanoparticles can serve as catalysts for advanced oxidation processes, contributing to the degradation of organic pollutants. When utilized as carriers for other phases, magnetic nanoparticles enhance the delivery and targeted release of treatment agents, allowing for controlled and localized treatment of contaminants. Whether acting independently or as carriers, the magnetic properties enable the recovery and recyclability of nanoparticles, ensuring a sustainable and cost-effective approach to water purification. Overall, the multifaceted capabilities of engineered magnetic nanoparticles

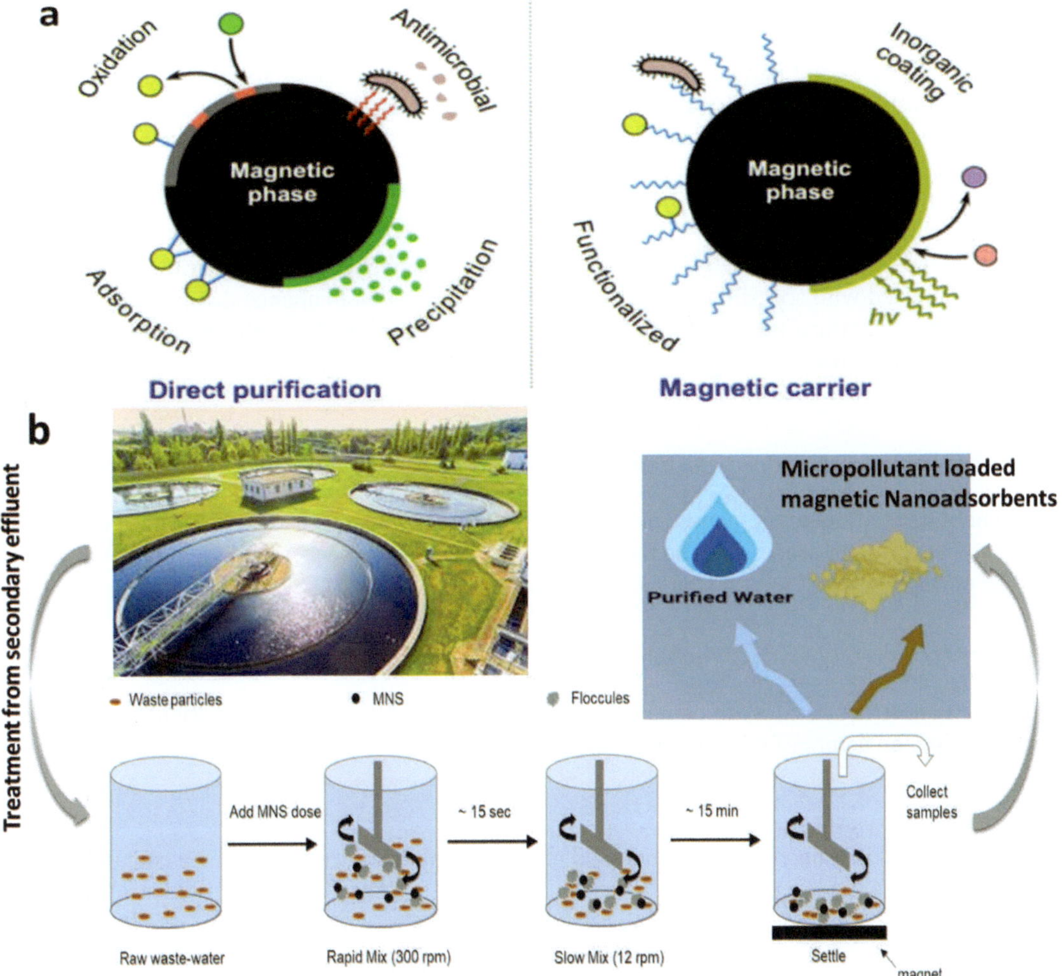

Figure 4. (a) Overview of potential water treatment mechanisms using engineered magnetic nanoparticles in direct and magnetic carrier applications (Martinez-Boubeta et al. 2019) Copyrights© (2019) Elsevier, (b) Visual representation of the utilization of magnetic nanoadsorbents and integration of magnetic separation within existing wastewater treatment facilities for the removal of micropollutants.

position them as versatile and efficient contributors to various water treatment mechanisms, holding considerable promise for addressing the complex challenges associated with water pollution.

This conceptual framework unfolds in three key stages. Initially, the discharge from the secondary treatment stage undergoes advanced treatment with carefully chosen magnetic nano-adsorbents. Subsequently, the treated effluent from this advanced treatment phase undergoes processing within an integrated magnetic separation system. In this step, the micro pollutant-laden magnetic nano-adsorbents are separated from the purified wastewater. Lastly, the purified water and micro pollutant-loaded magnetic nanoa-dsorbents follow separate streams for subsequent use and processing (Figure 4b). The micro pollutant-loaded magnetic nano-adsorbents are then regenerated to complete the cycle (Mudhoo et al. 2021).

Magnetic nanostructure offers several promising applications in wastewater treatment due to their unique physicochemical properties. Here are some ways in which these nanostructures can be utilized in addressing wastewater treatment challenges:

Adsorption of contaminants

Magnetic nanowires and nanotubes can be functionalized to have high surface areas and tailored surface chemistries, making them effective adsorbents for a variety of contaminants present in wastewater, such as heavy metals, organic pollutants, and dyes.

Targeted removal of pollutants

Magnetic nanostructures, typically based on iron oxide nanoparticles (e.g., magnetite or maghemite), are functionalized with specific ligands, coatings, or materials that enhance their affinity for target contaminants. Functionalized magnetic nanowires can be designed to selectively target specific pollutants in wastewater. This selective adsorption can enhance the efficiency of contaminant removal, reducing the need for additional treatment steps.

Introduction into wastewater

The functionalized magnetic nanostructures are introduced into the wastewater either directly or through pre-treatment processes. The nanoparticles disperse in the water, and the functional groups on their surfaces interact with the target contaminants.

Adsorption of contaminants

Magnetic nanostructures adsorb contaminants from the wastewater through various mechanisms, depending on the nature of the contaminants:

Chemisorption

It is the chemical bonding between functional groups on the nanostructure and the contaminant.

Physisorption

It refers to the physical interactions such as van der Waals forces, hydrogen bonding, or electrostatic interactions.

Enhanced contaminant interaction

The high surface area and specific functionalization of magnetic nanostructures enhance the interaction with contaminants. This allows for efficient adsorption of a wide range of pollutants, including heavy metals, organic compounds, pathogens, and more.

Photocatalytic degradation of adsorbed pollutants

The photocatalytic degradation of pollutants using magnetic nanostructures involves the activation of these nanostructures under light irradiation to generate reactive species that can break down contaminants. Here is a general mechanism for the photocatalytic degradation of pollutants using magnetic nanostructures (Figure 5a): Magnetic nanostructures, often based on materials like iron oxide nanoparticles (e.g., magnetite or maghemite), are synthesized and engineered to have photocatalytic properties. This may involve surface modification or doping with photocatalytic materials such as titanium dioxide (TiO_2) or zinc oxide (ZnO).

The magnetic nanostructures are functionalized to enhance their photocatalytic activity. This may involve the introduction of co-catalysts, sensitizers, or other materials that can improve the efficiency of the photocatalytic process.

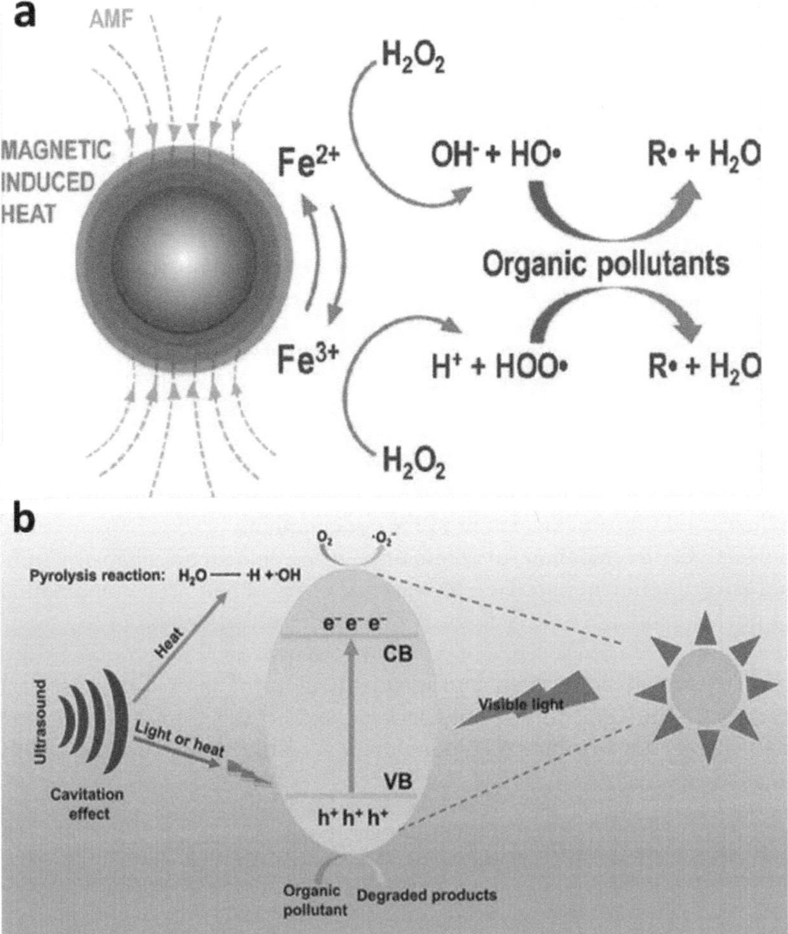

Figure 5. (a) An environmental catalytic process: Advanced oxidation of organic pollutants using iron oxide nanoparticles under an alternating magnetic field (AMF) (R: degradation products) (taken from an open access source), (b) Catalytic degradation scheme under light and ultrasound waves using magnetic nanomaterials (taken from an open access source).

The wastewater containing dispersed magnetic nanostructures is exposed to light irradiation, typically in the ultraviolet (UV) or visible light range. The nanostructures absorb the light energy and undergo electronic transitions, generating electron-hole pairs. Sometimes the reaction mixture is ultrasonically treated for enhanced photocatalysis (Figure 5b). The excited electrons and holes created by light absorption can initiate redox reactions and generate reactive species, such as:

Hydroxyl radicals ($\cdot$OH): Highly reactive and capable of oxidizing a wide range of pollutants.

Superoxide radicals ($\cdot O^{2-}$): Participate in redox reactions with contaminants.

The generated reactive species attack and interact with the pollutants present in the wastewater. This interaction leads to the degradation of pollutants into smaller, less harmful molecules.

Magnetic separation

An external magnetic field is applied to the wastewater or treatment system. The magnetic properties of the nanostructures facilitate their separation from the treated water, along with any residual contaminants.

Recovery and recycle

The recovered magnetic nanostructures can be reused in subsequent treatment cycles. If necessary, the nanostructures can undergo regeneration to remove any adsorbed or attached contaminants, allowing for multiple use cycles.

Monitoring and optimization

The efficiency of the photocatalytic degradation process is monitored using various analytical techniques. This information can be used to optimize the conditions, such as catalyst concentration, light intensity, and reaction time, for enhanced performance.

Final disposal or additional treatment

The treated water, now with reduced pollutant levels, can be discharged or subjected to additional treatment steps for polishing. The residual magnetic nanostructures may be disposed of or recycled, depending on their condition and the nature of the contaminants.

The photocatalytic degradation of pollutants using magnetic nanostructures provides a sustainable and energy-efficient approach to wastewater treatment. The combination of magnetic properties and photocatalytic activity enhances the overall effectiveness of the treatment process, allowing for the removal of a broad range of pollutants from water. This general mechanism outlines the fundamental steps involved in the use of magnetic nanostructures for contaminant removal in wastewater treatment. The versatility of magnetic nanostructures allows for customization based on the specific contaminants of concern, making them valuable tools in addressing a wide range of environmental challenges.

Potential advances in hybrid magnetic nanostructures

In recent years, there has been a growing interest in exploring potential advances in hybrid magnetic nanostructures and their application in wastewater treatment. These nanostructures, combining magnetic materials with various other components, exhibit unique properties that can significantly enhance the efficiency of wastewater treatment processes. One promising avenue involves the integration of magnetic nanoparticles with functional materials such as polymers or catalysts, creating hybrid structures with improved adsorption capabilities and catalytic activities. Additionally, the development of multifunctional hybrid magnetic nanostructures allows for the simultaneous removal of different pollutants, addressing the complexity of wastewater composition. Magnetic properties enable easy separation and recovery of the hybrid nanostructures after treatment, contributing to sustainable and cost-effective processes. Moreover, advancements in the design and synthesis of these materials offer tailored solutions for specific contaminants, optimizing their removal from diverse wastewater sources. As research progresses, the potential of hybrid magnetic nanostructures to revolutionize wastewater treatment by providing efficient, selective, and scalable solutions becomes increasingly evident.

Several specific advancements in hybrid magnetic nanostructures have shown promise in revolutionizing wastewater treatment. The integration of magnetic nanoparticles onto adsorbent materials like activated carbon or graphene oxide enhances the adsorption capacity for heavy metals, organic pollutants, and dyes in wastewater. Functionalizing magnetic nanoparticles with specific groups or ligands can improve their affinity towards target pollutants, enabling selective removal and efficient treatment of contaminated water. Hybrid structures incorporating magnetic nanoparticles with catalytic materials facilitate advanced oxidation processes, enabling the degradation of organic contaminants through Fenton-like reactions, thereby enhancing the efficiency of wastewater treatment. Smart materials that respond to changes in environmental conditions, such as pH or temperature, when combined with magnetic nanoparticles, enable tailored and controlled release of treatment agents, optimizing pollutant removal. Integration of magnetic nanoparticles into membrane systems enhances separation processes by improving fouling resistance and allowing for the magnetic recovery of membranes, extending their lifespan. Designing nanostructures with multiple functionalities, such as adsorption, catalysis, and magnetic separation, provides a comprehensive solution for the removal of diverse contaminants in a single treatment step. Hybrid magnetic nanozymes exhibit enzyme-like activities, facilitating the degradation of organic pollutants, and can be easily recovered using magnetic fields, making them promising for sustainable and reusable wastewater treatment. Incorporating magnetic nanoparticles into sensor systems allows for real-time monitoring of pollutant concentrations in wastewater, enabling timely and precise treatment interventions. Advancements in scalable synthesis methods and the development of large-scale hybrid magnetic nanostructures facilitate practical implementation in wastewater treatment plants, ensuring the technology's viability for broader applications. These specific advancements highlight the diverse applications of hybrid magnetic nanostructures in addressing various challenges associated with wastewater treatment, paving the way for more sustainable and effective water purification processes.

Conclusion

In conclusion, this book chapter underscores the promising prospects of employing magnetic nanostructures for wastewater treatment. The comprehensive overview of wastewater treatment fundamentals, magnetic nanostructures, their synthesis, hybridization strategies, and analytical characterization methods contributes to a nuanced understanding of the field. The elucidation of wastewater contaminants and the mechanisms of their removal by magnetic nanostructures emphasizes the versatility and efficacy of these materials in addressing water pollution challenges. The exploration of potential advances in hybrid magnetic nanostructures augurs well for the continuous evolution of wastewater treatment methodologies. By amalgamating theoretical insights with practical applications, this chapter serves as a valuable resource for researchers, engineers, and practitioners seeking to enhance the efficiency and sustainability of wastewater treatment processes through the integration of magnetic nanostructures.

References

Abada, A., Younes, A., and Manseri, A. 2023. Magnetic and structural properties of nanostructured FeSn, FeSnTi, FeSnV and FeSnTiV alloys elaborated via ball milling process. Journal of the Korean Physical Society, 1–11.

Ahmadi, F., Akbari, J., Saeedi, M., Seyedabadi, M., Ebrahimnejad, P., Ghasemi, S. et al. 2023. Efficient synergistic combination effect of curcumin with piperine by polymeric magnetic nanoparticles for breast cancer treatment. Journal of Drug Delivery Science and Technology, 104624.

Arabkhani, P., Sadegh, N., and Asfaram, A. 2023. Nanostructured magnetic graphene oxide/UIO-66 sorbent for ultrasound-assisted dispersive solid-phase microextraction of food colorants in soft drinks, candies, and pastilles prior to HPLC analysis. Microchemical Journal, 184: 108149.

Arfa, U., Alshareef, M., Nadeem, N., Javid, A., Nawab, Y., Alshammari, K.F. et al. 2023. Sunlight-driven photocatalytic active fabrics through immobilization of functionalized doped titania nanoparticles. Polymers, 15(13): 2775.

Ashiq, H., Nadeem, N., Mansha, A., Iqbal, J., Yaseen, M., Zahid, M. et al. 2021. G-C3N4/Ag@ CoWO4: A novel sunlight active ternary nanocomposite for potential photocatalytic degradation of rhodamine B dye. Journal of Physics and Chemistry of Solids, 110437.

Dadfar, S.M., Roemhild, K., Drude, N.I., von Stillfried, S., Knüchel, R., Kiessling, F. et al. 2019. Iron oxide nanoparticles: Diagnostic, therapeutic and theranostic applications. Adv. Drug Deliv. Rev., 138: 302–325. doi: 10.1016/j.addr.2019.01.005.

de Tuesta, J.L.D., Silva, A.S., Roman, F.F., Sanches, L.F., da Silva, F.A., Pereira, A.I. et al. 2023. Polyolefin-derived carbon nanotubes as magnetic catalysts for wet peroxide oxidation of paracetamol in aqueous solutions. Catalysis Today, 419: 114162.

Dongsar, T.T., Dongsar, T.S., Abourehab, M.A.S., Gupta, N., and Kesharwani, P. 2023. Emerging application of magnetic nanoparticles for breast cancer therapy. European Polymer Journal, 187: 111898. doi: https://doi.org/10.1016/j.eurpolymj.2023.111898.

Esther Nimshi, R., Judith Vijaya, J., Bououdina, M., John Kennedy, L., Al-Najar, B., and Lemine, O. 2023. Green synthesis of functional CuFe2O4@ TiO2@ rGO nanostructure for magnetic hyperthermia and cytotoxicity of human breast cancer cell line. Journal of Inorganic and Organometallic Polymers and Materials, 33(4): 1016-1027.

Fert, A., and Piraux, L. 1999. Magnetic nanowires. Journal of Magnetism and Magnetic Materials, 200(1): 338–358. doi: https://doi.org/10.1016/S0304-8853(99)00375-3.

Gregurec, D., Senko, A.W., Chuvilin, A., Reddy, P.D., Sankararaman, A., Rosenfeld, D. et a;. 2020. Magnetic Vortex nanodiscs enable remote magnetomechanical neural stimulation. ACS Nano, 14(7): 8036–8045. doi: 10.1021/acsnano.0c00562.

Hoang, T.K.N., Duong, H.Q., Nguyen, Q.B., and Nguyen, D.B.T. 2023. Colloidal stability and rheological properties of bio-ferrofluids of polymer-coated single-core and multi-core nanoparticles. Journal of Magnetism and Magnetic Materials, 579: 170838.

Huang, P., Wu, W., Li, M., Li, Z., Pan, L., Ahamad, T. et al. 2024. Metal-organic framework-based nanoarchitectonics: A promising material platform for electrochemical detection of organophosphorus pesticides. Coordination Chemistry Reviews, 501: 215534.

Kammar, S., Munnolli, C., Gaikwad, A., Shelke, S., Shirsath, S., Kadam, R. et al. 2023. Improved magnetic anisotropy of nano-crystalline Na substituted CaNb0. 5Ti0. 5O3 perovskite synthesized by sol–gel method. Materials Today: Proceedings.

Khan, S., Falahati, M., Cho, W.C., Vahdani, Y., Siddique, R., Sharifi, M. et al. 2023. Core-shell inorganic NPs@ MOF nanostructures for targeted drug delivery and multimodal imaging-guided combination tumor treatment. Advances in Colloid and Interface Science, 103007.

Li, X., Shi, C., Feng, Z., He, J., Zhang, R., Yang, Z. et al. 2023. Construction of Si nanowires/ZnFe2O4/Ag photocatalysts with enhanced photocatalytic activity under visible light and magnetic field. Journal of Alloys and Compounds, 946: 169467.

Li, Z. 2023. Bimetallic alloy Nanoparticles From Molecular Approaches: Synthesis, Characterization and their Magnetic and Electrocatalytic Applications.

Martinez-Boubeta, C., and Simeonidis, K. 2019. Chapter 20—Magnetic nanoparticles for water purification. pp. 521–552. In: Thomas, S., Pasquini, D., Leu, S.-Y., and Gopakumar, D.A. (Eds.). Nanoscale Materials in Water Purification. Elsevier.

Milian, Y.E., Claros, M., Ushak, S., and Vallejos, S. 2023. Silica based Janus nanoparticles: Synthesis methods, characterization, and applications. Applied Materials Today, 34: 101901.

Mudhoo, A., and Sillanpää, M. 2021. Magnetic nanoadsorbents for micropollutant removal in real water treatment: a review. Environmental Chemistry Letters, 19(6): 4393–4413. doi: 10.1007/s10311-021-01289-6.

Muhiuddin, G., Bibi, I., Nazeer, Z., Majid, F., Kamal, S., Kausar, A. et al. 2023. Synthesis of Ni doped barium hexaferrite by microemulsion route to enhance the visible light-driven photocatalytic degradation of crystal violet dye. Ceramics International, 49(3): 4342–4355.

Munir, T., Mahmood, A., Rasul, A., Imran, M., and Fakhar-e-Alam, M. 2023. Biocompatible polymer functionalized magnetic nanoparticles for antimicrobial and anticancer activities. Materials Chemistry and Physics, 301: 127677.

Nadeem, N., Yaseen, M., Rehan, Z.A., Zahid, M., Shakoor, R.A., Jilani, A. et al. 2021a. Coal fly ash supported CoFe2O4 nanocomposites: Synergetic Fenton-like and photocatalytic degradation of methylene blue. Environmental Research, 112280.

Nadeem, N., Zahid, M., Hanif, M.A., Bhatti, I.A., Shahid, I., Rehan, Z.A. et al. 2021b. Silver-doped metal ferrites for wastewater treatment. Silver Nanomaterials for Agri-Food Applications (pp. 599–622): Elsevier.

Nadeem, N., Zahid, M., Rehan, Z.A., Hanif, M.A., and Yaseen, M. 2021c. Improved photocatalytic degradation of dye using coal fly ash-based zinc ferrite (CFA/ZnFe2O4) composite. International Journal of Environmental Science and Technology. doi: 10.1007/s13762-021-03255-9.

Nadeem, N., Zahid, M., Tabasum, A., Mansha, A., Jilani, A., Bhatti, I.A., and Bhatti, H.N. 2020. Degradation of reactive dye using heterogeneous photo-Fenton catalysts: ZnFe2O4 and GO-ZnFe2O4 composite. Materials Research Express, 7(1)– 015519.

Nahar, B., Chaity, S.B., Gafur, M.A., and Hossain, M.Z. 2023. Synthesis of spherical copper oxide nanoparticles by chemical precipitation method and investigation of their photocatalytic and antibacterial activities. Journal of Nanomaterials, 2023.

Nemati, Z., Salili, S.M., Alonso, J., Ataie, A., Das, R., Phan, M.H. et al. 2017. Superparamagnetic iron oxide nanodiscs for hyperthermia therapy: Does size matter? Journal of Alloys and Compounds, 714: 709–714. doi: https://doi.org/10.1016/j.jallcom.2017.04.211.

Nithya, K., Sathish, A., and Sivamani, S. 2023. In situ synthesis of mesostructured iron oxide nanoparticles embedded in L. camara: adsorption insights and modeling studies. Biomass Conversion and Biorefinery, 13(9): 7827–7838.

Oyegoke, J. 2023. Exploring Novel Catalysts for the Chemical Vapor Deposition Synthesis, and Characterization of Zinc and Gallium Oxide Nanostructures.

Rehan, Z.A., Zahid, M., Kanwal, S., Nadeem, N., Hafeez, A., Jamil, A. et al. 2023. Optimization of carboxylated graphene oxide (C-GO) content in polymer matrix: Synthesis, characterization, and application study. Chemosphere, 310: 136900. doi: https://doi.org/10.1016/j.chemosphere.2022.136900.

Rubab, M., Bhatti, I.A., Nadeem, N., Shah, S.A.R., Yaseen, M., Naz, M.Y. et al. 2021. Synthesis and photocatalytic degradation of rhodamine B using ternary zeolite/WO3/Fe3O4 composite. Nanotechnology, 32(34): 345705. doi: 10.1088/1361-6528/ac037f.

Saeed, H., Nadeem, N., Zahid, M., Yaseen, M., Noreen, S., Jilani, A. et al. 2021. Mixed metal ferrite (Mn0.6Zn0.4Fe2O4) intercalated g-C3N4 nanocomposite: efficient sunlight driven photocatalyst for methylene blue degradation. Nanotechnology, 32(50): 505714. doi: 10.1088/1361-6528/ac2847.

Shah, M.M., Fatema, M., Ansari, D.A., and Gupta, D.K. 2023. Tuning the structural, magnetic, and electrochemical properties of Mo-doped NiO nanostructures prepared by coprecipitation method. Inorganic Chemistry Communications, 151: 110641.

Tabasum, A., Alghuthaymi, M., Qazi, U.Y., Shahid, I., Abbas, Q., Javaid, R. et al. 2021. UV-accelerated photocatalytic degradation of pesticide over magnetite and cobalt ferrite decorated graphene oxide composite. Plants, 10(1): 6.

Tabasum, A., Bhatti, I.A., Nadeem, N., Zahid, M., Rehan, Z.A., Hussain, T. et al. 2020. Degradation of acetamiprid using graphene-oxide-based metal (Mn and Ni) ferrites as Fenton-like photocatalysts. Water Science and Technology, 81(1): 178–189.

Yang, H.-M., Jang, S.-C., Hong, S.B., Lee, K.-W., Roh, C., Huh, Y.S. et al. 2016. Prussian blue-functionalized magnetic nanoclusters for the removal of radioactive cesium from water. Journal of Alloys and Compounds, 657: 387–393. doi: https://doi.org/10.1016/j.jallcom.2015.10.068.

Zahid, M., Nadeem, N., Hanif, M.A., Bhatti, I.A., Bhatti, H.N., and Mustafa, G. 2019. Metal ferrites and their graphene-based nanocomposites: synthesis, characterization, and applications in wastewater treatment. Magnetic Nanostructures (pp. 181–212): Springer.

Zahid, M., Nadeem, N., Tahir, N., Majeed, M.I., Naqvi, S.A.R., and Hussain, T. 2020. Hybrid nanomaterials for water purification, Multifunctional Hybrid Nanomaterials for Sustainable Agri-Food and Ecosystems (pp. 155–188): Elsevier.

Zahid, M., Saeeda, M., Nadeem, N., Shakir, H.M.F., El-Saoud, W.A. et al. 2023. Carboxylated Graphene Oxide (c-GO) Embedded ThermoPlastic Polyurethane (TPU) Mixed Matrix Membrane with Improved Physicochemical Characteristics. Membranes, 13(2): 144.

Zubair, U., Zahid, M., Nadeem, N., Ghazal, K., AlSalem, H.S., Binkadem, M.S. et al. 2022. The design of ternary composite polyurethane membranes with an enhanced photocatalytic degradation potential for the removal of anionic dyes. Membranes, 12(6): 630.

11

Nutrient Management through Nano-biofertilizers for Smart Agriculture

Tayyaba Samreen,[1,*] *Sehar Rasool,*[1] *Aimen Tahir,*[1] *Umair Riaz,*[2] *Muhammad Zulqernain Nazir,*[1] *Sehrish Kanwal*[3] and *Sidra-Tul-Muntaha*[4]

Introduction

Global population growth during the past ten years has compelled the agricultural industry to boost crop yield to meet the requirements of billions of emerging nations. Widespread nutrient shortage in soil has caused major nutritional value decline and substantial financial losses for farmers. It is estimated that over half of all chemical pesticides and fertilizers are lost due to leaching and mineralization before they can be removed from the environment. Soil has a crucial role in controlling food production needs to sustain the world's growing population. Increased usage of inorganic fertilizers and pesticides has exacerbated the situation for soil quality and the environment as the need for agricultural food supply has increased, necessitating more intense crop production and protection (WHO 2021). Soil health must be protected from pollution and lost output; hence, it is essential to restrain the increasing use of inorganic fertilizers (Sambangi et al. 2022). There will be 9.6 billion people round the globe by 2050, which will put more pressure on the amount of arable land that can be used for farming. As a result, for agriculture to be sustainable, smart agricultural techniques must be developed and put into use employing cutting-edge technology (Nongbet et al. 2022). This involves the requirement for the creation of novel fertilizers with cutting-edge technologies, great efficacy, and few drawbacks. In the current era, nanotechnology and its related applications have taken on enormous significance since this branch of technology has fundamentally changed modern science and is expanding at an exponential rate (Bhardwaj et al. 2022).

[1] Institute of Soil and Environmental Sciences, University of Agriculture, Faisalabad-38040, Pakistan.

[2] MNS-University of Agriculture, Multan, Pakistan.

[3] DDSDP (Data-Driven Smart Decision Platform), PMAS University of Arid Agriculture Rawalpindi, Rawalpindi 46000, Pakistan.

[4] Directorate of Pest Warning and Quality Control of Pesticides, Lahore, Punjab, Pakistan.

* Corresponding author: tybasamreen@gmail.com; tayyabasamreen@uaf.edu.pk

In the previous 10 years, there has been a lot of study performed on bio-fertilizers, micro biomes, and soil health because of rising awareness of the negative impacts of fertilizers (Chhipa 2017). Nanotechnology is applicable in different fields of study such as agriculture, chemistry, physics, pharmaceutical and medicinal science. Success in other fields opened several opportunities in the agricultural sector. According to the policies of the European Union (EU), precision agriculture (PA) is an agricultural technique that involves management of varying crops and the whole field to maximize the output using available resources. The increasing use of nanotechnology in contemporary agriculture has made precision farming a practical possibility (Duhan et al. 2017).

The novel and most technologically sophisticated method of providing mineral nutrients to crops is through the use of biological fertilizers, which include both bio and nano-fertilizers. The capacity of a biological process to tightly regulate the form of the particles would be a significant benefit. The external release of microorganisms provides the benefit of getting them abundantly in pure condition, devoid of other proteins associated with organisms, which allows for relatively easy downstream processing. It seems promising to employ certain enzymes released by fungus in the creation of nanoparticles. Biological fertilizers are more efficient in terms of fertilizer consumption because they supply plants with the nutrients they need while reducing the amount of nutrients lost through leaching (Subbarao et al. 2013, Malusa et al. 2016, Pandey and Chandra 2016). Therefore, developing environmentally acceptable, sustainable, and economically viable methods of manufacturing nanoparticles and microbial inoculants is of the utmost importance (Elkhatib et al. 2015, Hussain et al. 2016, Bagherzade et al. 2017, El-Ghamry et al. 2018). Using organisms for biosynthesis also adheres to the criteria of "green chemistry", which states that the bio-organism in question should be:

(1) Harmless to the environment

(2) A reducing agent

(3) A capping agent

This chapter examines the idea of nano bio-fertilizers, how they are made, how they function in crop protection and production, the problems they cause, and the ways to fix them.

Nano-biofertilizers

Nano-biofertilizers are formed by combining nanoparticles with bio-fertilizers. It is a method of encapsulating bio-fertilizers into a suitable nanomaterial. *Pseudomonas, Bacillus, Rhizobium, Azospirillum, Azotobacter*, and blue-green algae are all examples of organisms that may help plants to absorb phosphorus (Wu et al. 2005). The transformation of organic matter into simple chemicals by microorganisms enhances the soil nutrient status, improves soil habitat and boosts crop production and output. The effectiveness of using bio-fertilizers relies on the processing, storage, and application process (Jha and Prasad 2006). Reduced life span, sensitivity, desiccation and temperature issues during storage are some limitations in their use. The potential for encapsulating bio-fertilizer preparations with polymer-based nanoparticles to synthesize formulations that are resistant to desiccation has also been explored. For storage and distribution of microbes, water-in-oil emulsion is a renowned technique. It uses liquid formulations (Vander et al. 2006). The microbe's oil barrier prevents the water from falling and reduces evaporation. Extremely helpful for microbes that can be killed by desiccation. This technique boosts viability of cell and distribute kinetics by adding compounds to aqueous and oil phases. However, during the storage, sedimentation is a major cause for alarm. Silica nanoparticles improved cell viability by decreasing cell sedimentation by increasing the amount of oil during storage (Vandergheynst et al. 2007, Duhan et al. 2017).

Nano-biofertilizers also play an essential role by providing effective ways to improve the usage of various land and water resources while lowering environmental issues. Therefore, eco-friendly biological methods might provide an alternative to nanoscale chemical fertilizers.

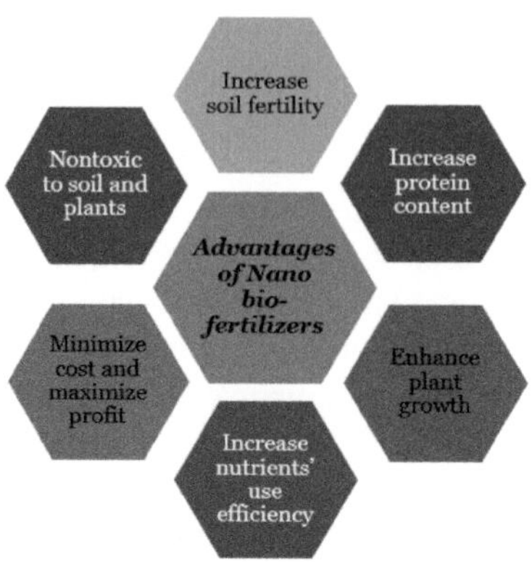

Figure 1. Nano-biofertilizers, also known as nano-fertilizers or nanobiotechnology-based fertilizers, have gained attention in recent years for their potential applications in agriculture. These innovative fertilizers utilize nanotechnology to enhance nutrient availability, plant growth, and overall crop productivity.

Azospirillum, Azotobacter, Rhizobium, Azotobacter, Cyanobacteria and nutrients (P and Zn, etc.) solubilizing microorganisms take part in different soil processes and plant growth. As a result, different inorganic and organic bio-nanoparticles are synthesized as environmentally friendly fertilizers by using microorganisms (bacteria, fungi, algae, actinomycetes, yeast) and plants. Utilizing nano-biofertilizers in modern plant nutrition explains the role of nanotechnology in the fields of agriculture and soil engineering (Sahu et al. 2022). As shown in Figure 1, these tiny bio-fertilizers have many additional important uses.

They regulate soil nutrient supply and buffer the effects of harsh environments. Nanomaterials are beneficial because they lessen the need for chemical fertilizers, raise nutrient availability and absorption, minimize environmental impact, and lower production costs (Pudake et al. 2019, Eliaspour et al. 2020). Nanoparticles influence plant-microbe interactions in two ways: directly, by increasing food availability in the rhizosphere, and indirectly, by improving the bacterial strains (Timmusk et al. 2018). Application of CeO_2-nanocomposite (75 ppm) significantly enhanced the length of *Trigonella foenum-graecum*, and the same effect was seen with extremely low concentrations (Sonali et al. 2022). To facilitate more successful crop establishment, nano-biofertilizers combine the benefits of bio inoculants with those of nanoparticles. They ensure that nutrients reach their proper targets. Still, nano-biofertilizers have not been aggressively supported due to less understanding of the interacting role of nanoparticles, bio-fertilizers, and plants (Akhtar et al. 2022, Elnahal et al. 2022).

Synthesis of nano-biofertilizers

Bio-fertilizers are mixtures of dormant or active strains of certain microorganisms that help plants to absorb nutrients and speed up other microbial activities. Such microbial activities, when immobilized on a carrier material, may help plants improve their nutrient absorption efficiency, as well as the cell count and surface area of such microorganisms. Table 1 describes different types of bio-fertilizers which will become nano-biofertilizers after combining with nanoparticles.

Nano-biofertilizers, also known as nanomaterials or nanoparticles, provide essential nutrients to plants at the nano scale, hence facilitating increased growth and plant yield (Khan and Rizvi 2017). Nano-biofertilizers may be categorized into three different categories: macro, micro, and nano-particulate. It is commonly known that some selected materials may be produced via nanotechnology at sizes sub 100 nm. These nanomaterials deliver one or more crucial plant nutrients to the soil by having nanoscale dimensions and specialized functionalities (Sahu et al. 2022). The similarities between nano- and bio-fertilizers are thus illustrated:

(1) Applying soil and foliar fertilizers to provide plants with the right nutrients for development.

(2) An inexpensive, environmentally acceptable source of plant fertilizers.

Table 1. Different types of bio-fertilizers.

Bio-fertilizers	Examples	Functions
N-fixing bio-fertilizers	*Azotobacter, Anabaena, Rhizobium, Anabaena azollae, Azospirillium*	• Fixes atmospheric nitrogen • Produces nitrogen by BNF process
P-solubilizing bio-fertilizers	*Bacillus subtilis, Pseudomonas striata, Aspergillus awamori*	• Secretes organic acids to increase solubility
P-mobilizing bio-fertilizers	*Glomus* spp., *Laccaria* spp., *Boletus* spp., *Amanita* spp., *Pezizella ericae*	• Increase resistance to root diseases and nematodes • Helps drought stressed plants
K-solubilizing bio-fertilizers	*Bacillus mucillagenous, Aspergillus terries, Aspergillus niger*	• Produces organic acids by glucose oxidation
For micronutrients	*Bacillus* spp., *Pseudomonas fluorescence, Thiobacillus*	• Iron uptake • Sulfur and zinc solubilizer • Sulfur oxidizing

(3) High fertilizing process efficiency.

(4) It serves as an adjunct to chemical fertilizers.

Furthermore, these biological fertilizers might be developing substitutes for traditional and chemical fertilizers and assist in reducing eutrophication and drinking water contamination. Therefore, it can be said that while there are several definitions of bio-fertilizers, there is a general understanding of what is a "nano-biofertilizer". Both biologically intermediated nano fertilizers and nano-biofertilizers have a number of comparable characteristics, most notably the ability to preserve and conserve agriculture while lowering the danger of environmental pollution and enhancing fertilization process efficiency (Mahanty et al. 2017). The combination of nano- and bio-fertilization must be thoroughly studied to produce food that is both safe and of the highest quality while also enhancing soil health (Sahu et al. 2022).

A combination of nanoparticles and bio-fertilizers is referred to as nano-biofertilizer. By gradually releasing nutrients, it can improve the efficiency with which plants utilize nutrients. By having long-lasting impacts on the physico-chemical and biological characteristics of the soil, it can enhance its attributes (Du et al. 2018). As a result of being protected from mechanical stress, the bacterial strains can thrive, and the product's efficacy is boosted by the sustained delivery of nutrients. Encapsulation is the process by which the bio-fertilizer cell is integrated into the nano-material capsule. Mixing starch with a harmless, biodegradable material like calcium alginate is involved. When starch is available, bacterial strains proliferate more rapidly (Vafa et al. 2021). The production of nano-biofertilizers entails three main steps: bio-fertilizer culture growth; its coating around nanoparticles, quality assessment, efficacy, purity, and shelf-life (Panichikkal et al. 2021a), just as the production of nano-fertilizers involves three approaches as shown in Figure 2. Microencapsulated nano-biofertilizers can also be manufactured. Two parts PGPR suspension to one part sodium alginate, three parts starch and four parts bentonite solution are required to make it. After the microcapsules have been cleansed with sterile deionized water, the solution is treated with the $CaCl_2$ solution used in crosslinking materials (Saberi-Rise and Moradi-Pour 2020). Also, the nanoparticles and salicylic acid have been used to develop a nano bio-fertilizer. For this method, bio-fertilizer is mixed with 2% sodium alginate, salicylic acid and 1.5 mM ZnO NPs (1 g/mL). Beads of 1 mm in diameter are formed, coated with a 3% solution of calcium chloride, allowed to air dry, and then incubated at 4°C in the solution (Panichikkal et al. 2021b). According to a review, organic waste from food, plants and animals may be blended with nanoparticles to synthesize potential nano-biofertilizer that improves fertility and nutrient status of soils. The organic waste was ground, rinsed with water to decontaminate and then decomposed or pyrolyzed. This pyrolyzed or partly decomposed garbage was combined with nanoparticles to make nano-biofertilizer (Singh et al. 2019, Akhtar et al. 2022).

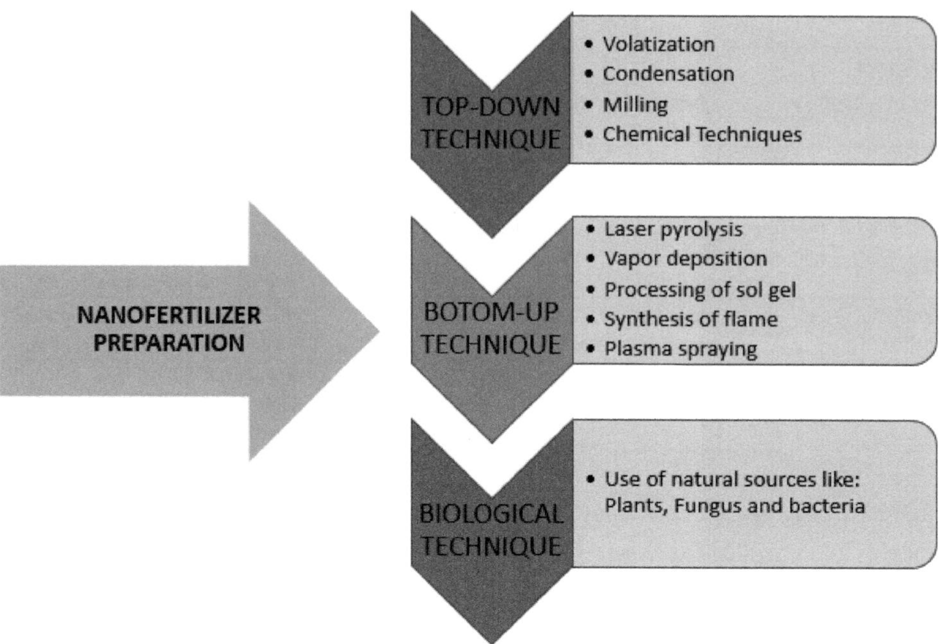

Figure 2. Preparation of nano-fertilizer.

Fate of nano-biofertilizers

It is commonly known that soil is essential for the survival of many terrestrial life forms, since it provides them with food, water, and a variety of other nutrients. Soil microbes are crucial to many nutrient biogeochemical cycles (such as C and N) and other soil minerals, yet soil is a non-renewable resource (Sathya et al. 2016). It's possible that the air, water, organic matter and minerals in this soil form a self-sustaining ecosystem (macro- and micro-organisms). Soil health, fertility and plant production are the primary factors influencing the sustainability of soil ecosystem (Seneviratne et al. 2017). Bio inoculants consist of several different beneficial microbial isolates, including biocontrol agents, nano-biofertilizers, and organic decomposers. These bacteria might give plants the nutrients they require, secrete various substances that encourage development, and give them resistance to a wide range of illnesses.

Nano-biofertilizers' fate and activity are profoundly influenced by the rhizosphere and vice versa. This suggests that both the positive (plant nutrition, protection, and agricultural yield) and the negative (rhizosphere biological state) elements of nano-biofertilizers have an impact on soil (Bhardwaj et al. 2014, Thul and Sarangi 2015, Dwiredi et al. 2016). The variety of microbial population, nutritional quality, soil fertility and crop yield are all impacted by bio- and nano-fertilizers. As a result, the agro-environment in which they are applied has a significant impact on the efficacy of both biologically mediated nano-fertilizers and bio-fertilizers. Numerous studies have been done on how nanoparticles interact with various environmental compartments, including the soil, plants, and microorganisms (Karimi and Mohseni 2017, Terekhova et al. 2017). Thus, the bioavailability, fate, translocation and phytotoxicity of nano-biofertilizers in soils are largely dependent on the soil's physio-chemical features (Benoit et al. 2013). Physical, chemical, and biological soil properties include things like clay content, salinity, pH, CEC, soil organic matter and microbial activity.

Typical reactions for nano-fertilizers in soils include agglomeration, transport, ionic metal dissolution, mobility, absorption, and nanoparticles sorption (Zhang et al. 2017). Soil properties must be regarded as a crucial determinant of the transport, dispersion, fractionation and consequent bioavailability of nanoparticles (or nano-biofertilizers) to plants in the soil. As previously noted, the features of the inoculant, the environment, and the soil characteristics all have an impact on the fate of the bio-fertilizers when applied (Akhtar et al. 2022). It is commonly revealing that nanotechnology is already being used for a variety of purposes, including wastewater treatment, targeted pesticide and fertilizer delivery, cancer treatments, focused medication delivery, cosmetic industries, electronics, and biosensors. The amount of money invested globally in the field of nanotechnology is predicted to have risen, reaching $10 billion in 2005 and $1 trillion in 2015 (Tripathi et al. 2017).

As previously indicated, handling or unintentional operations might cause produced nanoparticles to accidentally enter several environmental compartments such as soils, water, air and plants (Aziz et al. 2015, Prasad et al. 2016). According to previous research, these nanoparticles are primarily and ultimately absorbed by the soil, with the air serving as a secondary sink (Cornelis et al. 2014). If these nanoparticles are present in soil, they might pose a threat to the living creatures there as well as any plants that have been cultivated there. Many different nanoparticles may be transported by soil and absorbed by plants, leading to phytotoxicity (Tripathi et al. 2017). Contrarily, depending on the soil and fertilizer features, nano-fertilizers may be able to supply and retain the essential nutrient for plant nutrition. Therefore, as previously noted, a number of parameters will be used to limit the absorption, translocation, and phytotoxicity of nano-fertilizers. When compared to conventional fertilizers like urea, nano-fertilizers have been found to potentially release nutrients that are available for plant uptake over an extended length of time (Subramanian et al. 2009). Well-known microbial inoculants like bio-fertilizers aim to do one thing, and one thing only—provide nutrients to the soil and the plants that grow in it.

Soluble nutrients for plants are typically distributed by microbial communities, which may be aided by bio-fertilizers that convert inaccessible nutrient forms (Wong et al. 2015, Tomer et al. 2016, Mahanty et al. 2017). Direct and indirect processes regulate the nutrient uptake of bio- and nano-fertilizers. The beneficial functions of certain bacteria are the most frequent point of convergence between bio-fertilizers and biologically mediated nano-fertilizers. Therefore, it might be inferred that soil bacteria play an essential role in the soil rhizosphere through interactions with plant roots. These interactions include both direct effects (such as increased nutrient bioavailability in the rhizosphere) and indirect effects (such as improved nutrient absorption efficiency due to the stimulation of plant root development) (El-Ghamry et al. 2018). Nano-biofertilizers shows different functions in soils as shown in Figure 3.

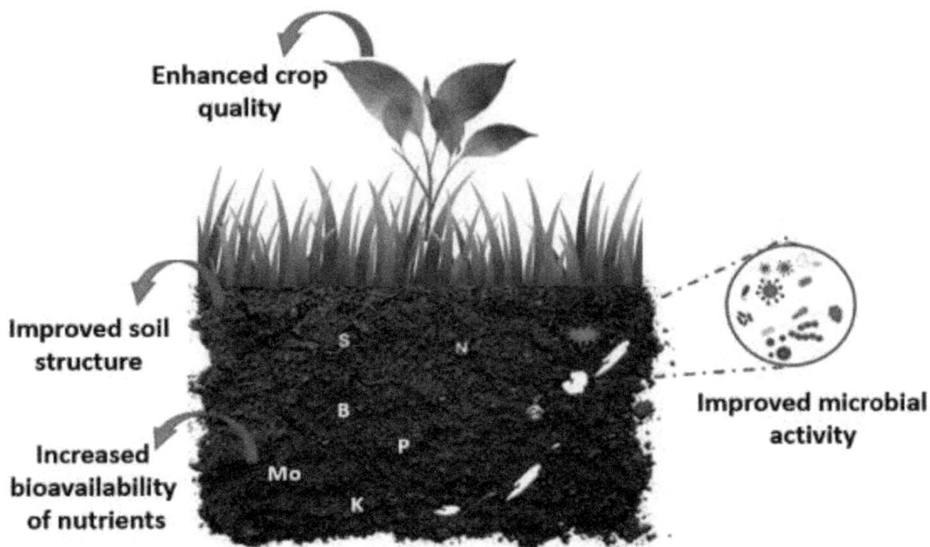

Figure 3. Nano-biofertilizers, also known as nano-fertilizers or nanobiotechnology-based fertilizers, offer several functions that can benefit plant growth, nutrient uptake, and overall crop productivity.

Role of nano-biofertilizers in crop production and protection

Improvement in crop production and nutrient status

Nano-biofertilizers increased crop yield and nutrient quality via many soil-plant interactions as shown in Table 2. Encapsulating nanomaterials like zeolite, chitosan and polymers provide a slow and steady delivery of essential nutrients to the plants (Qureshi et al. 2018). Nanoparticles coating on bio-fertilizer increased nutrient stability. The large surface area, nanoparticles size and better reactivity may allow more interaction and absorption of these nutrients to the plants with increased crop fertilization. This fertilizer has the potential to provide plants with sustainable bioavailable nutrients throughout their development. It reduces the application rate allowed for the delayed and continuous release of nutrients by nano-biofertilizer (Manjunatha et al. 2016). Bio-organic components (i.e., beneficial bacterial and fungal strains) improve the soil's nutrient status in several ways such as the atmospheric nitrogen fixation by the roots, siderophore production, metals chelation, phosphorus solubilization with the help of phosphorus-solubilizing bacteria and phyto-hyrdrolysis. Researchers have shown that the synergistic effects of the nanomaterial and the bio-fertilizer component of nano-biofertilizer lead to a more potent response, which manifests itself in increased crop growth, productivity, yield and quality features. Application of nano-biofertilizers has been credited with enhancing crop growth and quality because of the substantial advantages of these essential plant traits (Kumari and Singh 2020).

Table 2. Benefits of different nano-based fertilizers in the soil and plants.

Categories	Applications	Nano-particles	References
Soil Chemistry	Soil decontamination	Fe-NPs	Singh et al. 2018
	Soil aggregation stability	TiO_2, Au and Pt	Zhu et al. 2017
Soil microbiology	Increase microbial population (acidobacteria and cyanobacteria)	Ag-NPs	Grin et al. 2018
Plant growth and development	Plant growth stimulant	Fe-oxide NPs	Win et al. 2021
	Root growth and plant germination	Fe and Mn mono-and bimetallic NPs	Bettencourt et al. 2020
	Robust plant development	Carbon-based NPs	Ali et al. 2020
Bio-control	Prevent powdery mildew on grapevine leaves	Gold NPs	Rashad et al. 2021
	Manage bacterial infections in plants	Ag-NPs	Perumal et al. 2023
	Nematode control	Chitosan-NPs	Unsoy and Gunduz 2017
Abiotic stresses	Improve soil water retention	Si-nanosphere	Al-Mokadem et al. 2023
	Scavenge reactive oxidative stresses	Cu-NPs/ anti-oxidant enzymes	Verma et al. 2019
Bio-fortification	Nutritional bioavailability	Liposome, metal (Fe, Zn) NPs	Sambangi et al. 2022

Improvement in plant morphology and physiology

Application of nano-bio-fertilizer accelerates crop growth and enhances plant productivity by increasing photosynthesis, nutrient absorption, accumulation of photosynthates and nutrient movement to the economically important portions of the plant. They also enhance the plant growth through different biological mechanisms including siderophore production, mineral solubilization, i.e., phosphorus, phytohormone synthesis and biological nitrogen fixation in the form of metal chelates that are plant available. For this purpose, nitrogen fixing, phosphorus and potassium solubilizing bacteria are used so that these processes can speed up for the bioavailability and assimilation of nutrients to the plants (Sahu et al. 2022). The nano-biofertilizer used in agricultural crops was reportedly made by entrapping bio-fertilizer (such as *Bacillus subtilis, Pseudomonas fluorescence* and other growth promoting bacteria) between silver and gold nanoparticles, as reported by Dikshit et al. (2013). The crop-promoting abilities of this fertilizer were found to be dramatically enhanced. Some leguminous crops may benefit from the use of a neem cake + PGPR nanostructured fertilizer, which has been shown to increase agricultural production by improving the crop seedling germination and providing doped nutrients to crops more efficiently (Rahman and Zhang 2018).

Improvement in nutritional security of soil and plant system

Artificial fertilizers are continually depleting the essential nutrients from the soil where they are present naturally. One of the contributors to decreasing soil fertility is excessive soil acidity. Low fertility and nutrient imbalance are major concerns for farmers because of its correlation to diminishing crops yield and subpar food quality (Montreal et al. 2016). Nano-biofertilizers provide a long-term, cheap, and effective integrated nutrient management solution to these problems. In addition to lowering nutrient losses in the soil through gasification, leaching, and competition, it also increased plant nutrient absorption and assimilation (Janmohammadi et al. 2016). The nanoparticles coating of bio-fertilizer cause targeted nutrient release in a coordinated way with the crop demand (Morales-Diaz et al. 2017). The nano-biofertilizers also help nitrogen fixation, phosphorus solubilization and replenishment of soil nutrients.

Role in crop protection

About two million tons of pesticides are used annually across the world, with Europe accounting for about 45% of that total, the US for another 25%, and remaining 25% for rest of the countries (De et al. 2014). Unplanned and negligent use of pesticides increases the likelihood of pest and disease resistance, reduces the number of soil species, leads to pesticide biomagnification, reduces the number of pollinators and eliminates beneficial soil microorganisms. The benefits and applications of nanotechnology are just beginning to be explored. Examples include the use of nano-based herbicides and insecticides for insect control, and nano-encapsulated fertilizers for the slow and targeted release of nutrients to boost agricultural productivity (Duhan et al. 2017).

Nano-biofertilizer not only create resistance against insect and pathogens, but it also significantly improves crop protection (as seen in Figure 4), nutritional quality and water conservation potential. To determine the efficacy of nano-biofertilizers in mitigating effects of *Ralstonia solanacearum* on tomato crops, Gatahi et al. (2015) conducted a pest-resistant analysis. Nano-biofertilizer (including *Pseudomonas fluorescence, Bacillus subtilis*, and *Pseudomonas putida*) was reviewed by Gouda et al. (2018), who explained its protective effects against a variety of bacterial and fungal diseases in the rhizospheric region of leguminous crops. Nanobiofertilizers coated with titanium-NPs boosted beneficial bacteria to adhere onto the roots of oilseed, which in turn protected the crop against

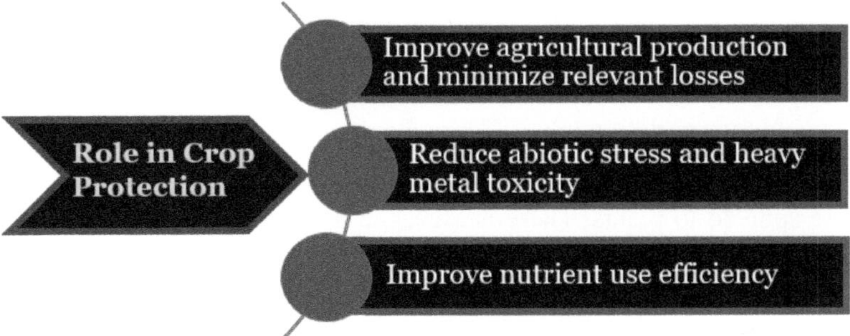

Figure 4. Nano-biofertilizers play a significant role in crop protection by helping to mitigate various biotic stresses such as plant diseases, pests, and pathogens.

damaging fungal infection, as shown by Mishra and Kumar (2009). Rabi crops can be protected against fungal-nematode diseases and abiotic stresses through application of nano-clay coated bio-fertilizers containing *Trichoderma* sp. and *Pseudomonas* sp. (Mukhopadhyay and De 2014).

Improving soil fertility

The unrestrained inorganic fertilizer usage nowadays has deteriorating effects on human health and health of ecosystem. It causes soil toxicity, eutrophication, and leaching, and it damages the structure of soil. For agriculture to be productive and sustainable, efforts must be made to control soil health. To satisfy the needs of crops without harming the environment, soil nutrients should be handled effectively (Mahapatra et al. 2022). For example, biofertilizers are involved in (a) replenishing nutrients through nitrogen fixation and phosphate solubilization (b) increasing the soil ability to absorb nutrients by the provision of suitable microorganisms (c) producing phytohormones and suitable solutes, making siderophores that chelate iron and sequester heavy metals to make them inaccessible to plants, preserving soil moisture, and fostering plant tolerance to biotic and abiotic stressors. (Sharma et al. 2021, Shafiel-Mosouleh 2022).

Constraints of nano-biofertilizers and their solutions

Effect on nutrient availability: Only the nutrients present are different in nano-biofertilizers made using organic nano materials. When compared to inorganic fertilizers, the nutrients' quantity is likewise insufficient. However, the uneven absorption of these nutrients by plants may be due to different factors such as nanoparticle size, soil pH and root interaction with the nanoparticles. All these attributes collectively affect the total plant growth and development. This problem can be solved by smart and targeted delivery system. It can reduce losses, enhance nutrient availability and adsorption by application of controlled-release nanoparticles and nano-fertilizers to certain plant parts in response to the plant needs and environment (Oh et al. 2023). Also, the nano-biofertilizers applied to the plants leach down into the soil. The fate of these nanoparticles depends on various soil factors such as properties, type and kind of soil.

Effect on soil microbes: Nano-biofertilizers based on synthetic, inorganic, macro- or micronutrients do not promote soil microbial life. They cause toxicity in the soil, affect the cell membrane, enzymatic and metabolic processes of soil microbes and thus kill the microbial population leading to decreased nutrient availability in soil and ultimately to the plants. These nanoparticles disrupt the ability of plant growth promoting rhizobacteria, affect the soil-microbes and plant-microbes symbiosis and inhibit the symbiotic nitrogen fixation etc. (Khodakovaskaya and Lahiani 2014, Sambangi et al. 2022). These limitations can be avoided by blending the two different types of nanomaterials. These nano-biofertilizers may be created by combining various nano-sized inorganic macro and micronutrients, depending on the kind and quantity of organic nanomaterials used. Metal nanoparticles display their antibacterial properties when used in nano-biofertilizers, such as mixed bio-fertilizers with metal nanoparticles. These live formulations must contain enough organic elements to support the bacterial growth. These are some of the difficulties in preserving microbial life. When the soil is unhealthy and deficient in minerals, these kinds of nano-biofertilizers are useless. Other restrictions include the shorter life span, a shortage of carrier material, sensitivity to high temperature and transportation issues (Thirughanasambandan 2018).

Effect on plants: Along with fertilizers, nutrients and insecticides, metal nanoparticles like gold nanoparticles are sprayed to plants. There is a chance that these nanoparticles will bind to metal nanoparticles when applied to plant proteins (Kumar et al. 2007). This can result in altered behavior and surroundings inside the plants. These outcomes might be positive, negative or both. Therefore, careful consideration and significant study are required before using metal nanoparticles on plants (Kanotop et al. 2014).

Effect on human: Nanomaterials' distinctive physio-chemical characteristics have hazardous effects. They can easily enter people through a variety of channels. These materials' tiny size and huge surface area make it simple for them to enter the human body and spread there, which causes nano toxicity. Their toxicity impact on human health is varied (Jain et al. 2018). Before using nanoparticles on a broad scale for agricultural uses, precautions should be followed. Animals and humans will eat these agricultural products. When consumed, they could have negative impacts on people's and animals' health due to nano toxicity (Thirughanasambandan 2018). Thus, a careful understanding of the selected nanoparticles and monitoring of methods used for the synthesis of nano-biofertilizers is required. The application rates and methods should also be improved in the addendum so that the soil properties and agricultural production is enhanced considering the environmental prospective (Sahu et al. 2022).

Effect on Cost: No doubt preparation of different type of nano-fertilizers may be costly due to limited resources in some areas making their use limited for agricultural purposes. Thus, nano-fertilizers may be manufactured by using possible cost-effective techniques that are scalable and commercially feasible for higher crop production (Su et al. 2022).

Effect on Environment: Numerous scientific research has proved that nanoparticles are harmful to plants due to their toxicity at higher dosage. This phytotoxicity is correlated with several factors such as nanomaterial content, size, surface charge, shape and amount of nanomaterials being used. This fact is also proved by different researchers. In order to improve and assure environmental security, a careful evaluation of feasible long-term effect of nano-fertilizers on soil health, plant growth, water quality and non-targeted creatures is needed. Thus, extensive research is needed to deal with all these nano-biofertilizer-ecosystem interactions (Vasseghian et al. 2022).

Summary

Given that nano-biofertilizers are currently in the growth stage, it is crucial in the current situation. In the future, these fertilizers will take the place of traditional fertilizers. Plant materials are diamagnetic; however, some plant nanoparticles have other characteristics (Theivasanthi 2014). The superparamagnetic property of certain plant nanoparticles is that they are temperature independent (*Acalypha indica, Cynodon dactylon, Terminalia chebula, Eugenia jambolana* and *Cassia auriculata*). Similarities may be seen in the behavior of nanomaterials generated from Curcuma longa and Cocos nucifera (size 20 nm and 18 nm respectively). However, the mixed vegetable powder (*Cyamopsis tetragonoloba, Vicia faba, Momordica charantia,* and *Abelmoschus esculentus*) and nanoparticles of Amorphophallus konjac (size 21 nm) do not exhibit superparamagnetic behavior (Thirughanasambandan 2018). Superparamagnetic nanoparticles can be used as carriers for proteins, oligonucleotides, liver enzymes, biomolecules, and antibodies among many other things. Additionally, these materials are used in many biological processes, such as magnetic cell and nucleic acid separation, as well as imaging procedures like magnetic resonance imaging (Hadef 2018).

These results showed that nanoparticles have a variety of properties, interact intimately with biomolecules, and could function inside the cells. These variances might result in unfavorable reactions or outcomes, difficulties, and negative environmental effects. These issues are the main challenges to the formation of such materials (Thirughanasambandan 2018). Before using such materials, these need to be adequately examined. It is possible to adhere to the standards (connected to nanomaterials) established by several organizations in order to discover the appropriate answers and take safety precautions. Additionally, techniques that are used in other domains may be applied when using nanoscale materials. An ISO definition of a nanomaterial is a material having internal and external dimensions, or surface structures all within the nanometer range of (1–100 nm) (Mansfield et al. 2017).

To label materials with a limited number of these particles as nanomaterials, however, would be inaccurate. The number of sizes may range between 1% and 50% statutory instances if the environment, health, safety, or competitiveness merits it (EU commission 2011). A concrete prepared by using Portland cementing agent that are smaller than 500 nm is referred to as nano-concrete. Cement particles typically have size range between a few nm to 100 mm. However, the micro cement's particle size is 5 m; additionally, size reduction is necessary to create nano cement (Balagura and Chong 2006). Nanomaterials are substances with particles smaller than 200 nm, according to the definition. The particle size of nanomaterials in cement, concrete, and construction-related products must be smaller than 500 nm. The range of nanomaterials' particle sizes, which is 1 to 100 nm, is explicitly stated in the set standards for nanomaterials (Norhasri et al. 2017).

Likewise, according to the EU Commission, the fraction of nanoparticles in nanomaterials can be reduced up to 50% if the environmental safety, human health or competitiveness call for it. According to the EU commission (2011), the unfavorable effects of nanomaterials can be reduced or avoided by reducing the proportion of nano sized particles in nano pesticides, nano bio-fertilizers, or nano-fertilizers or by blending them in to smaller amounts with conventional pesticides and bio-fertilizers or fertilizers. Nanomaterials' size varies between 1 and 100 nm. Particles larger than this size range show no change. Large size particles (those larger than 100 nm) are used in applications that are connected to building. Similarly, using insecticides, bio-fertilizers, and fertilizer materials (larger than 100 nm) will prevent any potentially harmful impacts of nanomaterials. Without any reservations, with large surface area they may be used in agriculture or for plant application like other traditional materials (Thirughanasambandan 2018).

Conclusion

Overusing fertilizers to increase agricultural yield has unintended environmental and ecological consequences, such as altered soil fertility, compromised plant nutrition, and a weakened ecosystem. It's crucial to enhance agricultural productivity with little environmental impact. Advanced nano-biotechnological tools and approaches can help make sustainable agriculture in the approaching era of agricultural mechanization. A nanoparticle loaded with organic fertilizers is an ideal "nutrient booster" because it permits the progressive and continuous nutrient release to plants, therefore achieving the nutrient supply to plants during the growth and development stages. Nano-bioertilizers may provide several advantages for plants, including better functional component stability, the use of micro dosages, reduced nutrient loss due to leaching and degradation, masked soil nutrient deficiencies, and enhanced crop output. Therefore, nano and bio based agricultural improvement has great potential and may prove to be an inexpensive and ecologically benign solution for the expansion of sustainable agriculture.

Future perspective

With limited space and resources, increasing agricultural productivity and dietary security requires an environmentally benign strategy, and this is where the cutting-edge perspective of biotechnology and nanotechnology for sustainable plant management originates. Nanotechnology provides a solution by nano-biofertilizers, a sector with prevailing future in the management of sustainable agriculture. In this way, nano-biofertilizers might act as a "nutrient booster", providing plants with a continuous supply of nutrients throughout their growth. Possibly, use of organic materials should be focused in tandem with nano-based fertilizers to enhance effective nutrient utilization and increased soil health.

Improved functional component stability, reduced nutrient loss through leaching and degradation, concealed soil nutrient deficiency, and enhanced crop production are some potential benefits of nano bio-slow-release fertilizer for plants. Because of the inherent low cost and little effect, nano-biofertilizers are often regarded as the wave of the future in environmentally responsible farming. However, the potential risks of using nanomaterials must be assessed before this technology can be properly applied in agriculture. Considering the positive and negative characteristics of nano-biofertilizers, it is crucial to make considerable efforts to promote cutting-edge research to appropriately compensate for the hazards associated with applications of nanoparticles and bioorganic materials.

- Overall, thorough research should be carried out to consider the benefits as well as the risks brought on by these nano-biofertilizers on soil and plant health. In this concern, a complete understanding of these nano-biofertilizers on various trophic levels and the process involved in their transformation should be kept under consideration.

- Any experiment designed to gauge the effects of nano-materials on the environment should be conducted in a natural setting for accuracy sake. Also, the viability of nano-biofertilizers in the agricultural industry cannot be completely advanced by laboratory experiments alone.

- Systematic and government-based security evaluations should be performed to confirm the allowed and safety limit of nanoparticle dosage. More study and clear definitions based on actual natural field settings are needed to manage the negative impacts of using organic waste.

- Biodegradability and bio magnification transfer effects must be accounted for to completely grasp the toxicity of micro bio-fertilizer applications on plants.

- Application of nanomaterials can significantly cause advancement in different sectors, but it requires an extensive research on the life cycle, disposal and environmental impact of these nanomaterials. For the sustainable and safe incorporation of nanomaterials in environment, possible understanding, responsible development and participation of the stakeholders is required.

References

Akhtar, N., Ilyas, N., Meraj, T.A., Pour-Aboughadareh, A., Sayyed, R.Z., Mashwani, Z.U.R. et al. 2022. Improvement of plant responses by nano-biofertilizer: a step towards sustainable agriculture. Nanomater., 12(6): 965.

Ali, M.A., Ahmed, T., Wu, W., Hossain, A., Hafeez, R., Islam Masum, M.M. et al. 2020. Advancements in plant and microbe-based synthesis of metallic nanoparticles and their antimicrobial activity against plant pathogens. Nanomater., 10(6): 1146.

Al-Mokadem, A.Z., Sheta, M.H., Mancy, A.G., Hussein, H.A.A., Kenawy, S.K., Sofy, A.R. et al. 2023. Synergistic effects of kaolin and silicon nanoparticles for ameliorating deficit irrigation stress in maize plants by upregulating antioxidant defense systems. Plants, 12(11): 2221.

Aziz, N., Faraz, M., Pandey, R., Shakir, M., Fatma, T., Varma, A. et al. 2015. Facile algae-derived route to biogenic silver nanoparticles: synthesis, antibacterial, and photocatalytic properties. Langmuir., 31(42): 11605–11612.

Bagherzade, G., Tavakoli, M.M., and Namaei, M.H. 2017. Green synthesis of silver nanoparticles using aqueous extract of saffron (*Crocus sativus* L.) wastages and its antibacterial activity against six bacteria. Asian Pac. J. Trop. Biomed., 7(3): 227–233.

Balaguru, P., and Chong, K. 2006. Nanotechnology and concrete: research opportunities. Proceedings of the ACI Session on Nanotechnology of Concrete: Recent Developments and Future Perspectives. 15–28.

Benoit, R., Wilkinson, K.J., and Sauvé, S. 2013. Partitioning of silver and chemical speciation of free Ag in soils amended with nanoparticles. Chem. Cent. J., 7(1): 1–7.

Bettencourt, G.M.D.F., Degenhardt, J., Torres, L.A.Z., de Andrade Tanobe, V.O., and Soccol, C.R. 2020. Green biosynthesis of single and bimetallic nanoparticles of iron and manganese using bacterial auxin complex to act as plant bio-fertilizer. Biocatal. Agric. Biotechnol., 30: 101822.

Bhardwaj, A.K., Arya, G., Kumar, R., Hamed, L., Pirasteh-Anosheh, H., Jasrotia, P. et al. 2022. Switching to nanonutrients for sustaining agroecosystems and environment: the challenges and benefits in moving up from ionic to particle feeding. J. Nanobiotechnology, 20(1): 1–28.

Bhardwaj, D., Ansari, M.W., Sahoo, R.K., and Tuteja, N. 2014. Bio-fertilizers function as key player in sustainable agriculture by improving soil fertility, plant tolerance and crop productivity. Microb. Cell Factories, 13(1): 1–10.

Bollag, J.M., Myers, C.J., and Minard, R.D. 1992. Biological and chemical interactions of pesticides with soil organic matter. Sci. Total Environ., 123: 205–217.

Chhipa, H. 2017. Nano-fertilizers and nano-pesticides for agriculture. Environ. Chem. lett., 15(1): 15–22.

Cornelis, G., Hund-Rinke, K., Kuhlbusch, T., Van den Brink, N., and Nickel, C. 2014. Fate and bioavailability of engineered nanoparticles in soils: a review. Crit. Rev. Environ. Sci. Technol., 44(24): 2720–2764.

De, A., Bose, R., Kumar, A., and Mozumdar, S. 2014. Targeted delivery of pesticides using biodegradable polymeric nanoparticles. New Delhi: Springer India, 10: 978–981.

Dikshit, A., Shukla, S.K., and Mishra, R.K. 2013. Exploring nanomaterials with PGPR in current agricultural scenario: PGPR with special reference to nanomaterials. Lap Lambert Academic Publishing.

Du, C., Abdullah, J.J., Greetham, D., Fu, D., Yu, M., Ren, L. et al. 2018. Valorization of food waste into bio-fertilizer and its field application. J. Clean. Prod., 187: 273–284.

Duhan, J.S., Kumar, R., Kumar, N., Kaur, P., Nehra, K., and Duhan, S. 2017. Nanotechnology: The new perspective in precision agriculture. Biotechnol. Rep., 15: 11–23.

Dwivedi, S., Saquib, Q., Al-Khedhairy, A.A., and Musarrat, J. 2016. Understanding the role of nanomaterials in agriculture. In Microbial Inoculants in Sustainable Agricultural Productivity, 2: 271–288.

El-Ghamry, A., Mosa, A.A., Alshaal, T., and El-Ramady, H. 2018. Nano-fertilizers vs. bio-fertilizers: new insights. Env. Biodivers. Soil Secur., 2: 51–72.

Eliaspour, S., Seyed Sharifi, R., Shirkhani, A., and Farzaneh, S. 2020. Effects of biofertilizers and iron nano-oxide on maize yield and physiological properties under optimal irrigation and drought stress conditions. Food Sci. Nutr., 8(11): 5985–5998.

Elkhatib, E.A., Mahdy, A.M., and Salama, K.A. 2015. Green synthesis of nanoparticles by milling residues of water treatment. Environ. Chem. Lett., 13: 333–339.

Elnahal, A.S., El-Saadony, M.T., Saad, A.M., Desoky, E.S.M., El-Tahan, A.M., Rady, M.M. et al. 2022. The use of microbial inoculants for biological control, plant growth promotion, and sustainable agriculture: A review. Eur. J. Plant Pathol., 1–34.

EU Commission. 2011. Recommendation on the definition of nanomaterial. Official Journal of the European Commission, 275: 38–40.

Gatahi, D.M., Wanyika, H., Kihurani, A.W., Ateka, E., and Kavoo, A. 2015. Use of bio-nanocomposites in enhancing bacterial wilt plant resistance, and water conservation in greenhouse farming. In: scientific conference proceedings, 41: 52.

Gouda, S., Kerry, R.G., Das, G., Paramithiotis, S., Shin, H.S., and Patra, J.K. 2018. Revitalization of plant growth promoting rhizobacteria for sustainable development in agriculture. Microbiol. Res., 206: 131–140.

Grun, A.L., Straskraba, S., Schulz, S., Schloter, M., and Emmerling, C. 2018. Long-term effects of environmentally relevant concentrations of silver nanoparticles on microbial biomass, enzyme activity, and functional genes involved in the nitrogen cycle of loamy soil. J. Environ. Sci., 69: 12–22.

Hadef, F. 2018. An introduction to nanomaterials. J. Environ. Nanotechnol., 1: 1–58.

Hussain, I., Singh, N.B., Singh, A., Singh, H., and Singh, S.C. 2016. Green synthesis of nanoparticles and its potential application. Biotechnol. Lett., 38(4): 545–560.

Jain, A., Ranjan, S., Dasgupta, N., and Ramalingam, C. 2018. Nanomaterials in food and agriculture: an overview on their safety concerns and regulatory issues. Crit. Rev. Food Sci. Nutr., 58(2): 297–317.

Janmohammadi, M., Navid, A., Segherloo, A.E., and Sabaghnia, N. 2016. Impact of nano-chelated micronutrients and biological fertilizers on growth performance and grain yield of maize under deficit irrigation condition. Biologija., 62(2).

Jha, M.N., and Prasad, A.N. 2006. Efficacy of new inexpensive cyanobacterial bio-fertilizer including its shelf-life. World J. Microbiol. Biotechnol., 22(1): 73–79.

Karimi, E., and Mohseni, F.E. 2017. Nanomaterial effects on soil microorganisms. In Nanoscience and plant–soil systems. Springer. Cham., 137–200.

Khan, M.R., and Rizvi, T.F. 2017. Application of nano-fertilizer and nano-pesticides for improvements in crop production and protection. In Nanoscience and plant–soil systems. Springer. Cham., 405–427.

Khodakovskaya, M.V., and Lahiani, M.H. 2014. Nanoparticles and plants: from toxicity to activation of growth. Handbook of Nano-Toxicology, Nanomedicine and Stem Cell Use In Toxicology, 121–130.

Konotop, Y.O., Kovalenko, M.S., Ulynets, V.Z., Meleshko, A.O., Batsmanova, L.M., and Taran, N.Y. 2014. Phytotoxicity of colloidal solutions of metal-containing nanoparticles. Cytol. Genet., 48(2): 99–102.

Kumar, S.A., Abyaneh, M.K., Gosavi, S.W., Kulkarni, S.K., Ahmad, A., and Khan, M.I. 2007. Sulfite reductase-mediated synthesis of gold nanoparticles capped with phytochelatin. Biotechnol. Appl. Biochem., 47(4): 191–195.

Kumari, R., and Singh, D.P. 2020. Nano bio-fertilizer: An emerging eco-friendly approach for sustainable agriculture. Proceedings of the National Academy of Sciences, India Section B: Biological Sciences, 90(4): 733–741.

Mahanty, T., Bhattacharjee, S., Goswami, M., Bhattacharyya, P., Das, B., Ghosh, A. et al. 2017. Bio fertilizers: a potential approach for sustainable agriculture development. Environ. Sci. Pollut., 24(4): 3315–3335.

Mahapatra, D.M., Satapathy, K.C., and Panda, B. 2022. Bio-fertilizers and nano-fertilizers for sustainable agriculture: Phycoprospects and challenges. Sci. Total Environ., 803: 149990.

Malusà, E., Pinzari, F., and Canfora, L. 2016. Efficacy of bio-fertilizers: challenges to improve crop production. In Microbial inoculants in sustainable agricultural productivity. Springer, New Delhi. 17–40.

Manjunatha, S.B., Biradar, D.P., and Aladakatti, Y.R. 2016. Nanotechnology and its applications in agriculture: A review. J farm. Sci., 29(1): 1–13.

Mansfield, E., Kaiser, D.L., Fujita, D., and Van de Voorde, M. (Eds.). 2017. Metrology and standardization for nanotechnology: protocols and industrial innovations. John Wiley and Sons.

Mishra, V.K., and Kumar, A. 2009. Impact of metal nanoparticles on the plant growth promoting rhizobacteria. Dig. J. Nanomater. Biostruct., 4(3): 587–592.

Monreal, C.M., DeRosa, M., Mallubhotla, S.C., Bindraban, P.S., and Dimkpa, C. 2016. Nanotechnologies for increasing the crop use efficiency of fertilizer-micronutrients. Biol. Fertil. Soils, 52(3): 423–437.

Morales-Díaz, A.B., Ortega-Ortíz, H., Juárez-Maldonado, A., Cadenas-Pliego, G., González-Morales, S., and Benavides-Mendoza, A. 2017. Application of nano-elements in plant nutrition and its impact in ecosystems. Advances in Natural Sciences: Nanoscience and Nanotechnology, 8(1): 013001.

Mukhopadhyay, R., and De, N. 2014. Nano clay polymer composite: synthesis, characterization, properties and application in rainfed agriculture. Global J. Bio. Biotechnol., 3(2): 133–138.

Nongbet, A., Mishra, A.K., Mohanta, Y.K., Mahanta, S., Ray, M.K., Khan, M. et al. 2022. Nano fertilizers: A Smart and Sustainable Attribute to Modern Agriculture. Plants, 11(19): 2587.

Norhasri, M.M., Hamidah, M.S., and Fadzil, A.M. 2017. Applications of using nano material in concrete: A review. Construction and Building Materials. 133: 91–97.

Oh, J.Y., Choi, E., Jana, B., Go, E.M., Jin, E., Jin, S. et al. 2023. Protein-precoated surface of metal-organic framework nanoparticles for targeted delivery. Small, 19: 2300218.

Pandey, V., and Chandra, K. 2016. Agriculturally important microorganisms as bio-fertilizers: Commercialization and regulatory requirements in Asia. In Agriculturally important microorganisms. Springer, Singapore, 133–145.

Panichikkal, J., Prathap, G., Nair, R.A., and Krishnankutty, R.E. 2021a. Evaluation of plant probiotic performance of *Pseudomonas* sp. encapsulated in alginate supplemented with salicylic acid and zinc oxide nanoparticles. Int. J. Bio. Macromol., 166: 138–143.

Panichikkal, J., Puthiyattil, N., Raveendran, A., Nair, R.A., and Krishnankutty, R.E. 2021b. Application of encapsulated Bacillus licheniformis supplemented with chitosan nanoparticles and rice starch for the control of *Sclerotium rolfsii* in *Capsicum annuum* (L.) seedlings. Curr. Microbiol., 78(3): 911–919.

Perumal, D., Khairuddin, N.F.M., Wong, J.H., and Abdullah, C.A.C. 2023. Plant Extract-Based Silver Nanoparticles and Their Bioactiviy Investigations. In Diversity and Applications of New Age Nanoparticles. IGI Global, 88–111.

Prasad, R., Pandey, R., and Barman, I. 2016. Engineering tailored nanoparticles with microbes: quo vadis. Wiley Interdisciplinary Reviews: Nanomedicine and Nanobiotechnology, 8(2): 316–330.

Pudake, R.N., Chauhan, N., and Kole, C. (Eds.). 2019. Nanoscience for sustainable agriculture. Springer International Publishing. 711.

Qureshi, A., Singh, D.K., and Dwivedi, S. 2018. Nano-fertilizers: a novel way for enhancing nutrient use efficiency and crop productivity. Int. J. Curr. Microbiol. App. Sci., 7(2): 3325–3335.

Rahman, K.M.A., and Zhang, D. 2018. Effects of fertilizer broadcasting on the excessive use of inorganic fertilizers and environmental sustainability. Sustainability.

Rashad, Y.M., El-Sharkawy, H.H.A., Belal, B.E.A., Abdel Razik, E.S., and Galilah, D.A. 2021. Silica nanoparticles as a probable antioomycete compound against downy mildew, and yield and quality enhancer in grapevines: field evaluation, molecular, physiological, ultrastructural, and toxicity investigations. Front. Plant Sci., 12: 763365.

Rudershausen, S., Grüttner, C., Frank, M., Teller, J., and Westphal, F. 2002. Multifunctional superparamagnetic nanoparticles for life science applications. Eur. Cells Mater., 3(2): 81–83.

Saberi-Rise, R., and Moradi-Pour, M. 2020. The effect of *Bacillus subtilis* Vru1 encapsulated in alginate–bentonite coating enriched with titanium nanoparticles against *Rhizoctonia solani* on bean. Int. J.biol. Macromol., 152: 1089–1097.

Sahu, S.K., Bindhani, S., Acharya, D.K., and Padhee, R.K. 2022. Development of nano bio fertilizer via chemical and biological synthesis. IRJMETS, 4(7): 1–17.

Sambangi, P., Gopalakrishnan, S., Pebam, M., and Rengan, A.K. 2022. Nano-bio fertilizers on soil health, chemistry, and microbial community: benefits and risks. Proceedings of the Indian National Science Academy, 1–12.

Sathya, A., Vijayabharathi, R., and Gopalakrishnan, S. 2016. Soil microbes: the invisible managers of soil fertility. In Microbial inoculants in sustainable agricultural productivity, Springer, New Delhi, 1–16.

Seneviratne, G., Weerasekara, M.L.M.A.W., Kumaresan, D., and Zavahir, J.S. 2017. Microbial signaling in plant—microbe interactions and its role on sustainability of agroecosystems. Agro-environmental Sustainability, 1: 1–17.

Shafiei-Masouleh, S.S. 2022. Use of magnetic nano-chitosan as bio-fertilizer to reduce production period in three cyclamen cultivars. J. Soil Sci. Plant Nutr., 22(1): 281–293.

Sharma, A.M.E.E.T.A., Patel, S.A.K.S.H.I., and Menghani, E.K.T.A. 2021. Synthesis, application and prospects of nano-bio fertilizers: A reappraisal. J. Phytol. Res., 34: 79–85.

Singh, P., Ghosh, D., Manyapu, V., Yadav, M., and Majumder, S. 2019. Synergistic impact of iron (iii) oxide nano-particles and organic waste on growth and development of *Solanum lycopersicum* plants: New paradigm in nano bio-fertilizer. Plant Arch., 19(1): 339–344.

Singh, V.K., Singh, A.L., Singh, R., and Kumar, A. 2018. Iron oxidizing bacteria: insights on diversity, mechanism of iron oxidation and role in management of metal pollution. Environ. Sustain., 1: 221–231.

Sonali, J.M.I., Kavitha, R., Kumar, P.S., Rajagopal, R., Gayathri, K.V., Ghfar, A.A. et al. 2022. Application of a novel nanocomposite containing micro-nutrient solubilizing bacterial strains and CeO2 nanocomposite as bio-fertilizer. Chemosphere, 286: 131800.

Su, Y., Zhou, X., Meng, H., Xia, T., Liu, H., Rolshausen, P. et al. 2022. Cost–benefit analysis of nanofertilizers and nanopesticides emphasizes the need to improve the efficiency of nanoformulations for widescale adoption. Nat. Food, 3(12): 1020–1030.

Subbarao, C.V., Kartheek, G., and Sirisha, D. 2013. Slow release of potash fertilizer through polymer coating. Int. J. Appl. Eng., 11(1): 25–30.

Subramanian, K.S., Sharmila, and Rahale, C. 2009. Synthesis of nano-fertilizers formulations for balanced nutrition. In: Proceedings of the Indian Society of Soil Science-Platinum Jubilee Celebration, 22–25 December, New Delhi.

Terekhova, V., Gladkova, M., Milanovskiy, E., and Kydralieva, K. 2017. Engineered nanomaterials' effects on soil properties: problems and advances in investigation. In Nanoscience and plant–soil systems, Springer, Cham. 115–136.

Theivasanthi, T. 2014. Transition to Superparamagnetism of Diamagnetic Nanoparticles.

Thirughanasambandan, T. 2018. Advances and trends in nano bio-fertilizers.

Thul, S.T., and Sarangi, B.K. 2015. Implications of nanotechnology on plant productivity and its rhizospheric environment. In Nanotechnology and Plant Sciences. Springer, Cham., 37–53.

Timmusk, S., Seisenbaeva, G., and Behers, L. 2018. Titania (TiO2) nanoparticles enhance the performance of growth-promoting rhizobacteria. Sci. Rep., 8(1): 1–13.

Tomer, S., Suyal, D.C., and Goel, R. 2016. Bio-fertilizers: a timely approach for sustainable agriculture. In Plant-microbe interaction: an approach to sustainable agriculture. Springer, Singapore, 375–395.

Tripathi, D.K., Singh, S., Singh, S., Pandey, R., Singh, V.P., Sharma, N.C. et al. 2017. An overview on manufactured nanoparticles in plants: uptake, translocation, accumulation and phytotoxicity. Plant Physiol. Biochem., 110: 2–12.

Unsoy, G., and Gunduz, U. 2017. Targeted drug delivery via chitosan-coated magnetic nanoparticles. In Nanostructures for Drug Delivery; Elsevier: Amsterdam, The Netherlands, 835–864.

Vafa, Z.N., Sohrabi, Y., Sayyed, R.Z., Luh Suriani, N., and Datta, R. 2021. Effects of the combinations of rhizobacteria, mycorrhizae, and seaweed, and supplementary irrigation on growth and yield in wheat cultivars. Plants, 10(4): 811.

Vander Gheynst, J.S., Scher, H., and Guo, H.Y. 2006. Design of formulations for improved biological control agent viability and sequestration during storage. Ind. Biotechnol., 2(3): 213–219.

Vandergheynst, J., Scher, H., Guo, H.Y., and Schultz, D. 2007. Water-in-oil emulsions that improve the storage and delivery of the biolarvacide *Lagenidium giganteum*. BioControl., 52(2): 207–229.

Vasseghian, Y., Arunkumar, P., Joo, S.W., Gnanasekaran, L., Kamyab, H., Rajendran, S. et al. 2022. Metal-organic framework-enabled pesticides are an emerging tool for sustainable cleaner production and environmental hazard reduction. J. Clean. Prod., 373: 133966.

Verma, M.L., Kumar, P., Sharma, D., Verma, A.D., and Jana, A.K. 2019. Advances in Nanobiotechnology with Special Reference to Plant Systems. pp. 371–387. *In*: Prasad, R. (Ed.). Plant Nanobionics. Nanotechnology in the Life Sciences; Springer International Publishing: Cham. Switzerland.

Win, T.T., Khan, S., Bo, B., Zada, S., and Fu, P. 2021. Green synthesis and characterization of Fe3O4 nanoparticles using Chlorella-K01 extract for potential enhancement of plant growth stimulating and antifungal activity. Sci. Rep., 11: 1Plant Nanobionics11.

Wong, W.S., Tan, S.N., Ge, L., Chen, X., and Yong, J.W.H. 2015. The importance of phytohormones and microbes in biofertilizers. In Bacterial metabolites in sustainable agroecosystem. Springer. Cham., 105–158.

World Health Organization. 2021. The State of Food Security and Nutrition in the World 2021: Transforming food systems for food security, improved nutrition and affordable healthy diets for all (Vol. 2021).

Wu, S.C., Cao, Z.H., Li, Z.G., Cheung, K.C., and Wong, M.H. 2005. Effects of bio-fertilizer containing N-fixer, P and K solubilizers and AM fungi on maize growth: a greenhouse trial. Geoderma., 125(1-2): 155–166.

Zhang, W., Musante, C., White, J.C., Schwab, P., Wang, Q., Ebbs, S.D. et al. 2017. Bioavailability of cerium oxide nanoparticles to *Raphanus sativus* L. in two soils. Plant Physiol. Biochem., 110: 185–193.

Zhu, X., Chen, H., Li, W., He, Y., Brookes, P.C., Whit, R. et al. 2017. Evaluation of the stability of soil nanoparticles: the effect of natural organic matter in electrolyte solutions. Eur. J. Soil Sci., 68.

12

Nanofibers with Functional Properties for Water Purification

Kamel A. Abd-Elsalam,[1,*] *Toka E. Abdelkhalek,*[2]
Rawan K. Hassan[2] and *Chandra S. Seth*[3]

Introduction

In the intricate fabric of our planet's ecosystems, water stands as an irreplaceable force, nurturing life and preserving a delicate balance (Kumar et al. 2014). However, amid rapid human activity and industrial growth, a looming threat shadows the purity of this invaluable resource (Badgar et al. 2022). This chapter opens with a poignant recognition of the escalating crisis of water pollution gripping our world (Botes and Cloete 2010). From pristine streams to vast oceans, pollutants infiltrate every corner of our water bodies (Ma and Hsiao 2018). Agricultural runoff carrying pesticides, industrial effluents laden with heavy metals, and unchecked urban waste collectively compromise the safety and integrity of our freshwater sources (Peydayesh and Mezzenga 2021). This backdrop of increasing contamination paints a stark reality, underscoring the urgent necessity for effective water purification solutions (Nalbandian et al. 2015).

As populations soar and industries expand, the pursuit of clean, drinkable water becomes increasingly urgent, emphasizing the need for innovative approaches to tackle this global challenge (Pasini et al. 2021). Amid these challenges, a glimmer of hope emerges in the form of nanofibers. The subsequent section shifts focus to explore the enigmatic realm of nanofibers and their profound potential in revolutionizing methods of water treatment (Liang et al. 2011). Despite their minuscule size, nanofibers possess exceptional versatility (Homaeigohar and Elbahri 2014). Their distinct structural characteristics and adaptable functional properties pave the way for groundbreaking applications, especially in the realm of water purification (Nalbandian et al. 2015). Throughout this chapter, we embark on a journey to unveil the essence of nanofibers, probing their inherent traits and burgeoning role in surmounting the obstacles in water treatment (Figure 1). Exploring the junction of escalating water pollution and the urgent need for resilient purification techniques

[1] Plant Pathology Research Institute, Agricultural Research Center, Giza 12619, Egypt.
[2] Biotechnology English Program, Faculty of Agriculture, Cairo University, Giza 12613, Egypt.
[3] Department of Botany, University of Delhi, New Delhi-110007, Delhi.
* Corresponding author: kamelabdelsalam@gmail.com

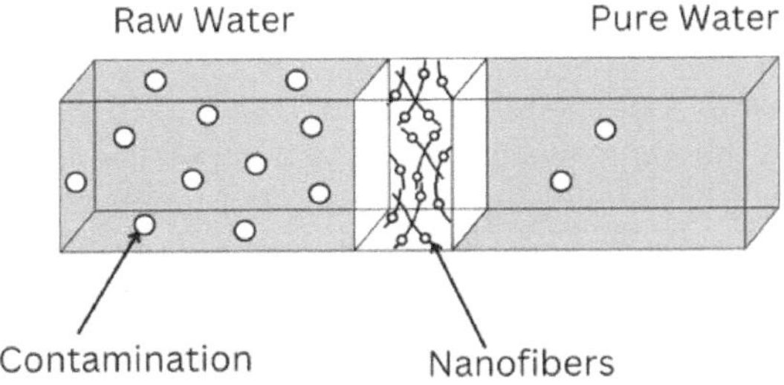

Figure 1. Water Purification Method by Nanofibers.

(Liang et al. 2011), we unravel the abundant opportunities that nanofibers offer in shaping a future characterized by sustainable and efficient water treatment solutions (Badgar et al. 2022).

The objective of this chapter is to explore the application of nanofibers in water treatment, emphasizing their unique properties. Specific objectives include introducing nanofibers, categorizing types, examining functional properties relevant to water purification, highlighting practical applications, addressing associated challenges, and concluding with key insights. Additionally, the chapter seeks to propose future research directions and contribute to a deeper understanding of nanofibers as innovative solutions for sustainable water treatment.

Types of nanofibers for water purification

The demand for continuous progress in filtration technology has led to limited exploration of advanced materials, specifically nanofiber membranes utilized in water distillation. The issue of organic matter and the accumulation of trace organics in wastewater stands as a formidable obstacle. Existing types of nanofibers, including polymeric nanofibers and composite nanofibers, are found to be inadequate in achieving the desired results (Figure 2). Furthermore, the heightened volume of sludge generated by these processes necessitates additional measures for processing and proper disposal (Tlili et al. 2019).

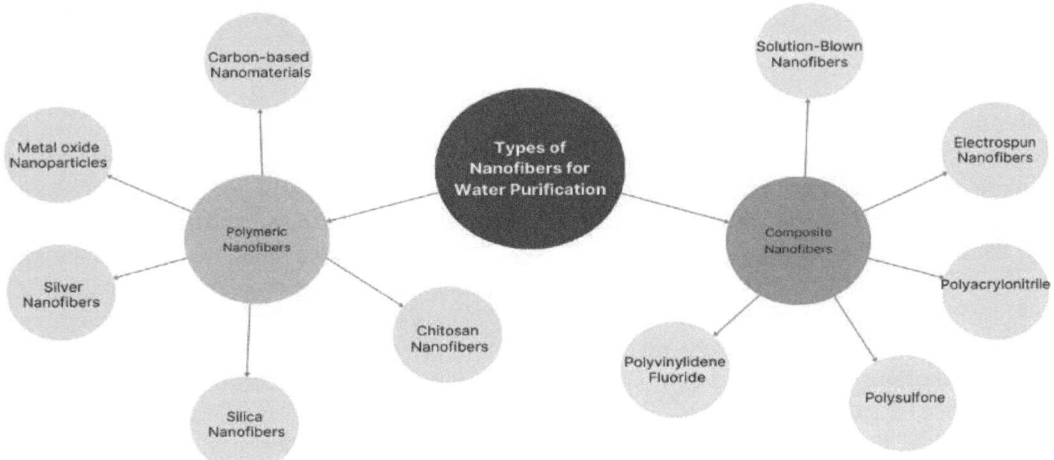

Figure 2. Different types of nanofibers used in water purification.

Polymeric nanofibers

Polymeric films coated with nanomaterials play a crucial role in advanced water treatment, addressing the removal of microorganisms, chemical compounds, heavy metals, and more. The integration of nanomaterials has proven instrumental in bolstering water resistance, mitigating the accumulation of pollutants, reducing contamination, improving reject efficiency, and enhancing mechanical properties and thermal stability (Salim et al. 2021). Due to the growing understanding of synthesizing and processing nanoscale materials, researchers globally favor polymeric nanomaterials for their novelty and significantly enhanced physicochemical properties. A primary reason for their widespread adoption is the recognition of polymeric membranes as valuable materials across various applications. These membranes are cost-effective, exhibit excellent separation selectivity, and deliver superior overall membrane performance (El-Aswar et al. 2022).

Polyvinylidene fluoride (PVDF)

Polyvinylidene fluoride (PVDF) stands at the forefront of polymer systems, renowned for its unique piezoelectric, pyroelectric, and ferroelectric properties. This chapter delves into recent advancements in nanotechnology, specifically the exploration of PVDF nanofibers, and unravels their potential applications in nanosensors, actuators, water purification, and energy generation, harnessing the remarkable piezoelectric capabilities of PVDF (Mokhtari et al. 2016). PVDF's appeal lies in its multifaceted properties. With heat resistance spanning –40 to +140°C, it exhibits durability, ductility, impact strength, and flexibility (Tansel 2021, Rakhmankulov et al. 2019).

Its pure composition, devoid of UV stabilizers and additives, aligns with eco-friendly practices, making it suitable for engineering and food applications. Additionally, PVDF's fire resistance, ease of processing, and adaptability through modification contribute to its reliability and longevity (Smejkalová et al. 2021, Kerbow et al. 1997). Critical processing parameters during electrospinning, such as applied voltage, solution flow rate, and collector distance, define PVDF's journey from polymer to nanofiber (Beachley and Wen 2009, Deitzel et al. 2001). Solution properties, including solvent ratio, viscosity, conductivity, solvent dielectric, and rates of solvent evaporation and cooling, intricately influence the electrospinning process, impacting the morphology and characteristics of electrospun PVDF fibers.

Polysulfone (PSf)

Polysulfone (PSF) membranes have emerged as pivotal components in wastewater treatment due to their intrinsic stability, mechanical robustness, and versatile modification capabilities. In recent explorations, a comprehensive investigation into the modification of PSF ultrafiltration (UF) and microfiltration (MF) membranes reveals significant strides in water purification. Techniques such as coating, grafting/blending, layer by layer, and deposition are detailed, showcasing marked improvements in water permeability, salt rejection, and anti-fouling properties when compared to conventional PSF membranes (Mamah et al. 2021). Within the broader scope of membrane modification for wastewater treatment, commonly employed methods, including physical blending, chemical grafting, coating, and layer by layer techniques, are integral to enhancing membrane performance (Alpatova et al. 2013).

Recognized as a preferred polymer for membrane production, PSF boasts excellent physicochemical properties. The chapter provides an insightful overview of recent developments in the application and modification of PSF membranes, addressing challenges related to membrane fouling and wetting, thereby guiding future advancements in PSF membrane applications (Kheirieh et al. 2018). Despite the hydrophobic nature inherent in PSF membrane surfaces, a critical review explores intelligent modifications. By incorporating metal nanoparticles into the polymer matrix, this strategy presents a promising avenue for reducing fouling potential and enhancing membrane performance in wastewater treatment applications (Heidi Lynn et al. 2021).

Polyacrylonitrile (PAN)

Polyacrylonitrile (PAN) stands as a significant synthetic resin obtained through the polymerization of acrylonitrile. A member of the acrylic resins family, PAN exhibits characteristics that make it a formidable material in various applications. PAN is described as a hard, rigid thermoplastic with resistance to solvents, slow combustibility, and low gas permeability (Britannica 2023). PAN-based membranes have garnered attention for combining chemical stability with effective performance in aqueous filtration. While commercially available PAN membranes are known to be brittle, innovative manufacturing processes, such as the one developed by GKSS, have addressed this limitation. The impregnation of substances like glycerol is no longer necessary, and these membranes maintain their filtration properties even when dry (Scharnagl and Buschatz 2001, Fester 1975).

The versatility of PAN extends to its applications in gray and black water treatment, as well as in the recycling of industrial waters, such as in the pulp and paper industry. Notably, PAN finds application in dead-end filtration processes, exemplified by the treatment of Elbe River water (Schamagl and Buschatz 1998, Schamagl and Buschatz 1999). PAN filtration membranes, produced with specified cut-offs and permeate fluxes, prove adaptable to various module types, showcasing their wide-ranging applicability. Furthermore, PAN membranes can undergo cleaning processes involving chemical agents like citric acid, sodium hypochlorite, acetic acid, weak sodium hydroxide, and hydrogen peroxide without suffering damage (Schamagl et al. 2000).

Electrospun nanofibers

In the realm of water treatment, electrospun polyvinylidene fluoride (PVDF) nanofiber membranes (ENMs) have demonstrated remarkable capabilities. A study revealed the successful desalination of saline water (6% NaCl content) for over 20 days without membrane wetting, showcasing the resilience and efficiency of PVDF ENMs (Feng et al. 2013). Beyond desalination, PVDF ENMs prove instrumental in the gas-stripping of chloroform, representing volatile organic compounds (VOCs). Notably, their mass transfer coefficients surpass those of hollow fiber membranes, indicating superior performance in VOC removal.

The versatility of ENMs extends further into membrane adsorption applications (Uyar et al. 2009, Sueyoshi et al. 2010). Chiral isomer separation and removal of hazardous substances underscore the potential of ENMs in diverse water purification contexts. In the realm of desalination, ENMs contribute significantly to membrane distillation. This highlights the broader applicability of ENMs in addressing various water treatment challenges (Feng 2009). Understanding the significance of ENMs requires insight into nanofiber fabrication techniques (Huang et al. 2003). Electro-spinning enables the production of nanofibers from a range of materials, enhancing the conducting and ceramic properties of inorganic materials like metal oxides. The electro-spinning process itself involves applying a high voltage to the solution passing through a spinneret. This unique method, distinct from conventional spinning, creates a repulsive force that shapes nanofibers with diverse configurations and assemblies (Gopal et al. 2006).

Composite nanofibers made of polycaprolactone, polyacrylic acid, and graphene oxide were created; PCL was taken into consideration since it possessed the right mechanical and physical qualities, together with PAA and GO, to maximize adsorption effectiveness (Rostami et al. 2024). Investigations were conducted on the absorption of lead heavy metal in water environments and the first use of this adsorbent nanocomposite to adsorb lead metal in apple juice and food systems (Figure 3).

Solution-blown nanofibers

Nanofibrous membranes have gained significant attention in microfiltration due to their unique properties. The solution blowing process, a novel nanofiber fabrication method with high productivity, was employed to successfully produce a polyvinylidene fluoride (PVDF) nanofibrous

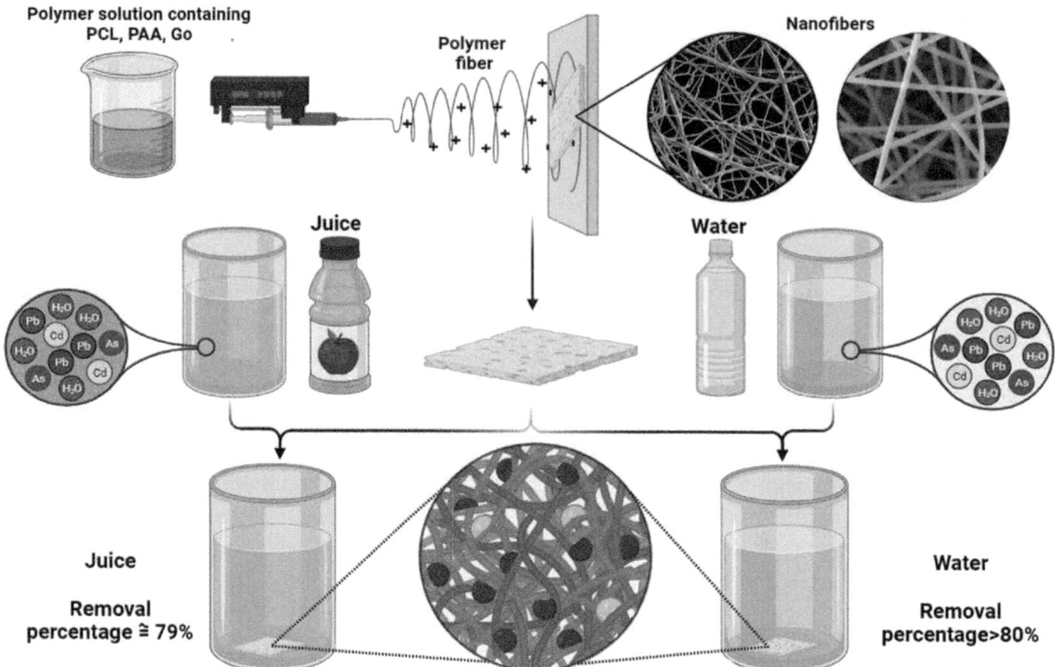

Figure 3. Diagram showing how PCL/PAA/GO nanofibers are made via electrospinning and how they are used to remove lead from apple juice and water (Reprinted from Rostami et al. 2024).

mat using a multiorifices die (Zhuang et al. 2013). The resulting fibers, ranging mostly from 60 to 280 nm in diameter, exhibited a three-dimensional curly structure, leading to a loose construction with high porosity (95.8%). Hot-pressing further enhanced the membrane's integrality, although it resulted in increased crystallinity, decreased porosity, and reduced pore size.

In the solution blowing process, spinning solution streams are blown into jets by high-velocity gas flow at the solution/gas interface. This method, akin to electrospinning but with different fiber formation driving forces, utilizes high-velocity gas flow to deform solution streams, evaporate the solvent, and solidify them into fibers (Zhuang et al. 2012). Filtration efficiency of commercially available filter media was significantly improved by incorporating electrospun nanofibers and ultrafine supersonically blown 20–50 nm nanofibers. Dual-coated filters with electrospun nanofibers deposited first and solution-blown 20–50 nm nanofibers on top proved most effective in removing below-200 nm Cu nanoparticles/clusters from aqueous suspensions. The smallest solution-blown nanofibers demonstrated higher efficiency in particle retention, attributed to attractive van der Waals forces (Sinha-Ray et al. 2015).

Solution blow spinning (SBS), a versatile method for fabricating nanofiber materials, combines the advantages of electrospinning and melt-blown spinning. Its simplicity, safety, high spinning efficiency, and suitability for a wide range of solution systems make it a valuable tool for controlling nanofiber structure and diameter (Daristotle et al. 2016). The evolution of nanofiber fabrication techniques, such as solution blowing and solution blow spinning, has paved the way for innovative applications in microfiltration and pollutant removal, showcasing the versatility and efficiency of nanofiber membranes in various fields (Khalid et al. 2017).

Composite nanofibers

Nanoscale polymer fibers have been successfully fabricated to create innovative membranes for water purification. These fibers possess several advantageous characteristics, such as high surface area and adjustable physical properties, which are crucial for effective filtration systems.

Important attributes include pore size, porosity, and hydrophilicity (Liu et al. 2019). Polymer membranes derived from nanofibers are commonly prepared using the electrospinning method, which utilizes the electrostatic charge present in the polymer solution jet. The resulting materials' overall physical properties are significantly influenced by factors like the diameter/thickness of each nanofiber strand and the incorporation of additives, such as fillers.

Carbon-based nanomaterials

Toxic metals and organic pollutants have emerged as significant environmental concerns due to their potential to cause irreparable harm to human health and aquatic organisms. Various materials have demonstrated considerable potential for removing conventional contaminants and purifying water. Among these materials, carbon nanofibers (CNFs) have garnered attention as promising candidates for water purification, thanks to their distinctive physical and chemical properties (Zhou et al. 2020a,b).

The electrospinning method was effectively used to create CNFs for the filtration and purification of water (Jamil et al. 2021). Following water filtration, Figure 4 displays the SEM and EDX pictures of the CNF filter, clearly displaying the contaminated particles.

Carbon-based nanomaterials, in general, have gained recognition for their excellent remediation capabilities and customizable properties in tackling pollutants effectively. Specifically, CNFs have shown high potential in eliminating pollutants from water due to their advantages, including ease of preparation, chemical stability, low cost, and environmental friendliness. CNFs exhibit a remarkable ability to remove toxic metals through multiple interactions such as π-π interaction, electrostatic interaction, and surface complexation (Wang et al. 2022). They also exhibit positive effects on the removal of organic compounds. To further enhance the application of CNFs, researchers have developed a series of CNF-based composites. CNFs, with their one-dimensional morphology characterized by a high aspect ratio and mechanical flexibility, have emerged as materials of interest. They possess a hexagonal carbon arrangement, where each carbon atom is bonded to four others.

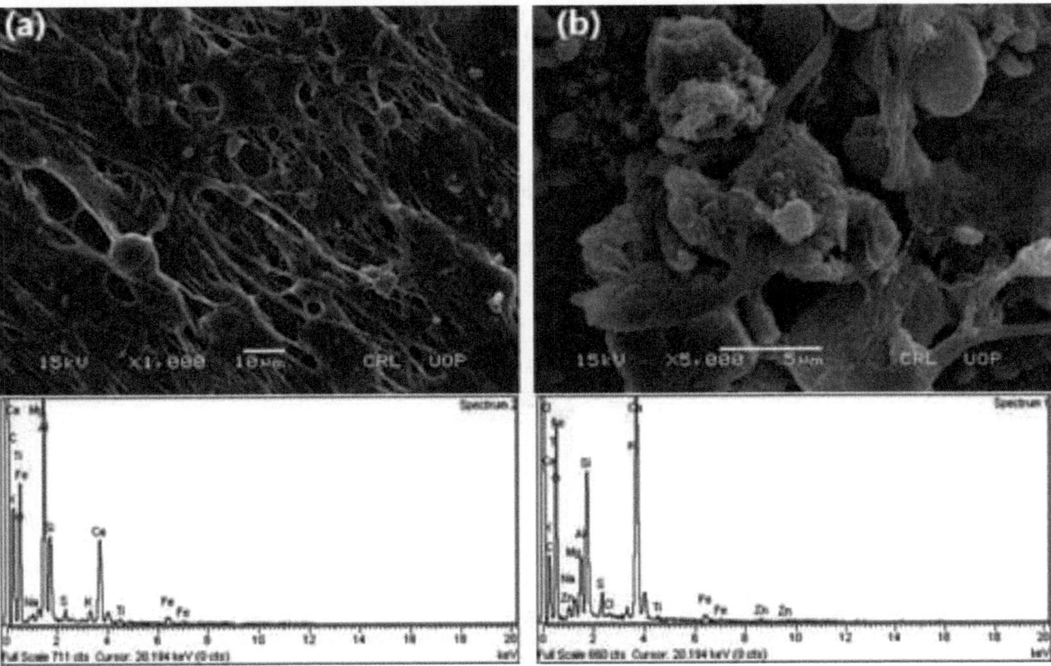

Figure 4. SEM and EDX outcomes of the CNF filter post-filtration are presented at two different resolutions: (a) captured at a resolution of 10 μm and (b) captured at a resolution of 5 μm (Cited from Jamil et al. 2021).

In this structure, three carbon atoms are sp2 hybridized with a bond length of 1.415 Å, while an additional carbon atom is connected through molecular accumulation, resulting in an interlayer spacing of 3.35 Å.

Metal oxide nanoparticles (e.g., TiO_2, ZnO)

By synthesizing nanostructured catalysts and optimizing light exposure wavelengths based on the selected semiconductor material's bandgap energy, the degradation of pollutants in wastewater can be improved. Photocatalytic oxidation processes using metal oxide nanomaterials and nanocomposites, such as zinc oxide, titanium dioxide, and zirconium dioxide, have gained significant research attention. These processes utilize ultraviolet (UV) and visible light, and even solar energy, to treat wastewater (Mondal et al. 2017).

The synthesis and engineering of a semiconductor-semiconductor (S-S) heterojunction with a low bandgap energy is being pursued for advanced wastewater treatment. This involves the use of nanostructured photocatalysts, primarily titanium dioxide (TiO_2) and zinc oxide (ZnO) nanoparticles, to enhance the degradation of refractory organic compounds in agricultural and industrial wastewater. By combining different treatment methods and optimizing light exposure wavelengths, the efficiency of pollutant removal can be improved. The photocatalysts utilize light energy to generate hydroxyl radicals, which completely oxidize the organic pollutants (Liu et al. 2010). The process also offers the potential for utilizing solar and UV spectra. A composite material, CTZO, is created by embedding TiO_2 nanoparticles within a polymeric matrix and depositing ZnO nanoparticles on the surface. This composite undergoes a calcination/annealing process to obtain a composite S-S heterojunction nanostructured material referred to as TZO. This approach demonstrates effective photo-degradation and antibacterial properties for wastewater treatment (Mousa et al. 2021).

Silver nanofibers

Among the various types of composite nanofibers, silver nanofibers have shown remarkable antimicrobial properties. The incorporation of silver nanoparticles or nanowires into the nanofiber structure enables the release of silver ions, which exhibit potent antimicrobial activity against a wide range of microorganisms, including bacteria, viruses, and fungi. This makes silver nanofibers highly effective in disinfecting water and preventing the growth of harmful pathogens (Chinthalapudi et al. 2021).

The fabrication of silver nanofibers can be achieved through different techniques, such as electrospinning or electrochemical deposition. These methods allow for the precise control of nanofiber morphology, size, and silver content, resulting in tailored properties for water purification applications. The inherent flexibility and high porosity of nanofibers also facilitate efficient water flow and contact between contaminants and the silver nanomaterial, enhancing the purification process (Zhang et al. 2016). Silver nanoparticles, hold great promise for advanced water purification systems. Their unique characteristics and antimicrobial properties make them suitable for various applications, ranging from point-of-use water filters to large-scale water treatment facilities, contributing to the provision of clean and safe drinking water.

Silica nanofibers

Silica-based composite nanofibers have excellent adsorption capabilities, making them highly effective for removing contaminants from water. These nanofibers have a large surface area and a porous structure that enhances their adsorption capacity and selectivity. They can be fabricated using techniques like electrospinning or sol-gel processing, allowing for tailored composition and structure (Beiranvand et al. 2023). Composite nanofibers, particularly those incorporating silica, offer a sustainable and cost-effective solution for water purification. They combine the strengths of polymers and silica nanoparticles, providing improved mechanical strength, chemical stability, and filtration efficiency. These nanofibers can remove a wide range of pollutants, including heavy metals and organic compounds, and their membranes can be customized for specific applications.

One of the most important silica nanofiber composites is the TiO_2-coated YSZ/silica composite nanofiber membrane, which demonstrated high separation ability and excellent adsorption/degradation efficiency for substances like humic acid, methylene blue, and tetracycline. The membrane also showed good reusability through multiple recycling tests (Huh et al. 2020).

Chitosan nanofibers

Chitosan exhibits a unique combination of properties, including enhanced mechanical strength and adsorption capabilities, making them ideal for removing contaminants from water. Silica's high surface area allows for the effective capture of pollutants like heavy metals and microorganisms. Techniques such as electrospinning are used to create nanofibrous membranes with controlled pore sizes, offering efficient and customizable filtration (Cárdenas Bates et al. 2021).

Composite nanofibers provide a sustainable and cost-effective approach to water treatment, suitable for various applications from point-of-use filters to large-scale industrial plants. The integration of silica represents a significant advancement, addressing water quality challenges with efficient, selective, and scalable solutions. The potential for regeneration and reusability further enhances the appeal of silica-based nanofibers in ensuring clean and safe drinking water (Jiang et al. 2018).

Functional properties of nanofibres for water purification

Nanofibers exhibit remarkable functional properties that make them highly effective in the domain of water purification. At the forefront of these properties is **adsorption**, a pivotal mechanism for the removal of contaminants. The multifaceted adsorption properties of nanofibers encompass sophisticated surface functionalization strategies and selective adsorption mechanisms.

Adsorption properties

Adsorption, a pivotal functional property of nanofibers, plays a crucial role in water purification, particularly in the removal of contaminants. The adsorption properties of nanofibers are multifaceted, encompassing surface functionalization strategies and selective adsorption mechanisms.

Surface functionalization for enhanced adsorption

CNF membranes offer a rapid and highly efficient water purification method due to their substantial adsorption capacity and swift adsorption rates for three specific contaminants, as demonstrated in the aforementioned batch adsorption data. The evaluation of filtration adsorption performance relied on the CNF-100 membrane, chosen for its exceptional flux and high adsorption capacity (Liang et al. 2010). Industries such as machinery manufacturing, fossil fuel combustion, oil refining, rubber and electronics production, and electroplating contribute significantly to heavy metal ion pollution in wastewater (Bhuyan et al. 2017). ENMs possess the capability to adsorb various types of metal ions through the presence of amino and hydroxyl groups on their surface. ENMs created via CS present an effective option for adsorbing metal ions (Habiba et al. 2017).

Toxic metal ions like copper (Cu2+), cadmium (Cd2+), lead (Pb2+), mercury (Hg2+), chromium (Cr4+), and nickel (Ni2+) are harmful and tend to accumulate in water, causing substantial disruptions in ecosystems and potential health issues for humans. However, these ions cannot undergo decomposition (Elwakeel et al. 2018). The removal or recovery of these heavy metal ions from aquatic environments is crucial. Various techniques, including chemical precipitation, membrane separation or adsorption, coagulation precipitation, and ion exchange, have been employed to treat heavy metals in wastewater from diverse sources (Chen et al. 2018, Wang et al. 2018).

The unmatched benefits of membrane adsorption for metal ions, such as ease of use, affordability, and high efficiency, are notable. Additionally, the extensive surface area of the membrane facilitates widespread adsorption of metal ions (Benzaoui et al. 2018). ENMs with high porosity, substantial specific surface area, good pore connectivity, and easily modifiable surface qualities are well-suited for application as membranes for heavy metal adsorption (Lee et al. 2018). A nanofiber membrane underwent functionalization, involving washing with water and drying at room temperature. This process included placing an electrospun PAN nanofiber membrane in a beaker with ethanol and sodium metal, resulting in the reduction of cyanide groups and formation of amine groups on the nanofiber surface. These functionalized membranes were utilized as nanoadsorbents for removing substances from water media (Patel et al. 2018).

Selective adsorption of contaminants

The open nanopores facilitate easy convection mass exchange within these nanonets, leading to a rapid adsorption characteristic and selective sieving performance. In contrast to the conventional solvent-evaporation-induced self-assembly process, the obtained BN network is formed by interconnected BN ultrathin nanofibers (Lian et al. 2013). Despite the ion-exchange capability of resins providing a high number of active centers, their selectivity is poor, making them susceptible to competition from coexisting interfering ions. In comparison, the nanofibrous adsorbents presented here exhibit adsorption capacity values approximately 2–3 times higher due to their high surface area (Martín et al. 2018).

The potential future applicability of removing toxic metals from industrial effluents was assessed by conducting adsorption of target heavy metal ions from a simulated water matrix containing a high background level of interfering ions (Lata et al. 2015). Interfering anions commonly present in wastewaters, such as chloride, nitrate, sulfate, or phosphate, can significantly reduce adsorption capacity by competing with metal ions for binding sites. Experimental studies were conducted separately for each interfering anion and then collectively as a group (Lata et al. 2015).

Nanofibers, characterized by high surface area, permeability, small, interconnected pores, and the ability to alter functionality at the nanoscale, possess unique structures (Zhu et al. 2021). The synthesis of selective nanofibers with controllable diameter, porosity, and hydrophobicity, along with the incorporation of specific functional groups, is feasible for targeting metal ion properties. This approach allows for easier addition of additives and practicality in creating the desired microstructure (Vo et al. 2020). Considering that industrial wastewater contains complex heavy metals combined with corrosive substances, the development of acid/base-resistant nanofibers is essential. Additionally, there is a need for research focusing on mechanically stable, strong, and reusable nanofibers, as current studies predominantly address static and single metal ion removal via nanofibers (Agrawal et al. 2021).

Filtration properties

Nanofiber membranes play a pivotal role in water treatment, particularly in achieving high filtration efficiency through size exclusion and particle removal mechanisms (Figure 5). The design of nanofibrous membranes considers the relationship between pore size distribution, fiber diameter, and bulk porosity within a non-woven layer structure. Research findings indicate that mean and maximum pore sizes are linearly scaled with the fiber diameter, regardless of the polymer material used.

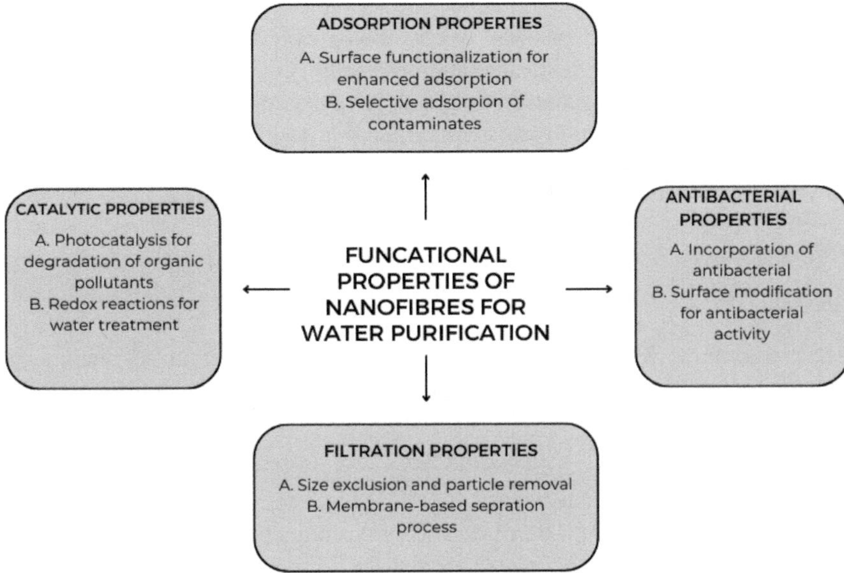

Figure 5. Different functional properties of nanofibers for water purification.

Size exclusion and particle removal

The design of the new nanofibrous membranes, aimed at achieving high filtration efficiency, is grounded in the concept that the pore size distribution within a non-woven layer structure correlates with both the fiber diameter and bulk porosity. Notably, the research revealed that the mean and maximum pore sizes can be linearly scaled with the fiber diameter, irrespective of the polymer material employed in the fabrication process (Chu et al. 2008). The investigation found that the mean pore size is approximately 3 ± 1 times the mean fiber diameter, while the maximum pore size is about 10 ± 2 times the mean fiber diameter. These correlations were established in nanofibrous scaffolds produced through electrospinning technology, utilizing various polymeric materials (Jin et al. 2002, Ryu et al. 2003, Lin et al. 2007).

The majority of membranes produced for drinking water production are composed of polymeric materials due to their significantly lower cost compared to other materials. Membranes for microfiltration (MF) and ultrafiltration (UF), employed in drinking water production, are frequently manufactured from the same materials but under distinct membrane formation conditions to produce varying pore sizes (Pinnau et al. 2000).

Membrane-based separation processes

Three membrane distillation (MD) membranes, each crafted from PTFE with distinct pore sizes and featuring a structured scrim support layer, were examined. The findings indicated that the distillation process utilizing the PTFE membrane exhibited significantly higher flux compared to PVDF microfiltration membranes under identical operational conditions (Khayet et al. 2011). The key structural parameters critical to water treatment applications in electrospun nanofiber membranes include surface area, porosity, and pore size. Generally, electrospun membranes demonstrate notably high surface area, substantial porosity, and adjustable pore size and distribution, effectively meeting the demands of water treatment processes (Ma et al. 2012).

The heightened porosity inherent in electrospun nanofibrous membranes proves advantageous for enhancing permeation flux, enabling water molecules to traverse the membrane with minimal hydraulic resistance. Moreover, this membrane system boasts interconnected pores formed by randomly deposited nanofibers, contributing to its advantageous channel structure (Chu et al. 2009). Surface properties of electrospun nanofibrous membranes exhibit partial dependence on material characteristics. The membrane surface functionality can be significantly expanded within the chosen family of electrospun nanofibrous nanofibers tailored for water purification (Agarwal et al. 2010, Wei et al. 2012, Agarwal et al. 2013).

Catalytic properties

Nanofiber membranes, with their catalytic prowess, are pivotal in advancing environmentally sustainable water treatment methods. In the realm of catalysis, two significant facets come to the forefront.

Photocatalysis for degradation of organic pollutants

Photocatalysis emerges as an economical and environmentally friendly process for eliminating organic impurities and non-biodegradable substances from wastewater. Several organic pollutants, such as humic acid, herbicides, pesticides, pharmaceuticals, and halophenols, have been efficiently eradicated from both water and wastewater using this method (Habibpanah et al. 2011, Zeltner et al. 2005, Cui et al. 2006, Alem et al. 2009).

The use of TiO_2 photocatalyst is primarily in particle suspension or immobilized on various supports by different methods. While suspension offers advantages, such as ease of use, it also poses disadvantages like separation issues and the need for additional steps to recover TiO_2, escalating operating costs. Moreover, suspended particles limit the depth of UV light penetration (Jung et al. 2008, Sabate et al. 1991). TiO_2, acting as a semiconductor, generates electron–hole pairs when exposed to UV radiation. These pairs interact with O2 or OH, producing active oxygen species that target DNA, cell membranes, and cellular proteins, ultimately causing cell death (Pant et al. 2013, Liua et al. 2010).

The catalyst amount significantly influences the efficiency of photocatalytic degradation. Increased catalyst amounts yield more active sites for generating highly reactive radicals but also escalate material costs. When designing highly efficient photocatalysis, factors like irradiation time and power intensity source should be considered (Mohamed et al. 2016).

Redox reactions for water treatment

RO demonstrates inherent OER activity and good electrical conductivity. However, its diminished OER overpotential limits its effectiveness in water treatment processes, affecting degradation efficiency (Yang et al. 2012, 2018, Kim et al. 2010). When it comes to electrospun nanofiber membranes produced from a single polymer, their ability to separate is typically lacking. To enhance membrane separation capabilities, adjustments are often made to the raw materials and the surface of the electrospun membrane. Surface modifications, including low-temperature treatments, plasma treatment, thermal crosslinking, and oxidation, have proven effective in improving membrane performance (Song et al. 2013, Higaki et al. 2015). Conducting polymers possess distinct optoelectronic characteristics like redox behavior, conductivity, and optical band gap, making them promising for applications. Consequently, covalent organic frameworks (COFs) containing redox units have potential applications as supercapacitor materials (Vyas et al. 2016).

Antibacterial properties

In the pursuit of effective water purification, the incorporation of antibacterial features into filtration processes has emerged as a transformative approach. Electrospun nanofibers (ENs) stand out as versatile agents, finding application in diverse water filtration processes such as desalination, metal removal, and filtration of organic matter and microparticles.

Incorporation of antimicrobial agents

Electrospun nanofibers (ENs) have found application in water filtration processes, serving in desalination, the removal of metals, filtration of organic matter and microparticles, as well as the elimination of microbes (Wang et al. 2013). Among various metallic nanoparticles, silver nanoparticles have received extensive attention and have demonstrated notable efficacy as antimicrobial agents (Rai et al. 2009, Kunlin et al. 2015). The antimicrobial effectiveness of resulting electrospun nanofibers using this method is reduced due to the aggregation of silver nanoparticles. Hence, the generation of silver nanoparticles typically involves reducing silver nitrate in a polymer solution before electrospinning, employing reducing agents like sodium borohydride or through atmospheric plasma treatment (De Faria et al. 2015, Shi et al. 2011). Composite nanofibers display remarkable antimicrobial properties, impacting the morphologies of electrospun nanofibers upon the addition of silver nanoparticles (Shi et al. 2011).

Surface modifications for antimicrobial activity

RO and NF membrane technologies serve various water purification needs, facing diverse feed water compositions with differing salinities and varying levels of natural organic and biological matter. The inherent properties of the membrane-forming materials prompt physicochemical interactions between the membrane surface and the feed solution, causing changes in membrane permeability and selectivity during operation. Membrane fouling, microbial attack, and concentration polarization contribute to fluctuations in membrane selectivity and flux (Zhang et al. 2015, Hoek et al. 2003). Water purification membranes exhibit the ability to eliminate a wide spectrum of contaminants, ranging from large colloids, algae, and bacteria with micrometer-scale dimensions to individual ions with angstrom-scale hydrated radii (Geise et al. 2010).

Reverse osmosis (RO) membranes, characterized as dense (nonporous), effectively remove salts from water, facilitating the desalination of brackish or seawater. Water transport through RO membranes follows a solution-diffusion mechanism (Baker et al. 2004, Lonsdale et al. 1965, Wijmans and Baker et al. 1995). When selecting membrane materials for water purification, factors like mechanical properties, ease of processing, thermal stability, and chemical resistance are crucial considerations. Several polymers, including polysulfone (PSf), poly(ether sulfone) (PES), and poly(vinylidene fluoride) (PVDF), hold importance in this regard (Liu et al. 2011).

Application of nanofibres in water purification

The application of nanofibers in water purification stands at the forefront of innovative solutions to address diverse water quality challenges. This section explores the multifaceted roles of nanofibers in mitigating pollution and enhancing water treatment methods.

Removal of organic pollutants

In the relentless battle against water pollution, the removal of organic pollutants stands as a pivotal challenge. This chapter explores the dynamic role of nanofibers in addressing this environmental concern through multifaceted approaches.

Dye removal

The widespread use of diverse dyes within the textile industry has resulted in significant contamination of surface water and groundwater due to the discharge of both toxic and colored effluents (Salem et al. 2000). Semiconductor photocatalysis stands as an advanced physicochemical process applicable to the photodegradation of environmental organic pollutants and toxins. In the elimination of dye pollutants, semiconductor metal oxides like TiO_2, ZnO, ZnS, CdS, and Fe_2O_3 nanoparticles (NPs) have shown considerable promise as the most effective materials in this domain (Poulios et al. 2003, Lizama et al. 2002).

The combined action of activated carbon and ZnO/TiO_2 nanocomposites has demonstrated a synergistic impact on the efficient degradation of certain organic compounds. This effect is closely linked to the augmentation of pollutant adsorption from water, facilitated by their exceptionally large surface area and high microporous volume. Recently, investigations have explored the antibacterial properties associated with ZnO nanomaterials (Yamamoto. 2001, Li et al. 2008). The synthesis of $MoS2/CdS/TiO_2$ nanocomposites and their integration into carbon nanofibers occurs simultaneously during thermal treatment. The amorphous carbon serves as a protective layer, preventing photocorrosion (Pant et al. 2014).

Pharmaceutical compound removal

Pharmaceutical remnants enter wastewater through the process of consumption, excretion, or the direct disposal of expired medications into toilets or sinks (Capodaglio et al. 2018). Initially intended to eliminate impurities and purify water, wastewater treatment plants (WWTPs) have demonstrated restricted efficacy in eliminating micropollutants like PPCPs. Evidence suggests that numerous molecules exhibit resilience to conventional methods such as filtration, chlorination, and fluorination (Lajeunesse et al. 2012). Xenobiotic compounds, including drugs, pose a significant challenge due to their poor degradability, raising concerns about their effective removal (Liu et al. 2020).

Traditional methods for eliminating xenobiotics in wastewater were utilized, but due to their hydrophilic nature, the effluents were discharged into combined environments. Over the years, numerous techniques have been developed, including advanced oxidation, electrochemical, and photochemical processes (Chakraborty et al. 2020). Nanomaterials possess remarkable characteristics, able to adapt their structures for specific functions, offering significant potential in wastewater treatment. A diverse range of nanomaterials has undergone testing, proving reliable, cost-effective, and environmentally sound for treating industrial effluents, soil waters, surface water, and drinking water (Kodama et al. 2015).

Removal of heavy metals

The adsorption capacity and selectivity of electrospun nanofibrous membranes (ENFMs) for heavy metals significantly depend on the surface's type and quantity of functional groups. Typically, a higher number of functional groups on the membrane surface correspond to an increased adsorption capacity (Zhu et al. 2021). Natural polymers and inorganic materials have been electrospun into nanofibrous membranes, serving as adsorbents to remove heavy metal ions from water. Certain natural polymers possess abundant functional groups with strong affinity for heavy metals, enabling their direct electrospinning into membranes for effective removal (Zhu et al. 2021).

Chitosan (CS), being widely available in nature, boasts multiple benefits including its ubiquity, affordability, biocompatibility, and facile biodegradability. Its molecular structure encompasses numerous amino and hydroxyl groups capable of forming stable chelates with various metal ions such as Hg2+, Ni2+, Cu2+, Pb2+, among others (Gupta et al. 2012). Presently, approved methods for eliminating heavy metal ions from aqueous solutions encompass chemical precipitation, electrochemical reduction, membrane separation, and adsorption (Toti et al. 2004, Al-Shannag et al. 2015). Chemical precipitation generates substantial waste residue and risks secondary pollution, while electrochemical reduction proves costly and inefficient. Meanwhile, membrane separation and adsorption stand out as promising methods due to their straightforward production, relatively lower costs, high selectivity, and potential for adsorbent reusability (Crini 2005, Wu et al. 2012, Haque et al. 2014).

Desalination and water softening

Desalination encompasses the treatment of seawater or inland brackish water, with seawater being more attainable compared to brackish water. Desalinating seawater provides a steady and ample supply of high-quality water (Saleem et al. 2020). The use of nanofiber membranes (NFMs), particularly in water treatment and desalination, is a relatively recent approach. Desalination methods include thermal-based and membrane-based technologies, with considerable interest directed towards membrane-based techniques like membrane distillation (MD), forward osmosis (FO), reverse osmosis (RO), and nanofiltration (NF) (Kugarajah et al. 2021).

Efforts are directed toward developing sustainable technologies capable of meeting the escalating water demands of future generations. Seawater desalination and wastewater treatment stand out as vital technologies for generating clean water. Traditional desalination methods, primarily reliant on energy-intensive thermal processes, contribute significantly to greenhouse gas emissions (Elimelech et al. 2011). In the realm of wastewater treatment and water desalination membranes, categorization includes inorganic, organic (polymeric), and hybrid membranes. Inorganic membranes, based on metallic or ceramic materials, find use in microfiltration (MF) and ultrafiltration (UF) membranes, owing to their high chemical and thermal stability suitable for corrosive and high-temperature environments. While ceramic membranes face limitations due to mechanical fragility and cost, metallic membranes are primarily employed in gas separation. Polymers remain the most widely used material for commercial water treatment membranes due to their effective separation performance (Setiawan et al. 2011).

Microbial and pathogen removal

The complexity of pathogen presence in drinking water is compounded by concerns regarding the spread of antibiotic-resistant bacteria, potentially posing significant public health implications (Xi et al. 2009). Approaches for eliminating water contaminants are broadly classified into three main categories: physical methods encompassing filtration and distillation; biological approaches involving the use of microorganisms as detoxifying agents; and chemical processes utilizing aqueous phase oxidation methods (Chiam et al. 2011). An ideal disinfectant should possess several key properties: (1) broad antimicrobial efficacy within a short duration; (2) safety for human health, without generating toxic byproducts during or after application; (3) cost-effectiveness and ease of application; (4) high water solubility without causing equipment corrosion; (5) manageability for safe disposal (Rutala et al. 2019).

Nanofiber membranes, with their high surface area to volume ratio, nano-sized pores, and significant porosity, have demonstrated enhanced efficiency compared to conventional materials in filtering and separating particulate matter (Aussawasathien et al. 2008). A class of nanomaterials known as nano-catalysts, including metal oxides and semiconductors, has garnered substantial attention among researchers in developing wastewater treatment technologies. These nanomaterials exhibit antimicrobial and chemical oxidative properties, targeting microbial removal in various water treatment applications (Chithra et al. 2022).

Emerging application and future prospects

Nanotechnology, particularly concerning nanomaterials like nanoparticles, polymeric nanomaterials, 2D nanosheets, or electrospun nanofibrous membranes, as well as nanocomposites designed for desalination or water treatment applications, represents a burgeoning concept that has swiftly garnered interest as a means to enhance water quality (Kugarajah et al. 2021). The utilization of nanofibers in water and wastewater treatment is rapidly gaining traction for its promising future potential. This is predominantly attributed to the myriad of intriguing and controllable properties and functionalities that can be tailored into nanofiber membranes. The upsurge in research related to nanofibers for desalination, water, and wastewater treatment over the past decade underscores its broad appeal across various sectors, be it academia or industry (Barhoum et al. 2019).

However, certain challenges need addressing and surmounting to optimize the pore size, porosity, and mechanical strength of Electrospun Nanomaterials (ENMs). There's an urgent necessity to establish a stable ultra-thin selection layer on ENM surfaces or create a nanostructure system to craft multifunctional composite ENMs possessing excellent permeability and corrosion resistance (Chen et al. 2020). Nanofibers, with their high surface-area-to-volume ratio, uniform pore distribution, and enhanced pore connectivity, showcase utility in desalination applications. In heavy-metal remediation, nanofibers have been extensively used as adsorbents, displaying high selectivity for target heavy metals amidst a background matrix. To serve as adsorbents for heavy-metal or CEC removal, nanofibers necessitate functional groups within their polymeric matrix, exhibiting selectivity for the targeted contaminants. Nanofibers with a high specific surface area efficiently incorporate these specific groups, aiding in contaminant removal and subsequent degradation through photocatalytic or electrochemical means (Agrawal et al. 2021). Nanofibers, as versatile and adjustable nanomaterials, find diverse environmental applications encompassing air and water filtration, as well as contaminant mitigation. Their extensive applicability arises from the adaptability of process parameters in their synthesis via electrospinning. By regulating the process, nanofibers with desired properties can be produced, fostering the development of both specific and nonspecific remediation techniques (Kugarajah et al. 2021).

Challenges and considerations

Nanofibers with functional properties represent a cutting-edge solution for water purification, showcasing remarkable characteristics such as an extensive surface area, high porosity, and customizable functionality. This innovative technology holds great promise for addressing water quality challenges. However, their application comes with inherent challenges and considerations described in Figure 6.

Scalability and cost-effectiveness

Nanocomposite membranes for water purification must balance pollutant removal efficacy and cost-effectiveness for large-scale projects. Unfortunately, cost details in the literature are often lacking. Estimating costs involve factors like precursor availability, processing techniques, and recyclability. Bio-based nanocomposite membranes, utilizing readily available and affordable components, offer advantages but require sustainable resource management (Al Harby et al. 2022).

A key challenge is the cost-effective procurement of nanomaterials essential for nanotechnology applications in water purification. The global membrane market's growth, fueled by nanomaterials, holds promise for industries due to their excellent selectivity, low energy consumption, and modular construction (Baten et al. 2013). Despite membrane filtration's anticipated dominance in water purification, researchers need to address operational costs, environmental impacts, and maintenance. Continuous improvement in membrane technology is crucial for sustainability, ensuring a more efficient and environmentally friendly approach to water treatment (Miller et al. 2015).

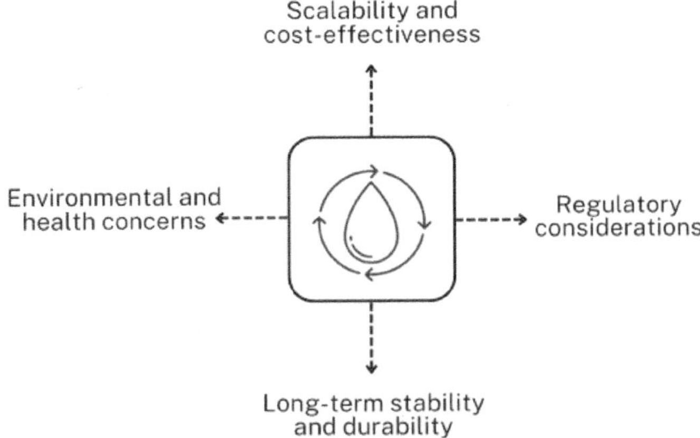

Figure 6. Challenges of nanofibers' usage for water purification.

Long-term stability and durability

Nanofibers with functional properties hold great potential for water purification, yet challenges and considerations, particularly regarding stability, persist. Nanotechnology is still evolving and understanding the environmental impact of nanoparticles (NPs) is crucial. Despite efforts by organizations like the OECD to collect data on NP impact, there remains a knowledge gap in comprehending their behavior in real-world settings (Peng et al. 2016).

The structural stability and fatigue resistance of nanocomposite membranes are pivotal for their efficacy in water purification, especially in challenging conditions. Stability involves the long-term assembly of components without leaching, aiming for consistent performance over time. Recognizing factors contributing to instability is essential for maintaining reliability, particularly through multiple uses (Chkirida et al. 2021). While performance and stability assessments often simulate real-world conditions in short-term experiments, a comprehensive understanding of long-term effects is lacking. Thorough investigations into the decontamination mechanisms and stability of nanocomposite membranes over extended periods are crucial for future scalability and implementation. Bridging the gap between laboratory findings and real-world applications is essential for advancing water purification technology effectively (Qu et al. 2013).

Environmental and health concerns

Nanofibers with functional properties for water purification are often promoted as environmentally friendly and non-toxic. However, the transition from these claims to practical sustainability assessments is hindered by a lack of research. The small size of nanoparticles poses potential dangers, as they can infiltrate organs and cells, causing adverse health effects. Occupational exposure during various phases of nanomaterials' lifecycle is a growing concern (Chkirida et al. 2021). While nanocomposite membranes claim to be biodegradable, there is limited research on their sustainability metrics. Leaching of nanoparticles during water purification raises health risks, and studies suggest potential adverse effects, including cell damage and DNA damage. Animal studies indicate the accumulation of nanoparticles in organs, highlighting ingestion-related health concerns. (Homaeigohar et al. 2014). To ensure public health, it is crucial to design nanocomposite membranes that do not release harmful nanoparticles. More research and a risk–benefit analysis are needed to verify non-toxicity claims. Balancing contaminant removal efficiency with potential adverse effects is essential for developing safe and sustainable water purification technologies. (Karthikeyan et al. 2020).

Regulatory considerations

The unique characteristics of nanofibers, including their size and potential environmental impacts, necessitate the establishment of regulatory frameworks to ensure safe and effective use. Current regulations are not specifically tailored to address nanofiber applications in water purification, requiring collaboration between researchers, industry stakeholders, and regulatory bodies. Guidelines should cover manufacturing, usage, and disposal, addressing safety, environmental impact, and potential leaching of nanoparticles or additives into treated water (Nalbandian et al. 2022). By addressing regulatory considerations, the integration of nanofibers into water treatment can proceed with clear standards, ensuring both effectiveness and environmental safety.

Conclusion

In this chapter, we delved into the promising realm of nanofibers with functional properties for water purification. Key points discussed include the unique characteristics of nanofibers, such as their high surface area, customizable functionality, and potential for enhancing water treatment efficiency. We explored the challenges and considerations associated with their application, ranging from regulatory concerns to environmental and health considerations.

Nanofiber technology emerges as a groundbreaking solution for advancing water purification. The exceptional attributes of nanofibers, particularly when incorporating functional properties, offer a versatile and efficient means of removing contaminants from water. The ability to tailor nanofiber membranes for specific applications, coupled with their scalability and cost-effectiveness, positions them as a transformative force in the field of water treatment.

Looking ahead, the prospects of nanofiber technology for water purification are promising. However, there is a crucial need for ongoing research to address regulatory gaps, assess long-term environmental impacts, and ensure the safety of these advanced materials. The continuous exploration of innovative fabrication techniques, consideration of sustainable materials, and a comprehensive understanding of the life cycle of nanofiber-based water treatment systems will be pivotal for their successful integration into mainstream practices.

In conclusion, nanofibers with functional properties represent a paradigm shift in water purification technology. The synthesis of knowledge and advancements discussed in this chapter underscores the transformative potential of nanofiber technology. As we move forward, interdisciplinary collaboration, robust regulatory frameworks, and sustained research efforts will be essential to unlock the full potential of nanofibers for ensuring clean and sustainable water resources globally.

References

Agarwal, S., Wendorff, J.H., and Greiner, A. 2010. Chemistry on electrospun polymeric nanofibers: merely routine chemistry or a real challenge?. Macromolecular Rapid Communications, 31(15): 1317–1331

Agarwal, S., Greiner, A., and Wendorff, J.H. 2013. Functional materials by electrospinning of polymers. Progress in Polymer Science, 38(6): 963-991.

Agrawal, S., Ranjan, R., Lal, B., Rahman, A., Singh, S.P., Selvaratnam, T. et al. 2021. Synthesis and water treatment applications of nanofibers by electrospinning. Processes, 9(10): 1779.

Al Harby, N.F., El-Batouti, M., and Elewa, M.M. 2022. Prospects of polymeric nanocomposite membranes for water purification and scalability and their health and environmental impacts: a review. Nanomaterials, 12(20): 3637.

Alem, A., Sarpoolaky, H., and Keshmiri, M. 2009. Titania ultrafiltration membrane: Preparation, characterization and photocatalytic activity. Journal of the European Ceramic Society, 29(4): 629–635. https://doi.org/10.1016/j.jeurceramsoc.2008.07.003.

Alpatova, A., Kim, E.S., Sun, X., Hwang, G., Liu, Y., and El-Din, M.G. 2013. Fabrication of porous polymeric nanocomposite membranes with enhanced anti-fouling properties: Effect of casting composition. Journal of Membrane Science, 444: 449–460.

Al-Shannag, M., Al-Qodah, Z., Bani-Melhem, K., Qtaishat, M.R., and Alkasrawi, M. 2015. Heavy metal ions removal from metal plating wastewater using electrocoagulation: Kinetic study and process performance. Chemical Engineering Journal, 260: 749–756.

Aussawasathien, D., Teerawattananon, C., and Vongachariya, A. 2008. Separation of micron to sub-micron particles from water: electrospun nylon-6 nanofibrous membranes as pre-filters. Journal of Membrane Science, 315(1-2): 11–19.

Barhoum, A., Pal, K., Rahier, H., Uludag, H., Kim, I.S., and Bechelany, M. 2019. Nanofibers as new-generation materials: From spinning and nano-spinning fabrication techniques to emerging applications. Applied Materials Today, 17: 1–35.

Baten, R., and Stummeyer, K. 2013. How sustainable can desalination be? Desalination and Water Treatment, 51: 44–52.

Badgar, K., Abdalla, N., El-Ramady, H., and Prokisch, J. 2022. Sustainable applications of nanofibers in agriculture and water treatment: a review. Sustainability, 14(1): 464.

Beachley, V., and Wen, X. 2009. Effect of electrospinning parameters on the nanofiber diameter and length. Materials Science and Engineering: C, 29: 663–668.

Beiranvand, R., and Sarlak, N. 2023. Electrospun nanofiber mat of graphene/mesoporous silica composite for wastewater treatment. Materials Chemistry and Physics, 309: 128311.

Benzaoui, T., Selatnia, A., and Djabali, D. 2018. Adsorption of copper (II) ions from aqueous solution using bottom ash of expired drugs incineration. Adsorption Science & Technology, 36: 114–129.

Bhuyan, M.S., Abu Bakar, M., Akhtar, A., Hossain, M.B., Ali, M.M., and Islam, M.S. 2017. Heavy metal contamination in surface water and sediment of the Meghna River, Bangladesh. Environmental Nanotechnology, Monitoring and Management, 8: 273–279

Botes, M., and Eugene Cloete, T. 2010. The potential of nanofibers and nanobiocides in water purification. Critical Reviews in Microbiology, 36(1): 68–81.

Britannica, T. Editors of Encyclopaedia. 2023, March 20. Polyacrylonitrile. Encyclopedia Britannica. [Link] (https://www.britannica.com/science/polyacrylonitrile)

Capodaglio, A.G., Bojanowska-Czajka, A., and Trojanowicz, M. 2018. Comparison of different advanced degradation processes for the removal of the pharmaceutical compounds diclofenac and carbamazepine from liquid solutions. Environmental Science and Pollution Research International. https://doi.org/10.1007/s11356-018-1913-6.

Cárdenas Bates, I.I., Loranger, É., Mathew, A.P., and Chabot, B. 2021. Cellulose reinforced electrospun chitosan nanofibers bio-based composite sorbent for water treatment applications. Cellulose, 28: 4865–4885.

Chakraborty, I., Sathe, S.M., Khuman, C.N., and Ghangrekar, M.M. 2020. Bioelectrochemically powered remediation of xenobiotic compounds and heavy metal toxicity using microbial fuel cell and microbial electrolysis cell. Materials Science and Energy Technology, 3: 104–115. https://doi.org/10.1016/j.mset.2019.09.01.

Chen, H., Huang, M., Liu, Y., Meng, L., and Ma, M. 2020. Functionalized electrospun nanofiber membranes for water treatment: A review. Science of the Total Environment, 739: 139944.

Chen, Q., Yao, Y., Li, X., Lu, J., Zhou, J., and Huang, Z. 2018. Comparison of heavy metal removals from aqueous solutions by chemical precipitation and characteristics of precipitates. Journal of Water Process Engineering, 26: 289–300.

Chiam, C.K., and Sarbatly, R. 2011. Purification of aquacultural water: conventional and new membrane-based techniques. Separation & Purification Reviews, 40(2): 126–160.

Chinthalapudi, N., Kommaraju, V.V.D., Kannan, M.K., Nalluri, C.B., and Varanasi, S. 2021. Composites of cellulose nanofibers and silver nanoparticles for malachite green dye removal from water. Carbohydrate Polymer Technologies and Applications, 2: 100098.

Chithra, A., Sekar, R., Kumar, P.S., and Padmalaya, G. 2022. A review on removal strategies of microorganisms from water environment using nanomaterials and their behavioural characteristics. Chemosphere, 295: 133915.

Chkirida, S., Zari, N., Achour, R., Hassoune, H., Lachehab, A., Qaiss, A.E.K. et al. 2021. Highly synergic adsorption/photocatalytic efficiency of Alginate/Bentonite impregnated TiO2 beads for wastewater treatment. Journal of Photochemistry and Photobiology A: Chemistry, 412: 113215.

Chu, B., Hsiao, B.S., and Yoon, K. 2008. Nanofiber and nanocomposite-fiber technology for environmental applications. AATCC Review, 8(2): 31–33.

Chu, B., and Hsiao, B.S. 2009. The role of polymers in breakthrough technologies for water purification. Journal of Polymer Science Part B: Polymer Physics, 47(24): 2431–2435.

Crini, G. 2005. Recent developments in polysaccharide-based materials used as adsorbents in wastewater treatment. Progress in Polymer Science, 30(1): 38–70.

Cui, P., Zhao, X., Zhou, M., and Wang, L. 2006. Photocatalysis-membrane separation coupling reactor and its application. Chinese Journal of Catalysis, 27(6): 752–754. https://doi.org/10.1016/S1872-2067(06)60041-7.

Daristotle, J.L., Behrens, A.M., Sandler, A.D., and Kofinas, P. 2016. A review of the fundamental principles and applications of solution blow spinning. ACS Applied Materials & Interfaces, 8(51): 34951–34963.

De Faria, A.F., Perreault, F., Shaulsky, E., Arias Chavez, L.H., and Elimelech, M. 2015. Antimicrobial electrospun biopolymer nanofiber mats functionalized with graphene oxide–silver nanocomposites. ACS Applied Materials & Interfaces, 7(23): 12751–12759.

Deitzel, J.M., Kleinmeyer, J., Harris, D.E.A., and Tan, N.B. 2001. The effect of processing variables on the morphology of electrospun nanofibers and textiles. Polymer, 42(1): 261–272.

El-Aswar, E.I., Ramadan, H., Elkik, H., and Taha, A.G. 2022. A comprehensive review on preparation, functionalization and recent applications of nanofiber membranes in wastewater treatment. Journal of Environmental Management, 301: 113908.

Elimelech, M., and Phillip, W.A. 2011. The future of seawater desalination: energy, technology, and the environment. Science, 333(6043): 712–717.

Elwakeel, K.Z., El-Bindary, A.A., Kouta, E.Y., and Guibal, E. 2018. Functionalization of polyacrylonitrile/Na-Y-zeolite composite with amidoxime groups for the sorption of Cu(II), Cd(II) and Pb(II) metal ions. Chem. Eng. J., 332: 727–736.

Feng, C.Y. 2009. Development of Novel Nanofiber Membranes for Seawater Desalination by Air-gap Membrane Distillation. Ph.D. Thesis, University of Ottawa.

Feng, C., Khulbe, K.C., Matsuura, T., Tabe, S., and Ismail, A.F. 2013. Preparation and characterization of electro-spun nanofiber membranes and their possible applications in water treatment. Separation and Purification Technology, 102: 118–135.

Fester, W. 1975. Polymer Handbook (2nd ed.). Wiley, New York.

Geise, G.M., Lee, H.S., Miller, D.J., Freeman, B.D., McGrath, J.E., and Paul, D.R. 2010. Water purification by membranes: the role of polymer science. Journal of Polymer Science Part B: Polymer Physics, 48(15): 1685–1718.

Gopal, R., Kaur, S., Ma, Z., Chan, C., Ramakrishna, S., and Matsuura, T. 2006. Electrospun nanofibrous filtration membrane. Journal of Membrane Science, 281: 581–586.

Gupta, N., Kushwaha, A.K., and Chattopadhyaya, M.C. 2012. Adsorptive removal of Pb2+, Co2+ and Ni2+ by hydroxyapatite/chitosan composite from aqueous solution. Journal of the Taiwan Institute of Chemical Engineers, 43(1): 125–131.

Habiba, U., Afifi, A.M., Salleh, A., and Ang, B.C. 2017. Chitosan/(polyvinyl alcohol)/zeolite electrospun composite nanofibrous membrane for adsorption of Cr6+, Fe3+ and Ni2+. J. Hazard. Mater., 322: 182–194.

Habibpanah, A.A., Pourhashem, S., and Sarpoolaky, H. 2011. Preparation and characterization of photocatalytic titania-alumina composite membranes by sol-gel methods. Journal of the European Ceramic Society, 31(14): 2867–2875. https://doi.org/10.1016/j.jeurceramsoc.2011.06.014.

Haque, E., Lo, V., Minett, A.I., Harris, A.T., and Church, T.L. 2014. Dichotomous adsorption behaviour of dyes on an amino-functionalised metal–organic framework, aminoMIL-101(Al). Journal of Materials Chemistry A, 2(1): 193–203.

Heidi Lynn, R., Priscilla GL, B., and Emmanuel, I. 2021. Metal nanoparticle modified polysulfone membranes for use in wastewater treatment: a critical review. Journal of Surface Engineered Materials and Advanced Technology, 2012.

Higaki, Y., Kabayama, H., Tao, D., and Takahara, A. 2015. Surface functionalization of electrospun poly(butylene terephthalate) fibers by surface-initiated radical polymerization. Macromolecular Chemistry and Physics, 216(10): 1103–1108.

Hoek, E.M., and Elimelech, M. 2003. Cake-enhanced concentration polarization: a new fouling mechanism for salt-rejecting membranes. Environmental Science & Technology, 37(24): 5581–5588.

Homaeigohar, S., and Elbahri, M. 2014. Nanocomposite electrospun nanofiber membranes for environmental remediation. Materials, 7(2): 1017–1045.

Huang, Z.M., Zhang, Y., Kotaki, Z., and Ramakrishna, S. 2003. A review on polymer nanofibers by electrospinning and their applications in nanocomposites. Composites Science and Technology, 63: 2223–2253.

Huh, J.Y., Lee, J., Bukhari, S.Z.A., Ha, J.H., and Song, I.H. 2020. Development of TiO2-coated YSZ/silica nanofiber membranes with excellent photocatalytic degradation ability for water purification. Scientific Reports, 10(1): 17811.

Jamil, T., Munir, S., Wali, Q., Shah, G.J., Khan, M.E., and Jose, R. 2021. Water purification through a novel electrospun carbon nanofiber membrane. ACS Omega, 6(50): 34744–34751.

Jiang, M., Han, T., Wang, J., Shao, L., Qi, C., Zhang, X.M., and Liu, X. 2018. Removal of heavy metal chromium using cross-linked chitosan composite nanofiber mats. International journal of biological macromolecules, 120: 213–221.

Jin, H.J., Fridrikh, S.V., Rutledge, G.C., and Kaplan, D.L. 2002. Electrospinning Bombyx mori silk with poly (ethylene oxide). Biomacromolecules, 3(6): 1233–1239.

Jung, Y.S., Kim, D.W., Kim, Y.S., Park, E.K., and Baeck, S.H. 2008. Synthesis of alumina-titania solid solution by sol-gel method. Journal of Physics and Chemistry of Solids, 69: 1464–1467. https://doi.org/10.1016/j.jpcs.2007.10.037.

Karthikeyan, P., Vigneshwaran, S., and Meenakshi, S. 2020. Al3+ incorporated chitosan-gelatin hybrid microspheres and their use for toxic ions removal: Assessment of its sustainability metrics. Environmental Chemistry and Ecotoxicology, 2: 97–106.

Kerbow, D.L. 1997. Modern Fluoropolymers. John Wiley & Sons: Chichester, UK, p. 301.

Khalid, B., Bai, X.P., Wei, H.H. et al. 2017. Direct blow-spinning of nanofibers on a window screen for highly efficient PM2.5 removal. Nano Letters, 17(2): 1140–1148.

Khayet, M., and García-Payo, M.C. 2011. Nanostructured flat membranes for direct contact membrane distillation. WO, 117, 12.

Kheirieh, S., Asghari, M., and Afsari, M. 2018. Application and modification of polysulfone membranes. Reviews in Chemical Engineering, 34(5): 657–693.

Kim, S., Choi, S.K., Yoon, B.Y., Lim, S.K., and Park, H. 2010. Effects of electrolyte on the electrocatalytic activities of RuO2/Ti and Sb–SnO2/Ti anodes for water treatment. Applied Catalysis B: Environmental, 97(1-2): 135–141.

Kodama, S., Matsumoto, S., Wang, D., Namihira, T., and Akiyama, H. 2015. Treatment of persistent organic pollutants in wastewater by nano-seconds pulsed discharge plasma. *In*: 2015 IIAI 4th International Congress on Advanced Applied Informatics, 695–698. https://doi.org/10.1109/IIAI-AAI.2015.215.

Kugarajah, V., Ojha, A.K., Ranjan, S., Dasgupta, N., Ganesapillai, M., Dharmalingam, S. et al. 2021. Future applications of electrospun nanofibers in pressure-driven water treatment: A brief review and research update. Journal of Environmental Chemical Engineering, 9: 105107.

Kumar, S., Ahlawat, W., Bhanjana, G., Heydarifard, S., Nazhad, M.M., and Dilbaghi, N. 2014. Nanotechnology-based water treatment strategies. Journal of Nanoscience and Nanotechnology, 14(2): 1838–1858.

Kunlin, S., Qinglin, W., Zhen, Z., Suxia, R., Tingzhou, L., and Quanguo, Z. 2015. Porous Carbon Nanofibers from Electrospun Biomass Tar/Polyacrylonitrile/Silver Hybrids as Antimicrobial Materials.

Lajeunesse, A., Smyth, S.A., Barclay, K., Sauve, S., and Gagnon, C. 2012. Distribution of antidepressant residues in wastewater and biosolids following different treatment processes by municipal wastewater treatment plants in Canada. Water Research, 46: 5600–5612. https://doi.org/10.1016/j.watres.2012.07.042.

Lata, S., Singh, P.K., and Samadder, S.R. 2015. Regeneration of adsorbents and recovery of heavy metals: a review. Int. J. Environ. Sci. Technol., 12: 1461–1478.

Lee, H., and Kim, I.S. 2018. Nanofibers: emerging progress on fabrication using mechanical force and recent applications. Polym. Rev., 58: 688–716.

Lian, G., Zhang, X., Si, H., Wang, J., Cui, D., and Wang, Q. 2013. Boron nitride ultrathin fibrous nanonets: one-step synthesis and applications for ultrafast adsorption for water treatment and selective filtration of nanoparticles. ACS Applied Materials & Interfaces, 5(24): 12773201512778.

Liang, H.W., Wang, L., Chen, P.Y., Lin, H.T., Chen, L.F., He, D. et al. 2010. Carbonaceous nanofiber membranes for selective filtration and separation of nanoparticles. Advanced Materials, 22(42): 4691–4695.

Liang, H.W., Cao, X., Zhang, W.J., Lin, H.T., Zhou, F., Chen, L.F. et al. 2011. Robust and highly efficient free-standing carbonaceous nanofiber membranes for water purification. Advanced Functional Materials, 21(20): 3851–3858.

Lin, K., Chua, K.N., Christopherson, G.T., Lim, S., and Mao, H.Q. 2007. Reducing electrospun nanofiber diameter and variability using cationic amphiphiles. Polymer, 48(21): 6384–6394.

Liu, F., Hashim, N.A., Liu, Y., Abed, M.M., and Li, K. 2011. Progress in the production and modification of PVDF membranes. Journal of Membrane Science, 375(1-2): 1–27.

Liu, J., Song, Y., Tang, M., Lu, Q., and Zhong, G. 2020. Enhanced dissipation of xenobiotic agrochemicals harnessing soil microbiome in the tillage-reduced rice-dominated agroecosystem. Journal of Hazardous Materials, 398: 122954. https://doi.org/10.1016/j.jhazmat.2020.122954.

Liu, R., Ye, H., Xiong, X., and Liu, H. 2010. Fabrication of TiO2/ZnO composite nanofibers by electrospinning and their photocatalytic property. Materials Chemistry and Physics, 121(3): 432–439.

Lizama, C., Freer, J., Baeza, J., and Mansilla, H.D. 2002. Optimized photodegradation of Reactive Blue 19 on TiO2 and ZnO suspensions. Catalysis Today, 76: 235–246.

Li, Q.L., Mahendra, S., Lyon, D.Y., Brunet, L., Liga, M.V., Li, D. et al. 2008. Antimicrobial nanomaterials for water disinfection and microbial control: potential applications and implications. Water Research, 42: 4591–4602.

Lonsdale, H.K., Merten, U., and Riley, R.L. 1965. Transport properties of cellulose acetate osmotic membranes. Journal of Applied Polymer Science, 9(4): 1341–1362.

Mamah, S.C., Goh, P.S., Ismail, A.F., Suzaimi, N.D., Yogarathinam, L.T. et al. 2021. Recent development in modification of polysulfone membrane for water treatment application. Journal of Water Process Engineering, 40: 101835.

Ma, H.Y., Chu, B., and Hsiao, B.S. 2012. Functional Nanofibers and their Applications. pp. 331–370. *In*: Huang, F., Wei, Q., and Cai, Y. (Eds.) (Chap. 15) Surface Functionalization of Polymer Nanofibers. *Wood Publishing, London.

Ma, H., and Hsiao, B.S. 2018. Current advances on nanofiber membranes for water purification applications. Filtering Media by Electrospinning: Next Generation Membranes for Separation Applications, 25–46.

Martín, D.M., Faccini, M., García, M.A., and Amantia, D. 2018. Highly efficient removal of heavy metal ions from polluted water using ion-selective polyacrylonitrile nanofibers. Journal of Environmental Chemical Engineering, 6(1): 236–245.

Mohamed, A., El-Sayed, R., Osman, T.A., Toprak, M.S., Muhammed, M., and Uheida, A. 2016. Composite nanofibers for highly efficient photocatalytic degradation of organic dyes from contaminated water. Environmental Research, 145: 18–25.

Miller, S., Shemer, H., and Semiat, R. 2015. Energy and environmental issues in desalination. Desalination, 366: 2–8.

Mokhtari, F., Latifi, M., and Shamshirsaz, M. 2016. Electrospinning/electrospray of polyvinylidene fluoride (PVDF): piezoelectric nanofibers. The Journal of The Textile Institute, 107(8): 1037–1055.

Mondal, K. 2017. Recent advances in the synthesis of metal oxide nanofibers and their environmental remediation applications. Inventions, 2(2): 9.

Mousa, H.M., Alenezi, J.F., Mohamed, I.M., Yasin, A.S., Hashem, A.F.M., and Abdal-Hay, A. 2021. Synthesis of TiO2@ ZnO heterojunction for dye photodegradation and wastewater treatment. Journal of Alloys and Compounds, 886: 161169.

Nabeela Nasreen, S.A.A., Sundarrajan, S., Syed Nizar, S.A., and Ramakrishna, S. 2019. Nanomaterials: Solutions to water-concomitant challenges. Membranes, 9(3): 40.

Nalbandian, M.J., Greenstein, K.E., Shuai, D., Zhang, M., Choa, Y.H., Parkin, G.F. et al. 2015. Tailored synthesis of photoactive TiO2 nanofibers and Au/TiO2 nanofiber composites: structure and reactivity optimization for water treatment applications. Environmental Science & Technology, 49(3): 1654–1663.

Nalbandian, M.J., Kim, S., Gonzalez-Ribot, H.E., Myung, N.V., and Cwiertny, D.M. 2022. Recent advances and remaining barriers to the development of electrospun nanofiber and nanofiber composites for point-of-use and point-of-entry water treatment systems. Journal of Hazardous Materials Advances, 8: 100204.

Pant, B., Pant, H.R., Barakat, N.A.M., Park, M., Jeon, K., Choi, Y. et al. 2013. Carbon nanofibers decorated with binary semiconductor (TiO2/ZnO) nanocomposites for the effective removal of organic pollutants and the enhancement of antibacterial activities. Ceramics International, 39(6): 7029–7035.

Pant, B., Barakat, N.A.M., Pant, H.R., Park, M., Saud, P.S. et al. 2014. Synthesis and photocatalytic activities of CdS/TiO2 nanoparticles supported on carbon nanofibers for high efficient adsorption and simultaneous decomposition of organic dyes. Journal of Colloid and Interface Science, 434: 159–166.

Pasini, S.M., Valerio, A., Yin, G., Wang, J., de Souza, S.M.G.U., Hotza, D. et al. 2021. An overview on nanostructured TiO2–containing fibers for photocatalytic degradation of organic pollutants in wastewater treatment. Journal of Water Process Engineering, 40: 101827.

Patel, S., and Hota, G. 2018. Synthesis of novel surface functionalized electrospun PAN nanofibers matrix for efficient adsorption of anionic CR dye from water. Journal of Environmental Chemical Engineering, 6(4): 5301–5310.

Peng, S., Jin, G., Li, L., Li, K., Srinivasan, M., Ramakrishna, S. et al. 2016. Multi-functional electrospun nanofibres for advances in tissue regeneration, energy conversion & storage, and water treatment. Chemical Society Reviews, 45(5): 1225–1241.

Peydayesh, M., and Mezzenga, R. 2021. Protein nanofibrils for next generation sustainable water purification. Nature Communications, 12(1): 3248.

Pinnau, I.F.B.D., and Freeman, B.D. 2000. Formation and modification of polymeric membranes: overview.

Poulios, E., Micropoulou, R., Panou, E., and Kostopoulou, E. 2003. Photooxidation of eosin Y in the presence of semiconducting oxide. Applied Catalysis B, 41(3): 345–355.

Qu, X., Brame, J., Li, Q., and Alvarez, P.J. 2013. Nanotechnology for a safe and sustainable water supply: enabling integrated water treatment and reuse. Accounts of Chemical Research, 46(3): 834–843.

Rai, M., Yadav, A., and Gade, A. 2009. Silver nanoparticles as a new generation of antimicrobials. Biotechnology Advances, 27(1): 76–83.

Rakhmankulov, A.A., and Davlatov, F.F. 2019. Research on the effect of dispersed graphite grade GMZ on the thermophysical properties and structure of polyvinylidene fluoride. International Science and Technology Journal, 87: 11–15.

Rostami, M., Jahed-Khaniki, G., aghaee, E.M., Shariatifar, N., Sani, M.A., Azami, M. et al. 2024. Polycaprolactone/polyacrylic acid/graphene oxide composite nanofibers as a highly efficient sorbent to remove lead toxic metal from drinking water and apple juice. Scientific Reports, 14(1): 4372.

Rutala, W.A., and Weber, D.J. 2019. Guideline for disinfection and sterilization in healthcare facilities, 2008. Update: May 2019.

Ryu, Y.J., Kim, H.Y., Lee, K.H., Park, H.C., and Lee, D.R. 2003. Transport properties of electrospun nylon 6 nonwoven mats. European Polymer Journal, 39(9): 1883–1889.

Sabate, J., Anderson, M.A., Kikkawa, H., Edwards, M., and Hill, C.G.J. 1991. A kinetic study of the photocatalytic degradation of 3-chlorosalicylic acid over TiO2 membranes supported on glass. Journal of Catalysis, 127: 167–177. http://www.osti.gov/energycitations/product.biblio.jsp?osti_id=5555878.

Salem, I.A., and El-Maazawi, M.S. 2000. Kinetics and mechanism of color removal of methylene blue with hydrogen peroxide catalyzed by some supported alumina surfaces. Chemosphere, 41(7): 1173–1180.

Saleem, H., Trabzon, L., Kilic, A., and Zaidi, S.J. 2020. Recent advances in nanofibrous membranes: Production and applications in water treatment and desalination. Desalination, 478: 114178.

Salim, S.H., Al-Anbari, R.H., and Haider, A.J. 2021. Polymeric Membrane with Nanomaterial's for Water Purification: A Review. IOP Conference Series: Earth and Environmental Science, 779.

Schamagl, N., and Buschatz, H. 1998. Proceedings, Wasseraufbereitung, Aachen, pp. A8/1-AS/12.

Schamagl, N., and Buschatz, H. 1999. Proceedings, Euromembrane 99, Leuven (B), 2: 266.

Schamagl, N., Bunse, U., and Peinemann, K.-V. 2000. Desalination, 131, 55.

Scharnagl, N., and Buschatz, H. 2001. Polyacrylonitrile (PAN) membranes for ultra- and microfiltration. Desalination, 139(1-3): 191–198.

Setiawan, L., Wang, R., Li, K., and Fane, A.G. 2011. Fabrication of novel poly(amide–imide) forward osmosis hollow fiber membranes with a positively charged nanofiltration-like selective layer. Journal of Membrane Science, 369: 196–205.

Shi, Q., Vitchuli, N., Nowak, J., Caldwell, J.M., Breidt, F., Bourham, M. et al. 2011. Durable antibacterial Ag/polyacrylonitrile (Ag/PAN) hybrid nanofibers prepared by atmospheric plasma treatment and electrospinning. European Polymer Journal, 47(7): 1402–1409.

Sinha-Ray, S., Sinha-Ray, S., Yarin, A.L., and Pourdeyhimi, B. 2015. Application of solution-blown 20–50 nm nanofibers in filtration of nanoparticles: The efficient van der Waals collectors. Journal of Membrane Science, 485: 132–150.

Sinha-Ray, S. 2017. Heavy metal adsorption on solution-blown biopolymer nanofiber membranes.

Smejkalová, T., Ţǎlu, Ş., Dallaev, R., Částková, K., Sobola, D., and Nazarov, A. 2021. SEM imaging and XPS characterization of doped PVDF fibers. In E3S Web of Conferences, 270: 01011. EDP Sciences.

Song, R., Yan, J., Xu, S., Wang, Y., Ye, T., Chang, J. et al. 2013. Silver ions/ovalbumin films layer-by-layer self-assembled polyacrylonitrile nanofibrous mats and their antibacterial activity. Colloids and Surfaces B: Biointerfaces, 108: 322–328.

Sueyoshi, Y., Fukushima, C., and Yoshikawa, M. 2010. Molecularly imprinted nanofiber membranes from cellulose acetate aimed for chiral separation. Journal of Membrane Science, 357: 90–97.

Tansel, T. 2021. Effect of electric field assisted crystallization of PVDF-TrFE and their functional properties. Sensors and Actuators A: Physical, 332: 113059.

Tlili, I., and Alkanhal, T.A. 2019. Nanotechnology for water purification: electrospun nanofibrous membrane in water and wastewater treatment. Journal of Water Reuse and Desalination, 9(3): 232–248.

Toti, U.S., and Aminabhavi, T.M. 2004. Different viscosity grade sodium alginate and modified sodium alginate membranes in pervaporation separation of water + acetic acid and water + isopropanol mixtures. Journal of Membrane Science, 228(2): 199–208.

Uyar, T., Havelund, R., Nur, Y., Hacaloglu, J., Besenbacher, F., and Kingshott, P. 2009. Molecular filters based on cyclodextrin functionalized electrospun fibers. Journal of Membrane Science, 332: 120–137.

Vo, T.S., Hossain, M.M., Jeong, H.M., and Kim, K. 2020. Heavy metal removal applications using adsorptive membranes. Nano Convergence, 7: 1–26.

Vyas, V.S., Vishwakarma, M., Moudrakovski, I., Haase, F., Savasci, G., Ochsenfeld, C. et al. 2016. Exploiting noncovalent interactions in an imine-based covalent organic framework for quercetin delivery. Advanced Materials, 28(39): 8749–8754.

Wang, J., Zhang, S., Cao, H., Ma, J., Huang, L., Yu, S. et al. 2022. Water purification and environmental remediation applications of carbonaceous nanofiber-based materials. Journal of Cleaner Production, 331: 130023.

Wang Wei, Q. (Ed.). 2012. Functional nanofibers and their applications. Elsevier.

Wijmans, J.G., and Baker, R.W. 1995. The solution-diffusion model: a review. Journal of Membrane Science, 107(1-2): 1–21.

Wu, N., Wei, H., and Zhang, L. 2012. Efficient removal of heavy metal ions with biopolymer template synthesized mesoporous titania beads of hundreds of micrometers size. Environmental Science & Technology, 46(1): 419–425.

Wang, R., Guan, S., Sato, A., Wang, X., Wang, Z., Yang, R. et al. 2013. Nanofibrous microfiltration membranes capable of removing bacteria, viruses and heavy metal ions. Journal of Membrane Science, 446: 376–382. https://doi.org/10.1016/j.memsci.2013.06.033.

Wang, Y.N., Goh, K., Li, X., Setiawan, L., and Wang, R. 2018. Membranes and processes for forward osmosis-based desalination: Recent advances and future prospects. Desalination, 434: 81–99.

Xi, C., Zhang, Y., Marrs, C.F., Ye, W., Simon, C., Foxman, B. et al. 2009. Prevalence of antibiotic resistance in drinking water treatment and distribution systems. Applied and Environmental Microbiology, 75(17): 5714–5718.

Xiangxiang Liu, Hongyang Ma, Benjamin S Hsiao ACS Applied Nano Materials, 2(6): 3606–3614, 2019.

Yamamoto, O. 2001. Influence of particle size on the antibacterial activity of zinc oxide. International Journal of Inorganic Materials, 3: 643–646.

Yang, K., Lin, H., Liang, S., Xie, R., Lv, S., Niu, J. et al. 2018. A reactive electrochemical filter system with an excellent penetration flux porous Ti/SnO 2–Sb filter for efficient contaminant removal from water. RSC Advances, 8(25): 13933–13944.

Yang, S.Y., Choo, Y.S., Kim, S., Lim, S.K., Lee, J., and Park, H. 2012. Boosting the electrocatalytic activities of SnO2 electrodes for remediation of aqueous pollutants by doping with various metals. Applied Catalysis B: Environmental, 111-112: 317–325.

Zeltner, W.A., and Tompkins, D.T. 2005. Shedding light on photocatalysis. ASHRAE Journal, 111: 523–534. https://doi.org/10.1038/nn.3723.

Zhang, W., Luo, J., Ding, L., and Jaffrin, M.Y. 2015. A review on flux decline control strategies in pressure-driven membrane processes. Industrial & Engineering Chemistry Research, 54(11): 2843–2861.

Zhang, S., Tang, Y., and Vlahovic, B. 2016. A review on preparation and applications of silver-containing nanofibers. Nanoscale Research Letters, 11: 1–8.

Zhou, X., Wang, Y., Gong, C., Liu, B., and Wei, G. 2020a. Production, structural design, functional control, and broad applications of carbon nanofiber-based nanomaterials: A comprehensive review. Chemical Engineering Journal, 402: 126189.

Zhou, X., Liu, B., Chen, Y., Guo, L., and Wei, G. 2020b. Carbon nanofiber-based three-dimensional nanomaterials for energy and environmental applications. Materials Advances, 1(7): 2163–2181.

Zhuang, X.P., Yang, X.C., Shi, L., Cheng, B.W., Guan, K.T., and Kang, W.M. 2012. Solution blowing of submicron-scale cellulose fibers. Carbohydrate Polymers, 90: 982–987.

Zhuang, X., Shi, L., Jia, K., Cheng, B., and Kang, W. 2013. Solution-blown nanofibrous membrane for microfiltration. Journal of Membrane Science, 429: 66–70.

Zhu, F., Zheng, Y.-M., Zhang, B.-G., d Dai, Y.-R. 2021. A critical review on the electrospun nanofibrous membranes for the adsorption of heavy metals in water treatment. J. Hazard. Mater., 401: 123608.

13

Utilizing Biomass-Sourced Carbon Quantum Dots
An Eco-Friendly Strategy for Wastewater Treatment

Muhammad Usama Ghafoor,[1] Kashif Ali,[1] Muhammad Zahid,[1,] Saima Noreen,[1] Ghulam Mustafa[2] and Zulfiqar Ahmad Rehan[3]*

Introduction

One of the most valuable resources on Earth is water, which is also necessary for human survival. However, the world is facing a serious environmental issue with limited water resources. The increasing population has put immense pressure on the already limited water supplies (Nasrollahzadeh et al. 2021a). Moreover, water pollution is a major problem affecting water quality and threatening the survival of aquatic life, human beings, and animals. Industrial wastewater is the principal source of environmental contamination, and it contains a mixture of multiple kinds of pollutants (Pirsaheb et al. 2018).

Water pollution causes

Water contamination predominantly arises from the activities of humanity. The progression of industrialization, urbanization, and population expansion has escalated the discharge of refuse and contaminants into the surroundings. As industries and urban regions grow at an unprecedented pace, they contribute to the emission of detrimental substances like heavy metals, natural compounds, and disease-causing microorganisms into the invaluable water bodies (Si et al. 2020). The wastewater produced by industries is a major source of pollution and comprises of complex mixture of hazardous

[1] Department of Chemistry, University of Agriculture Faisalabad, Faisalabad, Pakistan.
[2] Department of Chemistry, University of Okara, Okara, Pakistan.
[3] Department of Chemistry, Sultan Qaboos University, Oman.
* Corresponding author: rmzahid@uaf.edu.pk; zahid595@gmail.com

substances including organic dyes (Goswami et al. 2023), heavy metals (Ullah et al. 2018), and phenolic organics (Liang et al., 2020).

Chemical usage in agriculture presents another significant contributor to environmental pollution. To enhance crop productivity and safeguard against pests and diseases, farmers employ fertilizers, pesticides, and herbicides. Nonetheless, the excessive application of these chemicals can lead to soil and water contamination. Furthermore, through transportation facilitated by rainwater and wind, these chemicals can disseminate to distant regions, thereby causing pollution in other water bodies and ecosystems (Abbo et al. 2021).

Effects of water pollution

Contaminated water sources introduce hazardous substances, posing threats to public health, disrupting ecosystems, and compromising the well-being of living organisms (Ullah et al. 2018). The quality of water undergoes a profound impact, endangering the survival of aquatic organisms, humans, and animals equally. The discharged waste from textile plants contains residues of dyes and their byproducts, which are poisonous, carcinogenic, and detrimental to humans, aquatic life, and microbes. These pollutants contribute to a diverse range of diseases and health complications, encompassing, but not limited to, conditions such as cancer, skin irritations, and respiratory ailments. Notably, heavy metal pollutants prove particularly detrimental as they accumulate within the food chain, initiating long-term health issues (Al Ja'farawy et al. 2022, Saud et al. 2015). It also affects the ecosystem and biodiversity (Sharma et al. 2019). The water body contaminations can result in the killing of aquatic life, leading to a decline in biodiversity. Additionally, soil and air pollution affect the growth of plants and animals, contributing to a further decline in their population.

Methods used for water treatment

Conventional technologies for wastewater treatment include ultra-filtration, sedimentation, oxidation, adsorption, reverse osmosis, ion-exchange, membrane, flocculation, and advanced oxidation process (AOPs). Wastewater treatment approaches, including UV photolysis/photocatalysis, bio-processing, activated carbon adsorption, and ozonation have been employed to remove pollutants and organic contaminations from waters (Cai et al. 2018). The multifaceted nature of this treatment ensures the comprehensive purification and reclamation of the wastewater, safeguarding the environment and promoting sustainable water management practices (Nasrollahzadeh et al. 2021b). These methods aid in the elimination of various contaminants in wastewater, including colorants and organic compounds (Pirsaheb et al. 2018). Physical, biological, and chemical techniques are employed to decontaminate water and mitigate environmental pollution (Si et al. 2020). The treatment of wastewater typically involves three stages: primary, secondary, and tertiary treatments, each serving a specific function in removing pollutants from the wastewater. The primary stage separates suspended solids, the secondary stage eliminates biodegradable compounds, and the tertiary stage addresses non-biodegradable compounds. Conventional methods such as adsorption, coagulation, and membrane processes are commonly employed in wastewater treatment, but they possess certain limitations in terms of cost-effectiveness and efficiency. To overcome these drawbacks, nanotechnology, and nanomaterials offer promising solutions. These nanomaterials exhibit unique properties that effectively remove contaminants from water (Abbo et al. 2021). Nanoparticles made of graphene, carbon, and graphene-derived quantum nanomaterials, as well as carbon-based nanotubes, are particularly appealing options for wastewater treatment due to their exceptional characteristics (Guan et al. 2023).

Carbon quantum dots (CQDs)

Carbon quantum dots, also commonly known as carbon dots (abbreviated as CQDs, C-dots, or CDs), are a novel family of fluorescent carbon nanoparticles (CNPs) with sizes below 10 nm (Gayen et al. 2019). Due to their tunable optical and fluorescence emission properties, carbon quantum dots have become rising stars among carbon-based nanomaterials. Carbon quantum dots were first discovered by Xu et al. in 2004 accidentally during the purification and separation of single-walled carbon nanotubes (SWCNTs) through preparative electrophoresis prepared by the arc-discharge method (Xu et al., 2004). However, the name "carbon quantum dots" was coined by Sun et al. in 2006 for fluorescent carbon nanoparticles during the synthesis of carbon nanomaterials of different sizes (Sun et al. 2006).

As a sort of environmental-friendly carbon nanomaterials, carbon quantum dots are also featured with good water solubility, stable physicochemical properties, good biocompatibility, ease of synthesis, high photo response, high photostability, facile surface functionalization, low cytotoxicity, and good catalysis properties. Due to these characteristic features, carbon quantum dots are widely investigated in various fields such as drug delivery, sensing, bio-imaging, photovoltaic devices, solar cells, medical diagnosis, light-emitting diodes, and photocatalysis. Figure 1 illustrates the distinctive properties of carbon quantum dots.

Carbon quantum dots (CQDs) possess diverse functional groups including carboxyl groups on their surface. These groups enhance their water solubility, facilitating straightforward binding with other reactive groups for surface modification. This capability enables the alteration of biological, polymeric, inorganic, or organic materials (Sun et al. 2006). In summary, Carbon Quantum Dots (CQDs) present themselves as promising nanomaterials endowed with exceptional attributes that make them appealing for a variety of applications. This comprehensive article explores the fabrication of CQDs from biomass and delves into their applications in the realm of water treatment.

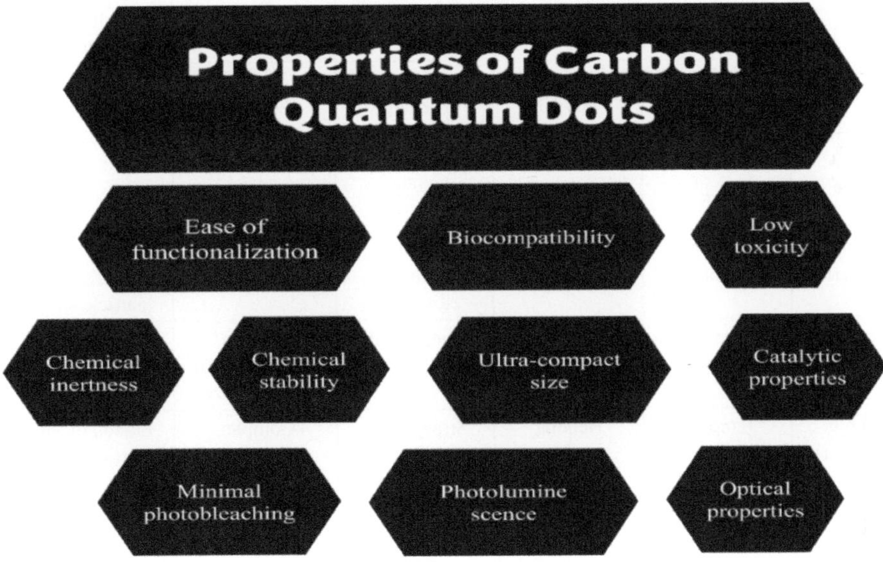

Figure 1. Features of CQDs.

Synthesis of CQDs

Carbon dots (CDs) follow two primary synthetic approaches: the "top-down" and "bottom-up" approaches. In the top-down approach, larger carbon structures are disintegrated into smaller entities, ultimately yielding CQDs, while in a bottom-up approach, smaller carbon precursors such as small molecules, biomolecules, or biomass combine to form carbon dots (Lim et al. 2015). The top-down approach involves arc discharge, laser ablation, and electrochemical oxidation, while hydrothermal/solvothermal, acidic oxidation, pyrolysis, and carbonization are summed up under the category of bottom-up approach (Luo et al. 2013). Researchers have used a variety of precursors, including citric acid, glucose, and natural resources such as papaya, garlic, and potato, to create CQDs through methods like carbonization, oxidation, hydrothermal, and electrochemical routes. Utilizing natural resources for CQD fabrication is eco-friendly and cost-effective, making it a highly accepted green synthesis method compared to chemical and physical methods. There is a growing interest in developing new routes using naturally available carbon sources to create CQDs (Chen et al. 2013; Liu et al. 2014a).

Overall, the use of natural carbon precursors for CQDs' synthesis provides a promising avenue for making ecologically friendly and economical CQDs. The development of new synthesis methods using naturally available resources can not only reduce the cost and environmental impact of CQDs synthesis but also expand their potential applications across several fields, including biomedicine, environmental cleanup, and energy conversion (Singh et al. 2013; Zhang et al. 2018b). Table 1 depicts top-down and bottom-up techniques (Zhu et al. 2020a).

Table 1. Comparison of various CQDs' synthesis methods.

	Carbon sources	Synthesis methods	Apparatus	Conditions	Disadvantages	Advantages	Product yields
Top-down methods	Carbon nanomaterials	Laser-ablation	Laser device	Steam at a high temperature, H2	High cost	Easy to use and adjustable size	Low
		Arc discharge	Arc discharge device	Electro discharge	Uneven distribution of particle size, complex purification procedure	Small size and excellent fluorescence performance	Low
	Carbon nanomaterials	Electronic scissoring	Electrolytic tank	Applied voltage, electrolyte	Complex operation	High purity, uniform particle size	Low
	Organic carbons	Chemical oxidation	Reflux unit	Heat, stirring	Uneven distribution of particle size	Simple operation and equipment	High
Bottom-up methods	Organic polymers	Template	Heating apparatus	Strong acid, elevated temperature, or alkali	Complex operation	Standardized particle size	Low
	Organics	Microwave irradiation	Microwave reactor	Microwave	More energy consumption	Simple operation,	Low
		Hydrothermal/ solvothermal treatment	Hydrothermal reactor	Heat, temperature control	Poor purity	Eco-friendly, low cost	High
		Pyrolysis	Tube furnace	High-temperature control	Uneven distribution of particle size	Simple operation, low cost	Low

Bio-based CQDs' formation mechanism via hydrothermal process

In recent years, carbon quantum dots (CQDs) have been synthesized utilizing both biomass (such as algae, lignocellulose, and municipal waste) and the model compositions of biomass (such as lignin, carbohydrate, and protein). These elements have distinct transformational pathways during the hydrothermal method, leading to the production of carbon dots with varying structures, yields, and characteristics. The mechanism of CQDs synthesis during the hydrothermal treatment from the view of biomass and biomass model compositions is discussed in the following section (Zhang et al. 2022).

Carbohydrate-based CQDs

Carbohydrate, characterized by a general formula $C_x(H_2O)_y$, constitutes the primary constituent of biomass including algae and lignocellulose. Numerous carbohydrate model substances, including starch, glucose, cellulose, sucrose, hemicellulose, and pectin, have been successfully used to make carbohydrate-based carbon quantum dots (Arias Velasco et al. 2021). Oligosaccharides and polysaccharides are believed to undergo hydrolysis, breaking down into individual sugar units (e.g., starch hydrolyzed to glucose). These units are then subjected to a complex dehydration process (see Figure 2a), ultimately leading to the formation of carbohydrate-based CDs (Arias Velasco et al. 2021, Meng et al. 2017).

According to several research studies, various bases or acids are first selectively added to ease the breakdown of carbohydrate into specific sugars. These substances act as catalytic agents, strategically facilitating the fragmentation of macromolecular chains (Liu et al. 2019, Zhang et al. 2018b). The transformation of one sugar unit, like glucose, into 5-hydroxymethylfurfural (HMF),

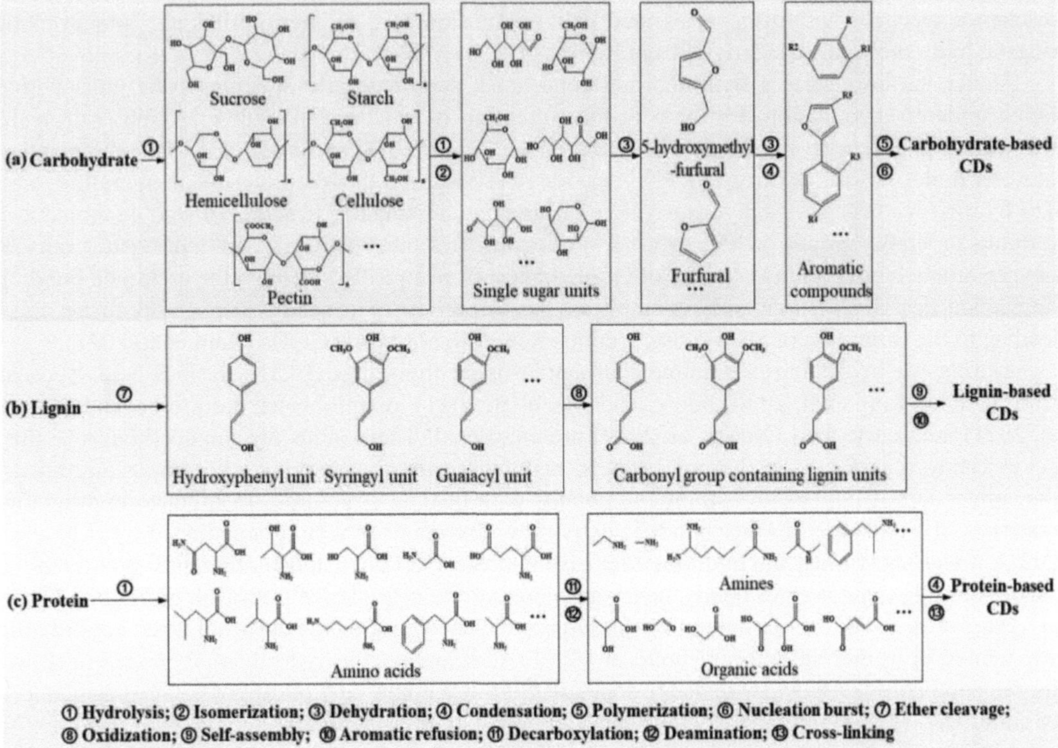

Figure 2. Proposed pathway for the formation of biomass-based CQDs using hydrothermal approach starting from (a) carbohydrate, (b) lignin, and (c) protein (Reproduced with permission from ref. (Zhang et al. 2022)).

followed by the subsequent dehydration of HMF into aromatic carbonaceous structures, is widely acknowledged as the process through which the carbon skeleton of CDs is formed (Shen et al. 2017).

According to (Papaioannou et al. 2018), the procedure starts with the isomerization of glucose into fructose, and then three water molecules are removed to create HMF. Subsequently, HMF undergoes dehydration and polymerization, resulting in the formation of small polymeric nanoclusters that gradually grow into spheres made of carbon. HMF's cycloaddition and condensation reactions lead to the generation of soluble polymers and the formation of aromatic clusters. Over time, these clusters reach a critical concentration, leading to a nucleation burst as they undergo aromatization and carbonization, ultimately increasing supersaturation (Liu et al. 2019). The solute then diffuses towards the particle surfaces, causing the nuclei of CDs to develop isotropically and uniformly, and the carbon compounds that undergo growth are oxidized, imparting improved properties to the resulting CQDs (Shen et al. 2017), such as outstanding solubility in aqueous solution (Hoang et al. 2019). According to (Zhao et al. 2020a), during the process, hemicellulose undergoes decomposition to produce furfural and furfuryl alcohol, meanwhile, the formation of cyclic compounds takes place through condensation, chain cleavage, and chain formation processes. These compounds are then subjected to polymerization and condensation, which produces aromatics with functional groups containing oxygen. These aromatics can be thought of as the building blocks for carbon dots (CDs).

Lignin-based CDs

Lignin, comprising roughly 10–45% of the lignocellulosic biomass (Carrott et al. 2007), plays a vital role in maintaining the structural stability of lignocellulosic biomass. The lignin polymer is abundant in aromatic ring formations, resulting from the intricate connections among hydroxyphenyl, syringyl, and guaiacyl units. The intriguing findings by Xu et al. (Xue et al. 2019) showcased the aromatic rings' identification within the syringyl and guaiacyl units using heteronuclear single-quantum coherence spectroscopy in two dimensions of alkali lignin-based CDs' nuclear magnetic resonance spectra. These discoveries shed light on the existence of interunit linkages and aromatic rings in hydrothermal CDs derived from lignin.

Under the influence of hydrothermal conditions, certain fragile phenylpropane units within lignin undergo degradation, resulting in the formation of smaller molecules through hydrolysis, subsequently dispersing into the aqueous phase (see Figure 2b) (Park et al. 2018). The dissociation of water molecules into hydrogen and hydroxide ions occurs as the hydrothermal temperature rises. The hydroxide ions can take hydrogen atoms from the main alcohol structure of soluble molecules, forming carbonyl groups in the process. Following this, the proximity of hydrogen atoms to oxygen atoms facilitates the creation of O-OH bonds. Consequently, through the hydrogen bonding interaction described above, self-assembly of the carbonyl group occurs in phenylpropane units, leading to the formation of nano-clusters comprised of lignin-based CQDs (Rani et al. 2021).

During the hydrothermal treatment of synthesizing lignin-based CDs, various acids such as sulfuric acid (Yang et al. 2020), nitric acid (Pei et al. 2021), o-aminobenzenesulfonic acid (Zhu et al. 2021), and citric acid (Xue et al. 2019) are employed. These acids are included because they act as catalysts in the hydrothermal process, making it easier to dissolve ether bonds and hasten the conversion of lignin into carbon compounds with distinct structures. To produce high-quality materials, the cleavage of ether bonds is extremely important. In a study conducted by (Liu et al. 2021), it was shown that the hydrothermal treatment with acid assistance efficiently promoted the scission or cleavage of ether bonds, hence accelerating the degradative transition of lignin. Ding et al. (Ding et al. 2018) further noted that the intensive cleavage of bonds occurred when alkali lignin was treated hydrothermally for 0.5 hours at 180°C. Additionally, the hydrothermal procedure allows for the doping of CQDs when acids are added with groups containing phosphorus (P), nitrogen (N), or sulfur (S). This doping process introduces effective active sites on the CQDs, enabling them to exhibit exceptional catalytic, optical, or electrochemical properties (Liu et al. 2021; Zhu et al. 2021).

Protein-based CDs

The majority of nitrogen in biomass is present in the protein form, serving as a fundamental building block of cells and actively contributing to their growth and metabolic activities. Proteins go through hydrolysis during the hydrothermal treatment, and produce amino acids. These amino acids are then further converted into amines and low-molecular-weight organic acids by deamination and decarboxylation processes (see Figure 2c). In further detail, amines and carbonic acid are generated when amino acids are decarboxylated, and ammonia and organic acids are simultaneously produced during the deamination of amino acids in the hydrothermal environment (Sato et al. 2004). According to (Liu et al. 2020), these two routes for protein hydrothermal processing favor decarboxylation more than deamination. On the other hand, Sato et al. (Sato et al. 2004) stated that the amount of deamination to decarboxylation varies according to the kind of amino acids involved. Subsequently, condensation and cross-linking of amines and organic acids, which act as CQDs' precursors, are followed to produce CQDs (Liu et al. 2020). However, there hasn't been a thorough explanation of the reaction pathway for these condensation and cross-linking processes.

Preparation of biomass-derived CQDs

Peanut shells, citrus fruit peel, rice husk, bamboo stalks, and sweet potato have emerged as valuable resources to produce carbonaceous materials. CQDs derived from sweet potatoes hold the potential to embody both carbon (C) and nitrogen (N) elements. However, nitrogen does not actively participate in the production process. Upon subjecting the sweet potato to prolonged high-temperature heating, the proteins undergo denaturation, resulting in undetectable nitrogen presence (Shen et al. 2017). Various biomass sources for synthesizing carbon quantum dots are described in a chart depicted in Figure 3.

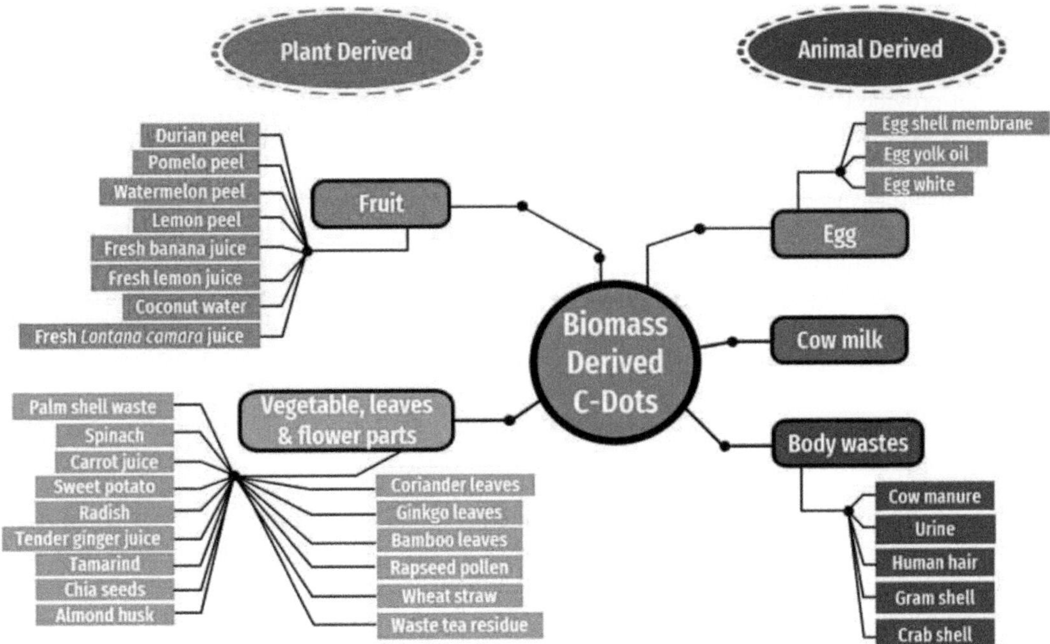

Figure 3. Carbon dots' precursors from biomass (Reproduced with permission from ref. (Khairol Anuar et al. 2021)).

Natural biomass feedstock

The large-scale manufacturing of CQDs has been explored using various affordable and sustainable carbon sources, including waste materials such as fruit peels, agricultural waste, and biomass. The utilization of waste materials not only renders the production process cost-effective but also contributes to waste management and environmental sustainability. Additionally, renewable carbon sources like cellulose and lignin have been investigated for carbon dots production. The development of such green and sustainable routes for CQDs synthesis is crucial for their potential applications in diverse fields, including optoelectronics, biomedical, and energy conversion (Yang et al. 2018). Biomass, such as plant matter, agricultural waste, and forestry residues, is an attractive source of carbon to produce CQDs owing to its abundance and low cost. In addition, biomass is a sustainable choice for the bulk manufacturing of CQDs since it is non-toxic and renewable (Shen et al. 2017). Biomass, as a rich source of carbon and oxygen, contains significant amounts of cellulose, hemicellulose, lignin, and inorganic substances. It comprises primary constituents, with cellulose accounting for 40%–50%, hemicellulose for 20%–30%, lignin for 20%–25%, and inorganic substances ranging from 1%–10%. Notably, these components exhibit remarkable water solubility, rendering them fascinating sources for CQDs (Mahat et al. 2020). Photoluminescent carbon quantum dots (CQDs) have been successfully synthesized using a diverse range of materials, including rice husk (Chaudhary et al. 2016), sugar cane bagasse (Chai et al. 2019), grape skin (Kang et al. 2023), wheat straw (Yuan et al. 2015), orange juice (Sahu et al. 2012), peanut shell (Ma et al. 2017), watermelon peel (Zhou et al. 2012), potato (Shen et al. 2017), durian (Wang et al. 2018a), mango peel (Jiao et al. 2019), onion (Hu et al. 2017), lychee (Sahoo et al. 2020), salvaged pine needles (Huang et al. 2023), lemon grass herb (Thota et al. 2018), rubber seed shells (Nizam et al. 2023), coffee grounds (Hsu et al. 2012b), broccoli (Arumugam et al. 2018), bamboo leaves (Liu et al. 2014b), rose (Feng et al. 2015), wool (Wang et al. 2016a), and prawn shell (Zhang et al. 2016). This conversion of biomass into useful goods accomplishes two goals at once: it effectively disposes of solid waste and lessens the growing environmental, resource, and energy problems (Liu et al. 2020).

Natural biomass is a great source of oxygen, carbon, and heteroatoms like sulfur, nitrogen, and silica that provide CQDs the ability to self-passivate and self-dope. Functional oxygenic groups, such as carboxyl, epoxide, alkyl, carbonyl, and hydroxyl are rich on the surface of biomass-derived CQDs, making them extremely water soluble (Jones et al. 2017). The degree of carbonization, size, and shape of carbon quantum dots can all be influenced by the synthetic paths used. Murugan and Sundramoorthy utilized borassus flabellifer as a precursor for CDs synthesis through pyrolysis treatment at temperatures of 200°C, 300°C, and 400°C, over a duration of 2 hours. The fluorescence intensity of the resulting CQDs demonstrated a notable decline with the rising temperature (Murugan et al. 2018). In summary, the combination of comparatively low temperature and extended heating duration tends to yield biomass-derived CQDs characterized by elevated crystallization and intense fluorescence.

Green production of CQDs using plant extracts

Various naturally occurring biomaterials have been utilized in the fabrication of CQDs. In particular, the aqueous leaf extract of Tulsi (Ocimum sanctum) has been utilized to create fluorescent CQDs through the eco-friendly hydrothermal method. These obtained CQDs showed outstanding photostability and a remarkably high quantum yield (QY) of about 9.3% (Kumar et al. 2017). These carbon dots worked well as sensors for Cr(VI), displaying remarkable sensitivity and specificity when detecting Cr(VI) in real water samples (Bhatt et al. 2018). In a separate investigation, versatile and remarkably luminescent CQDs were produced from Tulsi leaves using the hydrothermal method. When exposed to ultraviolet (UV) light, these CQDs displayed blue fluorescence. Their spherical form, which is around 5 nm in size, was revealed by transmission electron microscopy

(TEM) analysis. Fourier transform infrared spectroscopy (FTIR) confirmed the occurrence of hydroxyl, amino, carboxylic, and carbonyl groups on CQDs (Kumar et al. 2017). Without the usage of a chemical agent, henna leaf (Lawsonia inermis) extracts were used to create carbon quantum dots; when exposed to ultraviolet light, these quantum dots exhibited bright green fluorescence. The CQDs displayed well-dispersed 5-nanometer particles, an amorphous structure, and a quasi-spherical shape. Notably, they successfully combated Escherichia coli and Staphylococcus aureus using their antibacterial abilities. These carbon dots also showed the capacity to recognize methotrexate (MTX), a commonly used anticancer medication (Shahshahanipour et al. 2019). Using mango (Mangifera indica) leaf extracts in a one-pot pyrolysis procedure, CQDs were fabricated in a green manner. The reaction was supported for three hours at a temperature of 300 degrees Celsius (Jiao et al. 2019). The low-cost, photochemically produced carbon quantum dots (CQDs) showed intense blue emission with a distinct peak at about 525 nm. For the purpose of detecting Fe^{3+} in aqueous medium, mung bean seed-based CQDs were produced (Wang et al. 2018b). In another study, betel leaves (Piper betle) were utilized in the production of CQDs for the detection of Fe^{3+} in the presence of various metal ions (Raja et al. 2021). The N-doped CQDs demonstrated the removal of 37% of Cd^{2+} and 75% of Pb^{2+} (Sabet et al. 2019). CQDs derived from Prosopis juliflora leaf extract were employed in the development of a dual fluorescence biosensor. These carbon dots were utilized for the detection of Hg^{2+} and to mimic anti-poisoning drugs, as they emit blue fluorescence (Pourreza et al. 2019).

Green CQDs' synthesis from agro-industrial waste extracts

The agro-industry generates a considerable amount of waste with diverse properties. Agro-waste has accumulated to exceed 2 billion tons globally. Through the process of solid-state fermentation, these agro-industrial wastes serve as substrates for the production of antibiotics, enzymes, vitamins, animal feed, biofuels, and antioxidants. Various agricultural residues, including fruit peels like pomelo, are also included in this range (Lakshmipathy et al. 2015); papaya (Bhuvaneswari et al. 2015), watermelon, and orange (Wang et al. 2020), have been employed for the extraction of CQDs. Luminous carbon dots were created using agro-industrial waste products from tomato peels and seeds, grape pomace, sugar beet pulp, and olive marc. The synthesis process involved a mild thermal treatment assisted by γ-alumina (Brachi 2020). CQDs derived from the aqueous peel extract of Trapa bispinosa fluoresced bright green when exposed to UV light. At 450 nm, the CQDs showed the highest fluorescence intensity (Mewada et al. 2013). Unripe plum blossoms were utilized to fabricate highly fluorescent nanohybrids of carbon quantum dots (CQD) and silver nanoparticles (AgNPs). These nanomaterials demonstrated photocatalytic capability for the degradation of the dyes, like methylene blue and methyl orange (Atchudan et al. 2018).

Traditional synthesis method

CQDs can be produced using various bottom-up approaches from a range of inexpensive and renewable natural biomass sources (Lim et al. 2015). The template synthesis approach encompasses the formation of CQDs through the calcination process within a mesoporous spheres, often comprising silicon materials. Subsequently, the removal of carriers is achieved through etching. This method yields CQDs with adjustable particle sizes, a uniform distribution, and the ability to prevent agglomeration even under high temperatures (Liu et al. 2009; Zong et al. 2011). The pyrolysis technique is derived from the conventional production pathway of magnetic or semiconductor nanomaterials. It involves the carbonation and dehydration reactions of organic molecules during the thermal treatment. Dong et al. and Xu et al. successfully attained blue luminous carbon dots with soybeans and citric acid as precursors, respectively, at a low pyrolysis temperature of 200°C (Dong et al. 2012; Xu et al. 2016). The structure, size, and morphology of the resulting CQDs were influenced by the carbonization level determined by the pyrolysis temperature and retention time (Ren et al. 2019). Producing CQDs can be achieved easily and universally through chemical oxidation, involving the carbonization of precursors followed by extraction using oxidants like mixed H_2SO_4, H_2O_2, HNO_3, HAC, and others. Using a yield of around 15%, Wang et al. oxidized rice husks using H_2SO_4 and HNO_3, producing high-quality CQDs with a rich surface of oxygen groups (mainly carboxylic acids) (Wang et al. 2016c). The hydrothermal technique sterilizes objects in a hydrothermal reactor at a moderate temperature (about 200°C) without the use of hazardous or poisonous chemicals (Yuan et al. 2015). Sahu et al. used a hydrothermal approach to synthesize CQDs derived from orange juice, exhibiting a high fluorescence QY of 26% (Sahu et al. 2012). However, the majority of carbon dots that used water as their reaction medium were only able to generate short-wavelength blue fluorescence. The solvothermal treatment reportedly offers certain benefits for creating multi-color CQDs. Yuan et al. showed how to make green and red CQDs from biomass using an alkaline solution (Yuan et al. 2018).

In comparison to other bottom-up approaches, the hydrothermal approach has proven to be a more ecologically friendly method for synthesizing biomass-derived CQDs due to its low cost, easy operation, ease of modification, and moderate conditions (Prasannan et al. 2013; Zhu et al. 2020b). However, some CQDs made using conventional bottom-up techniques without any changes show low quantum yields (less than 15%) and constrained luminescence color (usually blue or green). In order to enable the large-scale manufacturing of high-quality CQDs from biomass waste, it is becoming more necessary to develop improved synthesis methodologies and modifications (Liu et al. 2020). The fabrication of carbon quantum dots from various biomass sources through different synthesis approaches is described in Figure 4.

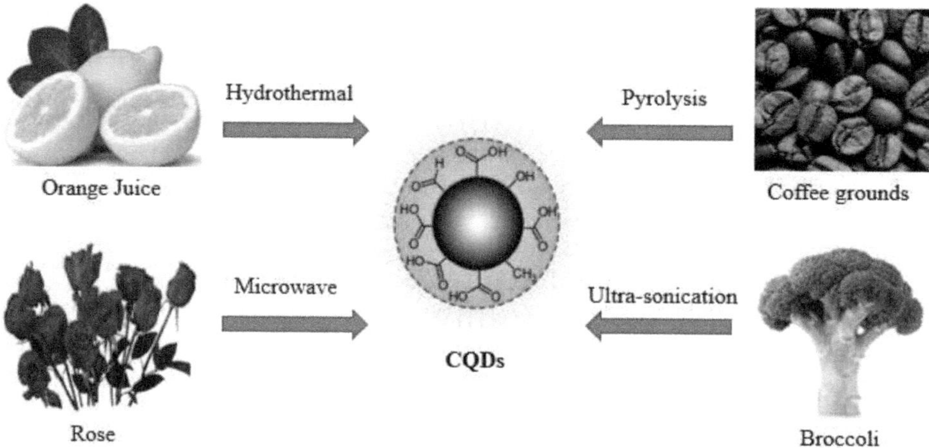

Figure 4. Synthesis of CQDs from orange juice (Sahu et al. 2012), coffee grounds (Hsu et al. 2012b), rose (Feng et al. 2015), and broccoli (Arumugam et al. 2018).

Advanced synthesis strategy

To overcome the constraints of conventional bottom-up approaches in generating high-quality carbon quantum dots (CQDs) from biomass, scientists have devised advanced preparation strategies that amalgamate the advantages of both bottom-up and top-down approaches. These approaches entail initially synthesizing sp^2 carbon flakes, like pyrolytic carbon and hydrochar, from various biomass materials through a bottom-up approach. Subsequently, the resulting carbon flakes are transformed into carbon dots using a top-down approach, often involving oxidation. This pioneering combination of bottom-up and top-down approaches exhibits significant potential for facilitating the large-scale production of CQDs from biomass.

The hydrothermal approach is a thermochemical procedure that transforms a substrate in water into a liquid product and hydrochar. This process typically occurs at relatively low experimental temperature, usually below 250°C. During hydrothermal treatment, the substrate is exposed to high-pressure water, which leads to the breakdown of its chemical bonds and the formation of new products. The resulting liquid product and hydrochar can be used as fuels, soil amendments, or precursors to produce carbon nanomaterials (Gao et al. 2015b).

Although hydrothermal treatment has the potential to produce carbon quantum dots (CQDs) from biomass, a notable disadvantage is the poor yield of CQDs caused by the generation of large volumes of hydrochar. However, researchers have developed strategies to convert hydrochar into value-added nanomaterials for improving the yield of biomass-derived carbon dots (Zhang et al. 2018a). For example, Zhou et al. achieved the production of multicolor CQDs by refluxing hydrochar derived from food waste with concentrated H_2SO_4 and HNO_3. However, the use of strong acids presents an environmental threat (Zhou et al. 2018). In contrast, Jing and colleagues devised a sustainable and efficient two-step approach to scale up the synthesis of biomass-derived CQDs through mild oxidation (Jing et al. 2019). Likewise, Zhao et al. successfully prepared lignin-derived carbon dots with a yield of 42.5 wt% by treating hydrochar with a low-concentration sodium hydroxide/oxygen solution (Zhao et al. 2020b).

The up-down joint methods have shown the capability of using low-value solid wastes, including hydrochar and pyrolytic carbon, to make high-value CQDs. However, because of its mild conditions and a notable improvement in QY, hydrothermal treatment combined with oxidation continues to be a better method for the large-scale manufacturing of biomass derived CQDs. For instance, using pyrolysis of the metal salt lignin mixture, nitric acid oxidation, and potassium hydroxide hydrothermal cutting, Temerov et al. synthesized lignin-derived GQDs from cobalt, nickel, and iron graphene capsules. These GQDs emit a yellowish light with QY varying from 11.7% to 12.4% (Temerov et al. 2019). To develop this discipline, further study on up-down joint techniques and their affecting variables is required.

Formation mechanism of CQDs derived from biomass

Biomass, as a renewable and abundant resource, is a promising precursor for the manufacturing of carbon quantum dots. The three primary constituents of natural biomass, namely hemicellulose, cellulose, and lignin, play crucial roles in determining the properties and functionalities of the resulting CQDs. D-glucose monomers constitute the basis of the polymer cellulose [$(C_6H_{10}O_5)n$], which is joined together by 1,4-glycosidic linkages to produce a homogenous glycan structure Hemicellulose is with amorphous and polysaccharide structures comprised of a variety of monosaccharide units, and a lot of side chains, which increase its high reactivity. A complex three-dimensional polymer, lignin contains polyaromatic units (p-hydroxyphenyl, syringyl, and guaiacyl) interconnected through β-O-4 ether linkages or C–C bonds (Zhao et al., 2018). The physical structure of biomass contains cellulose situated within the shell, while lignin forms the outer cell wall, and hemicellulose functions as an adhesion agent between them. Moreover, lignin and hemicellulose exist as a lignin-carbohydrate complex, while cellulose and hemicellulose are linked through H-bonding (Chai et al. 2019). Glycosidic and hydrogen bonds are broken during the hydrothermal treatment of biomass. Hydrolysis of hemicellulose and cellulose results in intermediate states composed of C/H/O components, including oligomeric-reducing sugars, primarily levoglucosan. Further dehydration and pyran ring-opening events occur, resulting in the creation of soluble small molecules such as 1,2,4-benzenetriol, hydroxyacetaldehyde, and hydroxymethylfurfural derivatives. Following that, intermolecular dehydration and aldol condensation of these intermediates induce polymerization and condensation reactions, resulting in the formation of condensed polyaromatic and furfural structures in CDs, as well as the presence of numerous oxygen-containing functional groups (Liu et al. 2020).

The hydrothermal process poses challenges in decomposing lignin due to its complex aromatic structure (Wang et al. 2019). Generally, the primary precursor materials for the synthesis of CQDs are recognized as polycyclic aromatic hydrocarbon molecules, which are formed via the acidolysis of lignin (Dong et al. 2012). However, through acid treatment, lignin can be broken down into smaller molecules, including alcohols, acids, aldehydes, and aromatic compounds. This breakdown occurs by cleaving β-O-4 ether linkages, C–C bonds, and hydrogen bonds. Acids act as both cutting agents and oxidizing agents in this process. The resulting functional fragments undergo transformations, including polymerization, condensation, aromatization, dehydration, and carbonization, facilitated by cycloaddition and aldol condensation reactions. Nucleation occurs when the concentration of carbon substances reaches a critical supersaturation point, followed by uniform and isotropic growth of nuclei on the particle surface due to solute diffusion. This leads to the creation of CQDs. CQDs derived from lignin exhibit sp2 hybridization and hold abundant surface functional groups, including carbonyl, epoxy, hydroxyl, ether, and carboxylic acid (Wang et al. 2019).

The potential mechanism for the formation of carbon quantum dots derived from biomass encompasses a complex sequence of chemical reactions. These reactions involve concurrent processes such as dehydration, polymerization, aromatization, condensation, decomposition, and nucleation (Hsu et al. 2012a; Wang et al. 2016b). But characterizing the intermediate products in each reaction step is difficult, leaving the synthesis method of CQDs from natural biomass uncertain (Shen et al. 2017). Further research is required since it is currently unclear how hemicellulose, cellulose, and lignin interact during hydrothermal carbonization. The number of carbon atoms, the surface functional groups, and the length of the carbon chain of the precursors are some of the variables that affect the structure and physicochemical characteristics of biomass-derived CQDs (Shen et al. 2018). Therefore, a comprehensive understanding of the preparation mechanism of CQDs from natural biomass is necessary to improve the quality.

Moreover, understanding the role of various reaction parameters, such as temperature, time, pH, and pressure, is crucial for the formation of high-quality CQDs with desirable properties (Kumar et al. 2018). For instance, the hydrothermal carbonization of lignin under acidic conditions at 200–300°C for 0.5–2 h has been reported to produce CQDs with high crystallinity and a little size variation. In addition to the above factors, the choice of catalysts and their concentration also play a critical role in the hydrothermal synthesis of CDs. Various metal ions, such as Fe(III), Co(II), and Ni(II), have been used as catalysts to promote the formation of CQDs from lignin. These metal ions can act as Lewis acids to catalyze the dehydration and condensation reactions of lignin-derived intermediates, leading to the formation of highly crystalline and uniform CQDs. However, excessive metal ions can also promote the formation of metal impurities and reduce the QY of CQDs. Therefore, optimizing the concentration of metal ions is essential for the synthesis of high-quality CQDs (Zhu et al. 2020b).

Optical properties

In general, the distinctive optical absorption spectra of carbon quantum dots in the UV region consist of two peaks, with a maximal absorption peak at about 260–270 nm corresponding to the π–π* transition (HOMO-LUMO) of the C=C bonding of the sp2-hybridized C-domains. The n–π* transition of the C=O bonding is responsible for the absorption peaks at 300 and 330 nm. The observable extending tail across the visible region is primarily due to nanoparticle (Carbonaro et al. 2019). Carbon quantum dots are notable for their photoluminescence, which enables their usage in biosensing, bioimaging, and the detection of numerous chemical species.

Unmodified carbon quantum dots often have a lower quantum yield; therefore, functionalization or surface passivation is used to improve optical properties, particularly fluorescence emission intensity.

Considering CQDs to be carbonaceous materials with both sp2 and sp3 domains, it is hypothesized that on a micro-level, CQDs are associated with defects in the form of non-perfect sp2 domains, resulting in energy gaps that contribute to CQD photoluminescence (Li et al. 2018c). Although the exact cause of photoluminescence is uncertain, three suggestions have been offered that may provide light on the mechanism: (1) quantum confinement caused by the carbon core's conjugated domains; (2) surface states caused by the presence of functional groups associated with the carbon backbone; and (3) a molecular state explained by photoluminescence from free and/or linked fluorescent molecules (Carbonaro et al. 2019).

Surface passivation is the process of treating carbon quantum dots with polymers and organic compounds to stabilize surface energy traps that contribute to fluorescence emission. Furthermore, coating the pristine CQDs with a thin layer of long-chain agents reduces impurity adhesion, which may affect particle stability and optical properties (Dimos 2016). As previously stated, adding heteroatoms to CQDs may increase various properties such as optical, electric, and chemical capabilities. Nitrogen has been commonly employed for carbon material doping because it has a similar atomic size to carbon and more electronegativity, as well as five electrons in its outer shell (Kang et al. 2020). In contrast, sulfur and phosphorus are used for heteroatom doping of CQDs due to their atomic properties, as these elements have a greater atomic radius and can easily produce disorders within the carbon backbone. Some of these processes may also cause red-shift fluorescence, or an emission shift toward longer wavelengths, which is mostly caused by the addition of electron-drawing groups to the surface of CQDs. It has been claimed that red-shift emission is more practical for analysis and sensing since UV light can be damaging to biological systems and shifting the excitation toward longer wavelengths can mitigate this risk (Meng et al. 2019).

Applications of biomass-derived CQDS in wastewater treatment

Carbon quantum dots (CQDs) are an exciting possibility for wastewater monitoring and treatment because of their distinctive qualities, including low toxicity, chemical inertness, and adjustable photoluminescence. In the field of wastewater treatment, CQDs have diverse applications at all steps, particularly in the monitoring of hazardous materials. Numerous studies have been conducted using functionalized CQDs for specific sites or purposes. One example is the use of fluorescent CQDs with a portable UV lamp for on-stream visual detection of heavy metal ions in wastewater discharge. This approach allows for real-time monitoring, providing a convenient and efficient method to identify the presence of heavy metal ions in wastewater (Li et al. 2018b).

Efforts have also been made to develop efficient CQD nanomaterials for the effective absorption of specific heavy metal ions such as Cd(II). The abundance of sites on the exterior surface of CQDs facilitates the adsorption of Cd(II) ions, enabling their removal from wastewater (Rahmanian et al. 2018). Overall, the utilization of CQDs in wastewater treatment offers promising opportunities for monitoring and addressing the presence of hazardous materials. Their unique properties, small size, and adaptability make them valuable tools for improving the efficiency and effectiveness of wastewater treatment processes (Manikandan et al. 2022).

Using a one-step alkali-assisted electrochemical technique, Li et al. developed carbon quantum dots with high up-conversion PL properties (Li et al. 2010). Additionally, they described the creation of TiO_2/CQDs and SiO_2/CQDs semiconductor nanocomposites, which showed a stable and strong visible-light response for the degradation of methylene blue (MB) dye. After 25 minutes of exposure to a halogen lamp (300 W), the highest photodegradation capacities of TiO_2/CQDs, SiO2/CQDs, and nanocomposites toward aqueous methylene blue contaminants were determined to be 100%. In contrast, negligible photodegradation of methylene blue (0%) was seen in control studies using pure CQDs and/or without the presence of either of the metal oxide components (SiO_2 or TiO_2). Additionally, a hydrothermal one-step method was employed to synthesize a CQDs/ZnO nanocomposite (about 20–30 nm) that was then used for the photodegradation of hazardous benzene and methanol (Yu et al. 2012). After 24 h, at room temperature under visible light, this nanocomposite demonstrated over 80% degradation efficiency.

Researchers have shown that nano-complex-based photocatalysts (such as WO_3/CQDs, N-ZnO/CQDs) are effective in photodegradation of a variety of organic pollutants (Chen et al. 2016, Muthulingam et al. 2016). A good example is Zhao et al.'s preparation and testing of CQDs-decorated Bi_2WO_6 nanocomposites for the photocatalytic elimination of gaseous VOCs under UV and visible light, which demonstrated increased photo-oxidation performance and stability towards toluene and acetone (Qian et al. 2016). Like this, Mendes et al. (Martins et al. 2016) looked at how well N-CQDs/TiO$_2$ (P25) nanocomposites performed in the photo-oxidation of NO pollutants under UV and visible-light irradiation, resulting in higher NO conversion rates than P25 alone.

Metal sulfides (e.g., CdS) with outstanding features, such as narrow band gap, good transport, and high thermal/chemical stability, are one of the candidates for improving CQDs-based photocatalysts. Liu and his co-workers prepared CQDs/CdS photocatalysts using a hydrothermal method and evaluated their ability to degrade RhB (Liu et al. 2013). About 50% of RhB was degraded with flower-shaped CdS because of the CQDs' effective electron trapping and slower recombination of the photoexcited carriers. However, the degrading efficiency was greatly increased to 90% by the addition of 1% CQDs/CdS. Alizarin red S (ARS) dye was successfully photodegraded by Kansal et al. (Kaur et al. 2016), utilizing CQDs/ZnS photo/nanocatalyst, which also displayed better photocatalytic activity for ARS degradation (89% after 250 min). Under comparable visible-light irradiation circumstances, it was discovered to be 1.4 times better than pristine ZnS (63%).

Polymers such as polyvinyl alcohol are used to make polymer dots (PDs), which contain CDs (Zhu et al. 2015). These PDs are made by hydrothermally treating a PVA precursor. The activities of the obtained semiconductor/CD nanohybrids for MO dye degradation in aqueous solutions were investigated. CDs/PDs-TiO$_2$ nanocomposites with Ti-O-C bands were created by grafting CDs/PDs with TiO$_2$ NPs using a simple hydrothermal procedure. The combined action of CDs/PDs and TiO$_2$ may greatly improve hazardous MO removal, and the nano/photocatalytic rate constants of PDs/TiO$_2$ are 9.5 and 3.6 times higher, respectively than commercially available pure P25 and TiO$_2$ (Figure 5a,b).

In a remarkable achievement, researchers have successfully produced hybrid nanocomposites comprising of carbon quantum dots (CQDs) and a diverse range of nanoparticles, such as zinc oxide, titanium dioxide iron oxide, and silica. Notably (Zhang et al. 2011), employed a hydrothermal process to synthesize Fe$_2$O$_3$/CQDs nanocomposites, exhibiting exceptional photocatalytic performance for the degradation of hazardous gases like benzene or methanol in the presence of visible light. The resulting magnetic nanohybrids possess a cubic morphology that combines the magnetic properties of iron oxide with the fluorescent characteristics of CQDs. Thanks to the impressive electron storage capacity of CQDs, excited photons from Fe$_2$O$_3$ nanoparticles can readily facilitate electron transportation within the CQD conducting system. Consequently, these nanohybrids exhibit a heightened ability to interact with adsorbed reductants/oxidants, generating an increased quantity of active oxygen radicals. As a result, they exhibit remarkable oxidation capabilities, effectively eliminating toxic gases.

Furthermore, in a groundbreaking study by (Markova et al. 2012), a remarkable procedure was employed to create core-shell photocatalysts with a magnetic and fluorescent nature. The ingenious approach involved electrostatic modifications of negatively charged NP-cores, such as biogenic magnetite, CNTs, and silver, using CQDs and betaine hydrochloride. As a result, a diverse array of core-shell fluorescent photocatalysts emerged, as magnetite and CNTs were coated with CQDs. These nanocomposites exhibited a captivating synergy between the mechanical, optical, and magnetic attributes of the NP cores and the fluorescence properties of CQDs.

Biomass-derived CQDs as photocatalyst

Carbon quantum dots produced from biomass have emerged as promising photocatalytic materials. These CQDs exhibit unique optical properties and are derived from sustainable sources such as carbohydrates, lignin, and proteins. Their use as photocatalysts is distinguished by their great efficiency in harvesting solar energy for catalytic reactions such as pollutant degradation and hydrogen evolution. The renewable nature of biomass sources, as well as the eco-friendly synthesis of CQDs, highlight their promise as environmentally sustainable photocatalysts with a wide range of applications in clean energy and environmental remediation (Rani et al. 2020).

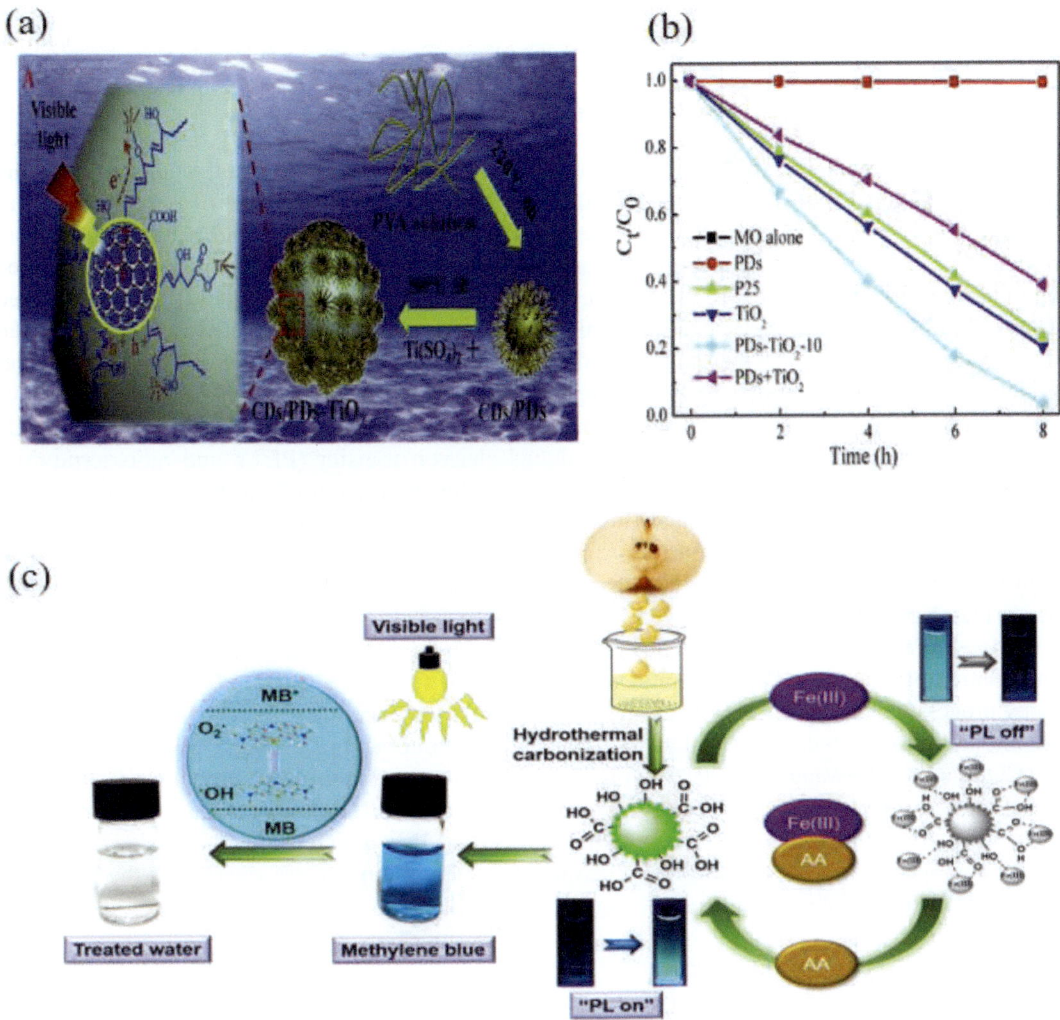

Figure 5. (a) Mechanistic representation of photocatalysis for CDs/PDs-TiO$_2$ in the presence of visible light. (b) MO degradation when exposed to UV/Visible light (220–800 nm) (Reproduced with permission from ref. (Li et al. 2018a), (c) Schematic illustration of the hydrothermal synthesis of CQDs from pear juice and their visible-light-induced dye degradation and sensing applications (Reproduced with permission from ref.(Das et al. 2019).

In recent years, there has been increasing attention to producing carbon quantum dots from a variety of biomass waste, presenting innovative and ecological methods for material synthesis. The manufacture of CQDs, which uses materials such as waste frying oil (Aji et al. 2016), orange peels (Prasannan et al. 2013), lignocellulosic waste (Achilleos et al. 2020), bitter apple peel waste (Aggarwal et al. 2020), spent coffee grounds (Jin et al. 2023), peer juice (Das et al. 2019) (Figure 5c), and lemon peel waste (Tyagi et al. 2016), not only addresses environmental concerns related to waste disposal but also capitalizes on the abundance of renewable resources. Because of their unique optical and electrical properties, biomass-derived CQDs are ideal for photocatalytic applications in wastewater treatment (Singh et al. 2023). The hydrothermal approach was utilized for organic synthesis of luminescent CQDs from fish scale waste. This study investigates the effect of CQDs on improved photocatalytic degradation of organic dyes including Methylene blue and Reactive Red 120 dyes, and metal ion detection (Alshammari et al. 2023). These CQDs have good photostability and a high quantum yield, which increases their efficiency as photocatalysts.

The diverse nature of biomass sources allows for the tailoring of CQD properties, ensuring versatility in their application.

Detection of heavy metals with CQDs

The development of optical sensors utilized to detect heavy metals is greatly facilitated by the optical absorption features of carbon quantum dots. To identify the presence of heavy metals and metalloids in water, these sensors make use of colorimetric, absorbance, and photoluminescence principles. The photo-excitation of charge carriers inside the carbon dots, which results in electrons moving from the valence band to the conduction band, is what gives CQDs their optical properties. This technology generates electron-hole pairs that can be utilized to detect metal ions based on the specific interactions between the ions and the surface ligands of the CQDs. Common sensing principles used when using CQDs for metal detection include fluorescence quenching, inner filter effects (IFEs), surface-enhanced Raman scattering (SERS), phosphorescence, fluorescence resonance energy transfer (FRET), photo-induced electron transfer (PET), and ratiometric dual emission. These processes depend on the intermolecular interaction of the target molecules and CQDs. When compared to traditional fluorescent chemosensors, CQD-based fluorescent devices provide selective and sensitive detection of target compounds, especially heavy metals. Figure 6 highlights the various sensing mechanisms used in the development of an optical chemosensor for heavy metals/metalloids in water.

Metal ions sensing in fluorescence quenching is based on the analyte-induced non-radiative annihilation of charge carriers in photoexcited CQDs. To selectively quench CQD optical probes, many processes and factors such as primary and secondary inner filter effects (IFEs), non-radiative recombination activated by the analyte ions, and chelating interactions between the analyte ions and the CQDs are used (Wang et al. 2010). This method is often used in the development of CQD-based metal ion sensing platforms, especially for ions like Cu(II) and Fe(III), which may efficiently cause quenching when attached to the surfaces of these materials. IFE, which happens when the absorbers' absorption spectra and the fluorophores' excitation or emission spectra overlap in the detection system, is closely related to fluorescence quenching in CQDs. Therefore, choosing the appropriate

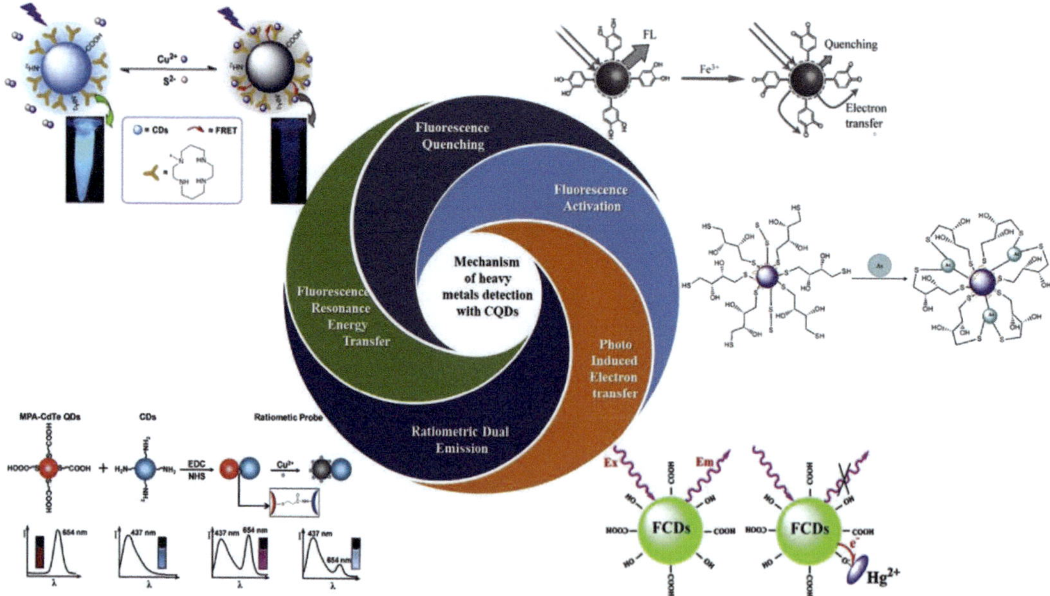

Figure 6. Heavy metal/metalloid detection mechanism using CQDs, from ref. (Devi et al. 2019).

absorber-fluorophore couple is essential in IFE-based fluorescent sensing systems (Mao et al. 2018). The analyte-induced fluorescence amplification of carbon dots is the second most frequently employed sensing approach. This is accomplished by passivating the surface of the trap states and adding a passivation layer on the carbon dots surface to facilitate analytes' binding. Gao et al. used this mechanism to detect Ag^+ ions using carbon dots, where the binding of Ag^+ ions increased the PL intensity by reducing Ag^+ to Ag on the carbon dots surface (Gao et al. 2015a).

The photoinduced electron transfer (PET) strategy is frequently utilized in the development of ion-induced "turn-on" and "turn-off" phosphorescence-based chemosensors. Through ion coordination, the presence of the analyte in PET often affects the emission of complexes. When the receptor attaches to the analyte, the generated complex inhibits (or induces) the PET, restoring (or weakening) the complex's emission. Systems based on phosphorescence rely on certain recognition sites to connect the analyte and receptor. The energy levels of the excited states of the CQDs are altered by the coupling between the receptor and analyte, changing the phosphorescent signal that may be employed for detection. Phosphorescent chemo sensors are based on the PET process, which is frequently accomplished via the "receptor-conjugated signaling unit approach." This method involves chelating the receptor and signaling unit to produce a complex that can display PET upon analyte binding. This method, for instance, has been successfully used to detect Cu^{2+} ions using CQDs. On the surface of CQDs, carboxyl groups can be coordinated with Cu^{2+} ions to achieve fluorescence quenching.

The fluorescence resonance energy transfer (FRET) mechanism, which relies on energy transfer between donor and acceptor units following analyte attachment, is another extensively used sensing approach. The simultaneous quenching of fluorescence from the donor unit and electronic stimulation of the acceptor unit characterize this occurrence. This can result in the creation of a low-energy photon or relaxation via a non-radiative pathway. The amount of energy transferred in FRET is affected by the spectral overlap between acceptor absorption and donor emission, the fluorescence QY of the donor unit, the comparative geometric orientation of the acceptor and donor units, and the comparative distance between the acceptor and donor molecules (Devi et al. 2019).

Toxic metals including lead, cadmium, mercury, and arsenic must be found and removed from water sources to maintain environmental sustainability. In addition, it's essential to degrade organic dyes, which are often used in the textile and dyeing industries, to stop water contamination. Titanium dioxide (TiO_2) is a widely studied and frequently employed semiconductor material owing to its excellent photocatalytic properties. TiO_2 nanoparticles exhibit high surface area, strong oxidizing power, and photoactivity under ultraviolet (UV) light irradiation. However, their wide bandgap limits their absorption to UV light only. To overcome this limitation, various strategies have been employed, including doping with metals or nonmetals, and coupling with other materials such as CQDs (Mehta et al. 2018). CQDs-TiO_2 suspensions had an 82% antibacterial activity, making them superior antimicrobial agents (Manikandan et al. 2022). For sensing and photocatalytic applications, the combination of CQDs and TiO_2 nanoparticles in the shape of a nanocomposite offers synergistic benefits. By extending the light absorption to visible wavelengths through energy transfer mechanisms, the CQDs can improve the photoactivity of TiO_2. In addition, the functional groups on the surface of carbon dots can offer locations for the adsorption of poisonous metals and organic dyes, improving the effectiveness of detection and removal (Mehta et al. 2018).

It has been thoroughly investigated how CQDs/TiO_2 nanocomposites may degrade organic dyes by photocatalysis. When exposed to UV or visible light, the nanocomposite generates electron-hole pairs, which initiate redox reactions with the adsorbed dyes, resulting in their degradation into harmless products. The distinctive features of CQDs, include their substantial surface area and efficient charge transfer abilities, which enhance the photocatalytic performance of TiO_2, leading to faster dye degradation rates. This approach offers a sustainable and green solution for the treatment of dye-contaminated wastewater. Nitrogen-doped carbon quantum dots generated from spent coffee grounds were used to decorate TiO_2. N-CQDs boost visible light absorption and facilitate the transfer and separation of photo-generated carriers. Over methylene blue, NCQDs/TiO_2 exhibit outstanding photocatalytic activity (Jin et al. 2023).

"Kill Waste by Waste" Concept

The "kill waste by waste" concept refers to the use of waste materials as precursors or catalysts in the synthesis of nanocomposites. This approach not only reduces production costs but also promotes the utilization of waste materials, thereby minimizing environmental impact. In the context of CQDs/TiO_2 nanocomposites, waste materials, such as biomass-derived carbon sources or industrial by-products, can be employed as carbon sources for CQD synthesis. Moreover, the incorporation of waste materials, such as discarded eggshells or seashells, can enhance the photocatalytic performance of TiO_2, reducing the reliance on expensive precursors (Mehta et al. 2018).

Conclusion

The tremendous challenge of water pollution poses a significant threat to both the natural environment and human well-being, demanding the implementation of highly effective and reliable water treatment technologies. The utilization of carbon quantum dots derived from biomass sources emerges as a viable and ecologically responsible strategy to address this global issue. The synthesis of CQDs from diverse biomass sources, including carbohydrates, lignin, and proteins, not only provides a compelling alternative to conventional methods but also embodies an eco-friendly and responsible approach. These biomass-derived CQDs exhibit unique structural and chemical properties, rendering them exceptionally suitable for various applications in water treatment. Advancements in the understanding of the intricate processes involved in CQD production from biomass have facilitated more precise synthesis methods and enhanced control over their inherent features. The incorporation of carbon quantum dots/titanium dioxide (TiO_2) nanocomposites in heavy metal detection and dye detoxification through photosensitivity represents a groundbreaking development with significant potential for advancing water treatment systems. Moreover, the groundbreaking concept of "kill waste by waste" introduces a transformative paradigm in waste management, capitalizing on the efficiency and efficacy of CQDs for precise and targeted elimination of pollutants. In essence, extensive research on biomass-derived CQDs and their diverse applications in water treatment provides invaluable insights, paving the way for sustainable solutions in mitigating water pollution. This ensures the continued availability of pristine and untainted water resources for generations to come.

References

Abbo, H.S., Gupta, K.C., Khaligh, N.G., and Titinchi, S.J. 2021. Carbon nanomaterials for wastewater treatment. ChemBioEng Reviews, 8(5): 463–489.

Achilleos, D.S., Kasap, H., and Reisner, E. 2020. Photocatalytic hydrogen generation coupled to pollutant utilisation using carbon dots produced from biomass. Green Chemistry, 22(9): 2831–2839. doi:10.1039/D0GC00318B.

Aggarwal, R., Saini, D., Singh, B., Kaushik, J., Garg, A.K., and Sonkar, S.K. 2020. Bitter apple peel derived photoactive carbon dots for the sunlight induced photocatalytic degradation of crystal violet dye. Solar Energy, 197: 326–331. doi:https://doi.org/10.1016/j.solener.2020.01.010.

Aji, M.P., Wiguna, P.A., Susanto, S., Rosita, N., Suciningtyas, S.A., and Sulhadi, S. 2016. Performance of photocatalyst based carbon nanodots from waste frying oil in water purification. Paper presented at the AIP Conference Proceedings.

Al Ja'farawy, M.S., Purwanto, A., and Widiyandari, H. 2022. Carbon quantum dots supported zinc oxide (ZnO/CQDs) efficient photocatalyst for organic pollutant degradation–A systematic review. Environmental Nanotechnology, Monitoring & Management, 18: 100681.

Alshammari, G.M., Al-Ayed, M.S., Abdelhalim, M.A., Al-Harbi, L.N., Qasem, A.A., and Abdo Yahya, M. 2023. Development of luminescence carbon quantum dots for metal ions detection and photocatalytic degradation of organic dyes from aqueous media. Environmental Research, 226: 115661. doi:https://doi.org/10.1016/j.envres.2023.115661.

Arias Velasco, V., Caicedo Chacón, W.D., Carvajal Soto, A.M., Ayala Valencia, G., Granada Echeverri, J.C., and Agudelo Henao, A.C. 2021. Carbon quantum dots based on carbohydrates as nano sensors for food quality and safety. Starch - Stärke, 73(11-12): 2100044. doi:https://doi.org/10.1002/star.202100044.

Arumugam, N., and Kim, J. 2018. Synthesis of carbon quantum dots from Broccoli and their ability to detect silver ions. Materials Letters, 219: 37–40. doi:https://doi.org/10.1016/j.matlet.2018.02.043.

Atchudan, R., Edison, T.N.J.I., Aseer, K.R., Perumal, S., Karthik, N., and Lee, Y.R. 2018. Highly fluorescent nitrogen-doped carbon dots derived from Phyllanthus acidus utilized as a fluorescent probe for label-free selective detection of Fe3+ ions, live cell imaging and fluorescent ink. Biosensors and Bioelectronics, 99: 303–311. doi:https://doi.org/10.1016/j.bios.2017.07.076.

Bhatt, S., Bhatt, M., Kumar, A., Vyas, G., Gajaria, T., and Paul, P. 2018. Green route for synthesis of multifunctional fluorescent carbon dots from Tulsi leaves and its application as Cr(VI) sensors, bio-imaging and patterning agents. Colloids and Surfaces B: Biointerfaces, 167: 126–133. doi:https://doi.org/10.1016/j.colsurfb.2018.04.008.

Bhuvaneswari, G., and Radjarejesri, S. 2015. Green synthesis and characterization of CdS quantum dots. Int. J. ChemTech. Res., 8(5): 104–108.

Brachi, P. 2020. Synthesis of fluorescent carbon quantum dots (CQDs) through the mild thermal treatment of agro-industrial residues assisted by γ-alumina. Biomass Conversion and Biorefinery, 10(4): 1301–1312. doi:10.1007/s13399-019-00503-4.

Cai, Z., Dwivedi, A.D., Lee, W.-N., Zhao, X., Liu, W., Sillanpää, M. et al. 2018. Application of nanotechnologies for removing pharmaceutically active compounds from water: development and future trends. Environmental Science: Nano., 5(1): 27–47.

Carbonaro, C.M., Corpino, R., Salis, M., Mocci, F., Thakkar, S.V., Olla, C. et al. 2019. On the emission properties of carbon dots: Reviewing data and discussing models. C, 5(4): 60.

Carrott, P., and Carrott, M.R. 2007. Lignin–from natural adsorbent to activated carbon: a review. Bioresource Technology, 98(12): 2301–2312.

Chai, X., He, H., Fan, H., Kang, X., and Song, X. 2019. A hydrothermal-carbonization process for simultaneously production of sugars, graphene quantum dots, and porous carbon from sugarcane bagasse. Bioresource Technology, 282: 142–147. doi:https://doi.org/10.1016/j.biortech.2019.02.126.

Chaudhary, S., Kumar, S., Kaur, B., and Mehta, S. 2016. Potential prospects for carbon dots as a fluorescence sensing probe for metal ions. RSC Advances, 6(93): 90526–90536.

Chen, B., Li, F., Li, S., Weng, W., Guo, H., Guo, T. et al. 2013. Large scale synthesis of photoluminescent carbon nanodots and their application for bioimaging. Nanoscale., 5(5): 1967–1971.

Chen, Y., Lu, Q., Yan, X., Mo, Q., Chen, Y., Liu, B. et al. 2016. Enhanced photocatalytic activity of the carbon quantum dot-modified BiOI microsphere. Nanoscale Research Letters, 11: 1–7.

Das, G.S., Shim, J.P., Bhatnagar, A., Tripathi, K.M., and Kim, T. 2019. Biomass-derived carbon quantum dots for visible-light-induced photocatalysis and label-free detection of Fe(III) and Ascorbic acid. Scientific reports, 9(1): 15084. doi:10.1038/s41598-019-49266-y.

Devi, P., Rajput, P., Thakur, A., Kim, K.-H., and Kumar, P. 2019. Recent advances in carbon quantum dot-based sensing of heavy metals in water. TrAC Trends in Analytical Chemistry, 114: 171–195.

Dimos, K. 2016. Carbon quantum dots: surface passivation and functionalization. Current Organic Chemistry, 20(6): 682-695.

Ding, Z., Li, F., Wen, J., Wang, X., and Sun, R. 2018. Gram-scale synthesis of single-crystalline graphene quantum dots derived from lignin biomass. Green Chemistry, 20(6): 1383–1390.

Dong, Y., Shao, J., Chen, C., Li, H., Wang, R., Chi, Y. et al. 2012. Blue luminescent graphene quantum dots and graphene oxide prepared by tuning the carbonization degree of citric acid. Carbon, 50(12): 4738–4743. doi:https://doi.org/10.1016/j.carbon.2012.06.002.

Feng, Y., Zhong, D., Miao, H., and Yang, X. 2015. Carbon dots derived from rose flowers for tetracycline sensing. Talanta, 140: 128–133. doi:https://doi.org/10.1016/j.talanta.2015.03.038.

Gao, X., Lu, Y., Zhang, R., He, S., Ju, J., Liu, M. et al. 2015a. One-pot synthesis of carbon nanodots for fluorescence turn-on detection of Ag+ based on the Ag+-induced enhancement of fluorescence. Journal of Materials Chemistry C, 3(10): 2302–2309. doi:10.1039/C4TC02582B.

Gao, Y., Yu, B., Wang, X., Yuan, Q., Yang, H., Chen, H. et al. 2015b. Orthogonal test design to optimize products and to characterize heavy oil via biomass hydrothermal treatment. Energy, 88: 139–148. doi:https://doi.org/10.1016/j.energy.2015.04.014.

Gayen, B., Palchoudhury, S., and Chowdhury, J. 2019. Carbon dots: A mystic star in the world of nanoscience. Journal of Nanomaterials, 2019: 1–19.

Goswami, J., Basyach, P., Purkayastha, S.K., Guha, A.K., Hazarika, P., and Saikia, L. 2023. Boosting photodegradation of pollutant dye with CuWO4/g-C3N4 heterojunction by introducing biomass derived N-CQD as electron mediator: Mechanism and DFT Calculation. New Journal of Chemistry.

Guan, J., Liu, X., Bai, N., Wang, F., Yang, Z., Zhang, J. et al. 2023. Luminescence properties of CQDs and photocatalytic properties of TiO2/ZnO/CQDs ternary composites. Journal of Materials Science: Materials in Electronics, 34(32): 2169.

Hoang, V.C., Nguyen, L.H., and Gomes, V.G. 2019. High efficiency supercapacitor derived from biomass based carbon dots and reduced graphene oxide composite. Journal of Electroanalytical Chemistry, 832: 87–96. doi:https://doi.org/10.1016/j.jelechem.2018.10.050.

Hsu, P.-C., and Chang, H.-T. 2012a. Synthesis of high-quality carbon nanodots from hydrophilic compounds: role of functional groups. Chemical Communications, 48(33): 3984–3986.

Hsu, P.-C., Shih, Z.-Y., Lee, C.-H., and Chang, H.-T. 2012b. Synthesis and analytical applications of photoluminescent carbon nanodots. Green Chemistry, 14(4): 917–920. doi:10.1039/C2GC16451E.

Hu, Y., Zhang, L., Li, X., Liu, R., Lin, L., and Zhao, S. 2017. Green preparation of S and N Co-doped carbon dots from water chestnut and onion as well as their use as an off–on fluorescent probe for the quantification and imaging of coenzyme A. ACS Sustainable Chemistry & Engineering, 5(6): 4992–5000.

Huang, F., Tan, D., Li, D., Guo, S., Yan, Y., and Zhang, W. 2023. Synthesis of broad spectrum-driven photocatalysts waste biomass-derived carbon quantum dots/g-C3N4 with superior energy bands for PPCPs restoration. Journal of Alloys and Compounds, 947: 169487. doi:https://doi.org/10.1016/j.jallcom.2023.169487.

Jiao, X.-Y., Li, L.-s., Qin, S., Zhang, Y., Huang, K., and Xu, L. 2019. The synthesis of fluorescent carbon dots from mango peel and their multiple applications. Colloids and Surfaces A: Physicochemical and Engineering Aspects, 577: 306–314. doi:https://doi.org/10.1016/j.colsurfa.2019.05.073.

Jin, Y., Tang, W., Wang, J., Ren, F., Chen, Z., Sun, Z. et al. 2023. Construction of biomass derived carbon quantum dots modified TiO2 photocatalysts with superior photocatalytic activity for methylene blue degradation. Journal of Alloys and Compounds, 932: 167627. doi:https://doi.org/10.1016/j.jallcom.2022.167627.

Jing, S., Zhao, Y., Sun, R.-C., Zhong, L., and Peng, X. 2019. Facile and high-yield synthesis of carbon quantum dots from biomass-derived carbons at mild condition. ACS Sustainable Chemistry & Engineering, 7(8): 7833–7843. doi:10.1021/acssuschemeng.9b00027.

Jones, S.S., Sahatiya, P., and Badhulika, S. 2017. One step, high yield synthesis of amphiphilic carbon quantum dots derived from chia seeds: a solvatochromic study. New Journal of Chemistry, 41(21): 13130–13139.

Kang, C., Huang, Y., Yang, H., Yan, X.F., and Chen, Z.P. 2020. A review of carbon dots produced from biomass wastes. Nanomaterials, 10(11): 2316.

Kang, K., Liu, B., Yue, G., Ren, H., Zheng, K., Wang, L. et al. 2023. Preparation of carbon quantum dots from ionic liquid modified biomass for the detection of Fe3+ and Pd2+ in environmental water. Ecotoxicology and Environmental Safety, 255: 114795. doi:https://doi.org/10.1016/j.ecoenv.2023.114795.

Kaur, S., Sharma, S., and Kansal, S.K. 2016. Synthesis of ZnS/CQDs nanocomposite and its application as a photocatalyst for the degradation of an anionic dye, ARS. Superlattices and Microstructures, 98: 86–95. doi:https://doi.org/10.1016/j.spmi.2016.08.011.

Khairol Anuar, N.K., Tan, H.L., Lim, Y.P., So'aib, M.S., and Abu Bakar, N.F. 2021. A review on multifunctional carbon-dots synthesized from biomass waste: Design/fabrication, characterization and applications. Frontiers in Energy Research, 9: 67.

Kumar, A., Chowdhuri, A.R., Laha, D., Mahto, T.K., Karmakar, P., and Sahu, S.K. 2017. Green synthesis of carbon dots from Ocimum sanctum for effective fluorescent sensing of Pb2+ ions and live cell imaging. Sensors and Actuators B: Chemical, 242: 679–686. doi:https://doi.org/10.1016/j.snb.2016.11.109.

Kumar, D.S., Kumar, B.J., and Mahesh, H. 2018. Quantum nanostructures (QDs): an overview. Synthesis of Inorganic Nanomaterials, 59–88.

Lakshmipathy, R., Palakshi Reddy, B., Sarada, N.C., Chidambaram, K., and Khadeer Pasha, S. 2015. Watermelon rind-mediated green synthesis of noble palladium nanoparticles: catalytic application. Applied Nanoscience, 5(2): 223–228. doi:10.1007/s13204-014-0309-2.

Li, G., Wang, F., Liu, P., Chen, Z., Lei, P., Xu, Z. et al. 2018a. Polymer dots grafted TiO2 nanohybrids as high performance visible light photocatalysts. Chemosphere, 197: 526–534. doi:https://doi.org/10.1016/j.chemosphere.2018.01.071.

Li, H.-Y., Li, D., Guo, Y., Yang, Y., Wei, W., and Xie, B. 2018b. On-site chemosensing and quantification of Cr(VI) in industrial wastewater using one-step synthesized fluorescent carbon quantum dots. Sensors and Actuators B: Chemical, 277: 30–38. doi:https://doi.org/10.1016/j.snb.2018.08.157.

Li, H., He, X., Kang, Z., Huang, H., Liu, Y., Liu, J. et al. 2010. Water-soluble fluorescent carbon quantum dots and photocatalyst design. Angewandte Chemie International Edition, 49(26): 4430–4434.

Li, L., and Dong, T. 2018c. Photoluminescence tuning in carbon dots: surface passivation or/and functionalization, heteroatom doping. Journal of Materials Chemistry C, 6(30): 7944–7970. doi:10.1039/C7TC05878K.

Liang, H., Tai, X., Du, Z., and Yin, Y. 2020. Enhanced photocatalytic activity of ZnO sensitized by carbon quantum dots and application in phenol wastewater. Optical Materials, 100: 109674.

Lim, S.Y., Shen, W., and Gao, Z. 2015. Carbon quantum dots and their applications. Chemical Society Reviews, 44(1): 362–381.

Liu, H., Ding, L., Chen, L., Chen, Y., Zhou, T., Li, H. et al. 2019. A facile, green synthesis of biomass carbon dots coupled with molecularly imprinted polymers for highly selective detection of oxytetracycline. Journal of Industrial and Engineering Chemistry, 69: 455–463. doi:https://doi.org/10.1016/j.jiec.2018.10.007.

Liu, R., Huang, H., Li, H., Liu, Y., Zhong, J., Li, Y. et al. 2014a. Metal nanoparticle/carbon quantum dot composite as a photocatalyst for high-efficiency cyclohexane oxidation. Acs Catalysis, 4(1): 328–336.

Liu, R., Wu, D., Liu, S., Koynov, K., Knoll, W., and Li, Q. 2009. An aqueous route to multicolor photoluminescent carbon dots using silica spheres as carriers. Angewandte Chemie International Edition, 48(25): 4598–4601.

Liu, W., Ning, C., Sang, R., Hou, Q., and Ni, Y. 2021. Lignin-derived graphene quantum dots from phosphous acid-assisted hydrothermal pretreatment and their application in photocatalysis. Industrial Crops and Products, 171: 113963. doi:https://doi.org/10.1016/j.indcrop.2021.113963.

Liu, X., Wang, M., Zhang, S., and Pan, B. 2013. Application potential of carbon nanotubes in water treatment: A review. Journal of Environmental Sciences, 25(7): 1263–1280. doi:https://doi.org/10.1016/S1001-0742(12)60161-2.

Liu, Y., Zhao, Y., and Zhang, Y. 2014b. One-step green synthesized fluorescent carbon nanodots from bamboo leaves for copper(II) ion detection. Sensors and Actuators B: Chemical, 196: 647–652. doi:https://doi.org/10.1016/j.snb.2014.02.053.

Liu, Y., Zhu, C., Gao, Y., Yang, L., Xu, J., Zhang, X. et al. 2020. Biomass-derived nitrogen self-doped carbon dots via a simple one-pot method: Physicochemical, structural, and luminescence properties. Applied Surface Science, 510: 145437. doi:https://doi.org/10.1016/j.apsusc.2020.145437.

Luo, P.G., Sahu, S., Yang, S.-T., Sonkar, S.K., Wang, J., Wang, H. et al. 2013. Carbon "quantum" dots for optical bioimaging. Journal of Materials Chemistry B, 1(16): 2116–2127.

Ma, X., Dong, Y., Sun, H., and Chen, N. 2017. Highly fluorescent carbon dots from peanut shells as potential probes for copper ion: The optimization and analysis of the synthetic process. Materials Today Chemistry, 5: 1–10. doi:https://doi.org/10.1016/j.mtchem.2017.04.004.

Mahat, N.A., and Shamsudin, S.A. 2020. Transformation of oil palm biomass to optical carbon quantum dots by carbonisation-activation and low temperature hydrothermal processes. Diamond and Related Materials, 102: 107660. doi:https://doi.org/10.1016/j.diamond.2019.107660.

Manikandan, V., and Lee, N.Y. 2022. Green synthesis of carbon quantum dots and their environmental applications. Environmental Research, 212: 113283.

Mao, M., Tian, T., He, Y., Ge, Y., Zhou, J., and Song, G. 2018. Inner filter effect based fluorometric determination of the activity of alkaline phosphatase by using carbon dots codoped with boron and nitrogen. Microchimica Acta, 185: 1–6.

Markova, Z., Bourlinos, A.B., Safarova, K., Polakova, K., Tucek, J., Medrik, I. et al. 2012. Synthesis and properties of core–shell fluorescent hybrids with distinct morphologies based on carbon dots. Journal of Materials Chemistry, 22(32): 16219–16223. doi:10.1039/C2JM33414C.

Martins, N.C.T., Ângelo, J., Girão, A.V., Trindade, T., Andrade, L., and Mendes, A. 2016. N-doped carbon quantum dots/TiO2 composite with improved photocatalytic activity. Applied Catalysis B: Environmental, 193: 67–74. doi:https://doi.org/10.1016/j.apcatb.2016.04.016.

Mehta, A., Mishra, A., Kainth, S., and Basu, S. 2018. Carbon quantum dots/TiO2 nanocomposite for sensing of toxic metals and photodetoxification of dyes with kill waste by waste concept. Materials & Design, 155: 485–493.

Meng, W., Bai, X., Wang, B., Liu, Z., Lu, S., and Yang, B. 2019. Biomass-derived carbon dots and their applications. Energy & Environmental Materials, 2(3): 172–192.

Meng, Y., Zhang, Y., Sun, W., Wang, M., He, B., Chen, H. et al. 2017. Biomass converted carbon quantum dots for all-weather solar cells. Electrochimica Acta, 257: 259–266. doi:https://doi.org/10.1016/j.electacta.2017.10.086.

Mewada, A., Pandey, S., Shinde, S., Mishra, N., Oza, G., Thakur, M. et al. 2013. Green synthesis of biocompatible carbon dots using aqueous extract of Trapa bispinosa peel. Materials Science and Engineering: C, 33(5): 2914–2917. doi:https://doi.org/10.1016/j.msec.2013.03.018.

Murugan, N., and Sundramoorthy, A.K. 2018. Green synthesis of fluorescent carbon dots from Borassus flabellifer flowers for label-free highly selective and sensitive detection of Fe 3+ ions. New Journal of Chemistry, 42(16): 13297–13307.

Muthulingam, S., Bae, K.B., Khan, R., Lee, I.-H., and Uthirakumar, P. 2016. Carbon quantum dots decorated N-doped ZnO: Synthesis and enhanced photocatalytic activity on UV, visible and daylight sources with suppressed photocorrosion. Journal of Environmental Chemical Engineering, 4(1): 1148–1155. doi:https://doi.org/10.1016/j.jece.2015.06.029.

Nasrollahzadeh, M., Sajjadi, M., Iravani, S., and Varma, R.S. 2021a. Carbon-based sustainable nanomaterials for water treatment: state-of-art and future perspectives. Chemosphere, 263: 128005.

Nasrollahzadeh, M., Sajjadi, M., Iravani, S., and Varma, R.S. 2021b. Green-synthesized nanocatalysts and nanomaterials for water treatment: Current challenges and future perspectives. Journal of Hazardous Materials, 401: 123401.

Nizam, N.U.M., Hanafiah, M.M., Mahmoudi, E., and Mohammad, A.W. 2023. Synthesis of highly fluorescent carbon quantum dots from rubber seed shells for the adsorption and photocatalytic degradation of dyes. Scientific Reports, 13(1): 12777. doi:10.1038/s41598-023-40069-w.

Papaioannou, N., Marinovic, A., Yoshizawa, N., Goode, A.E., Fay, M., Khlobystov, A. et al. 2018. Structure and solvents effects on the optical properties of sugar-derived carbon nanodots. Scientific Reports, 8(1): 6559.

Park, S.K., Lee, H., Choi, M.S., Suh, D.H., Nakhanivej, P., and Park, H.S. 2018. Straightforward and controllable synthesis of heteroatom-doped carbon dots and nanoporous carbons for surface-confined energy and chemical storage. Energy Storage Materials, 12: 331–340. doi:https://doi.org/10.1016/j.ensm.2017.10.008.

Pei, Y., Chang, A.Y., Liu, X., Wang, H., Zhang, H., Radadia, A. et al. 2021. Nitrogen-doped carbon dots from Kraft lignin waste with inorganic acid catalyst and their brain cell imaging applications. AIChE Journal, 67(5): e17132.

Pirsaheb, M., Asadi, A., Sillanpää, M., and Farhadian, N. 2018. Application of carbon quantum dots to increase the activity of conventional photocatalysts: a systematic review. Journal of Molecular Liquids, 271: 857–871.

Pourreza, N., and Ghomi, M. 2019. Green synthesized carbon quantum dots from *Prosopis juliflora* leaves as a dual off-on fluorescence probe for sensing mercury (II) and chemet drug. Materials Science and Engineering: C, 98: 887–896. doi:https://doi.org/10.1016/j.msec.2018.12.141.

Prasannan, A., and Imae, T. 2013. One-pot synthesis of fluorescent carbon dots from orange waste peels. Industrial & Engineering Chemistry Research, 52(44): 15673–15678. doi:10.1021/ie402421s.

Qian, X., Yue, D., Tian, Z., Reng, M., Zhu, Y., Kan, M. et al. 2016. Carbon quantum dots decorated Bi2WO6 nanocomposite with enhanced photocatalytic oxidation activity for VOCs. Applied Catalysis B: Environmental, 193: 16–21. doi:https://doi.org/10.1016/j.apcatb.2016.04.009.

Rahmanian, O., Dinari, M., and Abdolmaleki, M.K. 2018. Carbon quantum dots/layered double hydroxide hybrid for fast and efficient decontamination of Cd(II): The adsorption kinetics and isotherms. Applied Surface Science, 428: 272–279. doi:https://doi.org/10.1016/j.apsusc.2017.09.152.

Raja, D., and Sundaramurthy, D. 2021. Facile synthesis of fluorescent carbon quantum dots from Betel leafs (Piper betle) for Fe3+sensing. Materials Today: Proceedings, 34: 488–492. doi:https://doi.org/10.1016/j.matpr.2020.03.096.

Rani, U.A., Ng, L.Y., Ng, C.Y., and Mahmoudi, E. 2020. A review of carbon quantum dots and their applications in wastewater treatment. Advances in Colloid and Interface Science, 278: 102124. doi:https://doi.org/10.1016/j.cis.2020.102124.

Rani, U.A., Ng, L.Y., Ng, C.Y., Mahmoudi, E., Ng, Y.-S., and Mohammad, A.W. 2021. Sustainable production of nitrogen-doped carbon quantum dots for photocatalytic degradation of methylene blue and malachite green. Journal of Water Process Engineering, 40: 101816. doi:https://doi.org/10.1016/j.jwpe.2020.101816.

Ren, R., Zhang, Z., Zhao, P., Shi, J., Han, K., Yang, Z. et al. 2019. Facile and one-step preparation carbon quantum dots from biomass residue and their applications as efficient surfactants. Journal of Dispersion Science and Technology, 40(5): 627–633. doi:10.1080/01932691.2018.1475239.

Sabet, M., and Mahdavi, K. 2019. Green synthesis of high photoluminescence nitrogen-doped carbon quantum dots from grass via a simple hydrothermal method for removing organic and inorganic water pollutions. Applied Surface Science, 463: 283–291. doi:https://doi.org/10.1016/j.apsusc.2018.08.223.

Sahoo, N.K., Jana, G.C., Aktara, M.N., Das, S., Nayim, S., Patra, A. et al. 2020. Carbon dots derived from lychee waste: Application for Fe3+ ions sensing in real water and multicolor cell imaging of skin melanoma cells. Materials Science and Engineering: C, 108: 110429. doi:https://doi.org/10.1016/j.msec.2019.110429.

Sahu, S., Behera, B., Maiti, T.K., and Mohapatra, S. 2012. Simple one-step synthesis of highly luminescent carbon dots from orange juice: application as excellent bio-imaging agents. Chemical Communications, 48(70): 8835–8837.

Sato, N., Quitain, A.T., Kang, K., Daimon, H., and Fujie, K. 2004. Reaction kinetics of amino acid decomposition in high-temperature and high-pressure water. Industrial & Engineering Chemistry Research, 43(13): 3217–3222.

Saud, P.S., Pant, B., Alam, A.-M., Ghouri, Z.K., Park, M., and Kim, H.-Y. 2015. Carbon quantum dots anchored TiO2 nanofibers: Effective photocatalyst for waste water treatment. Ceramics International, 41(9): 11953–11959.

Shahshahanipour, M., Rezaei, B., Ensafi, A.A., and Etemadifar, Z. 2019. An ancient plant for the synthesis of a novel carbon dot and its applications as an antibacterial agent and probe for sensing of an anti-cancer drug. Materials Science and Engineering: C, 98: 826–833. doi:https://doi.org/10.1016/j.msec.2019.01.041.

Sharma, S., Dutta, V., Singh, P., Raizada, P., Rahmani-Sani, A., Hosseini-Bandegharaei, A. et al. 2019. Carbon quantum dot supported semiconductor photocatalysts for efficient degradation of organic pollutants in water: a review. Journal of Cleaner Production, 228: 755–769.

Shen, J., Shang, S., Chen, X., Wang, D., and Cai, Y. 2017. Facile synthesis of fluorescence carbon dots from sweet potato for Fe3+ sensing and cell imaging. Materials Science and Engineering: C, 76: 856–864. doi:https://doi.org/10.1016/j.msec.2017.03.178.

Shen, T., Wang, Q., Guo, Z., Kuang, J., and Cao, W. 2018. Hydrothermal synthesis of carbon quantum dots using different precursors and their combination with TiO2 for enhanced photocatalytic activity. Ceramics International, 44(10): 11828–11834. doi:https://doi.org/10.1016/j.ceramint.2018.03.271.

Si, Q.-S., Guo, W.-Q., Wang, H.-Z., Liu, B.-H., and Ren, N.-Q. 2020. Carbon quantum dots-based semiconductor preparation methods, applications and mechanisms in environmental contamination. Chinese Chemical Letters, 31(10): 2556–2566.

Singh, P., Raizada, P., Pathania, D., Sharma, G., and Sharma, P. 2013. Microwave induced KOH activation of guava peel carbon as an adsorbent for congo red dye removal from aqueous phase.

Singh, P., Rani, N., Kumar, S., Kumar, P., Mohan, B., Bhankar, V. et al 2023. Assessing the biomass-based carbon dots and their composites for photocatalytic treatment of wastewater. Journal of Cleaner Production, 413: 137474.

Sun, Y.-P., Zhou, B., Lin, Y., Wang, W., Fernando, K.A.S., Pathak, P. et al. 2006. Quantum-Sized Carbon Dots for Bright and Colorful Photoluminescence. Journal of the American Chemical Society, 128(24): 7756–7757. doi:10.1021/ja062677d.

Temerov, F., Belyaev, A., Ankudze, B., and Pakkanen, T.T. 2019. Preparation and photoluminescence properties of graphene quantum dots by decomposition of graphene-encapsulated metal nanoparticles derived from Kraft lignin and transition metal salts. Journal of Luminescence, 206: 403–411. doi:https://doi.org/10.1016/j.jlumin.2018.10.093.

Thota, S.P., Thota, S.M., Srimadh Bhagavatham, S., Sai Manoj, K., Sai Muthukumar, V.S., Venketesh, S. et al. 2018. Facile one-pot hydrothermal synthesis of stable and biocompatible fluorescent carbon dots from lemon grass herb. IET Nanobiotechnology, 12(2): 127–132.

Tyagi, A., Tripathi, K.M., Singh, N., Choudhary, S., and Gupta, R.K. 2016. Green synthesis of carbon quantum dots from lemon peel waste: applications in sensing and photocatalysis. RSC advances, 6(76): 72423–72432.

Ullah, N., Mansha, M., Khan, I., and Qurashi, A. 2018. Nanomaterial-based optical chemical sensors for the detection of heavy metals in water: Recent advances and challenges. TrAC Trends in Analytical Chemistry, 100: 155–166.

Wang, C., Shi, H., Yang, M., Yan, Y., Liu, E., Ji, Z. et al. 2020. Facile synthesis of novel carbon quantum dots from biomass waste for highly sensitive detection of iron ions. Materials Research Bulletin, 124: 110730. doi:https://doi.org/10.1016/j.materresbull.2019.110730.

Wang, G., Guo, Q., Chen, D., Liu, Z., Zheng, X., Xu, A. et al. 2018a. Facile and highly effective synthesis of controllable lattice sulfur-doped graphene quantum dots via hydrothermal treatment of durian. ACS Applied Materials & Interfaces, 10(6): 5750–5759.

Wang, H., Zhang, M., Song, Y., Li, H., Huang, H., Shao, M. et al. 2018b. Carbon dots promote the growth and photosynthesis of mung bean sprouts. Carbon, 136: 94–102. doi:https://doi.org/10.1016/j.carbon.2018.04.051.

Wang, L., Bi, Y., Hou, J., Li, H., Xu, Y., Wang, B. et al. 2016a. Facile, green and clean one-step synthesis of carbon dots from wool: Application as a sensor for glyphosate detection based on the inner filter effect. Talanta, 160: 268–275. doi:https://doi.org/10.1016/j.talanta.2016.07.020.

Wang, L., Li, B., Xu, F., Shi, X., Feng, D., Wei, D. et al. 2016b. High-yield synthesis of strong photoluminescent N-doped carbon nanodots derived from hydrosoluble chitosan for mercury ion sensing via smartphone APP. Biosensors and Bioelectronics, 79: 1–8. doi:https://doi.org/10.1016/j.bios.2015.11.085.

Wang, R., Xia, G., Zhong, W., Chen, L., Chen, L., Wang, Y. et al. 2019. Direct transformation of lignin into fluorescence-switchable graphene quantum dots and their application in ultrasensitive profiling of a physiological oxidant. Green Chemistry, 21(12): 3343–3352. doi:10.1039/C9GC01012B.

Wang, X., Cao, L., Yang, S.T., Lu, F., Meziani, M.J., Tian, L. et al. 2010. Bandgap-like strong fluorescence in functionalized carbon nanoparticles. Angewandte Chemie International Edition, 49(31): 5310–5314.

Wang, Z., Yu, J., Zhang, X., Li, N., Liu, B., Li, Y. et al. 2016c. Large-scale and controllable synthesis of graphene quantum dots from rice husk biomass: a comprehensive utilization strategy. ACS Applied Materials & Interfaces, 8(2): 1434–1439. doi:10.1021/acsami.5b10660.

Xu, M., Huang, Q., Sun, R., and Wang, X. 2016. Simultaneously obtaining fluorescent carbon dots and porous active carbon for supercapacitors from biomass. RSC Advances, 6(91): 88674–88682.

Xu, X., Ray, R., Gu, Y., Ploehn, H.J., Gearheart, L., Raker, K. et al. 2004. Electrophoretic Analysis and Purification of Fluorescent Single-Walled Carbon Nanotube Fragments. Journal of the American Chemical Society, 126(40): 12736-12737. doi:10.1021/ja040082h.

Xue, B., Yang, Y., Sun, Y., Fan, J., Li, X., and Zhang, Z. 2019. Photoluminescent lignin hybridized carbon quantum dots composites for bioimaging applications. International Journal of Biological Macromolecules, 122: 954–961. doi:https://doi.org/10.1016/j.ijbiomac.2018.11.018.

Yang, Q., Duan, J., Yang, W., Li, X., Mo, J., Yang, P., and Tang, Q. 2018. Nitrogen-doped carbon quantum dots from biomass via simple one-pot method and exploration of their application. Applied Surface Science, 434: 1079–1085. doi:https://doi.org/10.1016/j.apsusc.2017.11.040.

Yang, X., Guo, Y., Liang, S., Hou, S., Chu, T., Ma, J. et al. 2020. Preparation of sulfur-doped carbon quantum dots from lignin as a sensor to detect Sudan I in an acidic environment. Journal of Materials Chemistry B, 8(47): 10788-10796. doi:10.1039/D0TB00125B.

Yu, H., Zhang, H., Huang, H., Liu, Y., Li, H., Ming, H. et al. 2012. ZnO/carbon quantum dots nanocomposites: one-step fabrication and superior photocatalytic ability for toxic gas degradation under visible light at room temperature. New Journal of Chemistry, 36(4): 1031–1035. doi:10.1039/C2NJ20959D.

Yuan, B., Guan, S., Sun, X., Li, X., Zeng, H., Xie, Z. et al. 2018. Highly Efficient Carbon Dots with Reversibly Switchable Green–Red Emissions for Trichromatic White Light-Emitting Diodes. ACS applied materials & Interfaces, 10(18): 16005–16014. doi:10.1021/acsami.8b02379.

Yuan, M., Zhong, R., Gao, H., Li, W., Yun, X., Liu, J. et al. 2015. One-step, green, and economic synthesis of water-soluble photoluminescent carbon dots by hydrothermal treatment of wheat straw, and their bio-applications in labeling, imaging, and sensing. Applied Surface Science, 355: 1136–1144. doi:https://doi.org/10.1016/j.apsusc.2015.07.095.

Zhang, H., Kang, S., Wang, G., Zhang, Y., and Zhao, H. 2016. Fluorescence determination of nitrite in water using prawn-shell derived nitrogen-doped carbon nanodots as fluorophores. Acs Sensors, 1(7): 875–881.

Zhang, H., Ming, H., Lian, S., Huang, H., Li, H., Zhang, L. et al. 2011. Fe2O3/carbon quantum dots complex photocatalysts and their enhanced photocatalytic activity under visible light. Dalton Transactions, 40(41): 10822-10825. doi:10.1039/C1DT11147G.

Zhang, J., Xia, A., Zhu, X., Huang, Y., Zhu, X., and Liao, Q. 2022. Co-production of carbon quantum dots and biofuels via hydrothermal conversion of biomass. Fuel Processing Technology, 232: 107276.

Zhang, X., Jiang, M., Niu, N., Chen, Z., Li, S., Liu, S. et al. 2018a. Natural-product-derived carbon dots: from natural products to functional materials. ChemSusChem, 11(1): 11–24.

Zhang, Y., Zhao, Y., Duan, J., and Tang, Q. 2018b. S-doped CQDs tailored transparent counter electrodes for high-efficiency bifacial dye-sensitized solar cells. Electrochimica Acta, 261: 588–595. doi:https://doi.org/10.1016/j.electacta.2017.12.183.

Zhao, S., Liu, M., Zhao, L., and Zhu, L. 2018. Influence of interactions among three biomass components on the pyrolysis behavior. Industrial & Engineering Chemistry Research, 57(15): 5241–5249.

Zhao, S., Song, X., Chai, X., Zhao, P., He, H., and Liu, Z. 2020a. Green production of fluorescent carbon quantum dots based on pine wood and its application in the detection of Fe3+. Journal of Cleaner Production, 263: 121561. doi:https://doi.org/10.1016/j.jclepro.2020.121561.

Zhao, Y., Jing, S., Peng, X., Chen, Z., Hu, Y., Zhuo, H. et al. 2020b. Synthesizing green carbon dots with exceptionally high yield from biomass hydrothermal carbon. Cellulose, 27(1): 415–428. doi:10.1007/s10570-019-02807-0.

Zhou, J., Sheng, Z., Han, H., Zou, M., and Li, C. 2012. Facile synthesis of fluorescent carbon dots using watermelon peel as a carbon source. Materials Letters, 66(1): 222–224. doi:https://doi.org/10.1016/j.matlet.2011.08.081.

Zhou, Y., Liu, Y., Li, Y., He, Z., Xu, Q., Chen, Y. et al. 2018. Multicolor carbon nanodots from food waste and their heavy metal ion detection application. RSC Advances, 8(42): 23657–23662.

Zhu, L., Shen, D., Liu, Q., Wu, C., and Gu, S. 2021. Sustainable synthesis of bright green fluorescent carbon quantum dots from lignin for highly sensitive detection of Fe3+ ions. Applied Surface Science, 565: 150526. doi:https://doi.org/10.1016/j.apsusc.2021.150526.

Zhu, L., Shen, D., Wu, C., and Gu, S. 2020a. State-of-the-art on the preparation, modification, and application of biomass-derived carbon quantum dots. Industrial & Engineering Chemistry Research, 59(51): 22017–22039.

Zhu, S., Song, Y., Zhao, X., Shao, J., Zhang, J., and Yang, B. 2015. The photoluminescence mechanism in carbon dots (graphene quantum dots, carbon nanodots, and polymer dots): current state and future perspective. Nano Research, 8: 355–381.

Zhu, Z., Yang, P., Li, X., Luo, M., Zhang, W., Chen, M. et al. 2020b. Green preparation of palm powder-derived carbon dots co-doped with sulfur/chlorine and their application in visible-light photocatalysis. Spectrochimica Acta Part A: Molecular and Biomolecular Spectroscopy, 227: 117659. doi: https://doi.org/10.1016/j.saa.2019.117659.

Zong, J., Zhu, Y., Yang, X., Shen, J., and Li, C. 2011. Synthesis of photoluminescent carbogenic dots using mesoporous silica spheres as nanoreactors. Chemical Communications, 47(2): 764–766.

Index

A

agriculture 130, 131, 134, 136–138, 181, 184, 185, 190, 193, 195, 196, 198
Agriculture production 1, 2
Agroecosystem 35–37, 40, 43, 51, 52, 56
algae 81–85, 90, 91
animals 81, 89, 90, 96, 97, 99, 103, 104, 113
aquaporin 63, 69, 70
aquatic ecosystems 87, 89, 96, 113

B

Biomass-derived carbon dots 277

C

carbon nanotubes 63, 64
clean water 257
contamination 81, 87, 97, 109, 113

D

desalination 62–65, 67–69, 72, 73

E

ecosystem balance 244
environment 82, 83, 86, 87, 92–94, 113
environmentally friendly 230

G

graphene 63, 65, 66, 67, 71

H

Heavy metal detection 284
Hybrid materials 216, 218
hydrogel 130, 131, 133–139

I

industrial effluents 244, 252, 256
innovative materials 260

M

Magnetic nanostructures 208, 209, 213–219, 222–225
metal oxides 63, 69, 80, 82, 91, 104, 113

metalloids 89
Metals 80, 94, 97, 106, 108, 109, 111–113
Microbiome 163, 164, 166, 171, 172
microorganisms 96, 104, 105, 108, 111

N

Nano-agrochemicals 164, 166–168, 171, 173
Nano-biofertilizer 228–239
Nano-fertilizers 2–4, 13, 142, 143, 154
Nanofibers 243–256, 258–260
Nano-fungicides 142, 144
Nanomaterials 55, 62–64, 70–73, 94, 104, 105, 109, 112, 113, 178, 180, 181, 183, 194, 197
Nanoparticles 21–32, 79, 82, 89, 104, 112, 113, 163–173, 185, 188, 189, 195
Nano-pesticides 2, 5, 7, 13, 142, 144, 154, 155
Nanosensors 2, 8–10, 13, 35, 36, 40, 44, 45, 47, 48, 51
Nanotechnology 163, 168, 172, 173
Nutrient 40, 42, 54
nutrient release 235, 238

P

Photocatalysis 268, 269, 281, 283
Photoluminescence 278, 279, 282
Plant interfaces 146, 150
plant stress management 30
plants 79–81, 87–89, 104–113
Pollutant 39, 51
pollution 244, 251, 255, 257
purification methods 245, 251

S

sandy soil 130, 132, 133, 135, 137, 138
soil amendment 132, 33
soil carbon sequestration 32
soil ecosystems 79
soil health 228, 229, 231, 232, 236–238
soil health management 21–23, 26, 28
Soil-plant system 141, 147
soil pollution 178
sustainable agriculture 238
sustainable solutions 245, 250
Sustainable Water Management 208
Synthesis strategies 218, 219

W

water management 25, 26
Water pollution 267, 268, 284
water retention 131, 133, 135–138
water treatment 244–247, 250–258, 260

Z

zeolite 63, 67–70